Toughening Mechanisms in Quasi-Brittle Materials

NATO ASI Series

Advanced Science Institutes Series

A Series presenting the results of activities sponsored by the NATO Science Committee, which aims at the dissemination of advanced scientific and technological knowledge, with a view to strengthening links between scientific communities.

The Series is published by an international board of publishers in conjunction with the NATO Scientific Affairs Division

A **Life Sciences**	Plenum Publishing Corporation
B **Physics**	London and New York
C **Mathematical and Physical Sciences**	Kluwer Academic Publishers
	Dordrecht, Boston and London
D **Behavioural and Social Sciences**	
E **Applied Sciences**	
F **Computer and Systems Sciences**	Springer-Verlag
G **Ecological Sciences**	Berlin, Heidelberg, New York, London,
H **Cell Biology**	Paris and Tokyo
I **Global Environmental Change**	

NATO-PCO-DATA BASE

The electronic index to the NATO ASI Series provides full bibliographical references (with keywords and/or abstracts) to more than 30000 contributions from international scientists published in all sections of the NATO ASI Series.
Access to the NATO-PCO-DATA BASE is possible in two ways:

– via online FILE 128 (NATO-PCO-DATA BASE) hosted by ESRIN,
Via Galileo Galilei, I-00044 Frascati, Italy.

– via CD-ROM "NATO-PCO-DATA BASE" with user-friendly retrieval software in English, French and German (© WTV GmbH and DATAWARE Technologies Inc. 1989).

The CD-ROM can be ordered through any member of the Board of Publishers or through NATO-PCO, Overijse, Belgium.

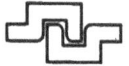

Series E: Applied Sciences - Vol. 195

Toughening Mechanisms in Quasi-Brittle Materials

edited by

S. P. Shah

NSF Science and Technology Center for Advanced Cement-Based Materials,
Robert R. McCormick School of Engineering and Applied Sciences,
Northwestern University, Evanston, Illinois, U.S.A.

Springer-Science+Business Media, B.V.

Proceedings of the NATO Advanced Research Workshop on
Toughening Mechanisms in Quasi-Brittle Materials
Evanston, Illinois, U.S.A.
16–20 July, 1990

Library of Congress Cataloging-in-Publication Data

Toughening mechanisms in quasi-brittle materials : preprints of the
  proceedings, NATO advanced research workshop, July 16-20, 1990 /
  S.P. Shah, editor ; sponsored by NATO Scientific Affairs Division
  ... [et al.].
        p.   cm. -- (NATO ASI series. Series E, Applied sciences ; vol.
  195)
    Includes indexes.
    ISBN 978-0-7923-1198-0   ISBN 978-94-011-3388-3 (eBook)
    DOI 10.1007/978-94-011-3388-3
    1. Ceramic materials--Fracture--Congresses. 2. Concrete-
  -Fracture--Congresses. 3. Fracture mechanics--Congresses.
  4. Micromechanics--Congresses.   I. Shah, S. P. (Surendra P.)
  II. North Atlantic Treaty Organization. Scientific Affairs
  Division. III. Series: NATO ASI series. Series E, Applied sciences
  ; no. 195.
  TA430.T68   1991
  620.1'126--dc20   91-2598

ISBN 978-0-7923-1198-0

Printed on acid-free paper

Table of Contents

PREFACE

A variety of ceramic materials has been recently shown to exhibit nonlinear stress-strain behavior. These materials include transformation-toughened zirconia which undergoes a stress-induced crystallographic transformation in the vicinity of a propagating crack, microcracking ceramics, and ceramic-fiber reinforced ceramic matrices. Since many of these materials are under consideration for structural applications, understanding fracture in these quasi-brittle materials is essential.

Portland cement concrete is a relatively brittle material. As a result mechanical behavior of concrete, conventionally reinforced concrete, prestressed concrete and fiber reinforced concrete is critically influenced by crack propagation. Crack propagation in concrete is characterized by a fracture process zone, microcracking, and aggregate-bridging. Such phenomena give concrete toughening mechanisms, and as a result, the macroscopic response of concrete can be characterized as that of a quasi-brittle material. To design super high performance cement composites, it is essential to understand the complex fracture processes in concrete.

A wide range of concern in design involves fracture in rock masses and rock structures. For example, prediction of the extension or initiation of fracture is important in: 1) the design of caverns (such as underground nuclear waste isolation) subjected to earthquake shaking or explosions, 2) the production of geothermal and petroleum energy, and 3) predicting and monitoring earthquakes. Depending upon the grain size and mineralogical composition, rock may also exhibit characteristics of quasi-brittle materials.

Recently, considerable interest has been developing in understanding and modeling the fracture processes in these quasi-brittle materials as well as in designing materials with improved toughness. The research activities in these groups of materials: ceramics, cement and rock can be substantially enhanced with the exchange of information between these three groups of investigators. Since the field is relatively new, it is likely that the researchers working with one set of materials are not aware of similar developments with other sets of materials. Although each material has its own set of specific characteristics, many common aspects can be shared among these quasi-brittle materials. These include: 1) application of nonlinear fracture mechanics, 2) experimental and theoretical considerations of strain localization, 3) microscopic observation of fracture process zone, 4) non-destructive evaluation of damage, 5) models to relate microstructure with macroscopic response, and 6) development of experimental and theoretical tools.

The purpose of this workshop was to bring together researchers addressing the problem of fracture in cement, ceramics, and rock so that they can share their knowledge and develop a more general syntheses of the problem.

This volume contains contributions from lecturers and reporters for each of the 9 sessions.

I hope that the efforts of all who have contributed to this workshop will produce lasting and worthwhile results.

Surendra P. Shah
January, 1991
Evanston, Illinois, USA

ix

ACKNOWLEDGEMENTS

This workshop was one of a series of NATO Advanced Research Workshops and was sponsored by the Scientific Affairs Division of the North Atlantic Treaty Organization under a program directed by G. A. Venturi and assisted by James Bombace. Additional financial support was provided by the U. S. Air Force Office of Scientific Research (Spencer T. Wu, Program Director), the U. S. Air Force WRDC/MLLN (Theodore Nicholas, Program Director), and the Materials Research Division of the U. S. Army Research Office (Edward Chen, Program Director).

Advice and assistance received from the following was greatly appreciated: Jerome Cohen, Dean of the Robert R. McCormick School of Engineering and Applied Sciences; Raymond Krizek, Chairman of the Department of Civil Engineering; Mary Lynne Williams, Administrative Assistant, and Auria Rosenberg, Executive Secretary for the NSF Science and Technology Center for Advanced Cement-Based Materials.

Thanks are due to Eric Landis for his work as Workshop Secretary. Additional assistance was provided by Chengcheng Ouyang, David Lange, Tianxi Tang, Sokhwan Choi, and Alan Pope.

The Editor would also like to acknowledge the support and patience of Tjaddie Ammerdorffer of Kluwer Academic Publishers.

Fracture of Ceramics with Process Zone

FRACTURE PROPERTIES OF SiC-BASED PARTICULATE COMPOSITES

K. T. Faber, W.-H. Gu, H. Cai, and R. A. Winholtz
Northwestern University
Department of Materials Science and Engineering
Evanston, Illinois 60208
U.S.A.

D. J. Magley
The Ohio State University
Department of Materials Science and Engineering
Columbus, Ohio 43210
U.S.A.

ABSTRACT. In order to evaluate the role of residual stresses in fracture toughening, a SiC-based particulate composite has been studied under uniform stressing conditions and in the near tip stress field of a pre-cracked specimen. First, residual stresses in a SiC-TiB2 composite before and after stressing have been measured using x-ray diffraction. Tensile residual stresses in the TiB2 drop by 50% after bending stresses of 250 MPa were applied. Likewise, the compressive residual stresses in the SiC phase decrease accordingly. Second, in the near tip stress field, a process zone of microcracks has been measured using transmission electron microscopy of thin foils taken from various locations from a fracture surface of a fracture mechanics specimen. Microcrack zones greater than 150 μm in height have been measured. Crack bridging sites of TiB2 particles operate more than a few millimeters behind a propagating crack. Hence, the toughening in this system is comprised of both stress-induced microcracking and crack bridging. The various contributions to the toughening are discussed.

1. Introduction

The toughening of ceramic materials has received a great deal of attention over the last decade. Mechanisms by which brittle materials can be toughened fall into two categories: process zone mechanisms and bridging mechanisms [1]. Process zone toughening, either by stress-induced phase transformations or stress-induced microcracking, provides shielding of a propagating crack by virtue of the microstructural changes which occur in the near vicinity of the crack. Bridging mechanisms operate behind a crack tip and provide closure forces which also act to reduce the crack tip stress intensity. In both cases, residual stresses may serve either as a source or as a consequence of

3

S. P. Shah (ed.), Toughening Mechanisms in Quasi-Brittle Materials, 3–17.
© 1991 *Kluwer Academic Publishers.*

the toughening process and should be considered in examining the
toughening increment.

Residual stresses in two phase materials have long been recognized
as having a significant influence on mechanical strength. In the
extreme, residual stresses can result in spontaneous microcracking on
cooling, destroying mechanical integrity. The conditions under which
such cracking occurs are now well established [2,3]. Cracking can be
avoided by maintaining a particle size distribution below some critical
size, b_c, described by

$$b_c = \eta \, K_{Ic}^2 \{(1+\nu_m)/2 + \beta(1-2\nu_p)\}/[E_m \, \Delta\alpha\Delta T]^2 \qquad (1)$$

where η is constant ranging from 2 to 8, K_{Ic} is the fracture toughness
of the microcracking site (often the interface), E is the elastic
modulus, ν is Poisson's ratio, β is the modulus ratio, E_m/E_p, and the
subscripts, m and p, refer to the matrix and particulate phases,
respectively. The product $\Delta\alpha\Delta T$ is the thermal mismatch strain, where $\Delta\alpha$
is the difference in thermal expansion coefficients of the matrix and
particulate phases, and ΔT is the difference between the temperature at
which relaxation stops and the temperature of interest. However, it is
below the threshold b_c where the present interest lies. It has been
postulated that a regime exists whereby residually-stressed particles in
the vicinity of an advancing crack will microcrack when a critical value
of applied stress is reached. Such stress-induced microcracking is
suggested to result in significant toughening by shielding the crack
from the applied loading [3-7]. It is the intent of this paper to
examine stress-induced phenomena in two-phase, non-transforming ceramics
under uniform and non-uniform loading. The system chosen for study is a
SiC-TiB2 composite where residual stresses arise from a high thermal
expansion mismatch.

2. The Silicon Carbide-Titanium Diboride System

Silicon carbide, although a highly refractory material, is limited by
its low fracture toughness ($K_{Ic} \cong 3.22$ MPa$\sqrt{m}$). Additions of TiB2 to SiC
have provided significant increases in the fracture toughness ($\Delta K_c \cong 5.0$
MPa$\sqrt{m}$) as measured by a few investigators [8-10], although the operative
toughening mechanisms were not identified. SiC-TiB2 has also been known
to exhibit R-curve behavior [10], a manifestation of both shielding and
bridging processes.

The materials studied here were prepared by pressureless sintering
at temperatures in excess of 2000°C and were nearly 99% dense. The
materials examined in this study contained TiB2 concentrations of 15.2%,
by volume. The average TiB2 grain size was measured to be approximately
4.5 μm, less than the critical size for spontaneous microcracking
calculated using Eqn. (1). Residual stresses in this system arise from
a high thermal expansion mismatch where α (SiC) $\cong 5.6 \times 10^{-6}$ $^{\circ}C^{-1}$ and α
(TiB2) $\cong 7.9 \times 10^{-6}$ $^{\circ}C^{-1}$ over the temperature range 25° to 1700°C. We
may approximate the residual stress by considering a spherical TiB2

particle in an isotropic SiC matrix by [11]:

$$\sigma^R = \frac{\Delta\alpha\Delta T}{(1+\nu_m)/(2E_m) + (1-2\nu_p)/E_p} \qquad (2)$$

For the appropriate thermal and elastic properties of SiC and TiB2, σ^R is of the order of 1.9 GPa.

3. Experimental Observations of Toughening Processes in SiC-TiB$_2$

In an effort to examine the potential toughening mechanisms associated with the high degree of residual stress in these SiC-based materials, a variety of experimental techniques have been utilized. First, if stress-induced microcracking is actually occurring, then some monitor of the stress relief on microcracking is warranted. Residual stress analysis via x-ray diffraction is chosen for such studies. This technique will provide conclusive, albeit indirect, evidence for the microcracking process. Direct observation of the microcrack zone via transmission electron microscopy is also employed. Finally, additional toughening behind the crack, also associated with residual stress, is examined through interrupted fracture experiments. The three will be described herein.

3.1 Microcracking Under Uniform Stress

To establish stress-induced microcracking in a uniform stress field, residual stresses were measured by x-ray diffraction prior to stressing and upon unloading. For this investigation, the SiC-TiB2 specimens were electrically discharge machined into bars 7.62 x 1.27 x 0.635 cm^3 and polished to a 0.25 μm finish using a sequential treatment of diamond pastes. A pressureless-sintered SiC, containing sintering aids of boron and carbon, was diamond machined with identical surface finishes and used as a standard. The reflections examined were the (121) for 6H SiC, the major SiC polytype present, and the (202) for TiB2 at approximately 148° 2θ and 144° 2θ, respectively. Further details of the x-ray diffraction experimental set up may be found in [12].

During x-ray diffraction the sample was oriented according to the scheme shown in Figure 1. A General Electric quarter circle was used for both the ψ-tilts and the ϕ-rotations. Diffraction measurements were made at six ϕ values between 0 and 300° and six ψ values between 0 and 45° for each ϕ setting. Peak positions at any ϕ, ψ rotation were measured by step scanning at 0.05° 2θ increments for 100 seconds. Because peaks have a tendency to broaden as the specimen is rotated about the ψ axis due to x-ray defocussing, longer counting times were employed at large values of ψ for greater accuracy. Peak positions were determined by fitting the measured values of intensity versus 2θ to a parabola.

After the x-ray measurements were made and the residual stresses determined, the SiC-TiB2 composite and the SiC standard were placed in a

6

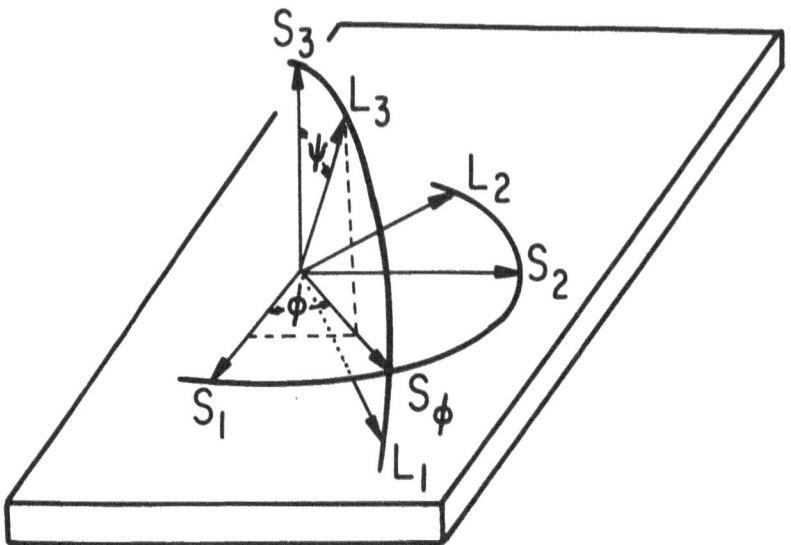

Figure 1. The coordinate system used in residual stress determinations. The S_1-S_2 surface represents the tensile surface of the SiC-TiB$_2$ bend bar.

four-point bend fixture and loaded until fracture occurred. At fracture the outer fiber tensile stress reached 249 and 266 MPa for the composite and the standard, respectively. X-ray measurements were repeated in areas which experienced the maximum tensile stress to monitor any changes in the residual stress profile following stressing. Although four-point bending can hardly be considered uniform loading, the depth of penetration of the x-rays (approximately 50 μm below the surface to account for 90% of the x-ray intensity) allows examination of the stress over a nearly uniformly stressed region.

The complete stress matrix can be determined from the stress analysis, as shown by Noyan [13] and Cohen [14]. The change in the interplanar spacing related to stresses in the coordinate system in Figure 1 is

$$\frac{d_{\phi\psi}-d_o}{d_o} = \epsilon_{\phi\psi} = \frac{S_2}{2}(\sigma_{11}\cos^2\phi + \sigma_{12}\sin2\phi + \sigma_{22}\sin^2\phi - \sigma_{33})\sin^2\psi$$

$$+ \frac{S_2}{2}\sigma_{33} - S_1(\sigma_{11} + \sigma_{22} + \sigma_{33})$$

$$+ \frac{S_2}{2}(\sigma_{13}\cos\phi + \sigma_{23}\sin\phi)\sin2\psi \qquad (3)$$

where d_0 is the unstressed lattice spacing, ϵ represents the strain, and S_1 and $S_2/2$ are the x-ray elastic constants $1+\nu/E$ and ν/E, respectively, for an isotropic solid. (The bulk values of ν and E were used to determine their values [15,16].) The stresses were determined from the measured d-spacings by a least-squares procedure [17]. The x-ray d-spacing measurements yield the total stresses, $^t\sigma_{ij}$, associated with each phase. These must be partitioned into microstresses, $^\mu\sigma_{ij}$, (those resulting from thermal expansion mismatch and macrostresses, $^m\sigma_{ij}$, (those from machining):

$$\left\langle ^t\sigma_{ij}^{\ SiC}\right\rangle = \left\langle ^\mu\sigma_{ij}^{\ SiC}\right\rangle + \left\langle ^m\sigma_{ij}\right\rangle \tag{4}$$

$$\left\langle ^t\sigma_{ij}^{\ TiB_2}\right\rangle = \left\langle ^\mu\sigma_{ij}^{\ TiB_2}\right\rangle + \left\langle ^m\sigma_{ij}\right\rangle \tag{5}$$

where the macrostresses are assumed to be the same in each phase. The total microstresses must also sum to zero, such that

$$(1-f)\left\langle ^\mu\sigma_{ij}^{\ SiC}\right\rangle + f\left\langle ^\mu\sigma_{ij}^{\ TiB_2}\right\rangle = 0 \tag{6}$$

where f is the volume fraction of the TiB$_2$ phase. Solving Eqns. (4), (5) and (6) simultaneously allows the microstresses and macrostresses of each phase to be resolved:

$$\left\langle ^\mu\sigma_{ij}^{\ SiC}\right\rangle = f\left[\left\langle ^t\sigma_{ij}^{\ SiC}\right\rangle - \left\langle ^t\sigma_{ij}^{\ TiB_2}\right\rangle\right] \tag{7}$$

$$\left\langle ^\mu\sigma_{ij}^{\ TiB_2}\right\rangle = (1-f)\left[\left\langle ^t\sigma_{ij}^{\ TiB_2}\right\rangle - \left\langle ^t\sigma_{ij}^{\ SiC}\right\rangle\right] \tag{8}$$

$$\left\langle ^m\sigma_{ij}\right\rangle = (1-f)\left\langle ^t\sigma_{ij}^{\ SiC}\right\rangle + f\left\langle ^t\sigma_{ij}^{\ TiB_2}\right\rangle \tag{9}$$

A difficulty of triaxial stress measurement by diffraction has been the determination of a precise valve of d_0. It has been shown that errors in d_0 give rise to an error in the hydrostatic microstresses

[18]. In contrast, errors of this kind will cancel out in the present measurements when comparing the pre-stressed with the post-stressed material.

Rather than examining the total stress matrices in order to compare the pre-loaded and post-loaded microstresses, the stress matrices are better analyzed by examining the hydrostatic component of the stress. However, one must first note that the stresses derived from x-ray diffraction analysis do not provide the true internal stresses. The σ_{33} component is influenced by the near surface effect, and should necessarily be zero if stresses only at the surface are measured. However, the σ_{33} measurement represents d-spacing data collected from up to 50 microns, or approximately 10 grain diameters below the surface. To supersede these difficulties, an "effective" hydrostatic stress (defined as $(\sigma_{11} + \sigma_{22})/2$) has been calculated and it is plotted in histogram form in Figure 2. It is clear that the data represents definitive evidence for residual stress relaxation on application of a stress in SiC-TiB2 as the microstresses drop by nearly 50%. The SiC standard, in contrast, shows no significant change in the average stress after loading. The present observations demonstrate indisputable evidence for stress-induced microstress relief in a non-transforming brittle material. These results are consonant with the model of stress-induced microcracking in SiC-TiB2.

3.2 Microcracking in the Near Crack Tip Stress Field

The near crack tip region in conventional fracture mechanics specimens provides a region to examine stress-induced process zone formation. The crack opening displacement of microcracks in SiC-TiB2 composites is too small to be detected by either optical microscopy or scanning electron microscopy; the resolution of transmission electron microscopy is necessary to explore any microcracking. Thin sections were cut from the fracture surface of the SiC-TiB2 composite prepared from four depths from the fracture surface of a double cantilever beam specimen. Details of the TEM sample preparation are given in [19].

Microcracks can be identified by Fresnel diffraction using the underfocus-overfocus process in the bright-field image [20]. Typical microcracks occur at the boundaries between TiB2 and SiC (Figure 3(a)), due to the thermal expansion mismatch described earlier, or between TiB2 particles (Figure 3(b)), due to thermal mismatch or thermal expansion anisotropy in the hexagonal TiB2 structure. The former, however, are the more common. The thickness enabling the image formation is about 0.1 μm and the crack opening displacement is about 26 nm. Therefore, it is necessary to rotate the thin foil through various stereo angles to observe every microcrack. The observable fraction of the solid angle for the double tilt stage which is allowed to rotate from $-30°$ to $+30°$ is 0.29.

The distribution of microcrack lengths at different distances from the fracture surface are similar in shape and are shown in Figure 4. The number of microcracks increases dramatically as the fracture surface is approached. The microcracks located at approximately 5 mm from the fracture surface represent cracks produced during sample preparation,

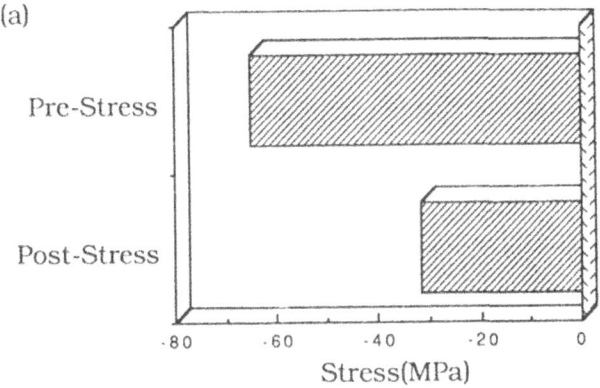

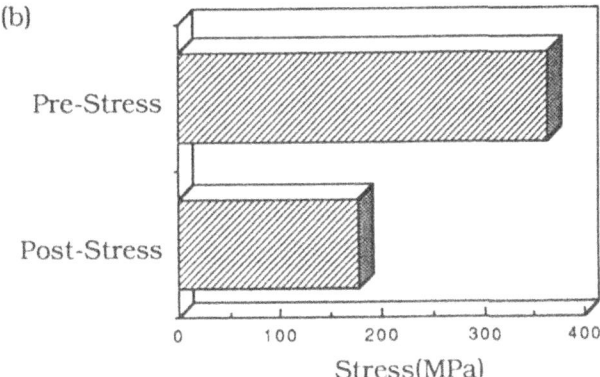

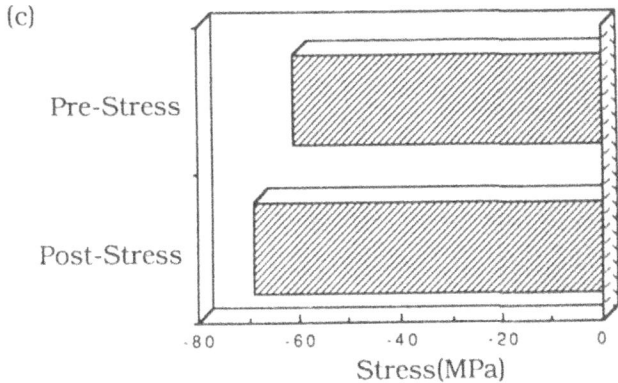

Figure 2. The effective hydrostatic stresses before and after loading for (a) the SiC phase in SiC-TiB₂, (b) the TiB₂ phase in SiC-TiB₂, and (c) the SiC standard.

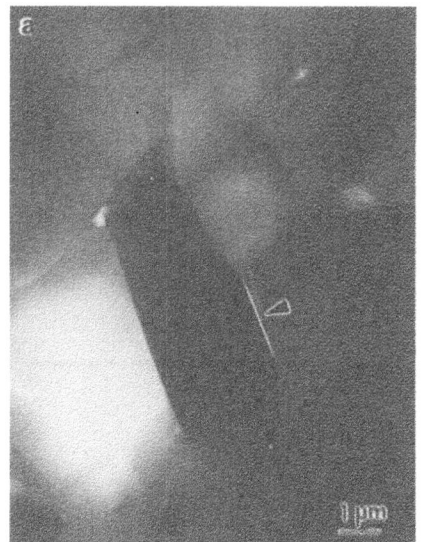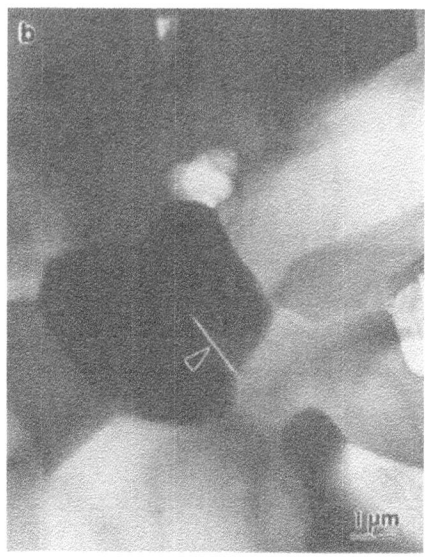

Figure 3. Transmission electron micrographs of (a) microcracks occurring on the boundary between TiB₂ and SiC and (b) a microcrack on the boundary between two TiB₂ particles.

either occurring spontaneously during cooling from the sintering temperature (based upon the conditions in Eqn. 1) or as a result of ion thinning, and will be treated as background. Since the TiB₂ particles are polyhedra and the microcracks are generally located along grain facets of the polyhedra, a microcrack can be reasonable treated as a penny-shaped crack. Hence, the microcrack density parameter, ϵ, may be defined as [21]:

$$\epsilon = \frac{3}{4\pi} N_A <\ell^2>$$ (10)

where N_A is the number of microcracks per unit area and ℓ is the length of the microcrack measured in the thin foil. The mean microcrack diameter is approximately 2.8 μm and is independent of distance from the fracture surface. This implies that the residual stress from the thermal expansion mismatch dominates the applied stress term. The profile of microcrack density perpendicular to the fracture surface is shown in Figure 5. These data are corrected for the fraction of the observable solid angle described above. The microcrack density is said to be saturated; that is, all TiB₂ particles have a least one microcrack associated with them, and ϵ equals 0.53 near the fracture surface. Furthermore, from the limited data, ϵ linearly decreases with increasing distance from the fracture surface. The microcrack density extrapolates to the background level at about 160 μm if a linear profile is assumed. Microcrack zones on the order of hundreds of microns in a similar SiC-TiB₂ have also been measured using small angle x-ray scattering [22].

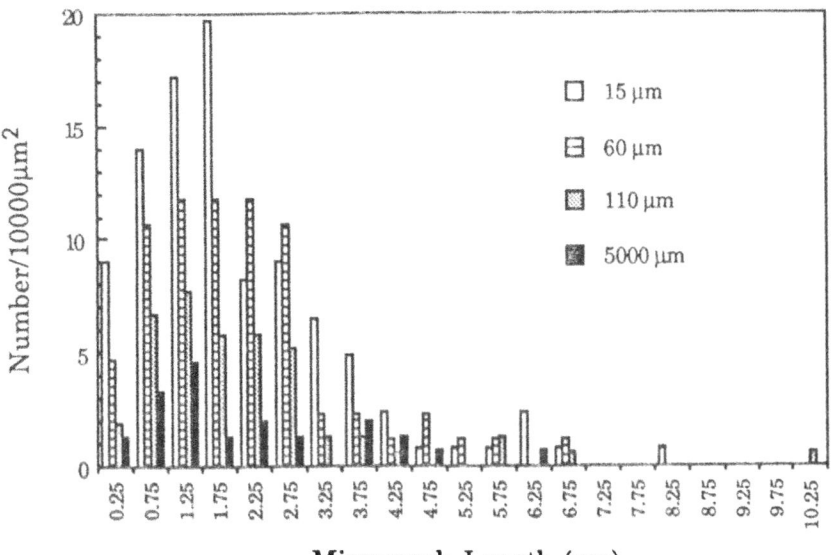

Figure 4. Microcrack distributions at different distances from the crack plane in a SiC-TiB2 double cantilever beam specimen.

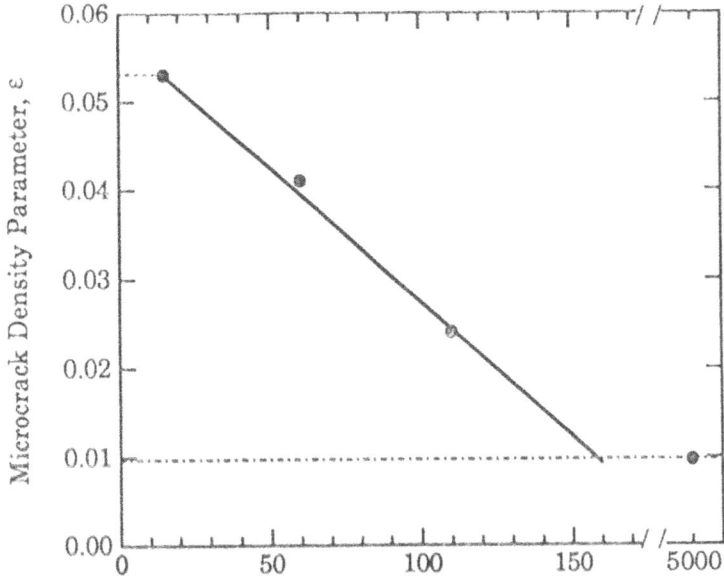

Figure 5. The microcrack density parameter, ϵ, as a function of the distance from the crack plane in a SiC-TiB2 double cantilever beam specimen.

This quantitative examination provides unambiguous information to evaluate microcrack toughening. It provides the first direct evidence of stress-induced microcracking in a particulate non-transforming ceramic composite by demonstrating the existence of a microcrack process zone.

3.3 Crack Bridging Behind an Advancing Crack

The fracture surface of SiC-TiB2 provides further insight into the mechanisms which give rise to enhanced toughening in this system. As in unreinforced sintered SiC [23], fracture in the SiC appears to be primarily transgranular. However, TiB2 particles rarely fail transgranularly at the crack tip in the materials studied. Upon examination of double cantilever beam specimens interrupted during testing, there are numerous examples of TiB2 particles acting as bridges spanning the primary crack (Figure 6). On examination of samples during testing and prior to catastrophic failure on unloading, the bridged grains were observed to be active nearly 3 mm from the crack tip. At large distances from the crack tip, breakaway generally occurs in the SiC matrix, or by complete pullout of the TiB2 grains. Given these microstructural observations, the contribution of crack bridging to the toughening by unbroken TiB2 grains must also be evaluated in addition to the contribution of stress-induced microcracking.

4. The Toughening Analysis

4.1 Stress-Induced Microcracking

In order to assess the contribution of stress-induced microcracking, we examine the model of Hutchinson [24]. Shielding from microcracking is comprised of two contributions: shielding due to the reduced moduli and shielding due to residual strain associated with the microcracking event. For conditions where the crack is assumed to be steadily growing with a zone of randomly-oriented microcracks induced by a critical mean stress, the shielding due to the reduced moduli can be written as

$$K_I/K_I^\infty = 1 - 0.608 \, \delta_1 + 0.707 \, \delta_2 \tag{11}$$

where K_I is the stress intensity factor at the crack tip, K_I^∞ is the applied stress intensity and δ_1 and δ_2 are given as

$$\delta_1 = 32 \, (5-\nu) \, \epsilon_s / \, [45 \, (2-\nu)]$$

and

$$\delta_2 = 16\nu \, (1-8\nu+3\nu^2) \, \epsilon_s / \, [45 \, (2-\nu)]$$

where ϵ_s is the saturated microcrack density.

The shielding contribution from the residual strain is of the form

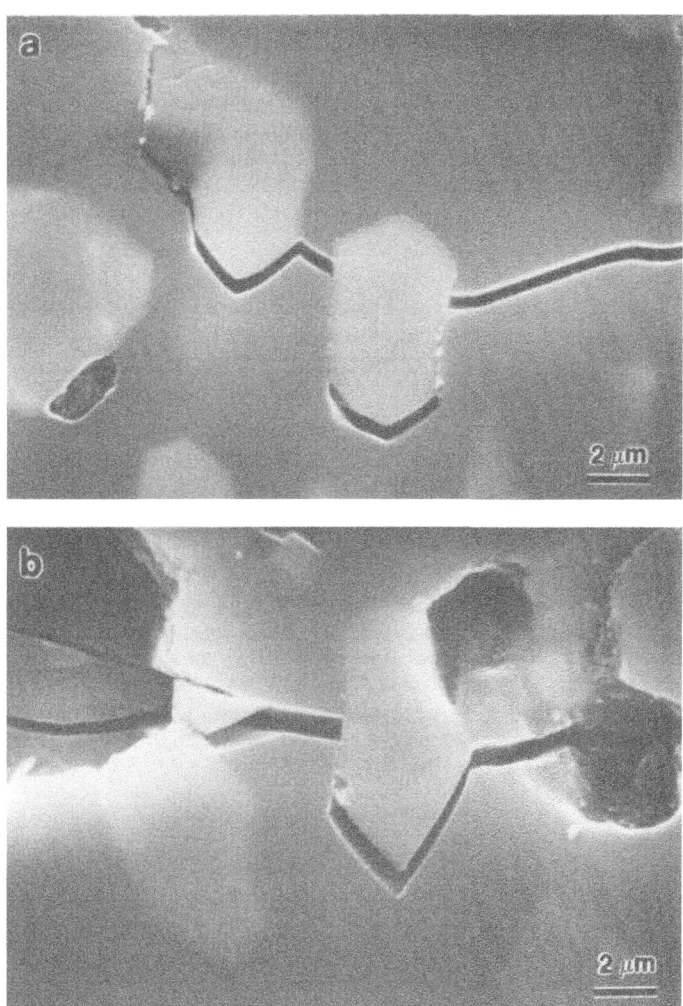

Figure 6. Scanning electron micrographs of bridging TiB2 grains behind the crack tip in a SiC-TiB2 composite. Micrographs were taken during an interrupted double cantilever beam test.

$$K_I - K_I^\infty = \frac{1}{3\sqrt{2\pi}} \frac{E}{(1-\nu)} \int \theta^M \frac{\cos(3\theta/2)}{r^{3/2}} \, dA \qquad (12)$$

where θ and r are the polar coordinates, θ^M is the dilatation strain associated with microcracking, and the integral is taken over the upper half of the microcrack zone. We approximate the microcrack which forms along a SiC-TiB2 facet by a prototypic microcrack which forms in the center of a particle under a residual tensile stress. The dilatation strain may then be written

$$\theta^M = (16/3) (1-\nu^2) \, \epsilon \, \sigma^R / E \qquad (13)$$

where σ^R is the residual stress. The elastic mismatch between SiC and TiB2 is small, and therefore, neglected.

Substituting Eqn (13) into Eqn (12), we obtain

$$K_I - K_I^\infty = \frac{16}{9\sqrt{2\pi}} (1+\nu) \, \sigma^R \int \epsilon \, \frac{\cos(3\theta/2)}{r^{3/2}} \, dA \qquad (14)$$

The above integral was evaluated numerically using the microcrack distribution information presented in Section 3.2, and coupled with calculations of σ^R. The amount of shielding due to the residual strain contribution for the SiC-TiB2 system is then computed to be -1.98 MPa√m. The combined effects of the modulus reduction and residual strain derive from Eqns. (11) and (14). However, to evaluate the toughening, one must recall that a crack must now propagate through a medium of microcracks. The reduced toughness of the microcracked material K_c^m ahead of the crack tip has been found to be of the form, [25]

$$K_c^m/K_c = 1 - \zeta \, \epsilon \qquad (15)$$

where K_c is the toughness of the microcrack free material and the value of ζ is 0.825. By setting $K_I = K_c^m$ in Eqn (14), and equating K_c to the toughness of monolithic SiC (3.22 MPa√m), K_I^∞ affords a prediction of the toughness due to stress-induced microcracking. For the SiC-TiB2 examined, the toughening due to stress-induced microcracking is approximately 2.18 MPa√m, on the order of half of the observed toughening.

4.2 Crack Bridging

We use the model of Campbell et al. [26] designed for whisker-reinforced materials to examine toughening due to the bridging of TiB2 grains. The steady state toughening from whisker reinforcements can be written in terms of the critical strain energy release rate, $\mathcal{G}_c$, as

$$\Delta \mathcal{G}_c/fd \approx S^2/E - Ee_T^2 + 4(\Gamma_i/R)/(1-f) + (\tau/d)\sum_i(h_i^2/R) \qquad (16)$$

where d is the debond length, e_T is the misfit strain, Γ_i is the interface fracture energy, R is the reinforcement radius, f is the volume fraction of reinforcing particles, S is the reinforcement strength, τ is the sliding resistance, and h_i is the pullout length. The four terms represent, respectively: the stored elastic strain energy in the reinforcement over the microcrack length prior to failure, the residual strain energy within the microcrack length, the energy needed to fracture the interface and the energy contribution to pullout.

For particles under residual tension, the last term can be neglected and we use Campbell et al.'s solution for reinforcements of arbitrary aspect ratio under tension which are oriented normal to the applied stress:

$$\Delta \mathcal{G}_c = fS^2R \; [(\lambda_1+\lambda_2d/R)^2 - (E_p e_T/S)^2(\lambda_3 + \lambda_4d/R)^2]/\{E_p(\lambda_1 + \lambda_2d/R)\}$$

$$+ 4f\Gamma_id/(1-f)R \qquad (17)$$

The parameters, λ_i, for i = 1 to 4 are coefficients used to describe the matrix crack opening for a given value of modulus mismatch between the reinforcement and the matrix. Any estimate of the toughening increment will represent the toughening afforded from bridging only from grains oriented parallel to the applied loading (i.e. no TiB_2 grains are subject to bending moments, and consequently, there is no effective frictional pullout contribution). Further, these estimates are limited by the reliability of our estimates of the bridge strength, S, and of the interfacial toughness, Γ_i.

For estimates of the necessary parameters, we make the following assumptions: First, the elastic strain energy component is highly sensitive to the choice of reinforcement strength, S. As the TiB_2 particles rarely fracture in the wake, one would anticipate that the interface is strength-controlling. The strength of the interfaces may then be estimated to be bounded by the average residual stress, 1.6 GPa [19], since the interfaces did not spontaneously crack on cooling. Second, the misfit strain component is reduced from its theoretical value due to microcracking ahead of the crack tip. We calculate an effective misfit strain based upon microstress results from x-ray studies. Hence, the strain e_T is approximately 50% of the microcrack-free value. Third, the interfacial fracture toughness (Γ_i) is estimated to be of the order of SiC, i.e. approximately 15 J/m^2, as an upper bound. Fourth, debonding only occurs over the grain facet length *less* the average microcrack length. TEM observations have provided data for the average microcrack length, 2.1 μm, in this case. From the above assumptions, it is clear that the energy contribution to fracture the SiC-TiB_2 interface is dominant, though not large ($\cong$ 5.6 J/m^2). The elastic strain energy and the residual strain components are of nearly equal magnitude, though opposite in sign, and nearly cancel. Converting $\Delta\mathscr{G}_c$ to ΔK_c, the toughening increment due to crack bridging is equal to approximately 2.0 MPa$\sqrt{m}$. This value is of the correct order of magnitude to account for the remainder of the fracture toughening. However, the effect of interlocking grains (not oriented parallel to the applied stress) and their effective frictional contribution to the pullout are not accounted for and could result in additional toughening.

5. Concluding Remarks

In the SiC-TiB$_2$ composites studies, residual stresses are of a large enough magnitude to cause stress-induced microcracking in the near vicinity of a propagating crack. Both x-ray residual stress measurements and direct transmission electron microscopy observations provide evidence for this phenomenon. The total toughening, however, cannot be explained totally in terms of stress-induced microcracking. Instead, contributions from crack bridging are of the same order of magnitude as those from microcracking. A likely scenario is one suggested by Amazigo and Budiansky [27] where the bridging and process zone mechanisms are likely to interact. Specifically, the bridged crack, which allows for a greater crack tip stress intensity, will afford larger process zones. Direct measurements of the microcracking stress are required to test this hypothesis.

Acknowledgments

Materials for this study were graciously provided by the Carborundum Company, Niagara Falls, NY. This work was supported by the National Science Foundation under Grant No. DMR-8896212.

References

1. Clarke, D. R. and Faber, K. T. (1987) 'Fracture of Ceramics and Glasses', J. Phys. Chem. Solids, 48, 1115-55.
2. Davidge, R. J. and Green, T. J. (1968) 'The Strength of Two-Phase Ceramic/Glass Materials', J. Mater. Sci. 3, 629-34.
3. Evans, A. G. and Faber, K. T. (1981) 'Toughening of Ceramics by Circumferential Microcracking', J. Am. Ceram. Soc. 64, 394-98.
4. Green, D. J., Nicholson, P. S. and Embury, J. D. (1973) 'Fracture Toughness of Partially Stabilized ZrO$_2$ in the System CaO-ZrO$_2$', J. Am. Ceram. Soc. 56, 619-23.
5. Hoagland, R. G., Embury J. D. and Green, D. J. (1975) 'On the Density of Microcracks Formed During the Fracture of Ceramics', Scripta metall. 9, 907-09.
6. Evans, A. G. (1976) 'On the Formation of a Crack Tip Microcrack Zone', Scripta metall. 10, 93-97.
7. Evans, A. G. and Fu, Y. (1985) 'Some Effects of Microcracks on the Mechanical Properties of Brittle Solids-II. Microcrack Toughening', Acta metall. 33, 1525-31.

8. Garnier, J. E., McMurtry, C. H. and Zangvil, A. 'A Microanalytical Study of a SiC-based TiB2 Particulate Composite', unpublished work.

9. McMurtry, C. H., Boecker, W. D. G., Seshadri, S. G., Zanghi, J. S. and Garnier, J. E. (1987) 'Microstructure and Material Properties of SiC-TiB2 Particulate Composites', Am. Ceram. Soc. Bull. 66, 325-29.

10. Magley, D. J. (1989) M.S. Thesis, The Ohio State University.

11. Selsing, J. (1961) 'Internal Stresses in Ceramics', J. Amer. Ceram. Soc. 44, 419.

12. Magley, D. J., Winholtz, R. A. and Faber, K. T. (1990) 'Residual Stresses in a Two Phase Microcracking Ceramic', J. Amer. Ceram. Soc. 73, 1641-44.

13. Noyan, I. C. (1983) 'Equilibrium Conditions for the Average Stresses Measured by X-Rays', Metall. Trans. A 14A, 1907-1914.

14. Cohen, J. B. (1986) 'The Measurement of Stresses in Composites', Powder Diffraction 1, 15-21.

15. Simmons, G. and Wang, H. (1971) Single Crystal Elastic Constants and Calculated Aggregate Properties, A Handbook, 2nd Ed., M.I.T. Press, Cambridge, MA.

16. Lyubimskii, V. M. (1977) 'Elastic Properties of Silicon Carbide Polytypes', Sov. Phys. Solid State 18, 1814-15.

17. Winholtz, R. A. and Cohen, J. B. (1988) 'Generalized Least-squares Determination of Triaxial Stress States by X-Ray Diffraction and the Associated Errors', Aust. J. Phys. 41, 189-199.

18. Winholtz, R. A. and Cohen, J. B. (1989) 'Separation of the Macro- and Micro-Stresses in Plastically Deformed 1080 Steel', Adv. in X-Ray Anal. 32, 341-53.

19. Cai, H., Gu, W.-H. and Faber, K. T. (1990) 'Microcrack Toughening in a SiC-TiB2 Composite', in Proc. Am. Soc. Composites, Fifth Technical Conference on Composite Materials, Technomic Publishing Company, Lancaster, PA, pp. 892-901.

20. Clarke, D. R., (1980) 'Observation of Microcracks and Thin Intergranular Films in Ceramics by Transmission Electron Microscopy', J. Amer. Ceram. Soc. 63, 104-06.

21. Budiansky, B. and O'Connell, R. J. (1976) 'Elastic Moduli of a Cracked Solid', Int. J. Solids Structures 12, 81-97.

22. Steinbrech, R., personal communication.

23. Faber, K. T. and Evans, A. G., (1983) 'Intergranular Crack-Deflection Toughening in Silicon Carbide', J. Amer. Ceram. Soc. 66, C94-96.

24. Hutchinson, J. W. (1987) 'Crack Tip Shielding by Micro-Cracking in Brittle Solids', Acta metall. 35, 1605-19.

25. Suh, T.-G. (1988) M.S. Thesis, Michigan State University.

26. Campbell, G. H., Rühle, M., Dalgleish, B. J. and Evans, A. G. (1990) 'Whisker Toughening: A Comparison Between Aluminum Oxide and Silicon Nitride Toughened with Silicon Carbide', J. Amer. Ceram. Soc. 73, 521-30.

27. Amazigo, J. C. and Budiansky, B. (1988) 'Interaction of Particulate and Transformation Toughening', J. Mech. Phys. Solids 36, 581-95.

CRACK BRIDGING PROCESSES IN TOUGHENED CERAMICS

P. F. BECHER
Metals and Ceramics Division
Oak Ridge National Laboratory
P. O. Box 2008
Oak Ridge, Tennessee 37831-6068

ABSTRACT. The fracture toughness of ceramics can be improved by the incorporation of a variety of brittle, discontinuous reinforcing phases. Observations of crack paths in these systems indicate that these reinforcing phases bridge the crack in the region behind the crack tip. Recent developments in toughening models based on crack bridging processes in such systems are discussed and compared to the experimentally observed toughening responses with second-phase whisker and self- (matrix grain) reinforcement. The bridging model then can be used to optimize the toughening effects based modification of the pertinent material characteristics (e.g., microstructure and physical properties).

1. Introduction

The brittle nature of ceramics has prompted us over the years to explore a variety of approaches to increasing their fracture toughness/resistance. Initially, the concern was to toughen these materials to improve their fracture strength and/or reduce the flaw size sensitivity of the fracture strengths. Then, it was recognized that resistance to damage in service was a further issue and that toughening these materials could enhance their damage resistance. While many issues still need to be addressed (e.g., cyclic fatigue resistance, crack size effects-R-curve behavior), improving the fracture toughness has been deemed, in general, to be quite beneficial.

One approach to toughening ceramics has been the incorporation of strong, discontinuous brittle phases (e.g., whiskers) [1] which is the subject of this paper. The mechanisms contributing to the increased fracture toughness are described herein in terms of crack bridging by the reinforcement. A crack bridging model is discussed which is found to accurately predict the observed toughening response in SiC whisker-reinforced ceramics [1]. The results reveal that debonding of the interface between the reinforcing phase and the matrix is required to achieve significant toughening. The bridging model also illustrates how some of the properties of the matrix, interface, and reinforcing phase influence the fracture resistance of the composite. The predictive capability of the whisker bridging model then allows us to develop other approaches to achieving toughness by crack

S. P. Shah (ed.), Toughening Mechanisms in Quasi-Brittle Materials, 19–33.
© 1991 *Kluwer Academic Publishers.*

bridging. These include crack bridging by other types of second phases (platelets) and by matrix grains (self-reinforced).

2. Crack Bridging By Discontinuous Reinforcements

Bridging of the crack surfaces behind the crack tip by a strong discontinuous reinforcing phase which imposes a closure force on the crack is, at times, accompanied by pullout of the reinforcement [1-6]. The extent of pullout (i.e., the pull-out length) of brittle, discontinuous reinforcing phases is generally quite limited due both to the short length of such phases and the fact that bonding and clamping stresses often discourage pull out. However, pullout cannot be ignored as even short pull-out lengths contribute to the toughness achieved. Crack deflection by such reinforcements has also been suggested to contribute to the fracture resistance. Often, out-of-plane (nonmode I) crack deflections are limited in length and angle and are probably best considered as means of debonding the reinforcement-matrix interface. Such interfacial debonding is important in achieving frictional bridging (bridging by elastic ligaments which are partially debonded from the matrix) and pull-out processes. Frictional bridging elastic ligaments can contribute significantly to the fracture toughness as is described herein.

3. Analysis of Toughening by Discontinuous Bridging Phases

Here we will concentrate on the toughening due to crack bridging by various brittle reinforcing phases where the reinforcement simply bridges the crack surfaces and effectively pins the crack and increases the resistance to crack extension. The bridging contribution to the toughness for is:

$$K^c = \left[E^c J^m + E^c \Delta J^{cb}\right]^{1/2}, \qquad (1)$$

where K^c is the overall toughness of the composite, J^m is the matrix fracture energy, and the term ΔJ^{cb} corresponds to the energy change due to the bridging process.

The energy change associated with the bridging process is a function of the bridging stress/traction, T_u, and the crack opening displacement, u, and is defined as:

$$\Delta J = \int_0^{u_{max}} T_u du, \qquad (2)$$

where u_{max} is the maximum displacement at the end of the zone [7], Fig. 1.

One can equate the maximum crack opening displacement at the end of the bridging zone, u_{max}, to the tensile displacement in the bridging brittle ligament at the point of failure:

$$u_{max} = e_f^t \ell_{db}, \tag{3}$$

where ϵ_f^t represents the strain to failure of the whisker and ℓ_{db} is the length of the debonded matrix-whisker interface, Fig. 2. The strain to failure of the whisker can be defined as:

$$e_f^t = \left(\sigma_f^t/E^t\right), \tag{4}$$

where E^t is the Young's modulus of the reinforcing phase. The interfacial debond length depends on the fracture criteria for the reinforcing phase vs that of the interface and can be defined in terms of fracture stress or fracture energy. The analysis of Budiansky et al. [8] yields:

$$\ell_{db} = \left(r\gamma^t/6\gamma^i\right), \tag{5}$$

where γ^t/γ^i represents the ratio of the fracture energy of the bridging ligament to that of the reinforcement-matrix interface.

From Eq. (3), one quickly notices that the tensile strain displacement achieved in the bridging reinforcement, and hence the maximum crack opening displacement at the end of the bridging zone increases as the debonded length/the gauge length of the reinforcing ligament increases. Consideration of Eqs. (4) and (5) shows that increasing the reinforcing phase strength and/or enhancing interface debonding will contribute to greater tensile displacement within the reinforcing ligament. Increases in the crack opening displacement supported by the bridging zone will enhance the toughening achieved by such reinforcements. Therefore, debonding of the matrix-reinforcement interface can be a key factor in the attainment of increased fracture toughness in these elastic systems. In fact, in ceramics reinforced by strong ceramic whiskers, debonding is observed only in those systems which exhibit substantial toughening. An example of interfacial debonding associated with a bridging whisker in the wake of the crack tip is seen in Fig. 3. In this case debonding is evidenced by the interfacial offsets at the leading and trailing sides of the bridging whisker.

For the case of a bridging stress which increases linearly from zero at the crack tip to a maximum at the end of the bridging zone and immediately decreases to zero, Eq. (2) can be reduced to $T_{max}(u_{max})/2$. The maximum closure stress, T_{max}, imposed by the reinforcing ligaments in the crack tip wake is the product of the fracture strength of the ligaments, Σ_f^t and the areal fraction of ligaments intercepting the crack plane, A^t:

$$T_{max} = \sigma_f^t A^t \sim \sigma_f^t V^t, \tag{6}$$

where A^t is approximated by the volume fraction, V^t, for ligaments which have large aspect ratios (e.g., $\ell/r \geq 30$ for whiskers). Reinforcement by frictional bridging introduces a change in energy equal to:

$$\Delta J^{fb} = \left[\sigma_f^t V^t\left(\sigma_f^t/E^t\right)\left(r\gamma^t/\gamma^i\right)\right]^{1/2}. \tag{7}$$

From these results, the toughness contribution from frictional bridging by the reinforcing phase in the crack tip wake is:

$$\Delta K^{fb} = \sigma_f^i [(r V^t/12)(E^c/E^t)(\gamma^t/\gamma^i)]^{1/2} . \tag{8}$$

The overall toughness of the composite then includes both the bridging contribution, Eq. (8), and the fracture resistance of the matrix per Eq. (1). It does not, however, include a contribution from whisker pullout. A recent extension of this does include a whisker pull-out component [3b].

4. Material Characteristics Influencing Toughness

The toughening contribution then can be enhanced by utilizing matrix-reinforcing phase combinations with comparable Young's moduli and by improving the strength of reinforcing phase and increasing the reinforcement content and diameter. There are obvious limits as to how large a diameter reinforcing phase can be used in systems employing a matrix with a thermal expansion coefficient greater than that of the reinforcement as the thermal contraction mismatch tensile stress intensity scales with increase in inclusion/reinforcing phase diameter. In the alumina-SiC whisker system, the larger thermal expansion coefficient of the matrix vs the whisker and the high elastic property values results in substantial hoop and longitudinal tensile strains in the matrix [3,9]. Larger diameter reinforcements can generate matrix crack during post-fabrication cooling and degrade the properties of such composite [10]. The maximum reinforcement diameter employed will depend on the elastic and thermal expansion properties of the matrix vs those of the reinforcing phase.

A critical factor in such toughening processes is interfacial debonding which can be achieved if the interfacial failure conditions are much less than those required to fracture the reinforcement. In fact, substantial toughening by such crack bridging is obtained only when the reinforcement-matrix interface debonds before or just as the main crack tip reaches the interface. The formation of a debonded interface spreads the strain displacement imposed on the bridging reinforcement ligament over a longer gauge section generating a larger crack opening displacement per unit of stress supported by the ligament. As a result, the bridging traction/stress supported by the reinforcement increases more slowly with distance behind the crack tip. The resultant increase in crack opening displacement with distance behind the crack tip due to interfacial debonding, Eqs. (3)–(5), significantly enhances the fracture resistance/toughness of the composite.

At this point, this model of the frictional bridging contribution by discontinuous brittle reinforcing ligaments provides a very useful means of designing such composites and analyzing their response. One can, at least, characterize those properties which are most important when selecting materials, and then systematically dissect the toughening response of composites to either uncover problem areas or to develop advanced systems. The bridging ligament model can be further refined by including a pull-out contribution and by addressing the response and contribution of whisker which are inclined to the crack plane. In fact, the simple crack bridging model described here and the effects of reinforcement by brittle whiskers have been successfully applied to a variety of oxide (including glasses) and nonoxide matrix ceramics.

5. Observed Toughening By Crack Bridging Processes

Several types of discontinuous brittle reinforcements have been successfully employed to form toughened ceramics including second-phase whiskers [1–6] and platelets [3b, 11–13] and both elongated [14–17], plate-like [18], and large [3, 19–22] matrix grains. Studies of cracks in such materials reveal that, within the wake of the crack tip, the reinforcement does bridge the crack. The following sections will describe the observed toughening response in whisker-reinforced ceramics, ceramics with both elongated grains and larger grains, and when such bridging processes are combined in a composite.

5.1 CRACK BRIDGING BY BRITTLE WHISKERS

The experimental fracture toughness results obtained to date confirm the various features of the model for crack bridging by these discontinuous brittle reinforcements [2] as shown in Fig. 4 which compares experimental data with predicted curves based on Eq. (8). These results are based on a specific SiC whisker of a given strength and diameter. Thus, Fig. 4 reveals several features. First, the whisker bridging toughening contribution, $\Delta K^{wr} = \Delta K^{ftb}$, does increase with volume/areal content of the reinforcing phase as predicted. Second, the toughening contribution also increases as the ratio of the composite's Young's modulus to that of the whisker increases. This is best illustrated by the increase in ΔK^{wr} with increase in E^c at a given whisker content. For the examples here, E^c values were obtained by rule of mixtures [$E^c = E^m(1 - V_f) + E^w V_f$]; thus, at a constant volume fraction of whiskers, E^c increases in the order from glass ($E^m = 80$ GPa) to mullite ($E^m = 210$ GPa) to alumina ($E^m = 400$ GPa) vs SiC ($E^w = 500$ GPa). One can see that the use of Eq. (8) underestimates the actual toughness for the alumina and mullite-based composites. This indicates that whisker pullout does contribute in these cases. Experimental observations confirm the occurrence of whisker pullout in these two systems. More recent analysis shows that inclusion of pullout yields an accurate prediction of the toughening behavior [3b].

These same experimental observations [2] also show that the whisker bridging toughening contribution, ΔK^{ftb}, increases as r (the whisker radius)$^{1/2}$ increases as predicted by Eq. (8). For example, the toughness of alumina composites containing 20 vol % SiC whiskers increased from ∼6.5 to ∼9 to ∼12 MPa m$^{1/2}$ when the mean diameter of the SiC whiskers increased from 0.4 to 0.75 to 1–1.5 microns, respectively. From the toughening model, we also expect the toughness to increase as the matrix-whisker interface fracture energy (strength) decreases with respect to that of the whisker (γ^w substituted for γ^f). While values of the ratio of the whisker to interface fracture energy (γ^w/γ^i) are not available, there are two observations which support the predicted behavior. First, whisker-matrix interfacial debonding and crack bridging by the whiskers are only observed in the composites exhibiting significant toughening. Second, the length of the whiskers protruding above the fracture surface increases with increased toughening and this length can be related to the interfacial debond length. These findings indicate that the bridging contribution does indeed increase with increasing whisker diameter and when the fracture energy (and) of the interface decreases with respect to that of the whisker.

5.2 MATRIX GRAIN BRIDGING: GRAIN SHAPE EFFECTS

Crack bridging phenomena and toughening effects which are very comparable to those observed in whisker-reinforced ceramics are also found in ceramics containing other reinforcing phase geometries. For example, in the development of a more thermal shock-resistant electrical insulator, alumina ceramics with a microstructure which contains large (~100–200 μm across by ~10 μm thick) plate-like alumina grains in a medium sized (~5 micron) equiaxed grained matrix were developed. These materials had excellent thermal shock resistance. In fact, their thermal shock resistance was much greater than any of the variety of ceramics tested including zirconias, various other oxides, silicon nitrides, and aluminas with equiaxed grains. Further examination showed that fracture toughness values were ~7 MPa m$^{1/2}$ for samples containing ~25 vol % of these large, single-crystal alumina plates [18]. Aluminas prepared at the same time but with only equiaxed grains which were ~5 μm in size had toughness values of only 4–4.5 MPa m$^{1/2}$. Observations of the crack paths in the alumina containing the plate-like grains revealed that cracks deflected along the interface between the matrix and the large plate-like grains. This produced plates which bridged the main crack and contributed to the high toughness in much the same manner as do SiC whiskers.

The logical extension of this is to consider whether or not crack bridging by second-phase platelets contributes to fracture toughness. Composites consisting of an equiaxed polycrystalline matrix of TiO$_2$ in which alumina platelets are dispersed also exhibit increased fracture resistance as described by Hori et al. [11]. This work shows that under conditions where the platelet dimensions remained fairly similar, toughness increased with platelet content leading to nearly a three-fold increase at 30 vol % of alumina platelets. Initial studies also reveal that SiC platelets can produce similar increases in toughness in alumina as do SiC whiskers [12]. Each of these composites give evidence for crack bridging by the reinforcement.

In this same vein, reinforcement of Si$_3$N$_4$ [14–16] and SiAlON [17] ceramics by the *in situ* growth of elongated or whisker-like grains is also a potent toughening approach resulting in toughness values of ≥10 MPa m$^{1/2}$. Such materials have been labeled as self-reinforced, and from the crack observations of Li and Yamanis [15] crack bridging by these grains contributes to the improved toughness. Sufficient additional experimental results exist to begin to test how well the current crack bridging model describes the toughening effects of such elongated grains. First, Tajima et al. results show that the toughening contribution, ΔK^{fb}, increases with increase in volume content of the elongated grains [23].

More recent observations also reveal that ΔK^{fb} increases with increase in the cross section of the elongated grains, Table 1. In fact, the authors plotted the data in the form of ΔK^{fb} vs the square root of the diameter of the elongated grains [24]. The resulting plot exhibits excellent fit to the behavior predicted by Eq. (8). The diverse sources of observations then would support crack bridging by the elongated grains as the toughening process in these silicon nitride ceramics.

5.3 MATRIX GRAIN BRIDGING: GRAIN SIZE EFFECTS

In the present discussion, grain size effects on toughness are related to bridging ligaments formed by matrix grains which are left behind the crack tip [3, 19, 21, 25]. The toughening analysis is analogous to that for the whisker reinforcement described above. However, here the bridging stress supported by ligaments formed by microcracking along grain boundaries is the product of the frictional stress required to pull out each bridging grain times the fraction of bridging grains, $f_{gb} T_{gb}$. The grain bridging zone length is dictated by equating the crack opening displacement at the end of the zone u to that required to completely pull out the bridging grains. Assuming that half the grain must be pulled out to disrupt a ligament, u will be equal to one half the grain size (d), and the incremental increase in fracture toughness due to grain bridging ΔK^{gb} is:

$$\Delta K^{gb} = \left[f_{gb} \tau_{gb} E^c (d/2) \right]^{1/2} , \tag{9}$$

yielding a grain bridging toughening contribution consistent with experimental observations [3, 18, 21] at grain sizes below those resulting in spontaneous microcracking [18]. As noted in Figure 5, the grain size dependence of the fracture toughness of alumina ceramics is consistent with this behavior.

TABLE 1. Fracture toughness of silicon nitride ceramics with elongated grain structures[*]

Diameter of Elongated Grains μm	Fracture Toughness MPa m$^{1/2}$
2.8	5.7
3.5	6.4
4.5	7.0
7	8.3
8.7	9.0
10-11	10-11

6   Conclusions

Reinforced ceramics including reinforcement by strong whiskers initiate crack bridging processes to achieve improved fracture resistance. Similar toughening processes and effects are achieved by changes in grain size in noncubic ceramics, and/or by altering grain shape (e.g., formation of elongated grains in Si_3N_4 and SiAlON and plate-like grains in alumina ceramics). These reinforcing phases can contribute considerable toughening to brittle ceramics; factors of three increases in the fracture toughness are not uncommon.

The bridging processes involve frictional bridging where the matrix-reinforcement interface debonds which allows the reinforcement to elastically stretch over some finite gauge length hindered only by frictional sliding against the matrix. The contribution of pullout of these reinforcements to the toughness is rather limited (but not negligible), in part, due to their limited pull-out dimension. Enhanced interfacial debonding leads to greater toughening

[*]Data taken from results of H. Okamoto and T. Kawashima, NKK Corporation, Kawasaki, Japan.

effects in these systems by promoting the crack opening displacement supported by the bridging zone. The amount of toughness realized is dependent upon the properties and characteristics of the reinforcing phase and the interface as described by the micromechanics models developed for these systems. The model for frictional crack bridging reveals that the bridging contribution to the toughness is a function of the whisker strength, diameter, and content, as well as the ratio of the whisker to interface fracture resistance, and the ratio of the composite to whisker Young's moduli. The predicted effect of these parameters is supported by experimental observations for SiC whisker-reinforced ceramics.

Extension of the micromechanics model of toughening by crack bridging reinforcements illustrate the importance of considering other reinforcements including second phases and changes in matrix microstructure. Experimental results confirm various aspects of the toughening response due to crack bridging resulting from grain size and grain shape changes in alumina and silicon nitride ceramics. These findings suggest a variety of approaches may be possible to obtain improved fracture toughness in ceramic and other brittle systems by incorporating reinforcing phases which can generate crack bridging mechanisms. Such processes can be combined with each other or with other toughening mechanisms to develop synergistic toughening effects. The approach described here offers a means of developing these materials by considering the material characteristics/parameters which control the crack bridging contribution.

7 Acknowledgements

The contributions of T. N. Tiegs, C. H. Hsueh, K. B. Alexander, P. Angelini, S. B. Waters, W. H. Warwick, and L. M. Evans of ORNL and E. R. Fuller, Jr. of National Institute of Standards and Technology are gratefully acknowledged. This research was sponsored by the Division of Materials Sciences, U.S. Department of Energy, under contract DE-AC05-84OR21400 with Martin Marietta Energy Systems, Inc.

8 References

1. Becher, P. F. and Wei, G. C. (1984) 'Toughening Behavior in SiC-Whisker Reinforced Alumina,' *J. Am. Ceram. Soc.* 67(12), C267–69.

2. Becher, P. F., Hsueh, C. H., Angelini, P., and Tiegs, T. N. (1988) 'Toughening Behavior in Whisker Reinforced Ceramic Matrix Composites,' *J. Am. Ceram. Soc.* 71(12), 1050–61.

3. a. Becher, P. F. (1990) 'Recent Advances in Whisker Reinforced Ceramics,'in R. A. Huggins (ed.), *Annual Review of Materials Science, Vol. 20*, Annual Review, Palo Alto, Calif.

    b. Becher, P. F. (1990) 'Microstructural Design of Toughened Ceramics,' R. B. Sosman Lecture, in *Abstracts of the 92nd Annual Meeting of the American Ceramic Society*, American Ceramic Society, Westerville, Ohio, to be published.

4. Claussen, N. and Petzow, G. (1986) *Tailoring of Multiphase and Composite Ceramics, Vol. 20, Materials Science Research*, R. E. Tressler, G. L. Messing, C. G. Pantano, R. E. Newnham (eds.), Plenum Publishing, New York.

5. Becher, P. F., Tiegs, T. N., and Angelini, P. 'Whisker Toughened Ceramic Composites,' K. S. Mazdiyasni (ed.), *Fiber Reinforced Ceramics*, Noyes Publications, Park Ridge, N. J., in press.

6. Warren, R. and Sarin, V. K., 'Fracture of Whisker Reinforced Ceramics,' in K. Friedrich (ed.) *Applications of Fracture Mechanics to Composite Materials*, Elsevier, Amsterdam, in press.

7. Rice, J. R. (1968) 'Mathematical Analysis in the Mechanics of Fracture,' H. Liebowitz (ed.) *Fracture, Vol. II*, Academic Press, New York.

8. Budiansky, B., Hutchinson, J. W., and Evans, A. G. (1986) 'Matrix Fracture in Fiber-Reinforced Ceramics,' *J. Mech. Phys. Solids* 34(2), 167–89.

9. Angelini, P., Mader, W., and Becher, P. F. (1987) 'Strain and Fracture in Whisker Reinforced Ceramics,' P. F. Becher, M. V. Swain, and S. Somiya (eds.), *MRS Proceedings: Advanced Structural Ceramics, Vol. 78*, Materials Research Society, Pittsburgh.

10. Tiegs, T. N. (1989) 'Tailoring of Properties of SiC Whisker-Oxide Matrix Composites,' V. J. Tennery (ed.), *Ceramic Materials and Components for Engines*, American Ceramic Society, Westerville, Ohio.

11. Hori, S., Kaji, H., Yoshimura, M., and Somiya, S. (1987) 'Deflection-Toughened Corundum-Rutile Composites,' P. F. Becher, M. V. Swain, and S. Somiya (eds.), *MRS Proceedings: Advanced Structural Ceramics, Vol. 78*, Materials Research Society, Pittsburgh.

12. Alexander, K. B., Becher, P. F., and Waters, S. B. 'Characterization of Silicon Carbide Platelet-Reinforced Alumina,' *Proceedings of the 12th International Congress on Electron Microscopy*, San Francisco Press, San Francisco, in press.

13. Cambier, F. J., 'Processing and Properties of SiC-Platelet Silicon Nitride Ceramics,' in *Proceedings of the International Workshop for Fine Ceramics: New Processing and Properties of Fine Ceramics*, Nagoya, Japan, March 15–16, 1990, to be published.

14. Lange, F. F. (1979) 'Fracture Toughness of Si_3N_4 as a Function of the Initial a-Phase Content,' *J. Am. Ceram. Soc.* 62(7–8), 428–30.

15. Li, C. W. and Yamanis, J. (1989) 'Super-Tough Silicon Nitride with R-Curve Behavior,' *Ceram. Eng. Sci. Proc.* 10(7–8), 632–45.

16. Himsolt, G., Knoch, H., Huebner, H., and Kleinlein, F. W. (1979) 'Mechanical Properties of Hot-Pressed Silicon Nitride with Different Grain Structures,' *J. Am. Ceram. Soc.* 62(1), 29–32.

17. Lewis, M. H. (1986) 'Microstructural Engineering of Ceramics for High-Temperature Application,' in R. C. Brandt, A. G. Evans, D. P. H. Hasselman, and F. F. Lange (eds.), *Fracture Mechanics of Ceramics*, Plenum Publishing, New York.

18. Tiegs, T. N. and Becher, P. F. (1985) 'Particulate and Whisker Toughened Alumina Composites,' in *Proceedings of 22nd Automotive Technology Development Contractors' Coordination Meeting*, Society of Automotive Engineers, Inc., Warrendale, Pa.

19. Rice, R. W., Freiman, S. W., and Becher, P. F. (1981) 'Grain-Size Dependence of Fracture Energy in Ceramics: I, Experiment,' *J. Am. Ceram. Soc.* 64(6), 345–50.

20. a. Swanson, P. L., Fairbanks, C. J., Lawn, B. R., and Mai, Y. W. (1987) 'Crack-Interface Grain Bridging as a Fracture Resistance Mechanisms in Ceramics: I, Experimental Study of Alumina,' *J. Am. Ceram. Soc.* 70(4), 279–89.

    b. Mai, Y. W. and Lawn, B. R. (1987) 'Crack-Interface Grain Bridging as a Fracture Resistance Mechanism in Ceramics: II, Theoretical Fracture Mechanics Model,' *J. Am. Ceram. Soc.* 70(4), 289–94.

21. Mussler, B., Swain, M. V., and Claussen, N. (1982) 'Dependence of Fracture Toughness of Alumina on Grain Size and Test Technique,' *J. Am. Ceram. Soc.* **65**(11), 566–72.

22. Becher, P. F., Fuller, Jr., E. R., and Angelini, P. 'Matrix Grain Bridging in Whisker Reinforced Ceramics,' to be published.

23. Tajima, Y., Urashima, K., Watanabe, M., and Matsuo, Y. (1988) 'Fracture Toughness and Microstructure Evaluation of Silicon Nitride Ceramics,' G. L. Messing, E. R. Fuller, Jr., and H. Hausner (eds.), *Ceramic Transactions, Vol. 1: Ceramic Powder Science-IIB*, American Ceramic Society, Westerville, Ohio.

24. Okamoto, H. and Kawashima, T., unpublished results.

25. Wu, C. Cm., Freiman, S. W., Rice, R. W., and Mecholsky, J. J. (1978) 'Microstructural Aspects of Crack Propagation in Ceramics,' *J. Mater. Sci.* **13**:2659–70.

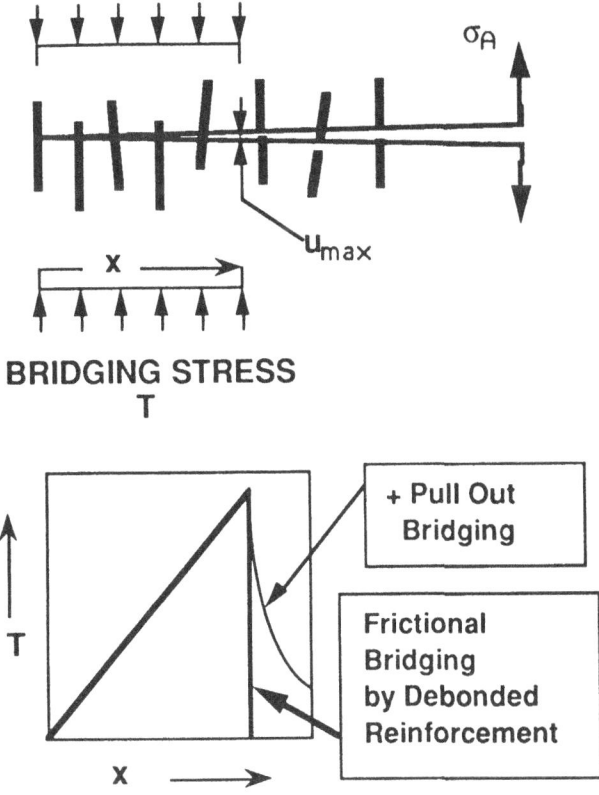

Figure 1. Crack bridging by discontinuous brittle reinforcing phases imposes a closure or bridging stress in the wake of the crack tip and enhances the fracture resistance of the brittle matrix.

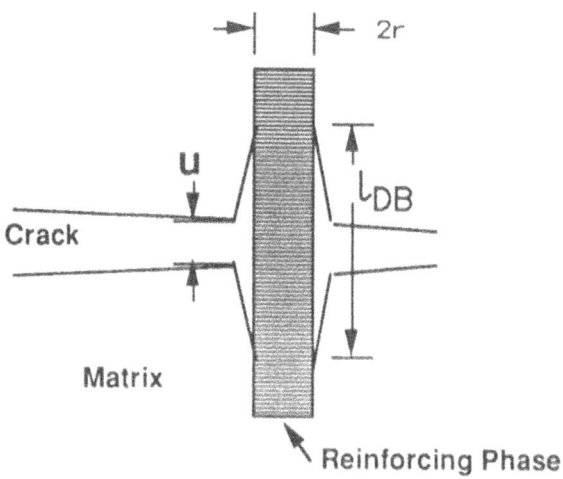

CRACK OPENING u ($=u_{max}$) AT THE END OF BRIDGING
ZONE EQUATED TO MAXIMUM WHISKER TENSILE
DISPLACEMENT ($l_{DB} \, \epsilon_f^w$).

Figure 2. The formation of the bridging zone behind the crack tip requires that the reinforcing phase-matrix interface separate/debond (a) during fracture. The crack opening displacement associated with the bridging zone then is related to the tensile displacement in the bridging ligaments (b). At the end of the bridging zone the maximum crack opening is equivalent to the displacement in the ligament corresponding to its fracture stress.

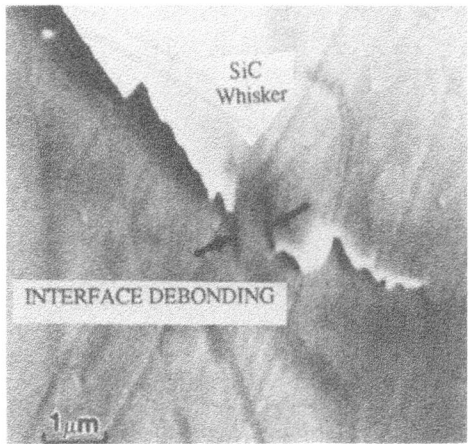

Figure 3. Debonded whisker-matrix interfaces are associated with whisker bridging in region immediately behind the crack tip in a polycrystalline aluminum oxide matrix.

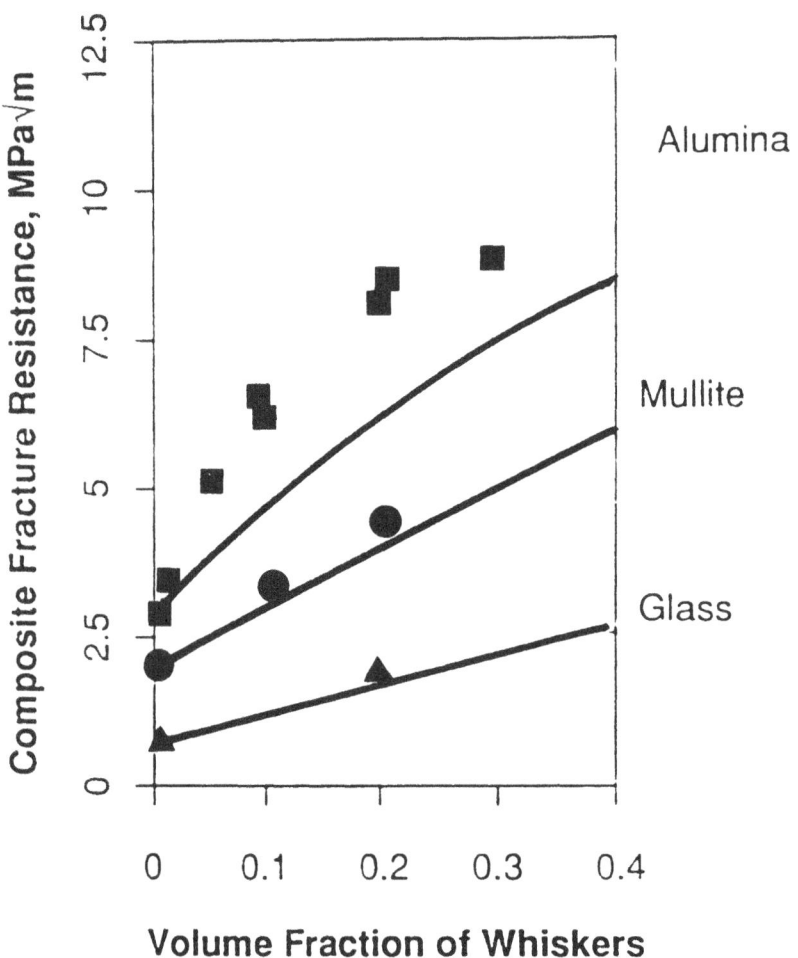

Figure 4. The measured fracture toughness of SiC whisker reinforced ceramics increases with increasing whisker content. The toughening response predicted by Eq. (8) for frictional bridging ligaments (solid lines) is quite consistent with the experimental observations for the glass and mullite matrices. It does, however, underestimate the response in the alumina based composite as whisker pullout also makes a significant contribution to the observed toughness in this case.

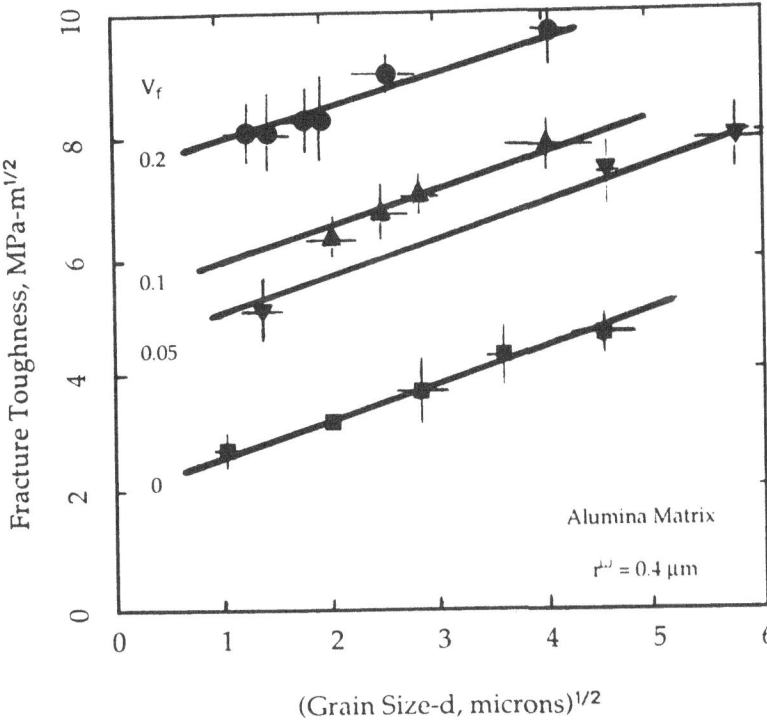

Figure 5. The fracture toughness of alumina ceramics and whisker reinforced alumina composites are enhanced by increases in the matrix grain size. The composites represented in this figure were prepared using the same SiC whisker source for various volume fraction V_f whisker additions.

FRACTURE PROCESS ZONE IN CONCRETE AND CERAMICS - A MATTER OF SCALING

Albert S. Kobayashi,* Neil M. Hawkins** and Richard C. Bradt***
* University of Washington, Department of Mechanical Engineering, Seattle, Washington 98195
* * University of Washington, Department of Civil Engineering, Seattle, Washington 98195
* * * University of Nevada-Reno, Mackay School of Mines, Reno, Nevada 89557

ABSTRACT. The crack closure stress versus crack opening displacement (COD) associated with the fracture process zone (FPZ) of a subcritically growing crack in concrete was determined through a hybrid experimental-numerical procedure. COD measurements, which were obtained by moire interferometry, were used interactively with an elastic finite element model of the concrete specimen to back out the above constitutive relation by a direct analysis. A constitutive relation, which is similar in form but different in magnitude, was also used to characterize the FPZ associated with stable crack growth in SiC_w/Al_2O_3-matrix ceramic composite. Similarity in the genesis of the FPZ's in concrete and ceramics is discussed.

1. Introduction

A decade ago, Hillerborg and his colleagues showed that a postulated fracture process zone (FPZ) ahead of a macro crack tip,* in which microcracks form and coalesce, will conveniently reproduce the strain softening phenomenon in direct tension tests of unnotched concrete specimens [1]. Papers [2-6], which followed Reference [1], continued to support indirectly the existence of the FPZ in concrete but its existence only became evident with the recent multiple notching experiments of three point bend specimens [7]. On the other hand, the existence of a small FPZ in polycrystalline Al_2O_3 fracture specimens was demonstrated by a

---

* Macro crack tip is defined as the tip of a traction free crack.

S. P. Shah (ed.), Toughening Mechanisms in Quasi-Brittle Materials, 35–46.

multiple notching experiment much earlier in 1982 [8] and led to the recent detailed study of the FPZ as a dominant energy dissipation mechanism in Al_2O_3 [9,10].

For mode I fracture, the behavior within the FPZ can be described by a crack closure stress, σ, vs crack opening displacement (COD), w, relation. This constitutive relation can be determined from the decreasing strain-softening portion of a direct tension test of concrete specimens [2,5]. When incorporated into a finite element model, this σ-w curve can be used to predict the overall fracture strength and to simulate the progressive damage process of a fracturing concrete structure. Crack closure stress has also been identified as the primary toughening mechanism of a ceramic composite [10-13], but a constitutive relation, which quantified the crack closure stress versus COD relation, was not derived.

The authors and their colleagues have been studying the FPZ associated with stable crack growth in concrete and ceramic composite using either a direct or an inverse analysis based on finite element models of fracture specimens of concrete and ceramics. These results are discussed in the following.

2. FPZ in concrete fracture specimens

The FPZ model used in previous studies was a nonsingular-FPZ model which had no stress singularity at the micro-crack tip presumably due to crack shielding by the frontal process zone. This assumed non-singular crack tip state appeared to be verified by the reversed curvature in the crack tip profile which was determined by Moire interferometry of stably growing cracks in statically loaded concrete specimens [14-16]. Subsequent dynamic Moire interferometry data [17] showed unequivocally that at high strain rate, the crack opening profile was parabolic with a $1/\sqrt{r}$ stress singularity. In retrospect, the low static stress intensity factor was obscured by the relatively large crack closure stress in the nonsingular-FPZ model [14-16] and thus did not generate a sufficiently large parabolic crack tip profile which could be detected by Moire interferometry. The holographic interferometry technique [18], on the otherhand, successfully detected the parabolic crack tip opening shape and led to the two parameter fracture criterion for concrete [19] with a critical stress intensity factor. The fracture criterion proposed in this paper closely follows that of [19].

Concrete specimens of two different geometries, i.e. a crack-line wedge-loaded, double-cantilever beam (CLWL-DCB) specimen shown in Figure 1(a) and a three-point bend specimen shown in Figure 1(b), were statically loaded to failure by stable crack growth. Figures 2 and 3 show typical moire fringe patterns in the CLWL-DCB and the three point bend specimens, respectively. Figure 4 shows the wedge load versus COD, $2V_1$, relations of four CLWL-DCB specimens. Figure 5 shows the load versus load-line displacement relations of four three-point bend specimens.

Figure 6 shows the σ versus w relation which was obtained by prescribing the COD data obtained through moire interferometry as input boundary condition to the finite element model for direct analysis. The three-line segmented model, which was used in previous studies [4,6,14-16], was replaced with the two-line segmented σ-w relation as shown in Figure 6.

Figures 7 and 8 show typical comparison between the COD's, which were computed by using the σ-w constitutive relation of Figure 6, and the COD obtained by moire interferometry. The computed load versus load line displacements and the wedge load versus COD relations are shown as data points in Figures 4 and 5, respectively. The excellent agreements between the measured and computed values validate the finite element model of the concrete fracture specimens with the FPZ constitutive relation of Figure 6.

Figure 9 shows the stress intensity factors associated with the crack extensions in the CLWL-DCB and three pont bend specimens. This flat resistance curve is typical of a strain softening material where the fully developed FPZ follows the extending microcrack tip.

3. FPZ of ceramic composite

The specimen for SiC-whisker/alumina-matrix (SiC_w/Al_2O_3) ceramic composite was a wedge-loaded, double cantilever beam (WL-DCB) specimen with a chevron notch along the entire length of its remaining ligament as shown in Figure 10. This fracture specimen is inherently rigid and is thus suitable for stable crack growth study of brittle materials where crack growth is confined to the entire chevron notch plane without the annoying crack curving which is prevalent in ceramic DCB specimens without side grooves. The friction force at the loading pin and the location of the crack tip, which is not visible, in addition to COD variation in the FPZ, are the unknowns which cannot be measured directly.

A hybrid experimental-numerical, inverse procedure was used to determine σ versus w relation of the FPZ as well as the energy release and dissipation rates of a stably growing crack. Moire interferometry was used to determine the displacement field on the side surface of the WL-DCB specimen. Figure 11 shows a typical moire pattern representing the lateral displacements on the side surface of the chevron-notched specimen. The measured horizontal displacement at the loading pin, together with the measured pin-load, were input to a three dimensional finite element model of the WL-DCB specimen. An assumed crack front, which is estimated from the moire fringe pattern, is used in the first iteration and this crack front was adjusted in subsequent numerical analyses until a reasonable match between the computed and measured side-surface COD, which was obtained from the moire fringes, was obtained.

Figure 12 shows the σ versus w relation obtained from this inverse analysis. Figure 13 shows the COD measurements and the computed COD's, one without the FPZ, which is identified as (LEFM) in this figure, and one with the FPZ, which is identified as (FPZ), on the exposed side surface of the fracture specimen. The notable difference between the computed COD's with, and without, the FPZ and the discrepancy with the measured surface displacement, is noted. These results validate the FPZ constitutive relation of Figure 12.

Figure 14 shows the typical variations in energy release rate and the energy dissipation rate in the FPZ with crack extension. The relatively flat energy release rate and the energy dissipation rate curves indicate that the FPZ is completely developed in this ceramic composite. The closeness of the resistance and dissipation rate curves also indicates that the dominant energy dissipation mechanism is the FPZ in this brittle material.

4. Conclusions

The FPZ constitutive relations for concrete and $SiCw/Al_2O_3$ ceramic composite are similar in form. The crack closure stress for SiC_w/Al_2O_3 ceramic composite was one order higher than that of concrete while the COD's were comparable in magnitude. The length of typical FPZ's in concrete was an order of magnitude larger than that of SiC_w/Al_2O_3 ceramic composite.

5. Acknowledgement

FPZ studies in concrete fracture, which was conducted by Dr. J. Du and Mr. Z.-K. Guo, was supported by National Science Foundation Grant MSM 831063 and Air Force Office of Scientific Research Contract 86-0204. Studies in ceramic fracture, which were conducted by Mr. C. T. Yu were supported by the Office of Naval Research Contract N00014-87-K-0326.

6. References

1. Hillerborg, A., Modeer, M. and Petersson, P.E. (1978) 'Analysis of Crack Formation and Crack Growth in Concrete by Means of Fracture Mechanics Finite Elements,' Cement and Concrete Research, 6, 773-782.
2. Petersson, P. E. (1981) 'Crack Growth and Development of Fracture Zones in Plain Concrete and Similar Materials,' Doctoral Dissertation, University of Lund, Sweden, TVBM-1006.
3. Wecharatana, M. and Shah, S. P. (1983) 'Prediction of Non-Linear Fracture Process Zone in Concrete,' ASCE, J. Eng. Mech., 109 (5), 1231-1246.
4. Cho, K. Z., Kobayashi, A. S., Hawkins, N. M., Barker, D. B. and Jeang, F. L. (1984) 'Fracture Process Zone of Concrete Cracks,' J. Eng. Mech., ASCE, 110 (8), 1174-1184.
5. Reinhardt, H. W., Cornelissen, A. W. and Hordjik, D. A., (1986) 'Tensile Tests and Failure Analysis of Concrete,' J. Struct. Eng., 112 (11), 2462-2477.

6. Liaw, B. M., Jeang, F. L., Du, J.J., Hawkins, N. M. and Kobayashi, A. S.,. (1990) 'Improved Non-Linear Model for Concrete Fracture,' ASCE, J. Eng. Mech., 116 (2), 429-445.

7. Hu, X.Z. and Wittman, F.H. (1989) 'Fracture Process Zone and Kr-Curve of Hardened Cement Paste and Mortar,' in S.P. Shah, S.E.Swartz and B. Barr (eds), Fracture of Concrete and Rock: Recent Developments, Elsevier Applied Science, London, 307-316.

8. Knehans, R. and Steinbrech, R. (1982) 'Memory Effect of Crack Resistance During Slow Crack Growth in Notched Al_2O_3 Bend Specimens,' Journal of Materials Science Letters. 1, 327-329.

9. Swanson, P.I., Fairbanks, C.L., Lawn, B.R., Mai, Y.-W. and Hockey, B.J., (1987) 'Crack-Interface Grain Bridging as a Fracture Resistance Mechanism in Ceramics: I Experimental Study on Alumina,' J. American Ceramic Society 70, 279-289.

10. Mai, Y.-W. and Lawn, B.R. (1987) 'Crack-Interface Grain Bridging as a Fracture Resistance Mechanism in Ceramics: II Theoretical Fracture Mechanics Model,' J. American Ceramic Society 70, 289-294.

11. Mori, T. and Mura, T. (1984) 'An Inclusion Model for Crack Arrest in Fiber Reinforced Materials,' Mech. of Mater. 3, 193-198.

12. Budiansky, B., Hutchinson, J.W. and Evans, A.G. (1986) 'Matrix Fracture in Fiber-Reinforced Ceramics,' J. Mech. Phys Solids, 34, 167-189.

13. Hori, M. and Nemat-Nasser, S. (1987) 'Toughening by Partial or Full Bridging of Cracks in Ceramics and Fiber Reinforced Composites,' Mech. of Mater., 6, 24269.

14. Du, J.J., Kobayashi, A.S., and Hawkins, N.M. (1989) 'Fracture Process Bond of a Concrete Fracture Specimen,' in J. P. Shah and S.E. Swartz (eds.), Fracture of Concrete and Rock, Springer-Verlag, 199-204.

15. Du, J.J., Kobayashi, A.S. and Hawkins, N.M., (1989) 'A Hybrid Procedure for Determining the Fracture Process Zone in Concrete,' in S.P. Shah, S.E. Swartz and B. Barr (eds.), Fracture of Concrete and Rock: Recent Developments, Elsevier Applied Science, 297-306.

16. Du, J.J., Kobayashi, A.S. and Hawkins, N.M. (1990) 'An Experimental-Numerical Analysis of Fracture Process Zone in Concrete Fracture Specimens,' Engineering Fracture Mechanics, 35 (1/2/3), 15-28.

17. Du, J.J., Yon, J.-H., Hawkins, N.M., Arakawa, K. and Kobayashi, A.S. (1990) 'Fracture Process Zone for Concrete for Dynamic Loading,' to be pubished in Material J. American Concrete Institute.

18. Miller, R.A., Shah, S.P. and Bjelkhagen, H.I. (1988) 'Crack Profiles in Mortar Measured by Holographic Interferometry,' Experimental Mechanics, 28, 388-394.

19. Jenq, Y.-S. and Shah, S.P. (1985) 'Two parameter Fracture Model for Concrete,' J. Engineering Mechanics, ASCE 116, 1227-1241.

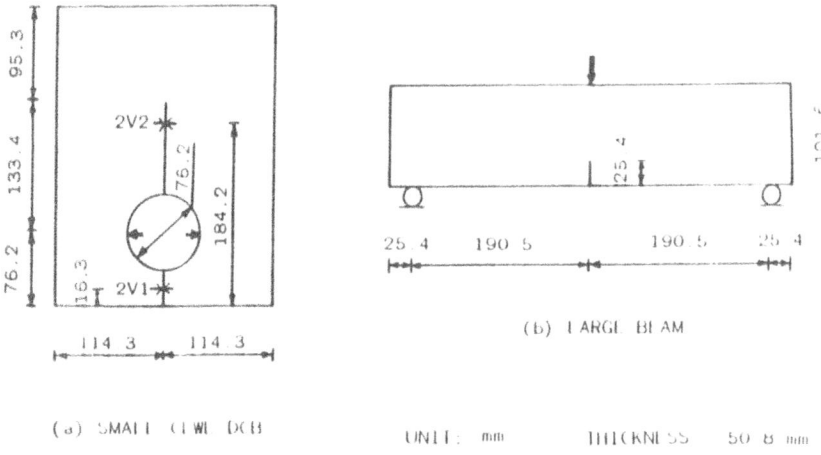

Figure 1. CLWL-DCB and Three-Point Bend Concrete Specimens.

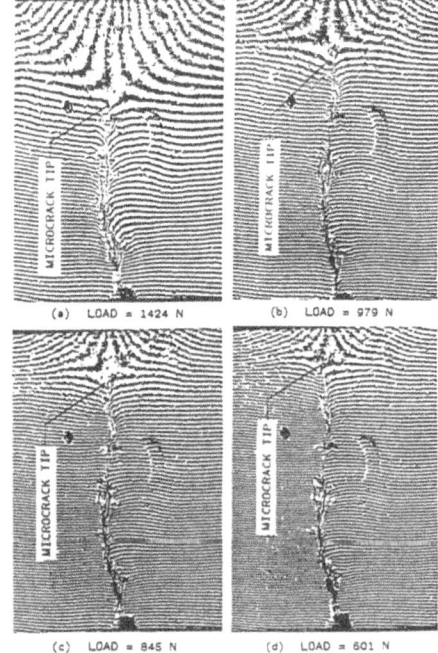

Figure 2. Sequential Moire Fringe Patterns of CLWL-DCB Specimen.
Specimen No. SD-2.

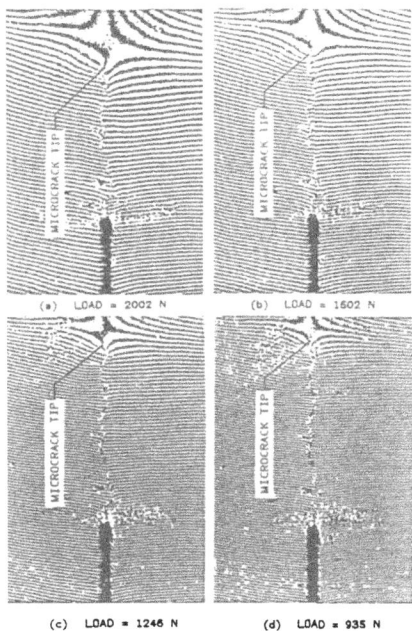

Figure 3.   Sequential Moire Fringe Patterns of Three-Point Bend Specimens. Specimen No. LB-1

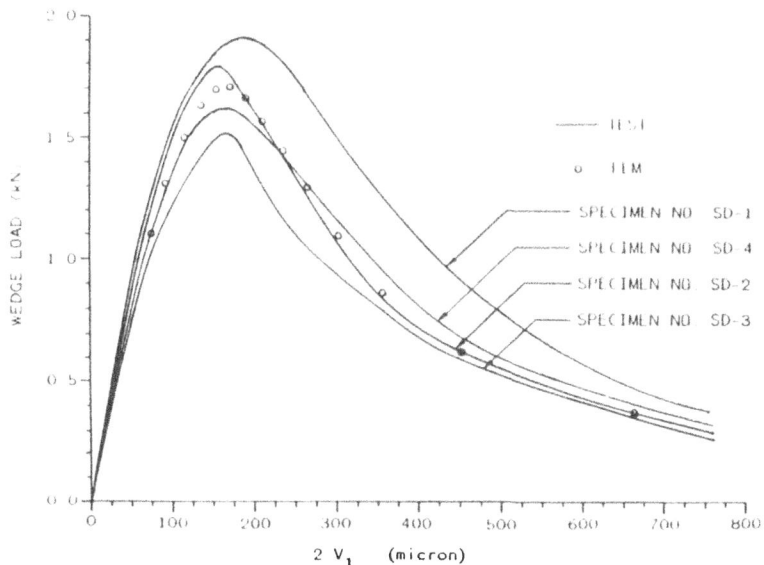

Figure 4.   Wedge Load Versus $2V_1$ CLWL-DCB Specimens.

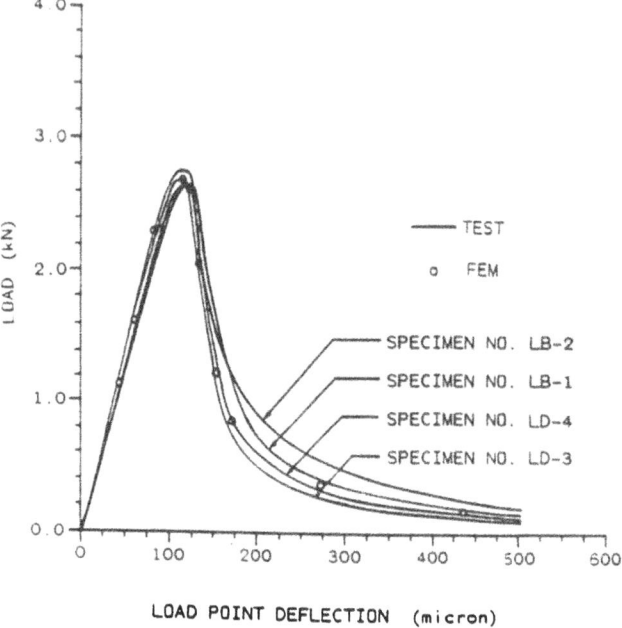

Figure 5.   Load Versus Load-Line Deflection Curves, Three-Point Bend
Specimens.

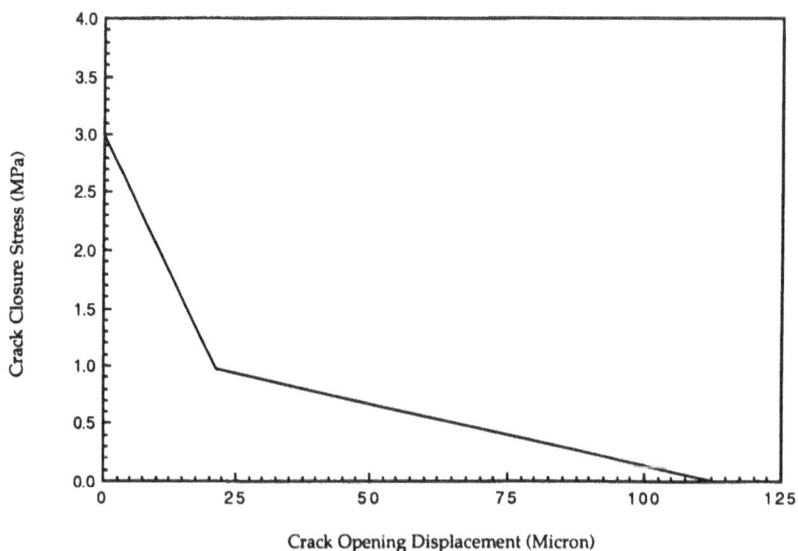

Figure 6.   Crack Closure Stress Versus COD Along FPZ.

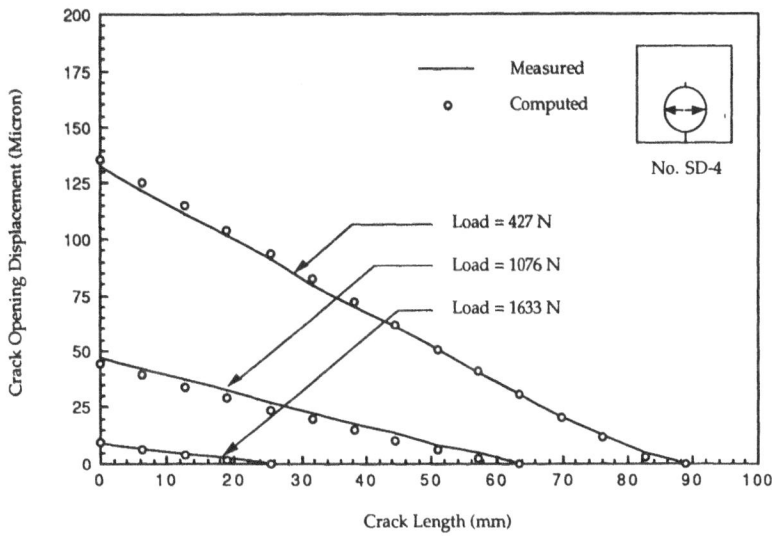

Figure 7.   COD Variations along Fracture Process Zone.
            CLWL-DCB Specimen No. SD-4.

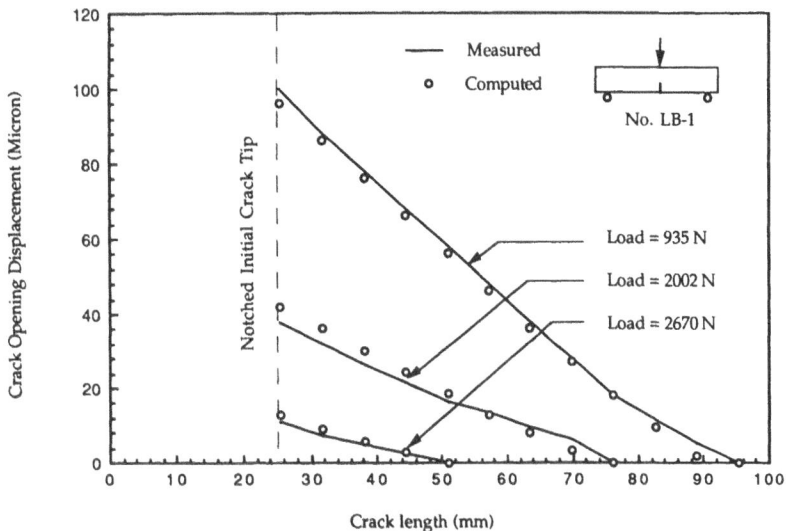

Figure 8.   COD Variations along Fracture Process Zone.
            Three-Point Bend Specimen No. LB-4.

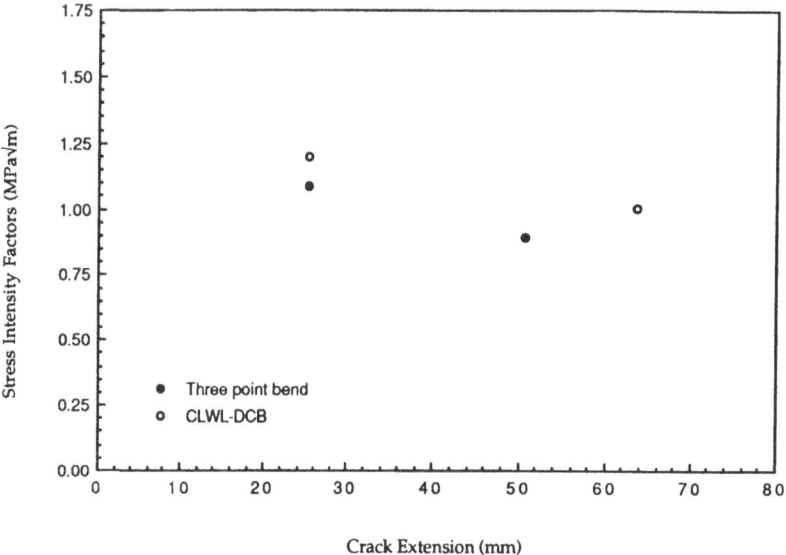

Figure 9.   Stress Intensity Factor versus Crack Extension.

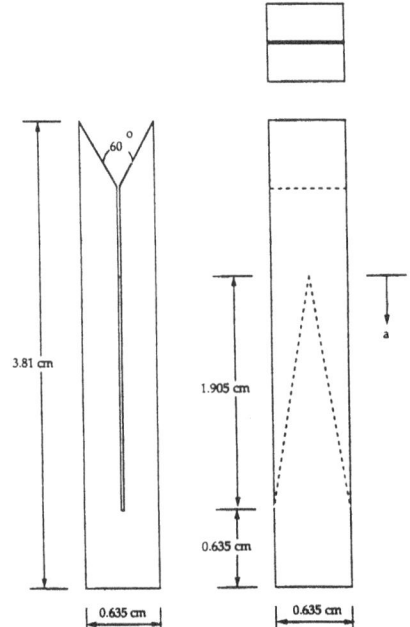

Figure 10.   Chevron Notched DCB Specimen.

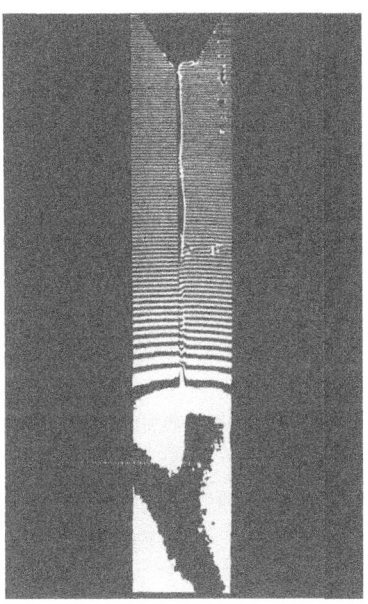

Figure 11. Moire Fringe Pattern of Chevron Notched DCB Specimen.

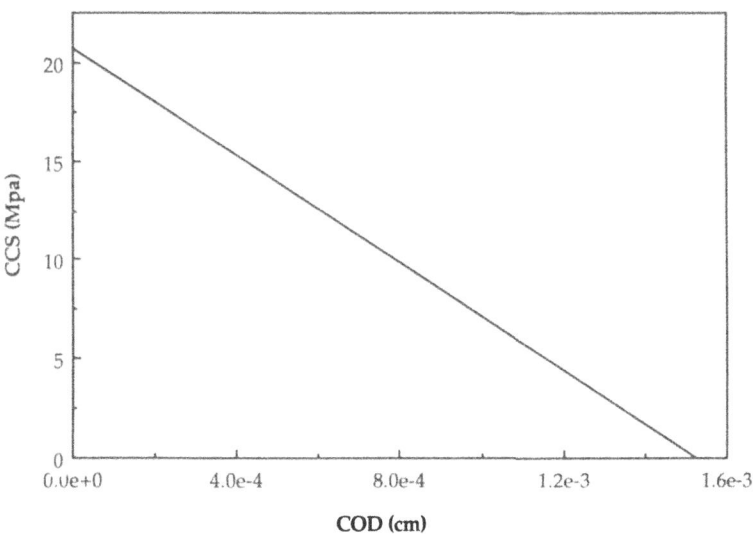

Figure 12. Crack Closure Stress Versus COD Along FPZ of Chevron Notched
DCB Specimen.

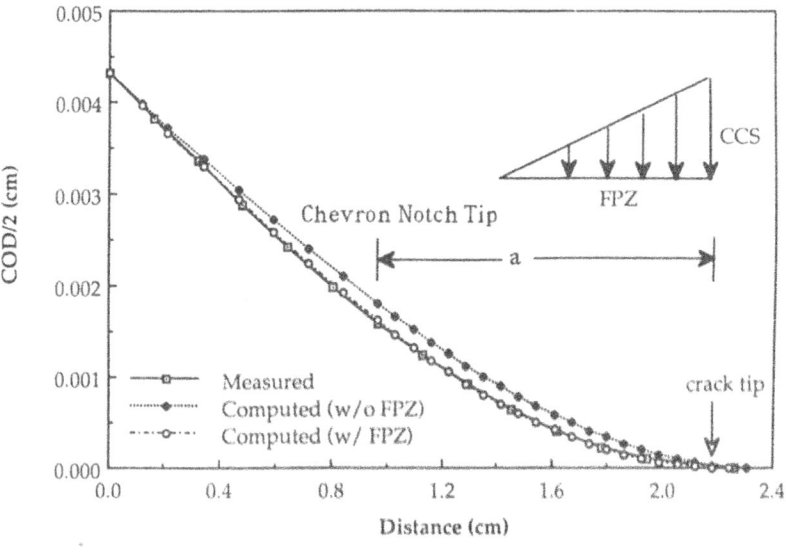

Figure 13. Crack Opening Displacement on the Surface of SiC$_w$/Al$_2$O$_3$ Chevron Notched DCB Specimen.

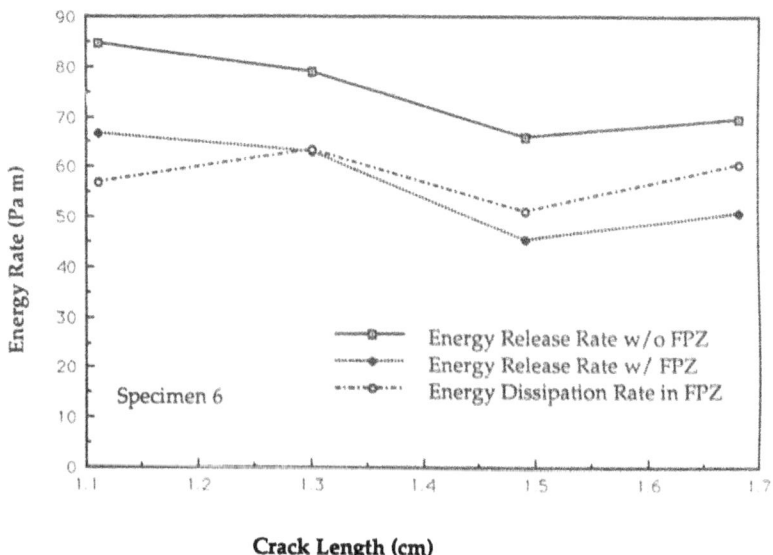

Crack Length (cm)

Figure 14. Variations in Energy Release and Dissipation Rates with Stable Crack Growth in SiC$_w$/Al$_2$O$_3$ Chevron Notched Specimen.

REPORT ON SESSION 1: FRACTURE OF CERAMICS WITH PROCESS ZONE

H. Hübner
Technische Universität Hamburg–Harburg
Eissendorfer Str. 42
2100 Hamburg 90
Germany

In the three foregoing lectures, the subject of toughness increase of ceramic materials by the incorporation of fibers and platelets and by the formation of a process zone was dealt with. The various toughening mechanisms acting in brittle materials were presented and their quantitative predictions were compared with experimental observations. The papers demonstrated that there exists agreement with respect to a systematic classification of the fundamental ceramic toughening mechanisms discussed today in the literature. Consequently, the discussion of the papers which is reported here turned out to follow the same classification which shall briefly be reviewed in the following.

When dealing with ceramic toughening mechanisms a general distinction is made between process zone mechanisms and bridging mechanisms. The former occur in the stressed region ahead the tip of the advancing crack, and they are ascribed to essentially two basic effects, i.e. stress–induced phase transformation and stress–induced microcracking. Both mechanisms affect the fracture toughness of the material by shielding the propagating crack from the applied stress intensity, thereby increasing the stress intensity factor necessary to drive the crack. The second group of mechanisms acts behind the crack tip. Among them, effects such as elastic bridging, debonding with subsequent frictional bridging, and pull–out bridging are distinguished. Whereas elastic bridging contributes to toughening through the build–up of a bridging stress across the crack planes which tends to close the crack, debonding and pull out can be understood most easily as processes which provide additional energy consumption as the main crack advances.

Likewise, the summary of the discussion of Session 1 will be given by following the schematic classification outlined above. First, questions concerning the relation and interaction of process zone toughening and bridging zone toughening are reported. The most part of the discussion, however, was concentrated about the details and specific aspects of debonding phenomena. Finally, the problems addressed with respect to pull out effects are dealt with.

As to the synergetic action between the process zone and the bridging zone the question came up if microcracking and crack bridging are additive processes or not. This was strongly affirmed by one of the speakers (K.T.F.) who pointed out that justification of this assumption is obtained from the comparison of model

47

S. P. Shah (ed.), Toughening Mechanisms in Quasi-Brittle Materials, 47–49.
© 1991 Kluwer Academic Publishers.

calculations and experimental results. Computer simulations permitted to look at both groups of mechanisms, and assuming their interaction gave the right order of magnitude of the toughness increase ΔK when compared with measurements. Furthermore, the question about the difference between the two zones was responded by several participants. It was stated that usually no distinction between the zones is made, neither in cements nor in ceramics. A separation between the regions ahead and behind the crack tip would require an exact determination of the crack tip location, which was considered to be a difficult task. Moreover, the identification of the crack tip was thought not to be even necessary, since both zones have similar effects on the toughness.At this point of the discussion it was criticized that in some models debonding is counted twice, i. e. as the real debonding and also as microcracking. This may lead to errors in the evaluation of the R curve if the microcrack length is not corrected by the debonding crack length.

Fiber debonding phenomena turned out to be the most interesting feature in the discussion of the foregoing lectures. A considerable number of contributions was concentrated around the question where the debonding of fibers starts. There were two opposite opinions given by several speakers, i. e. those who assumed debonding to start at the end of the fibers and others who believed that debonding initiated at the intersection of the crack with the fibers. Experimental evidence stemming from TEM observations was cited which showed that debonding begins at the tip of the whiskers in whisker–reinforced metals and in fiber–reinforced ceramics. Other speakers queried any end–face debonding to occur, remembering TEM work on whisker–reinforced ceramics and glass–fiber composites which clearly demonstrated that debonding started near the intersection of cracks with whiskers and that no discontinuity at the end of the fibers could be seen. It was also mentioned that end face–induced debonding instead of intersection–induced debonding should result in a much smaller contribution to the toughness increase than is determined experimentally. As resumed by the chairman a marked conflict between these two opinions could be observed among the scientists present. However, there was broad agreement that in any case the fracture process is strongly influenced by interfacial debonding independently of the location where debonding starts.

The question of the beginning of debonding was also regarded from a theoretical point of view. The stress analysis of the problem of a short fiber embedded in a matrix containing a crack perpendicular to the fiber axis permits the interfacial shear stress along the fiber axis to be calculated. As a result, the shear stress can be shown to increase at both the end face of the fiber and at its intersection with the crack. In terms of this analysis debonding is expected to start at a site where the shear stress first reaches the interface shear strength of the composite, and this should depend on three parameters, i. e. the modulus ratio E_f/E_m, the aspect ratio of the fibers, and the interface shear strength ratio τ_s/τ_i where E_f and E_m are the Young's moduli of the fiber and the matrix and τ_i and τ_s are the interface shear strengths before and after debonding, respectively. The analysis implies further that the crack may intersect the fiber without causing any debonding at all, but this consequence of the model was questioned by other participants.

A second area of interest in context with debonding phenomena was the effect of fiber orientation on toughening. It was stated that, particularly in the case of short fibers, small angles may frequently occur between fibers and the direction of crack propagation when short fibers are randomly oriented in the matrix, and this could lead to a compressive zone behind the crack tip. It was pointed out that it would be at least difficult to say which type of stress is built up when the main crack is close to the fiber axis. The subject was further addresses in another comment where it was pointed out that the theoretical analysis usually deal with the simple assumption that the fibers are perpendicular to the crack plane. When the fibers are inclined, however, much more energy dissipation may take place during de-bonding and pull–out due to fiber bowing. Thus, it was concluded that inclined fibers may give a very different (and enhanced) contribution to toughness. Finally, attention was called to the fact that not only inclined fibers but also inclined paricles are supposed to have a much greater effect on toughness than parallel particles since the frictional component of energy dissipation has to be included and contributes much to ΔK.

Apart from these statements on the effect of fiber orientation on pull–out pro-cesses, only few comments were made on the mechanisms contolling fiber or grain pull–out. Only limited knowledge exists about the details of the fiber–matrix interaction once debonding has occurred, i. e. on interface processes such as adhe-sion, friction and energy dissipation during the relative movement between fibers and the matrix walls or between platelets and the matrix grains. The only comment was made on the changing situation at the interface when the tempera-ture is increased. In microstructures containing small glass films between the whiskers or fibers and the surrounding matrix grains the rising temperature causes the amorphous phase to soften. Therefore, the amount of bonding between fibers and matrix and hence the strength becomes strain–rate dependent. At high strain rates, brittle behavior still prevails whereas at small strain rates some limited plastic deformation becomes possible due to the reduced viscosity of the glassy phase, thereby causing considerable stress relaxation to take place. It was agreed, however, that such high–temperature phenomena are beyond the scope of the workshop.

In his final word the chairman pointed out that "we had some very good questions and this is a very good beginning".

Fracture in Concrete and Rock

MICROCRACKING AND DAMAGE IN CONCRETE

D.FRANCOIS.
CNRS. URA 850
Ecole Centrale de PARIS. F92295
Chatenay-Malabry CEDEX

KEYWORDS/ABSTRACT : Concrete/microcracking/damage mechanics/fatigue.

This review attempts to show how the observation of the physical processes of damage in concrete allow to formulate sound damage mechanics models. Microcracks opening and propagation by successive steps provide concrete with a non linear behaviour and increase the fracture toughness with respect to the hardened cement paste. Several criteria were proposed for the propagation of microcracks. Damage mechanics is an efficient tool to modelize the behaviour of concrete. The damage is related to the strain. Various expressions were proposed for this relation. This evolution is related to the opening and to the propagation of microcracks through their influence on the compliances of the material. Extension of these models can be given in the case of fatigue, where the viscous behaviour of concrete must be taken in consideration. A new model is proposed. Lastly numerical models provide a better understanding of the behaviour of concrete and of the effect of structural parameters.

1 - Microcracking in concrete

The cracking of concrete was studied and described by many authors (for instance Slate and Hover, 1984). Let's simply recall that before any loading this material contains many defects and microcracks. In the cement paste porosities are present. Their sizes are distributed over a wide range and when they are observed at different scales, the appearence remains the same. The cement paste is a fractal object whose dimension is equal to 2.90 to 2.97. (Regourd, 1986). The density is lower near the aggregate interface, in the "auréole de transition", and moreover this feature is accentuated under the aggregates.

S. P. Shah (ed.), Toughening Mechanisms in Quasi-Brittle Materials, 53–65.
© 1991 *Kluwer Academic Publishers.*

Microcracks exists at the interface between the aggregates and the mortar, mostly underneath the aggregates. They are due to bleeding and to internal stresses coming from the mortar shrinkage.

There are also longer cracks at the surface which are produced by the shrinkage from drying.

When the concrete is loaded, at some critical stress the cracks at the aggregate interface, propagate through the mortar until they meet an aggregate boundary where they are stopped. It is not until further loading that they can propagate again and produce the final fracture. It is this two steps propagation which somewhat lessen the brittleness of concrete, if compared with the mortar. As the cracks are stopped before final fracture, their opening increases the deformation, producing a non linear stress strain curve and offering a way to reduce the stress concentrations at the macroscale. Furthermore this diffuse cracking absorbs energy and this increases the toughness of the material. At a smaller scale the same can be said of the mortar compared to the cement paste.

This behaviour of concrete leads quite naturally to a description in terms of damage mechanics.

If a connection is sought between the macroscopic behaviour and the physical processes of crack propagation at the microscale, a crack propagation criterion is needed.

2 - Crack propagation criteria

Cracks in concrete are loaded in mixed mode I and II. For a crack which propagates in its own plane, with no deviation, several propagation criteria can be written.

The Griffith criterion, which states that the strain energy release rate G_{is} equal to a critical value G_{Ic} yields :

$$G = G_{Ic} \text{ or } K_I{}^2 + K_{II}{}^2 = K_{IC}{}^2$$

K_I and K_{II} being the stress intensity factors in mode I and II respectively. It is the criterion which is the most satisfactory from a thermodynamics point of view.

The normal stress criterion, which is based on the idea that the propagation will start when the normal stress on the propagation plane reaches a critical value yields simply

$$K_I = K_{IC}$$

because the mode II loading does not produce any normal stress on the plane of the crack.

The maximun principal stress criterion yields :

$$K_I + K_{II} = K_{IC}$$

The criterion of the strain energy density proposed by Sih (1973) yields :

$$K_I^2 + \frac{K_{II}^2}{1 - 2\nu} = K_{IC}^2$$

where ν in the Poisson ratio. It is equivalent to a maximum strain criterion in that case.

The question becomes more confused when the crack is allowed to branch. The Griffith criterion is now very difficult to use because there is no simple relation between the strain energy release rate and the branching angle.

The normal stress criterion is the most likely. It gives now the following equations (Di Tommaso, 1984)

$$K_I \sin \theta_0 + K_{II} \left(3 \cos \theta_0 - 1\right) = 0$$

$$\cos \frac{\theta_0}{2} \cdot \left[K_I \cos^2 \frac{\theta_0}{2} - \frac{3}{2} K_{II} \sin \theta_0 \right] = K_{IC}$$

where q_0 is the crack branching angle.

The simplest and maybe the most likely criterion is to consider that the crack alrways propagates in the direction of pure mode I.

It can be shown that pure mode II toughness cannot be measured because the crack faces touch and friction forces must be overcome. This is a result of the higher order terms in the development of the crack faces displacement. In the preceding expressions the K_{IIc} fracture toughness which applies at relatively small K_{II}/K_I ratios is thus much lower than the one which might be measured in mode II.

When the normal stress on the crack is compressive, it can still propagate if the shear stress is high enough with respect to the friction force on the crack faces. The condition can be written :

$$\left| K_{II} \right| - f \left| K_I \right| = K'_{IIC}$$

f being the friction coefficient, with an appropriate value for K'_{IIc}.

Another complication comes from the cracks lying at the interface between two dissimilar materials. A solution to this problem exists (Di Tommaso, 1984). Depending on the ratio of the shear moduli and of the stress applied on the interface, the crack propagates in the interface or branches into either the aggregate or the mortar. The tendency is for the cracks to propagate into the less rigid material. Thus in normal concrete they will not cross the aggregates, whereas this can happen in light weight concrete or in high strength concrete.

3 - Damage mechanics models.

The microcracks decrease the effective area of the material which carries the load. A damage parameter can be introduced, defined as the ratio between the effective area and the total area. It is an increasing function of the number of cracks and of their sizes. As there are cracks allready present in the concrete before any loading, there is an initial damage present. It then increases when loads are applied.

Another definition is to relate the damage parameter to the decrease of the elastic moduli, owing to the opening and to the propagation of microcracks.

In any case, the development of the cracks is not isotropic and it is obvious that the damage must be rigorously defined as a tensor.

It seems better to relate the evolution of the damage to the strain rather then to the stress. The crack propagation, whatever the criterion used, is produced by a local stress, included in the K_I and K_{II} stress intensity factors. However on a macro scale, microcracks can propagate even when K_I and K_{II} are equal to zero. It is the case in a compression test where cracks open parallel to the load axis.

This of course is due to local stress fluctuations at the microscale and this would never happen in a perfectly homogeneous material. Back to the macroscale this damage can be related to the Poisson expansion of the concrete under compression.

But there are some thermodynamics arguments to support this idea. If the thermodynamics potential is ψ, the damage evolution force Y, i.e. the damage associated variable, is given by :

$$Y = \frac{\partial \psi}{\partial \mathbb{D}}(\varepsilon, \mathbb{D})$$

For an elastic material such as concrete :

$$\psi = \frac{1}{2}\,\varepsilon : \sigma = \frac{1}{2}\,\mathbb{M} : \varepsilon : \varepsilon = \frac{1}{2}\mathbb{M}_0 (1 - \mathbb{D}) : \varepsilon : \varepsilon$$

where $\mathbb{M}$ represents the matrix of the elastic moduli. The damage $\mathbb{D}$ is defined as an evolution of this matrix. In such a case :

$$Y = \frac{-1}{2}\,\mathbb{M}_0 : \varepsilon : \varepsilon$$

is purely related to the deformation and so it must be the same for the evolution of damage which is a function of Y.

Another way to look at it is to consider the strain energy release rate G which is the microcracks driving force. It can be expressed either as a function of the load or of the deformation. In the first case G is damage dependent, whereas, this parameter does not appear in the expression of G as a function of the deformation.

In a way this reflects the interaction between microcracks, an effect which is seldom taken in consideration.

The different behaviour of microcracks according to the sign of the normal stress shows that it is not possible to use a single law of evolution for the damage.

In the most general case there could be 21 components of the damage tensor as there are 21 elastic moduli. Simplifications are introduced the most stringent being to simply use a scalar.

Considering that the concrete is an orthotropic material, reduces the number of components to nine (Collombet, 1985; Pijaudier Cabot, 1985), To identify and to use such a model remains costly.

More simply Karihaloo and Fu (1990) proposed to use two parameters D_{11} and D_{22} to describe the damage in two dimensions, in the principal strains directions. Using the equivalence of complementary energy of the damaged and the equivalent elastic material these authors give the effective compliance tensor S^* as :

$$S^*{}_{1111} = \frac{1}{E(1 - D_{11})^2}$$

$$S^*{}_{1212} = S^*{}_{2121} = \frac{\nu}{E(1 - D_{11})(1 - D_{22})}$$

$$S^*{}_{2222} = \frac{1}{E(1 - D_{22})^2}$$

They further give the evolution law as :

$$D_{11} = \left[A + B\left(\sigma_{22}/\sigma_{11}\right)\right]\left(\varepsilon_{11} - \varepsilon_{11}^{th}\right)^C$$

$$D_{22} = \left[A + B\left(\sigma_{22}/\sigma_{11}\right)\right]\left(\varepsilon_{22} - \varepsilon_{22}^{th}\right)^C$$

where A = 150 ± 10, B = 40 ± 5, and C = 0.8 ± 0.02 ε_{11}^{th} and ε_{22}^{th} are thresholds, which depend on the concrete and below which there is no damage. According to experiments of Kaplan (1963) who measured the strain at first cracking (in micro strain)

$$\varepsilon_{11}^{th} = 79.5 \, (\pm 6) + 43.4 \, V - 406.1 V^2 + 707.2 V^3 - 1150 V^4 + 763 V^5$$

where V is the volume fraction of coarse mix. This model of damage cannot describe cases where one of the stress component is compressive.

Mazars (1984) took this problem in consideration but considering the damage as isotropic i.e. as a scalar. In a first version (Mazars, 1986 a) an equivalent deformation threshold was introduced, this equivalent deformation being defined as $\sqrt{<\varepsilon_i>^2}$ where ε_i are the principal positive strains. The evolution of damage differs in tension and in compression. Another version (Mazars, 1986) introduces an unilateral damage by partitioning the stress tensor in two parts, σ^+ built with the positive eigen values and σ^- with the negative ones.

Keeping only two damage scalars D_t and D_c the behaviour of the damaged concrete is described as :

$$\varepsilon = \frac{1}{E_o(1-D_t)}\left[(1+\nu_o)\sigma^+ - \nu_o(tr\sigma)^+ 1\right] + \frac{1}{E_o(1-D_c)}\left[(1+\nu_o)\sigma^- - \nu_o(tr\sigma)^- 1\right]$$

The corresponding damage extension forces are :

$$Y_t = \frac{1}{6E_o(1-D_t)^2}\left[(1+\nu_o)\left[3\sum_i \left(\sigma_i^+\right)^2 - \left(tr\sigma^+\right)^2\right] - (1+2\nu_o)\left(tr\sigma^+\right)^2\right]$$

and the same expression for Y_c replacing σ^+ by σ^-.

The evolutions of the damage parameters are given on loading by :

$$\frac{\dot{D_t}}{\dot{Y_t}} = \frac{\sqrt{Y_t^o}\,(1 - a_t)}{2\,(Y_t)^{3/2}} + \frac{a_t\,b_t}{2\sqrt{Y_t}\,\exp\left[b_t\left(\sqrt{Y_t} - \sqrt{Y_t^o}\right)\right]}$$

and a similar expression for D_c, where Y_t^o and Y_c^o are thresholds, a_t, b_t, a_c, b_c material parameters.

The damage parameters remain constant when unloading. In such a model the material keeps the memory of the previous damage in cyclic loading.

4 - Relation between the damage and the microcracks.

The relation between the damage parameters and the microcracks comes from the influence of their propagation and of their opening on the compliances of the material.

For a penny shaped crack of radius a under a normal stress σ, the crack discontinuity b is given by :

$$b = \left(a^2 - r^2\right)^{1/2} \cdot \frac{8\left(1 - v^2\right)}{\pi E} \cdot \sigma$$

whose average is :

$$\bar{b} = \frac{16\left(1 - v^2\right)}{3\pi E} \cdot \sigma a$$

This shows that the presence of such a crack introduces an extra strain $\pi a^2 \bar{b} / V$, V being the volume of the material. The compliance is increased proportionnally to the cube of the crack size. If it propagates a non linear effect appears.

Karihaloo and Fu (1969) remind that the effective moduli of a material containing voids can be computed as :

$$C^*_{ijkl} = C_{ijmn}\left[I_{mnkl} - f_p\left(I_{mnkl} - S_{mnkl}\right)^{-1}\right]$$

where f is the volume fraction of voids, and S_{ijkl} Elshelby's tensor, which depends on their shape. This expression can give the evolution of damage as a function of strain when there is no propagation of the defects nor nucleation of new voids. In the case of an orthotropic material under plane stress condition they find a damage evolution which displays a similar behaviour as the one given before.

$$D_{ii} = \left[A + B\sigma_{ii} + C\left(\sigma_{22} / \sigma_{33}\right)\right]\left(1 - \varepsilon_v^{th} / \varepsilon_v\right)^D$$

ε_v being the volumetric strain.

In the case of an evolution of the population of voids, the effective moduli can be computed, as a function of f and of S_{ijkl}.

In a similar fashion Krajcinovic and Fanella (1986) built a model in which initial cracks occupy a fraction of the aggregate mortar interfaces. Their sizes and orientations are randomly distributed. When the load is increased a few cracks will suddenly grow to a size equal to that of the aggregate. Then more and more cracks jump in this way until one crack will propagate in the mortar, this being considered as the final failure. It is possible to calculate the damaged compliances which are a function of the size distribution of the aggregates and of the applied load. Similar calculations were proposed by Nobile (1986).

5 - Extension to the fatigue of concrete.

The previous models are based on the opening and on the propagation of microcracks in concrete on loading. In those models, unloading is purely elastic. When loading again the microcracks would recover exactly the same configuration as before and the cycles could then be repeated over and over without any evolution. The only way to explain the fatigue of concrete is to take in consideration its viscous behaviour. Under constant load this material creeps. This could be due to water migration in the mortar and to slow growth of the microcracks. At each cycle of loading some irreversible deformation accumulates.

Tait and Garrett (1986) showed for example that the fatigue crack propagation in mortar could be correlated to the static crack growth rate under constant stress, checking a relation established by Evans and Fuller (1974).

Hu (1990) used this idea to study the evolution of fatigue damage in concrete. He writes the evolution of the sizes of a microcrack population obeying a Weibull statistics. Each crack grows according to the law :

$$da/dt = AK_I{}^q$$

In this way the evolution of the fracture probability can be evaluated as a function of time. This can be done either with a constant stress or with a cyclic stress.

However this model is based on the concept of the weakest link and it should apply to a single phase material, such as hardened cement paste containing a population of cracks. For concrete the detailed description of the microcrack behaviour could be the following : slow growth of the cracks at the interfaces of the mortar and the aggregates until they reach a critical size at which time they jump accross the whole interface and stop. This critical size is a function of the local stress. This first phase produces a decreasing damage rate because as time goes on more and more cracks have reached the critical size. The next step will be the branching of the microcracks into the mortar and their slow growth into an orientation which favours mode I. When they reach a critical size, they propagate quickly until they meet another aggregate where they stop and grow slowly. This phase will correspond to an increasing damage rate as the K_I factor on each crack keeps increasing because K_I can be written :

$$K_I = \alpha\sigma\sqrt{a}$$

Each crack grows according to the law :

$$a_0^{-\left(\frac{q-2}{2}\right)} - a^{-\left(\frac{q-2}{2}\right)} = \frac{q-2}{2} \cdot A\alpha^q \sigma^q t$$

The time for an individual jump of a crack to the next aggregate is thus proportionnal to σ^{-n}. This would give a viscous behaviour for concrete characterized by :

$$\dot{\varepsilon} = \left(\frac{\sigma}{\sigma_0}\right)^q$$

We can now build a fatigue model for concrete having such a viscous behaviour. We consider that the material is decomposed in n elements each having a fracture strain ε_R which is distributed according to a cumulative law $P_R(\varepsilon)$. For a strain ε the proportion of broken elements is :

$$n_R / n = P_R(\varepsilon)$$

We can take the damage parameter as this ratio n_R/n. We can now write the stress strain law as :

$$\dot{\varepsilon} = \left[\sigma / \sigma_0 (1 - D)\right]^q = \left(\sigma / \sigma_0\right)^q \left[1 - P_R(\varepsilon)\right]^{-q}$$

During one fatigue cycle :

$$\int_{\varepsilon}^{\varepsilon + \Delta\varepsilon/\Delta N} \left[1 - P_R(\varepsilon)\right]^q d\varepsilon = \int_{t}^{t+T/2} \left(\sigma / \sigma_0\right)^q dt$$

If the stress amplitude and the frequency remain unchanged this quantity is constant. After N cycles we will have a strain accumulation such that :

$$\int_{0}^{\varepsilon} \left[1 - P_R(\varepsilon)\right]^q d\varepsilon = N \int_{t}^{t+T/2} \left(\sigma / \sigma_0\right)^q dt$$

Or else : $(dN / d\varepsilon)(d\varepsilon / dN)_0 = \left[1 - P_R(\varepsilon)\right]^q \left[1 - P_R(\varepsilon_0)\right]^{-q}$

where $(d\varepsilon / dN)_0$ is the strain increment per cycle at the beginning of the test.

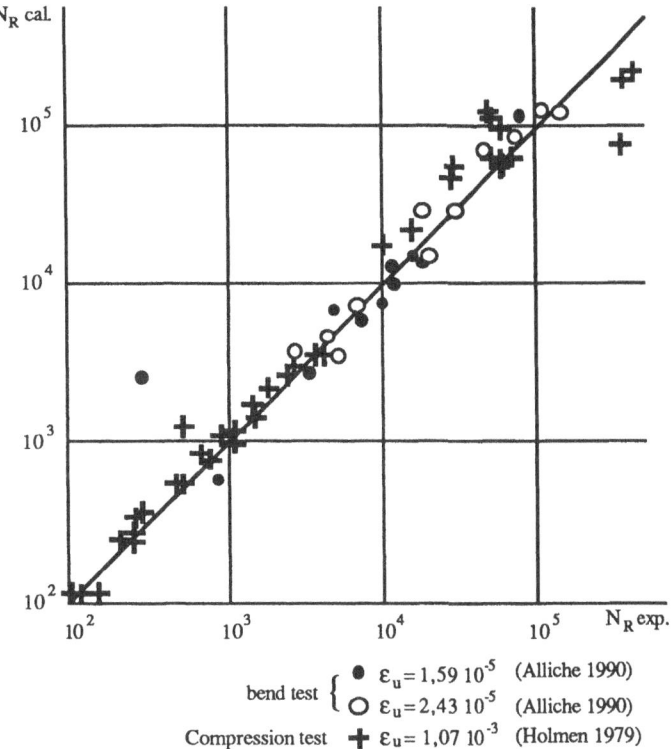

bend test $\left\{\begin{array}{l} \bullet \quad \varepsilon_u = 1,59 \ 10^{-5} \quad \text{(Alliche 1990)} \\ \bigcirc \quad \varepsilon_u = 2,43 \ 10^{-5} \quad \text{(Alliche 1990)} \end{array}\right.$

Compression test ✚ $\varepsilon_u = 1,07 \ 10^{-3}$ (Holmen 1979)

FIG 1 : Comparaison betwen the experimental and the theoretical fatigue life for different concretes.

Thus the increase of the deformation per cycle increases progressively and goes to infinity when $1 - P_R(\varepsilon_r) = 0$, i e for the ultimate damage $D = 1$. In other words concrete possess a certain strain capacity ε_r and it breaks when this value is reached whatever the loading, be it monotonic or cyclic.

To make things more precise we can for instance assume that the individual fracture strains are distributed according to a Weibull law such that :

$$D = n_R / n = 1 - \exp\left[- (\varepsilon / \varepsilon_u)^m\right]$$

The evolution of the deformation is now given as

$$(dN / d\varepsilon)(d\varepsilon / dN)_0 = \exp\left[(\varepsilon / \varepsilon_u)^{mq}\right] / \exp\left[(\varepsilon_0 / \varepsilon_u)^{mq}\right]$$

If m and q are large, which is the case for concrete, this quantity remains equal to 1 for a large number of cycles. A sudden instability occurs when $\varepsilon = \varepsilon_u$. So with an excellent approximation

$$N_R (d\epsilon / dN)_0 / \epsilon_u = \int_0^1 \exp(-x^{mq}) dx \cong 1$$

Experimental results obtained by Alliche (1990) and by Holmen (1979) obey rather well such a relation, provided ϵ_u is adjusted (fig 1). It is interesting to note that the tests performed by Alliche included two different R ratios for each concrete.

A simplified version consists in using a damage parameter the increase of which for one cycle is proportionnal to the increase in strain.(Alliche 1990). This enables to calculate the behaviour of a beam in a bend test.

In applying this model, the initial portion of the deformation per cycle curve which corresponds to a first phase of the microcracks propagation, must be eliminated, and $(d\epsilon / dN)_0$ must be measured once a steady state has been reached.

6 - Numerical models.

The complex structure of concrete led many authors to use numerical simulation to study the evolution of damage in concrete. The best known example is due to Zaitsev and Wittmann (1983). Other computation were published (Hu, Cotterell and Mai, 1986; Zaitsev, Kondrashenko and Arshabov, 1986; Roelfstra and Sadouki, 1986). They start from a certain distribution of initial cracks at the interfaces of the mortar and the aggregates and by introducing crack propagation criteria they follow the growth of the damage in a 2D structure. Roelfstra and Sadouki (1986) put special elements at the interfaces to simulate crack initiation.

A somewhat different numerical analysis consists in introducing random fracture resistances in the finite element mesh (Rossi and Richer, 1987).

A somewhat different approach was used by Schorn (1986) who modelized the concrete by a serie of struts with random fracture resistances. This kind of model is also developped by Hermann, Hansen and Roux (1989) so as to study the size effect. They found a master load-displacement curve if both the load and the displacement are divided by the size L of the elements to the power 3/4.(in a 2D strut system). More details can be found in the contribution of D. Krajcinovic to the present workshop.

Besides this possibility to test the size effect, the better understanding which the visualization of damage can bring, numerical concretes constitute good models to test the influence of various parameters.

In this way for instance Zaitsev, Kondrashenko and Arshabov (1986) found that a large heterogeneity of the structure lowers the strength of light-weight concrete, that it decreases drastically when the resistance of the interfaces drops below the resistance of the mortar, and, most of all they showed the detrimental effect of the porosity of the light-weight aggregates.`

7 - Conclusion.

A great deal of progress has been achieved in the understanding and in the modelization of such a complex, multiscale system as concrete. Damage mechanics, its relation with the influence of voids and microcracks on the compliances of the material, numerical modelling can help to predict its behaviour and the influence of various structural parameters. It seems that the models that are needed must incorporate a damage tensor the evolution of which is a function of strain, the viscous behaviour of concrete (or the slow crack growth) a statistical distribution of strength or of microcracks.

Experiments are still needed to better understand the fracture criteria of an individual microcrack in mixed mode, depending on its exact location, on the nature of the aggregate. A more exact knowlege of slow microcrack growth needs also to be obtained. This would allow more precise modelizations.

REFERENCES

ALLICHE N.: (1990) Thèse. Université d'Alger.

COLLOMBET F.: (1985) "Modélisation de l'endommagement anisotrope. Application au comportement du beton sous sollicitations multiaxiales". Thèse de 3ème cycle. Université Paris VI.

DI TOMMASO A.: (1984) "Evolution on Concrete Fracture"
in "Fracture Mechanics of Concrete", A.Carpinteri, A.R. Ingraffea ed. Martinus Nijhoff, Pub. The Hague, pp.31-65.

EVANS A.G. and FULLER E.R. : (1974) Met. Trans, $\underline{5}$, pp. 27-33.

HERMANN H.J., HANSEN A. and ROUX S. (1989) Phys. Rev. B. $\underline{39 \ p}$. 637

HOLMEN J.O. (1979). "Fatigue of concrete by constant and variable Amplitude loading". Norwegian Institute of Technology F.C.B. Trondheim.

HU Xiao-Zhi : (1990) "Flaw Analysis in Time Dependent Fracture for Cemetitious Materials".

HU X-Z., COTTERELL B. and MAI Y.N. : (1986) "Computer Simulation Models of Fracture in Concrete" in Fracture Toughness and Fracture Energy of Concrete. F.HWittmann ed. Elsevier, Amsterdam, pp. 91-100.

HU X-Z., MAI Y.N. and COTTERELL B. (1989). "A. Statistical Theory of Time-Dependent Fracture for Cementious Materials Subjected to cycle Loading" J. of Mater; Science $\underline{24}$ pp. 3118-3122.

KAPLAN M.F. : (1963) "Strains and Stresses of Concrete at Initiation of Cracking and Near Failure". J. Am. Conc. Inst., 60, pp. 853-880.

KARIHALOO B.L. and FU D. : (1989) "A Damaged Based Constitutive law for Plain Concrete in Tension". Eur. J. Mech. A/Solids 8 pp 373-384

KARIHALOO B.L. and FU D. : (1990) "An Anisotropic Damage Model for Plain Concrete".
In Proc. Int. Conf. on Fracture and Damage of Concrete and Rock. Vienna, July 1988, Pergamon Press, Engng. Fract. Mech., 35, 205-209.

KARIHALOO B.L. and FU D. : (1990) "An Orthotrophic Damage Model for Plain Concrete in Tension". ACI Materials J. pp. 62-67.

KRAJCINOVIC D. and FANELLA D. : (1986) "A Micromechanical Damage Model for Concrete", Engng. Fract. Mech., 25, 585-596.

MAZARS J. : (1981) "Mechanical Damage and Fracture of Concrete Structures", in Advances in Fracture Research. D.François et al, Eds. Pergamon, 4, 1499-1506.

MAZARS J. : (1984) "Application de la mécanique de l'endommagement au comportement non linéaire et à la rupture du béton de structure", Thèse de doctorat d'Etat, Université Paris VI.

MAZARS J. (a) : (1986) "A Description of Micro and Macroscale Damage of Concrete Structures". Engng. Fract. Mech., 25, pp. 729-737.

MAZARS J. (b) : (1986) "A Model of Unilateral Elastic Damageable Material and its Application to Concrete" in Fracture Toughness and Fracture Energy of Concrete, F.H.Wittmann, Ed. Elsevier, Amsterdam, pp. 61-71.

MAZARS J. , BERTHAUD Y. and RAMTANI S. : (1990) "The Unilateral Behaviour of Damaged Concrete", Engng. Fract. Mech., 35, 629-635.

MAZARS J. , and LEMAITRE J. : "Application of Continuous Damage Mechanics to Strain and Fracture of Concrete", Proc. of the NATO Advanced Research Worskhop on Application of Fracture Mechanics to Cementious Composites. Evanston, pp.375-388.

NOBILE L. : (1986) "A Microcrack Model of Concrete's Behaviour under Applied Load" in Fracture Toughness and Fracture Energy of Concrete, F.H.Witmann ed. Elsevier, Amsterdam, 73-80.

PIJAUDIER-CABOT G. : (1985) "Caractérisation et modélisation du comportement du béton par un essai multiaxial automatique", thèse de 3ème cycle, Université Paris VI.

REGOURD M. : (1986) "Microstructure of Cement Blends Including Fly Ash, Silica Fume, Slag and Fillers" Proc. of the Fall Meeting Materials Research Soc., Boston.

ROELFSTRA P.E. and SADOUKI H.: (1986) "Fracture Process in Numerical Concrete" in Fracture Toughness and Fracture Energy of Concrete. F.H.Wittmann ed. Elsevier, Amsterdam, pp. 105-116.

ROSSI P. and RICHER S. : (1987) "Stochastic Modelling of Concrete Cracking", in Constitutive Laws for Engineering Materials, C.S.Desai, ed.Tucson.

SCHORN H. : (1986) "Numerical Simulation of Composite Materials as Concrete" in Fracture Toughness and Fracture Energy of Concrete. F.H.Wittmann, ed. Elsevier, Amsterdam, pp. 177-188.

SIH G.C. : (1973) "Some Basic Problems in Fracture Mechanics and New Concepts", Engng. fract. Mech., 5,.

SLATE F.O. and HOVER K.C. : (1984) "Microcracking of Concrete" in Fracture Mechanics of Concrete, ed. A.Carpinteri, A.R. Ingraffea, p.137, Martinus Nijhoff, Pub. the Hague.

TAIT R.B. and GARRETT G.G. : (1986) "A Fracture Mechanics Evaluation of Static and Fatigue Crack Growth Rate in Cement Mortar" in Fracture Toughness and Fracture Energy of Concrete. F.H.Wittmann, ed. Elsevier, Amsterdam, 21-30.

WITTMANN F.H. : (1981) "Mechanisms and Mechanics of Fracture of Concrete" in Advances in Fracture Research, ed. D.François et al, 4, 1467, Pergamon.

ZAITSEV Y.V. , KONDRASHENKO V.I. and ARSHABOV A.A. : (1986) "Simulation of Behaviour of Light-Weight Concretes under Load and its Experimental Trial" in Fracture Toughness and Fracture Energy of Concrete. F.H.Wittmann, ed. Elsevier, Amsterdam, pp. 101-104.

ZAITSEV Y.V. and WITTMANN F.H. : (1983) "Fracture Mechanics of Concrete" Elsevier Scien. Pub., Amsterdam, 251-299.

CRACKING, DAMAGE AND FRACTURE IN STRESSED ROCK: A HOLISTIC APPROACH

P.G. MEREDITH, M.R. AYLING, S.A.F. MURRELL & P.R. SAMMONDS
Rock & Ice Physics Laboratory
Department of Geological Sciences
University College London
London WC1E 6BT, U.K.

ABSTRACT. A brief review is given of time-dependent cracking and failure in rock subjected to both tensile and compressive stresses. Recent results from our experimental program to measure changes in an integrated suite of rock physical properties contemporaneously with mechanical parameters during deformation experiments are also presented. We suggest that the observed changes in these properties provides additional information regarding the evolution of damage, and the rate of damage accumulation, within the rock sample. Future developments required for a complete holistic description of the development of crack-related damage leading to failure are also considered.

1. Introduction

The quasi-brittle behaviour of crustal rocks has a number of aspects in common with the processes of crack growth, fracture and failure in other quasi-brittle materials such as concrete and polycrystalline ceramics. Most of these similarities arise due to the intrinsic heterogeneous, polycrystalline and commonly polyphase microstructure of these materials. However, there are also a number of fundamental differences that affect rock fracture, and which arise primarily due to their burial in the environment of the Earth's crust. Important among these are: (1) that under all but exceptional circumstances, rocks are subjected to compressive triaxial stresses in the crust; (2) that in the overwhelming majority of situations in the crustal environment, pre-existing microcracks, pores and void spaces contain fluids (especially water) which are also commonly under pressure; and (3) the average rate of natural deformation is very slow.

Since crack growth is necessarily a dilatant process (involving volume increase) it is pressure dependent, and therefore the application of compressive stresses acts to stabilize crack growth and hence to strengthen or toughen the material. Conversely, the presence of pressurized pore fluids acts to weaken rocks, both by reducing the confining effect of any applied compressive normal stress via the principle of *effective* stress (e.g., see Jaeger & Cook 1976), and also through weakening fluid-rock chemical reactions that allow subcritical crack growth processes such as *stress corrosion* to proceed. Subcritical crack growth becomes a fundamentally important deformation process in rocks because the

S. P. Shah (ed.), Toughening Mechanisms in Quasi-Brittle Materials, 67–89.
© 1991 *Kluwer Academic Publishers.*

inferred rate of natural tectonic deformation in the crust (strain rates of the order of 10^{-14}/s) is much lower than typical engineering rates of strain. This leads to a time-, environment- and deformation rate- dependence of the mechanical properties of rocks. For example, the compressive fracture stress of unconfined rock can decrease by a factor of two or three as the strain rate is reduced from those ordinarily used in laboratory tests (10^{-4}- 10^{-6}/s) to that commonly associated with natural tectonic deformation. Most rocks containing pore fluids also exhibit brittle creep behaviour (time-dependent cracking at constant stress, or *static fatigue*), especially if the applied stress is a significant fraction of the short-term fracture stress (Kranz & Scholz 1977, Kranz 1979, Costin 1987).

Furthermore, the underlying rationale behind the methodology for determining basic fracture mechanics parameters is different for rocks than for some other quasi-brittle materials (e.g. concrete), and this is reflected in the different codes of practice recently developed by ISRM and RILEM (ISRM 1988, RILEM 1985, Hillerborg 1989). In general, the size of rock masses being considered in deep engineering problems and in crustal geophysics can be considered to be essentially infinite, and under these circumstances an LEFM analysis is appropriate. On the other hand, the realistic size of laboratory rock samples is normally many orders of magnitude smaller. Hence the ISRM approach has been to attempt to obtain valid fracture toughness values by testing sub-size samples and then making a correction for non-LEFM behaviour. There is, therefore, the additional problem of the scaling effect to be considered when modelling and analysing rock fracture problems using laboratory derived data.

By contrast, the RILEM approach has been to abandon LEFM altogether and to determine the fracture energy of concrete from tension-softening measurements. This stems, at least in part, from the relatively small scale difference between test samples and real engineering structures.

In summary then, the major part of this paper is concerned with an integrated discussion of those aspects of the fracture and failure of rocks that are different from other quasi-brittle materials, i.e. compressive stress fields, the presence of pore fluids, time-dependent properties, and scaling relations. However, it is well recognised (e.g. Kranz 1983, Meredith 1990) that even under compressive loading rock deformation proceeds by the growth, interaction and linkage of many tensile microcracks. It is apposite therefore to consider briefly the fracture of rocks under simple tensile loading before proceeding to the more complex case of compressive loading.

2. Fracture Under Tensile Loading

There have been a number of comprehensive reviews published in recent years detailing different aspects of the propagation of tensile cracks in rocks (e.g. Atkinson 1982, 1984, Swanson 1984, Atkinson & Meredith 1987a, 1987b, Meredith 1989) so that only a brief summary is provided here to avoid needless repetition.

2.1 CRITICAL FRACTURE

Meredith (1989) has shown that mode I fracture toughness measurements for rocks exhibit similar trends as for many other quasi-brittle materials. The fracture toughness of polycrystalline rocks is generally found to be much higher than that of any of the rock's constituent minerals (Atkinson & Meredith 1987b, Meredith 1990), and Figure 1 also illustrates that, for rocks of similar mineralogical composition, the fracture toughness tends to increase with increasing grain size. For rocks with a large variation in grain size, it is the maximum grain size that is the controlling influence. Large grains or phenocrysts appear to act as crack *stoppers* in a similar manner to the large inclusions that are used to artificially toughen some polycrystalline ceramics.

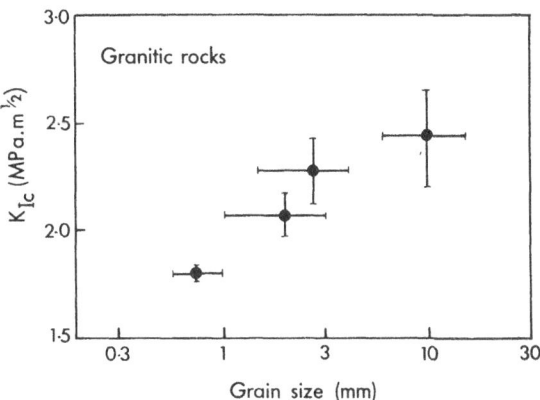

Figure 1. Mode I fracture toughness (K_{Ic}) as a function of grain size for granitic rocks. (After Meredith 1989).

In addition, Figure 2 shows for Merrivale granite the type of rising R-curve behaviour that is typical of many rocks (e.g. Schmidt & Lutz 1979, Ingraffea 1987, Swanson 1987, Meredith 1989) and polycrystalline ceramics (e.g. Hubner & Jillek 1977, Cook et al.

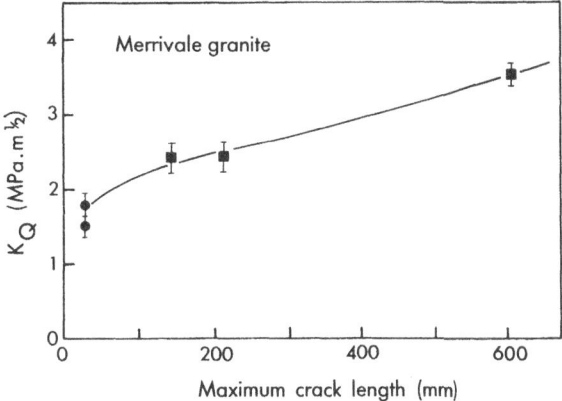

Figure 2. Illustration of R-curve behaviour in samples of Merrivale granite. Fracture resistance (K_Q) is plotted against maximum crack length at critical failure. (After Meredith 1989).

1985). The existence of R-curve behaviour is linked directly with the development of an inelastic fracture process zone that comprises microcracking ahead of, and crack interface tractions (frictional interlocking due to the rough three-dimensional nature of crack surfaces, and ligamentary bridging between opposing crack walls) behind a macrocrack tip (Labuz et al. 1985, Ingraffea 1987, Swanson 1987, Freiman & Swanson 1990). Crack extension resistance increases as crack length increases until the process zone is fully developed and there is at least a portion of the crack that is traction-free. Only at or beyond this critical crack length is a steady state reached, and only then can a fracture toughness value that is truly representative of the bulk material be determined.

2.2 SUBCRITICAL CRACK GROWTH

In systems where rocks are subjected to long-term loading, the classical LEFM approach does not provide an adequate description of crack growth. This is especially so at elevated temperatures and in the presence of reactive environmental species. A considerable body of evidence supports the idea that cracks can propagate in a stable, quasi-static manner at stress intensities that are substantially below the fracture toughness, albeit at velocities that are orders of magnitude lower than the terminal velocity associated with critical fracture. This phenomenon of subcritical crack growth has been observed experimentally in many quasi-brittle materials, including glass (Wiederhorn 1978), ceramics

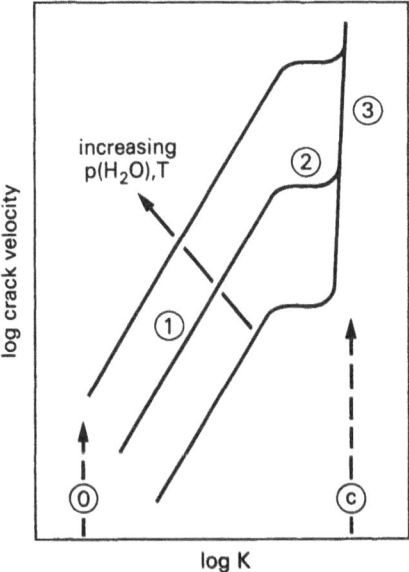

Figure 3. Schematic diagram showing the dependence of crack velocity on stress intensity (K) between the subcritical crack growth limit (0) and critical rupture (c). The influence of temperature and partial pressure of water is also indicated. The different behaviour in regions 1, 2 and 3 is described in the text.

(Wiederhorn 1974), and minerals and rocks (Atkinson 1974, Atkinson & Meredith 1987a, 1987b).

The overwhelming body of experimental and observational evidence suggests that extension of pre-existing cracks and flaws by the mechanism of stress corrosion is likely to dominate subcritical crack growth in rocks at low homologous temperatures. Stress corrosion proceeds by the preferential weakening of strained bonds at crack tips through reactions with chemical species in the environment. Although a whole range of surfactants can contribute to stress corrosion (e.g. Dunning et al. 1984), the most active reagent for the process appears to be water.

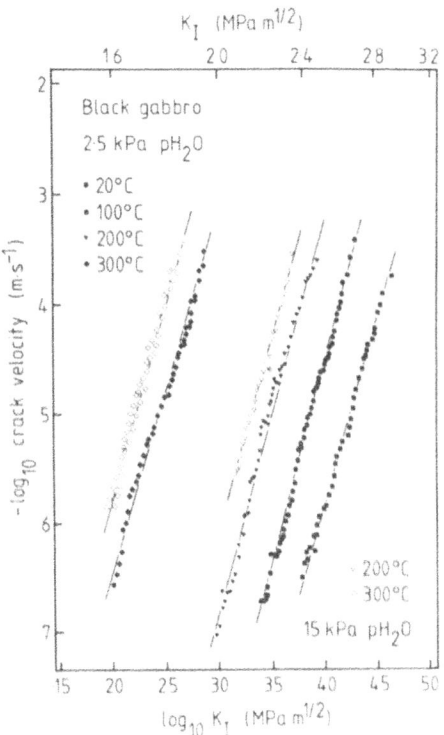

Figure 4. Stress intensity factor - crack velocity diagram for Black gabbro cracked at temperatures from 20° to 300°C under water vapour pressures of 2.5 and 15kPa. Solid lines are least squares fits to the data points. (After Meredith & Atkinson 1985).

The subcritical crack growth behaviour of many materials approximates to the classical trimodal pattern exhibited by glass in aqueous environments. This pattern of behaviour is illustrated schematically in Figure 3, where the logarithm of stress intensity (K) is plotted against the logarithm of crack velocity. In region 1, the crack extension velocity is controlled by the rate of stress corrosion reactions at crack tips. In region 2, crack growth is controlled by the rate of transport of reactive species to crack tips (Lawn &

72

Wilshaw 1975). In region 3 the curve becomes asymptotic to the critical value (denoted by c); crack growth is primarily due to thermally activated bond rupture and is relatively insensitive to the chemical environment (Freiman 1984). A lower limiting threshold is thought to exist, below which no crack growth will occur (denoted by 0). The existence of such a subcritical crack growth limit has been demonstrated for various glass/water systems (e.g. Wiederhorn & Johnson 1973, Simmons & Freiman 1980), but has not yet been confirmed for polycrystalline ceramics or rocks. Note that because stress corrosion is a chemically-enhanced and thermally activated process, the subcritical crack growth curve should be shifted to a higher velocity for the same value of K if either the partial pressure of the active environmental reagent or the temperature is increased.

Figure 4 illustrates experimentally derived subcritical crack growth data for Black gabbro under a range of environmental conditions. All of the data of Figure 4 relate to region 1 type behaviour and may be described by a power law (Charles 1958) of the form:

$$V = V_c(K/K_c)^n \tag{1}$$

where n is known as the stress corrosion index (for the data of Figure 4, values of n lie in the range 33 - 40).

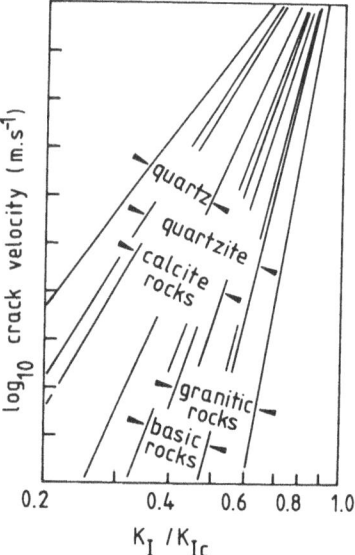

Figure 5. Synoptic diagram illustrating the variation in subcritical crack behaviour with rock type. In general, polyphase rocks exhibit higher values of n than monominerallic rocks, with single mineral phases having the lowest values. Arrows indicate the range of data. (After Atkinson 1984).

Atkinson & Meredith (1987b) provide a compilation of subcritical crack growth data for a wide range of rocks and minerals, and for polycrystalline rocks values of n in equation (1) generally lie in the range 20 - 60. Comparison of subcritical crack growth behaviour for different rock types is problematical, but an attempt is made to do this in Figure 5 (after Atkinson 1984) by normalizing stress intensity factors by the fracture toughness of each rock type. Figure 5 therefore illustrates in a broad way the relative susceptibility of different rock types (and quartz) to subcritical crack growth, bearing in mind that it is only constructed from existing experimental data. The overall trend is that the more complex the microstructure, the lower is a material's susceptibility to subcritical crack growth. Similar conclusions have been reported by Swanson (1984). These observations imply that the more heterogeneous rocks behave in a more brittle manner, at least during slow deformation. This is not what one might intuitively expect, and appears to contradict the conclusions of Cox & Scholz (1988) and Cox & Paterson (1990) that heterogeneous rocks with large initial flaw distributions tend to behave in a less brittle manner than more homogeneous rocks.

3. Fracture Under Compressive Stresses

Fracture of quasi-brittle materials subjected to compressive stresses is generally much more complex than for the tensile case, since compressive failure generally involves the nucleation, propagation and interaction of a distributed array of microcracks. Direct observation of such stress-induced microcracks in rocks (e.g. Peng & Johnson 1972, Tapponier & Brace 1976, Kranz 1979, 1980, 1983, Wong & Biegel 1985) suggests strongly that they nucleate from pre-existing flaws (pores, inclusions, microcracks, etc.), and propagate in a direction parallel or sub-parallel to the maximum principal compressive stress.

The minimum principal stress, or *confining stress*, acts as a stabilizing influence, and its influence on stress-strain curves is illustrated schematically in Figure 6. Where there is no confining stress (uniaxial compression) failure may be unstable, and is dominated by the propagation of a small number of axial cracks. By contrast, the application of even a modest compressive confining stress causes individual cracks to extend stably. A microcrack will extend to relieve the local stress concentration caused by an increase in the compressive stress difference, and will then arrest. In terms of fracture mechanics, neglecting for the moment any subcritical crack growth, microcracks extend once the critical tensile stress intensity (K_{Ic}) is exceeded locally. However, since K is a *decreasing* function of crack length under these conditions (Tada et al. 1973, Costin 1987), individual cracks only extend until equilibrium is reached at $K = K_{Ic}$. As the compressive stress difference is increased, an increasing population of microcracks extend until their density and average size is such that they interact to produce macroscopic failure.

At moderate confining stresses, the failure mode is controlled by the interaction of numerous microcracks in a relatively narrow fracture zone to form a macroscopic shear failure (i.e. a fault); and at much higher confining stresses by distributed, near-homogeneous microcracking to cause pseudo-ductile deformation by cataclastic flow. Hence the compressive strength of rocks is very pressure dependent.

Furthermore, Main, Peacock & Meredith (1990) have noted that both seismic data from crustal earthquakes and acoustic emission data from laboratory-scale rock fracture experiments (reported in a later section) are consistent with a feedback model of

74

compressive failure. Where a population of microcracks exists, the overall situation appears to be initially one of negative feedback, or *anti-persistency*. Once a crack has grown to relieve the stress locally in a high stress zone, it becomes a relatively low stress zone. It is then more likely that further stress relief will be accommodated by growth of a different crack than by further extension of the same crack. Eventually, under conditions of increasing stress, a proportion of the original population of cracks will have grown stably until their lengths are comparable to their spacing, whereupon the locally perturbed stress fields due to the presence of the cracks interact in a cooperative manner. The situation can then flip from one of negative feedback to one of positive feedback (*persistency*), leading to instability and failure. In terms of stress/strain relations this corresponds to a change from strain hardening to strain softening behaviour during dilatancy.

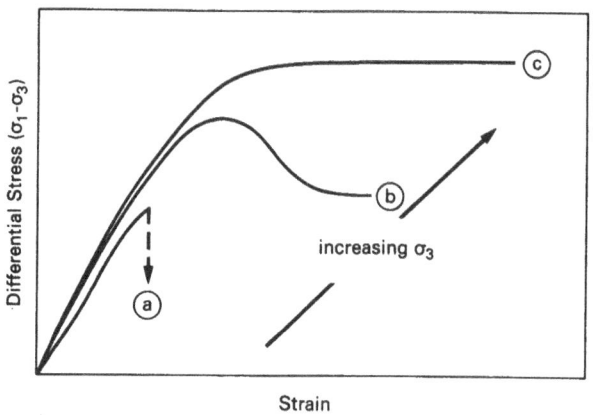

Figure 6. Schematic stress-strain curves for rock undergoing deformation over a range of confining stresses. (a) At zero confining stress the material fails catastrophically in a brittle manner by axial splitting. (b) At intermediate confining stresses the material has a higher strength and fails in a semi-brittle manner by localization of deformation on a shear fault. (c) At very high confing stresses the material is even stronger and fails in a pseudo-ductile manner by cataclastic flow. The material may flow at constant stress or exhibit strain hardening.

So far we have neglected any influence of time-dependent processes such as stress corrosion, but like tensile fracture, failure under compressive loading has long been recognised as a process that is dependent upon both environment and deformation rate. Therefore if we wish to use our understanding of the behaviour of individual tensile cracks to address the problem of fracture in compression it is necessary to consider the combined effects of arrays of multiple cracks in order to account for bulk material behaviour. In an attempt to address this problem, modified Griffith theories were developed to predict the shapes of failure envelopes in normal stress - shear stress space (*Mohr diagrams*) from consideration of the action of microcracks (modelled as elliptical flaws) at different orientations (e.g. McLintock & Walsh 1962, Murrell & Digby 1970). The essential features of this approach are illustrated in Figure 7 (after Murrell 1990).

In particular, the figure shows conditions for compaction (due to crack closure), dilatancy (following crack initiation), macroscopic shear fracture, and the transitions from fracture to cataclastic flow or plastic flow (at elevated temperatures). Each of these features is associated with a particular state of crack damage.

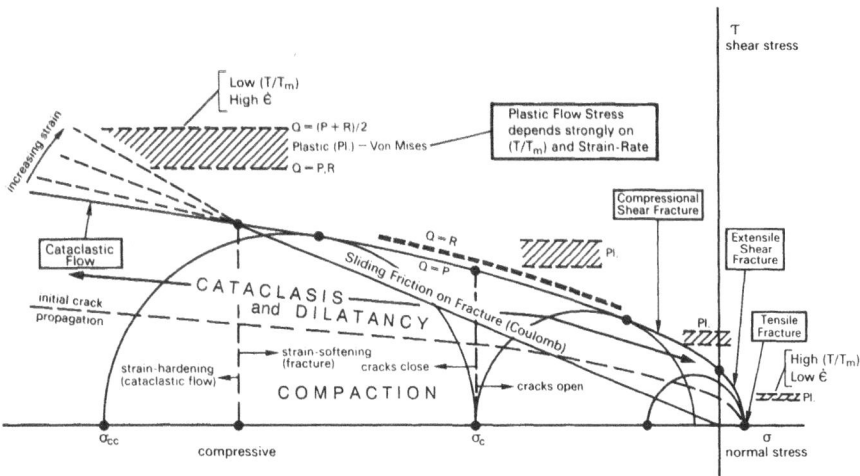

Figure 7. Mohr diagram (tensile stresses taken as positive) showing envelopes for crack propagation, fracture and sliding friction, and for plastic and cataclastic flow. Q is the intermediate principal stress. Cataclasis with dilatancy associated with new crack propagation occurs under stress conditions lying between the fracture and crack propagation envelopes, and in the cataclastic flow region of the diagram. (After Murrell 1990).

However, a full treatment of brittle failure requires consideration of all phases of crack development including crack nucleation, crack extension and crack interaction. In recent years a number of authors have contributed to the development of a new body of theory known as *damage mechanics* that takes account of all of these phases of cracking in an attempt to explain the various aspects of non-linear, time-dependent mechanical behaviour by utilising the concept of a single variable or set of variables to describe the changing microstructural state of a material as it is deformed (e.g. Costin 1983, 1985, Horii & Nemat-Nasser 1986, Ashby & Hallam 1986, Sammis & Ashby 1986, Kemeny & Cook 1987). A brief review and comparison of the predictions of these various damage models has been provided by Gueguen et al. (1990), and so is not repeated here. Costin (1987) has pointed out that there are three basic elements required for any damage theory: (1) a definition of the state of damage; (2) an equation to describe damage evolution; and (3) a constitutive law that predicts the relation of damage to stress and strain.

It is trivial to note that any predictions derived from the models will depend crucially upon the assumed initial flaw distribution. In his comprehensive review of microcracks in rocks, Kranz (1983) points out that while a great deal of theory exists on the role of microcrack populations in rock fracture processes,there is a paucity of supporting

76

observational data. Common assumptions include collinearity or randomness in orientation and location of the initial microcrack population, and initial length-frequency distributions that can be uniform, Gaussian, exponential or power law in form. However, the assumption of randomness in the cracking process and stochastic independence between cracks becomes increasingly untenable as stress levels increase and cracks interact and coalesce as failure is approached. Furthermore, since cracks are quasi-planar features, the growth of microcracks that are aligned with respect to principal stress directions will result in even an initially isotropic material being rapidly transformed into a mechanically anisotropic material.

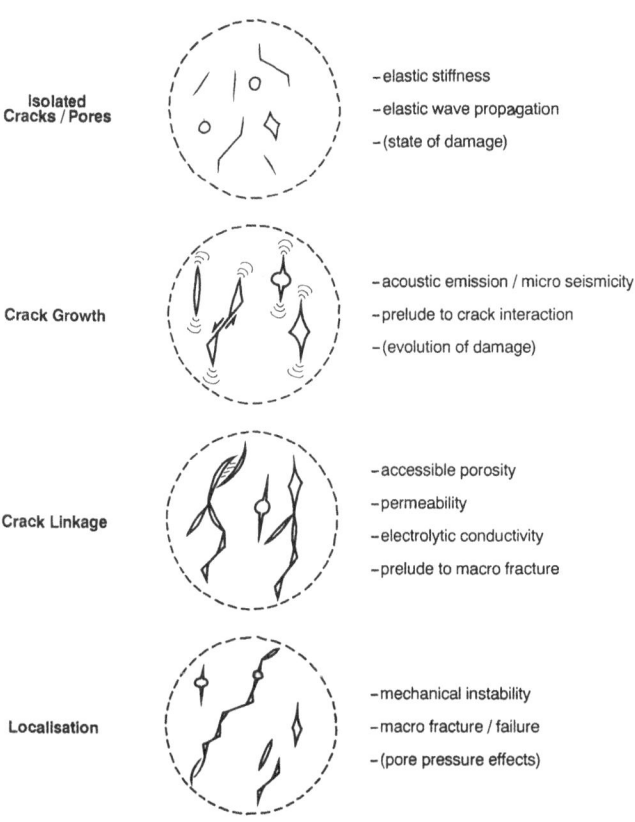

ROCK PHYSICS - the role of cracks / pores

Isolated Cracks / Pores
- elastic stiffness
- elastic wave propagation
- (state of damage)

Crack Growth
- acoustic emission / micro seismicity
- prelude to crack interaction
- (evolution of damage)

Crack Linkage
- accessible porosity
- permeability
- electrolytic conductivity
- prelude to macro fracture

Localisation
- mechanical instability
- macro fracture / failure
- (pore pressure effects)

Figure 8.   Schematic diagram illustrating the relationship between rock physical properties, damage development and the various stages of cracking leading to failure in brittle rock.

4. Indirect Monitoring of Damage Development by Integrated Rock Physical Property Measurements

Where the experimental work described in this section differs from, but aims to build upon, earlier theoretical damage mechanics studies is by incorporating the diagnostic results of contemporaneous measurements of rock physical property changes during deformation experiments into the modelling process, in order to provide a physically more realistic description of crack damage development. The underlying rationale of this approach is illustrated schematically in Figure 8. In essence: (1) changes in the velocity of elastic waves pulsed through an experimental sample occur in response to changes in the cumulative density of cracks and pores, and hence indirectly describes the changing state of damage; (2) acoustic emission statistics relate to the contemporary rate of crack extensions, and hence can provide information about the rate of damage evolution; and (3) measurement of changes in transport properties such as fluid permeability or electrolytic conductivity provides information about the interaction and linkage of dilatant microcracks leading to macroscopic fracture and failure. Currently, we are making simultaneous measurements of elastic wave velocity changes and variations in acoustic emission parameters during our triaxial rock deformation experiments. The contemporaneous measurement of changes in connected pore volume and fluid permeability is in an active state of development.

4.1   ELASTIC WAVE VELOCITY MEASUREMENTS

Results of measurements of both P- and S-wave velocities made during triaxial deformation of dry sandstone in our laboratory have previously been reported in Sammonds et al. (1989) for all of the failure modes illustrated in Figure 6. More recent results for the case of failure by localized shear faulting is shown in Figure 9. Note that the velocity of P and S waves increases during the initial quasi-linear phase of loading, with maximum velocity occurring at about half the peak stress difference. This is interpreted as being due to grain-scale compaction and the closure of cracks in response to the increasing applied stress. The most favourably oriented (i.e. those oriented normal or sub-normal to the axial stress) and most open cracks close first, and crack closure then becomes progressively more difficult. Above about half the peak stress, new dilatant cracks begin to propagate, and hence both wave velocities start to decrease. However, note that during the initial phase of deformation, the relative increase in V_p is substantially higher than the increase in V_s, but that V_s decreases more rapidly than V_p in response to dilatant crack growth. Since axial P waves are more sensitive to cracks normal to the axial stress, and S waves more sensitive to cracks parallel to the axial stress, this observation supports the concept that dilatancy occurs by the growth of tensile cracks oriented parallel to the maximum compressive stress. It also shows that simultaneous measurement of changes in both velocities provides a guide to the progressive development of crack anisotropy as deformation proceeds.

These continuous changes in the velocities of pulsed elastic waves have been used to calculate the changes in effective dynamic elastic constants (Young's modulus and Poisson's ratio) during deformation. In addition, the relative amplitudes of received waveforms have been used to determine the *seismic quality factors* (Q_p and Q_s). Both sets of data are also shown in Figure 9. These quality factors are merely the reciprocal of the attenuation coefficient. The change in effective Young's modulus exhibits a very similar trend to that of the wave velocities, indicating that the material becomes stiffer as cracks close and more compliant during dilatancy. The quality factors also increase during compaction and crack closure, since transmissibility is improved. Currently, we have no good explanation for why Q_p and Q_s peak at different values of strain.

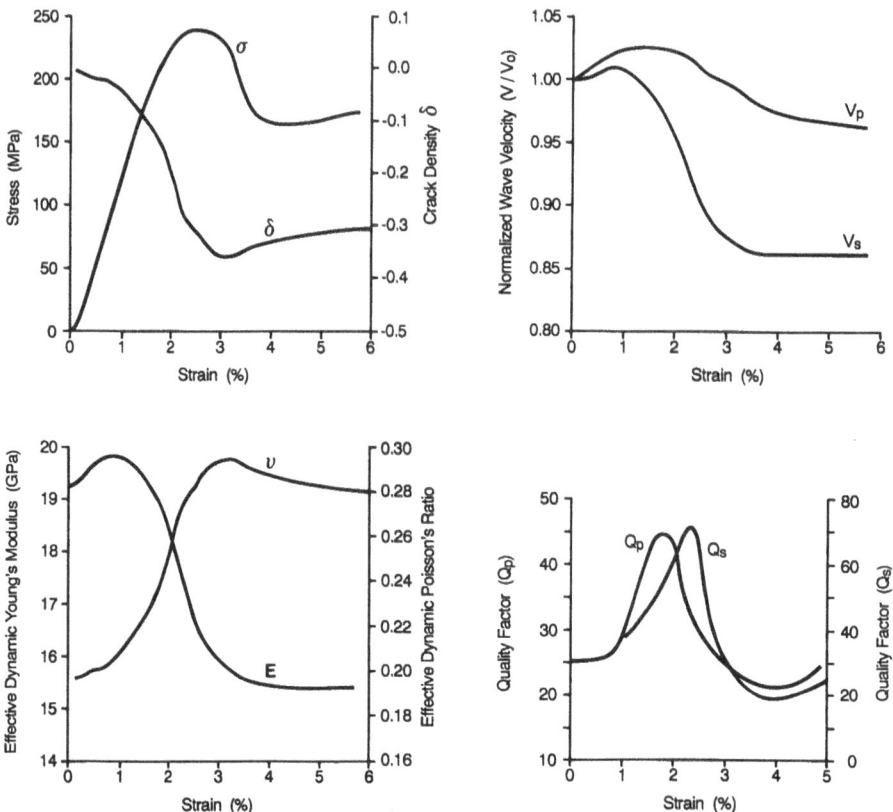

Figure 9. Stress-strain data and simultaneous measurements of compressional (P) and shear (S) wave velocities for Darley Dale sandstone deformed under a confining stress of 50MPa. Effective dynamic elastic moduli are calculated from standard relationships between V_p and V_s. Seismic quality factor is calculated by the spectral ratio method of Toksoz; and the crack density parameter is calculated according to the model of O'Connell & Budiansky.

A number of theories have been proposed to relate changes in elastic wave velocities and changes in effective elastic moduli in order to predict changes in microcrack concentration or density. One of the first of these was by Walsh (1965) who calculated the excess strain energy due to the presence of an isolated penny-shaped crack in an infinite medium under stress. More recently, O'Connell & Budiansky (1974) developed a self-consistent model which assumes that the $(N + 1)$th crack is imbedded in a homogeneous medium whose effective elastic moduli are those of the cracked material containing N cracks. Not surprisingly, this latter model predicts a much more rapid decrease in moduli with increase in crack density and average crack length. However, neither model takes account of stress field interactions between adjacent cracks, and are therefore only strictly applicable to low densities of relatively short cracks.

O'Connell & Budiansky's model predicts that not only both V_p and V_s, but also the ratio V_p/V_s would decrease with increasing crack density. However, the experimental results of Figure 9, and those reported by Gupta (1973) and Sammonds et al. (1989) do not support this latter prediction. For example, Sammonds et al. note that for their experimental results, the V_p/V_s ratio initially increases with increase in stress difference. This initial increase is followed by a period where the ratio remains essentially constant, before again increasing close to peak stress and beyond. Gupta (1973) reports a similar result.

Finally in Figure 9 we show the results of calculations of the change in crack density from the initial unstressed state determined according to O'Connell & Budiansky's model. The crack density parameter (d) is given by

$$d = N\langle a^3\rangle \qquad (2)$$

where N is the number of cracks per unit volume of the solid and $\langle a\rangle$ is the mean major axis of the ellipsoidal cracks. d is further defined as a function of the Poisson ratio (v) according to

$$d = [45(v_o - v)(2 - v)]/[16(1 - v^2)(10v_o^2 - 3v_o - v)] \qquad (3)$$

where v_o is the Poisson ratio for the initial state, and v is the effective Poisson ratio of the cracked solid. From elasticity theory,

$$v = [(V_p/V_s)^2 - 2]/[2(V_p/V_s)^2 - 1] \qquad (4)$$

and hence variations in d may be determined from variations in V_p and V_s. Rather than an increase in crack density, the model erroneously predicts a decreasing crack density even during the dilatancy that accompanies the post-elastic strain hardening and strain softening phases of deformation. This result stems directly from the observation that the ratio V_p/V_s does not decrease during deformation as predicted by the model.

The explanation for these discrepancies between model predictions and experimental observations is that the model not only does nor take account of interactions, but also does not take account of the development of any crack anisotropy. As rocks deform, their compliance parallel to the axis of compression decreases due to compaction and crack closure, and that normal to the axis of compression increases due to the growth of dilatant tensile axial cracks.

More recently, Hudson (1981) and Crampin (1984) have taken crack anisotropy into account by calculating theoretically the variations in P and S wave velocities through solids containing aligned cracks. However, once again, their formulations are only valid for crack concentrations that are low enough for no interaction to occur, and where the wavelengths of the pulsed waves are large compared with the length of the cracks.

To date, therefore, no systematic theoretical treatment exists that can take full account of both crack interactions and the development of anisotropy.

4.2 ACOUSTIC EMISSION MEASUREMENTS AND GEOMETRICAL SCALING RELATIONS

It is well documented that, for subcritical crack growth of tensile cracks, the rate of acoustic emission (AE) activity is directly related to the rate of crack propagation (e.g. Atkinson & Rawlings 1981, Meredith & Atkinson 1983, Meredith 1990), and that the magnitude or amplitude of an individual AE event is related to the source dimension and the increment of crack extension that generated the emission. Meredith & Atkinson (1983) have previously reported data describing the distribution of event amplitudes as a function of stress intensity factor, from tensile tests on a range of crystalline rock types. The parameter that characterizes the amplitude distribution is the *seismic b-value* (c.f. the Gutenberg-Richter frequency-magnitude relation for earthquakes), where b is an empirical constant in the relation

$$\log N = a - bm \tag{5}$$

where N is the number of times an event of magnitude (log scale) or peak amplitude (in dB) m occurs, and a is a constant. The data are shown in Figure 10. For these tensile experiments, the b-values are seen to lie in the range 1 - 3, and are negatively correlated and linearly related to K/K_c, the normalized stress intensity. Physically, a high b-value characterizes crack growth dominated by a large number of relatively small events, whereas a low b-value indicates a larger number of relatively large events. Note also that the b-value is higher in a "wet" as opposed to a "dry" environment, especially at low values of K (low crack velocity). Changes in b may therefore be due either to changes

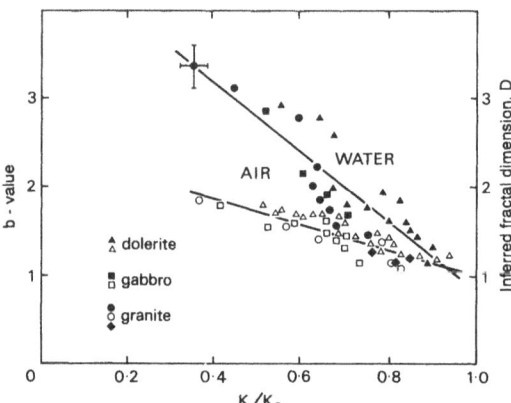

Figure 10. Synoptic diagram of the variation in seismic b-value and the inferred fractal dimension (D) with normalized stress intensity (K/K_c) and crack tip humidity for tensile crack propagation in a variety of crystalline rocks. Solid lines are least squares fits to the data points and converge approximately at the point ($K/K_c = 1$, $b = 1$, $D = 1$).

in stress intensity or to changes in the humidity of the crack tip environment.

This may be related to the observation of a trend from dominantly transgranular cracking at high crack velocity and low humidity, to dominantly intergranular cracking at low velocity and high humidity, since grain boundaries provide the main conduits for the access of water and will therefore stress corrode preferentially. At values of K close to the fracture toughness, crack growth is very rapid, and hence the environment has little effect on either crack growth rate or the b-value. For either environment, the critical b-value for dynamic failure (at $K = K_c$) is unity.

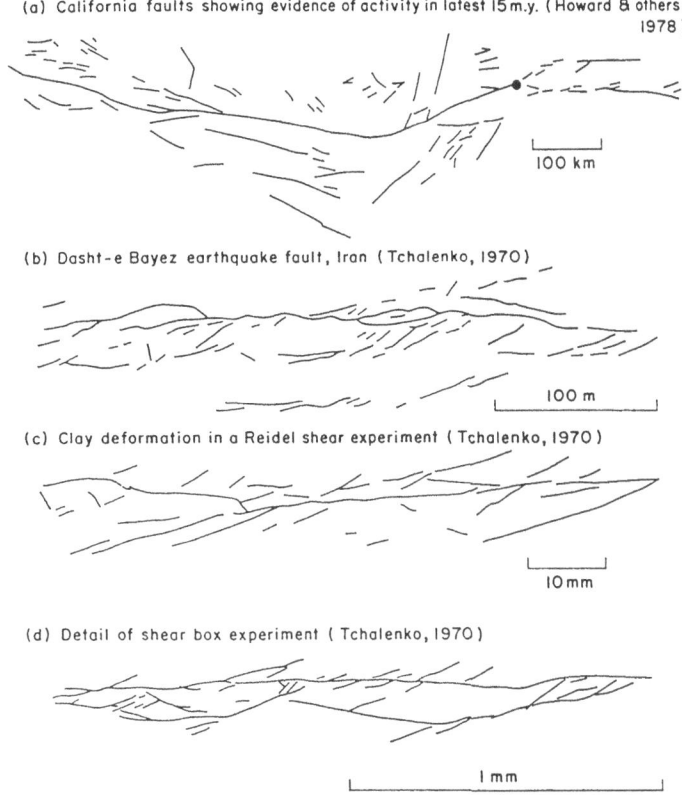

Figure 11.   Geometry of shear faulting over a range of scales (after Shaw & Gartner 1986). On the scales shown, each example is dominated by onr major throughgoing fault.

The power law distribution of AE event amplitudes described by equation (5) implies that a power law distribution of flaw sizes of the form

$$N(a) = C \ a^{-D} \tag{6}$$

is necessary to generate this range of event sizes, where C is a constant, a is the crack length, and D is the power law exponent. Furthermore, if the distribution is geometrically scale-invariant between the minimum and maximum lengths, then D is strictly defined as one aspect of the fractal dimension of the system (Mandelbrot 1982). Several authors (e.g. Caputo 1976, Aki 1981) have shown that

$$D = 3b/c \qquad (7)$$

where c is a constant that depends on the relative time constants of the event and the recording system. Main et al. (1989) have shown that, for the data of Figure 10, $c = 3$, so that $b = D$; whereas for the triaxial compression experiments to be reported below, $c = 3/2$ and therefore $b = D/2$. The crack length fractal dimension (D) inferred from the above argument is also indicated on Figure 10.

It is well known that many geological structures and fabrics are often scale-invariant within well-defined characteristic limits, otherwise it would not be necessary to include scale bars or the ubiquitous geological hammer on photographs of rock samples and rock outcrops. Figure 11 (after Shaw & Gartner 1986) shows convincing evidence for the scale-invariance of shear fault systems with one dominant throughgoing fault, over a range of scales from fractions of a *mm* to hundreds of *km*. Without the scale bars and annotation it would be very difficult to distinguish laboratory shear-box experiments from plate-rupturing faults. Figure 12 superposes the size distribution of the four fault systems

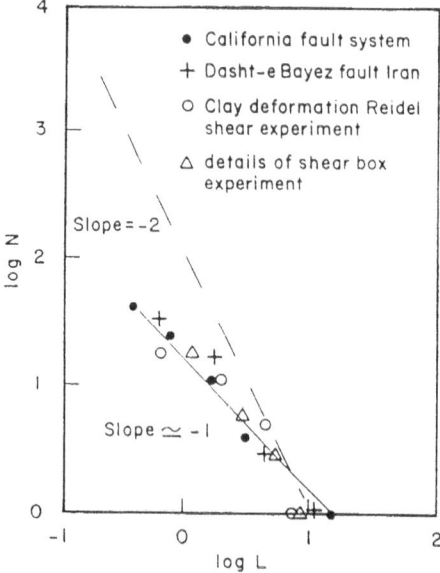

Figure 12.   Normalized discrete frequency-length distribution of the faults shown in Figure 11. All four data sets are consistent with a power law of negative slope $D = 1$ (solid line).

of Figure 11, with each normalized to the length of the dominant throughgoing fault. Two aspects emerge immediately: (1) all the fault systems have the same relative size distribution of subsidiary faults; and (2) the size distribution is a power law with a negative exponent $D = 1$, corresponding to critical rupture.

The most appropriate mathematical description of the size distributions of Figure 11 is the Cantor set (Mandelbrot 1982, Turcotte 1989, Main et al. 1990) which describes a dominant straight line drawn on a two-dimensional plane, with more and more identical replicas in a cascade of smaller scales, and with a power law frequency distribution of lengths. Extension to three dimensions would require a dominant fault plane with a cascade of smaller area fault planes (Main et al. 1990).

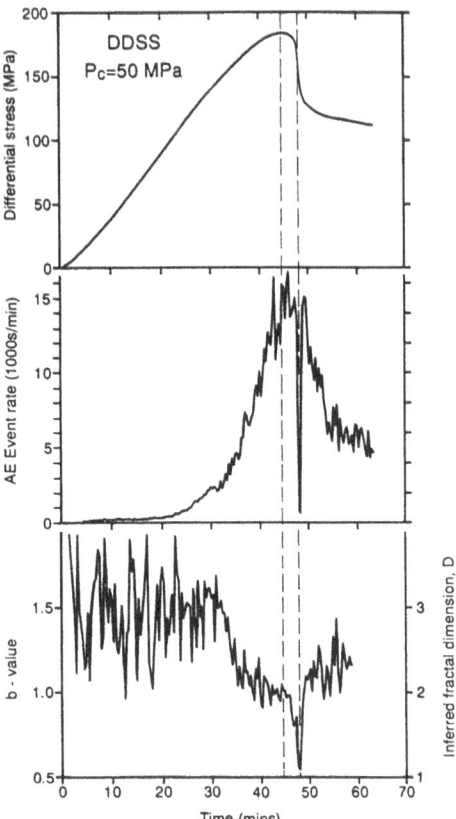

Figure 13. Contemporaneous measurements of (a) stress difference, (b) AE event rate, and (c) b-value and inferred fractal dimension, as functions of time for a water saturated sample of Darley Dale sandstone deformed at a nominally constant strain rate of 10^{-5}/s under a confining stress (P_c) of 50MPa. This sample failed by localized shear faulting after a dynamic stress drop. Peak stress and dynamic failure are marked by dashed vertical lines.

AE results from laboratory-scale triaxial compression experiments on samples of Darley Dale sandstone are shown in Figures 13 and 14. All data are plotted against time, but this is equivalent to strain for these constant strain rate tests. Details of the experimental arrangement and methodology are given in Meredith et al. (1990).

Figure 13 shows data from a test on a water-saturated sample deformed under a confining stress of 50MPa. The stress-time curve is markedly non-linear, with peak stress followed by a significant decrease in stress precursory to dynamic failure on a well-defined fault plane, and finally, stable sliding on the fault. Macroscopic dynamic failure occurs when the negative slope of the stress/time curve is a maximum (i.e. when the rate of stress drop is a maximum).

The AE rate increases rapidly with the onset of dilatancy, which in this case occurs at little more than 50% of the peak stress. We can usefully separate dilatancy into two phases; the first dominated by the growth of new microcracks in the period up to peak stress, and the second dominated by the interaction and coalescence of these microcracks to form a throughgoing fault in the period of post-peak strain softening. Around peak stress, the AE rate at first flattens out and then falls dramatically to a short period of apparent *quiescence*. This quiescence associated with dynamic failure is thought to be caused by saturation of the monitoring system due to a cascade of events which become indistinguishable from each other at critical crack coalescence. Similar apparent quiescence has been reported to occur close to dynamic failure by a number of other workers (Gowd 1980, Kikuchi et al. 1981, Sondergeld et al. 1984), but *only* when a strain softening phase preceded failure. Finally, following instability the AE rate recovers before decaying in a manner analogous to an earthquake aftershock sequence (i.e. according to Omori's law)

b-value data for this experiment exhibit the following important features: (1) during the early quasi-elastic phase of loading, where there is a low level of AE activity, the b-value remains essentially constant, with a high value close to 1.5 (and by inference $D = 3$). At higher levels of stress, the major trend is of a decreasing b-value, correlated with increasing AE rate and the growth of new microcracks, and which flattens out at around peak stress ($b = 1$, $D = 2$). This is followed by an inflection point leading to a much shorter time-scale b-value anomaly leading to dynamic failure close to the expected value of $b_c = 0.5$ ($D_c = 1$). Post-failure the b-value recovers as expected. Note that the error in measurement of b-values is proportional to $1/N^{1/2}$, where N is the total number of AE events in each time period for which a b-value is determined. So the variability in b-values is much greater during the relatively quiet low-stress phase of loading than during the dilatant phase where the AE rate is much increased.

Finally, Figure 14 shows the same suite of data from a test on a water-saturated sample deformed under a much higher confining stress of 200 MPa. In this case there is no stress drop, and deformation takes place quasi-statically by cataclastic flow rather than by dynamic rupture on a localized fault. Under these conditions, the AE rate again increases with the onset of microcracking, but remains at a consistently high level throughout the phase of cataclastic flow, with no indication of any period of quiescence or decay. Here, the b-value is simply negatively correlated with the level of stress, and falls from its background level of 1.5 to reach a constant value of unity (with an inferred instantaneous value of $D = 2$) associated with distributed microcracking during the phase of cataclastic flow. The value never falls to anywhere near the critical value of 0.5, since there is no critical stress concentration.

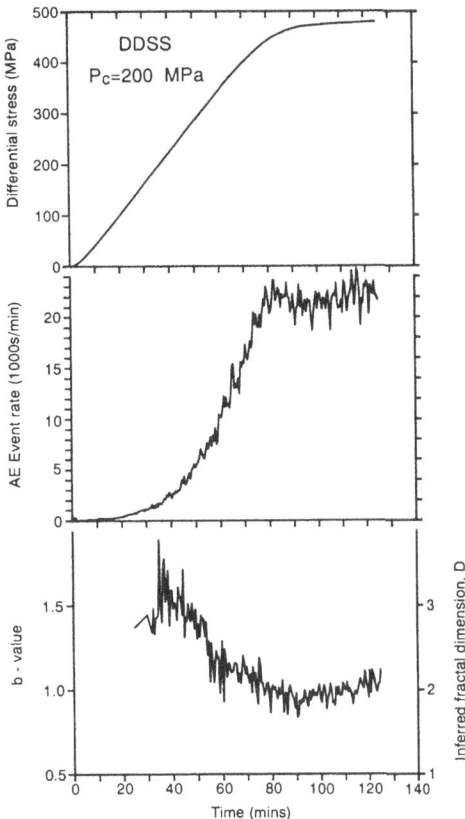

Figure 14. As for Figure 13, but for a sample of Darley dale sandstone deformed under a confining stress of 200MPa. In this case deformation occurred by distributed cataclastic flow, and there was no dynamic failure or stress drop.

5. Concluding Remarks

It is clear that the geometrical distribution of flaws on all scales plays a crucial role in controlling the mechanical and physical behaviour of heterogeneous, polycrystalline, quasi-brittle materials such as rocks. It is therefore vital that theoretical models describing the evolution of crack damage utilise physically realistic flaw distributions if predictions of the deformation response and failure of these materials are to be useful.

A whole plethora of evidence now supports the concept of a power law distribution of flaw sizes in rocks, with an exponent that varies as deformation proceeds, but that reaches a critical value for catastrophic rupture. Encouragingly, this also implies that, within certain limits, the process of fracturing is scale-invariant, which suggests that data derived from laboratory-scale samples may be able to be applied usefully to much larger scale rupture.

For the future, more detailed damage models will be required that accommodate time-dependent cracking processes and more realistic descriptions of both initial flaw distributions and the developing state of damage. This is likely to be an iterative process requiring input from the techniques that indirectly monitor damage development during progressive deformation into enhanced theoretical models.

References

Aki K. (1981) A probabilistic synthesis of precursory phenomena, In: *Earthquake Prediction: An International review*, Maurice Ewing Series 4, Amer. Geophys. Union, Washington, 566-574.

Ashby M.F. & Hallam S.D. (1986) The failure of brittle solids containing small cracks under compressive stress states, *Acta Metall.*, **34**, 497-510.

Atkinson B.K. (1982) Subcritical crack propagation in rocks: theory, experimental results and applications, *J. Struct. Geol.*, **4**, 41-56.

Atkinson B.K. (1984) Subcritical crack growth in geological materials, *J. Geophys. Res.*, **89**, 4077-4114.

Atkinson B.K. & Meredith P.G. (1987a) The theory of subcritical crack growth with application to minerals and rocks, In: *Fracture Mechanics of Rock*, (ed: B.K. Atkinson), Academic Press, London, 111-166.

Atkinson B.K. & Meredith P.G. (1987b) Experimental fracture mechanics data for rocks and minerals, In: *Fracture Mechanics of Rock*, (ed: B.K. Atkinson), Academic Press, London, 477-525.

Atkinson B.K. & Rawlings R.D. (1981) Acoustic emission during stress corrosion cracking in rocks, In: *Earthquake Prediction: An International Review*, Maurice Ewing Series 4, Amer. Geophys. Union, Washington, 605-616.

Caputo M. (1976) Model and observed seismicity represented in a two-dimensional space, *Ann. Geophys. (Rome)*, **4**, 277-288.

Charles R.J. (1958) Static fatigue of glass, *J. Appl. Physics*, **29**, 1549-1560.

Cook R.F., Lawn B.R. & Fairbanks C.J. (1985) Microstructure-strength properties of ceramics, I: Effect of crack size on toughness, *J. Am. Ceram. Soc.*, **68**, 604-615.

Costin L.S. (1983) A microcrack model for the deformation and failure of brittle rock, *J. Geophys. Res.*, **88**, 9485-9492.

Costin L.S. (1985) Damage mechanics in the post-failure regime, *Mech. Mat.*, **4**, 149-160.

Costin L.S. (1987) Time-dependent deformation and failure, In: *Fracture Mechanics of Rock*, (ed: B.K. Atkinson), Academic Press, London, 167-215.

Cox S.J.D. & Paterson L. (1990) Damage development during rupture of heterogeneous brittle materials: a numerical study, In: *Deformation Mechanisms, Rheology and Tectonics*, (eds: R.J. Knipe & E.H. Rutter), Geol. Soc. Lond. Spec. Pub., (in press).

Cox S.J.D. & Scholz (1988) On the formation and growth of faults: an experimental study, *J. Struct. Geol.*, **10**, 413-430.

Crampin S. (1984) Effective anisotropic elastic constants for wave propagation through cracked solids, *Geophys. J. R. Astr. Soc.*, **76**, 135-145.

Dunning J.D., Petrovski D., Schuyler J. & Owens A. (1984) The effect of aqueous chemical environments on crack propagation in quartz, *J. Geophys. Res.*, **89**, 4115-4123.

Freiman S.W. (1984) Effects of chemical environments on slow crack growth in glasses and ceramics, *J. Geophys. Res.*, **89**, 4072-4076.

Freiman S.W. & Swanson P.L. (1990) Fracture of polycrystalline ceramics, In: *Deformation Processes in Minerals, Ceramics and Rocks*, (eds: D.J. Barber & P.G. Meredith), Unwin Hyman, London, 72-83.

Gowd T.N. (1980) Factors affecting the acoustic emission response of triaxially compressed rock, *Int. J. Rock Mech. Min. Sci. & Geomech. Abstr.*, **17**, 219-223.

Gueguen Y., Reuschle T. & Darot M. (1990) Single crack behaviour and crack statistics, In: *Deformation Processes in Minerals, Ceramics and Rocks*, (eds: D.J. Barber & P.G. Meredith), Unwin Hyman, London, 48-71.

Gupta I.N. (1973) Seismic velocities in rocks subjected to axial loading, *J. Geophys. Res.*, **65**, 1083-1102.

Hillerborg A. (1989) Existing methods to determine and evaluate fracture toughness of aggregative materials - RILEM recommendation on concrete, In: *Fracture Toughness and Fracture Energy: Test Methods for Concrete and Rock*, (eds: H. Mihashi, H. Takahashi & F.H. Wittmann), Balkema, Rotterdam, 145-152.

Horii H. & Nemat-Nasser S. (1986) Brittle failure in compression: splitting, faulting and brittle-ductile transition, *Phil. Trans. Roy. Soc. Lond.*, **Ser.A**, **319**, 337-374.

Hubner H. & Jillek W. (1977) Sub-critical extension and crack resistance in polycrystalline alumina, *J. Mat. Sci.*, **12**, 117-125.

Hudson J.A. (1981) Wave speeds and attenuation of elastic waves in materials containing cracks, *Geophys. J. R. Astr. Soc.*, **64**, 133-150.

Ingraffea A.R. (1987) Theory of crack initiation and propagation in rock, In: *Fracture Mechanics of Rock*, (ed: B.K. Atkinson), Academic Press, London, 71-110.

ISRM Commission on Testing Methods (1988) Suggested method for determining the fracture toughness of rock, *Int. J. Rock Mech. Min. Sci & Geomech. Abstr.*, **25**, 71-96.

Jaeger J.C. & Cook N.G.W. (1976) *Fundamentals of Rock Mechanics*, 2nd. edition, Chapman and Hall, London.

Kemeny J.M. & Cook N.G.W. (1987) Crack models for the failure of rocks in compression, *Proc. 2nd. Int. Conf. on Constitutive Laws for Engineering Materials*, Tucson, Arizona.

Kikuchi M., McNally K. and Tittman B.R. (1981) Machine stiffness appropriate for experimental simulation of earthquake processes, *Geophys. Res. Lett.*, **8**, 321-323.

Kranz R.L. (1979) Crack growth and development during creep of Barre granite, *Int. J. Rock Mech. Min. Sci. & Geomech. Abstr.*, **16**, 23-35.

Kranz R.L. (1980) The effect of confining pressure and stress difference on static fatigue of granite, *J. Geophys. Res.*, **85**, 1854-1866.

Kranz R.L. (1983) Microcracks in rock: a review, *Tectonophysics*, **100**, 449-480.

Kranz R.L. & Scholz C.H. (1977) Critical dilatant volume at the onset of tertiary creep, *J. Geophys. Res.*, **82**, 4893-4898.

Labuz J.F., Shah S.P. & Dowding C.H. (1985) Experimental analysis of crack propagation in granite, *Int. J. Rock Mech. Min. Sci. & Geomech. Abstr.*, **22**, 85-98.

Lawn B.R. & Wilshaw T.R. (1975) *Fracture of Brittle Solids*, Cambridge University Press, Cambridge, 204pp.

Main I.G., Meredith P.G. & Jones C. (1989) A reinterpretation of the precursory seismic b-value anomaly using fracture mechanics, *Geophys. J.*, **96**, 131-138.

Main I.G., Meredith P.G., Sammonds P.R. & Jones C. (1990) Influence of fractal flaw distributions on rock deformation in the brittle field, In: *Deformation Mechanisms, Rheology and Tectonics*, (eds: R.J. Knipe & E.H. Rutter), Geol. Soc. Lond. Spec. Pub., (in press).

Main I.G., Peacock S. & Meredith P.G. (1990) Scattering attenuation and the fractal geometry of fracture systems, *Pure & Appl. Geophys.*, **133**, 283-304.

Mandelbrot B.B. (1982) *The Fractal Geometry of Nature*, Freeman, New York.

McClintock F.A. & Walsh J.B. (1962) Friction on Griffith cracks in rocks under pressure, *Proc. 4th. U.S. Nat. Congr. Appl. Mech.*, **Vol. II**, Am. Soc. Mech. Eng., New York, 1015-1021.

Meredith P.G. (1989) Comparative fracture toughness testing of rocks, In: *Fracture Toughness and Fracture Energy: Test Methods for Concrete and Rock*, (eds: H. Mihashi, H. Takahashi & F.H. Wittmann), Balkema, Rotterdam, 265-278.

Meredith P.G. (1990) Fracture and failure of brittle polycrystals: an overview, In: *Deformation Processes in Minerals, Ceramics and Rocks*, (eds: D.J. Barber & P.G. Meredith), Unwin Hyman, London, 5-47.

Meredith P.G. & Atkinson B.K. (1983) Stress corrosion and acoustic emission during tensile crack propagation in Whin Sill dolerite and other basic rocks, *Geophys. J. R. Astr. Soc.*, **75**, 1-21.

Meredith P.G. & Atkinson B.K. (1985) Fracture toughness and subcritical crack growth during high-temperature tensile deformation of Westerly granite and Black gabbro, *Phys. Earth & Planet. Ints.*, **39**, 33-51.

Meredith P.G., Main I.G. & Jones C. (1990) Temporal variations in seismicity during quasi-static and dynamic rock failure, *Tectonophysics*, **175**, 249-268.

Murrell S.A.F. (1990) Brittle to ductile transitions in polycrystalline non-metallic materials, In: *Deformation Processes in Minerals, Ceramics and Rocks*, (eds: D.J. Barber & P.G. Meredith), Unwin Hyman, London, 109-137.

Murrell S.A.F. & Digby P.J. (1970) The theory of brittle fracture initiation under triaxial stress conditions, Parts I and II, *Geophys. J. R. Astr. Soc.*, **19**, 309-334 and 499-512.

Peng S.S. & Johnson A.M. (1972) Crack growth and faulting in cylindrical specimens of Chelmsford granite, *Int. J. Rock Mech. Min. Sci.*, **9**, 37-86.

O'Connell R.J. & Budiansky B. (1974) Seismic velocities in dry and saturated cracked solids, *J. Geophys. Res.*, **79**, 5412-5426.

RILEM Draft Recommendation (1985) Determination of the fracture energy of mortar and concrete by means of three-point-bend tests on notched beams, *Materials and Structures*, **106**, 285-290.

Sammis C.G. & Ashby M.F. (1986) The failure of brittle porous materials under compressive stress states, *Acta Metall.*, **34**, 511-526.

Sammonds P.R., Ayling M.R., Jones C., Meredith P.G. & Murrell S.A.F. (1989) A laboratory investigation of acoustic emission and elastic wave velocity changes during rock failure under triaxial stresses, In: *Rock at Great Depth*, (eds: V. Maury & D. Fourmaintraux), Balkema, Rotterdam, Vol.1, 233-240.

Scmidt R.A. & Lutz T.J. (1979) K_{Ic} and J_{Ic} of Westerly granite - effects of thickness and in-plane dimensions, In: *Fracture Mechanics Applied to Brittle Materials*, ASTM Spec. Tech. Pub., STP 678, 166-182.

Shaw H.R. & Gartner A.E. (1986) On the graphical interpretation of palaeoseismic data, *U.S.G.S. Open File Report, 86-394*.

Simmons C.J. & Freiman S.W. (1983) Effect of corrosion processes on subcritical crack growth in glass, *J. Am. Ceram. Soc.*, **64**, 683-686.

Sondergeld C.H., Granryd L.A. & Estey L.H. (1984) Acoustic emissions during compression testing of rock, In: *Proc. 3rd. Conf. on Acoustic Emiision/Microseismic Activity in Geologic Structures and Materials*, (eds: H.R. Hardy & F.W. Leighton), Trans Tech, Clausthal, 131-145.

Swanson P.L. (1984) Subcritical crack growth and other time- and environment-dependent behavior in crustal rocks, *J. Geophys. Res.*, **89**, 4137-4152.

Swanson P.L. (1987) Tensile fracture resistance mechanisms in brittle polycrystals: an ultrasonics and in-situ microscopy investigation, *J. Geophys. Res.*, **92**, 8015-8036.

Tada H., Paris P.C. & Irwin G.R. (1973) *The Stress Analysis of Cracks Handbook*, Del Research Corp'n., Hellertown, PA.

Tapponier P. & Brace W.F. (1976) Development of strss-induced microcracks in Westerly granite, *Int. J. Rock Mech. Min. Sci. & Geomech. Abstr.*, **13**, 103-112.

Tchalenko J.G. (1970) Similarities between shear zones of different magnitudes, *Geol. Soc. Am. Bull.*, **81**, 1625-1640.

Turcotte D.L. (1989) Fractals in geology and geophysics, *Pure & Appl. Geophys.*, **131**, 171-196.

Walsh J.B. (1965) The effect of cracks on the compressibility of rocks, *J. Geophys. Res.*, **70**, 381-389.

Wiederhorn S.M. (1974) Subcritical crack growth in ceramics, In: *Fracture Mechanics of Ceramics*, (eds: R.C. Bradt, D.P.H. hasselman & F.F. Lange), Vol.2, Plenum, New York, 613-646.

Wiederhorn S.M. (1978) Mechanisms of subcritical crack growth in glass, In: *Fracture Mechanics of Ceramics*, (eds: R.C. Bradt, D.P.H. Hasselman & F.F. Lange), Vol.4, Plenum, New York, 549-580.

Wiederhorn S.M. & Johnson H. (1973) Effect of electrolyte pH on crack propagation in glass, *J. Am. Ceram. Soc.*, **56**, 192-197.

Wong T.-F. & Biegel R. (1985) Effects of pressure on the micromechanics of faulting in San Marcos gabbro, *J. Struct. Geol.*, **7**, 737-749.

TEST METHODS FOR DETERMINING MODE I FRACTURE TOUGHNESS OF CONCRETE

B.L. KARIHALOO and P. NALLATHAMBI
School of Civil and Mining Engineering
University of Sydney
N.S.W. 2006
Australia

ABSTRACT. This lecture will review the various test methods considered by sub-committees A and B of RILEM TC 89-FMT for the determination of mode I fracture toughness of plain concrete. These methods have been categorized depending on the geometry of the test specimen. Thus, sub-committee A considered only the notched beam specimen, while sub-committee B considered three different compact specimen geometries. Moreover, as linear elastic fracture mechanics is fully applicable only to very large structures, three separate models have been proposed in relation to the application of LEFM to laboratory-scale notched beam specimens and these will be discussed . A further method considered by an earlier RILEM committee (TC-50 FMC) will also be included in the review not only because reference will often be made to this method but also to complete the presentation of all existing test methods for mode I.

1. Introduction

One of the earliest methods for studying the fracture behaviour of concrete was pioneered by Hillerborg et al. (1976). In this method which is based on the so-called cohesive crack model, it is assumed that fracture under monotonically increasing mode I loading occurs when the maximum (tensile) principal stress reaches the (uniaxial) tensile strength of the material f_t. It is further assumed that fracture is localized in the so-called process zone such that there is no energy dissipation in the bulk of the structure. The process zone (Fig. 1) is modelled by a displacement discontinuity with the proviso that the faces of the discontinuity are capable of transmitting certain cohesive stresses, less than f_t, such that $\sigma = F(w)$ with $F(0) = f_t$, and $F(w) \geq 0$, where $F(w)$ describes the tensile softening behaviour. In practice, $F(w)$ was approximated by a linear or bi-linear relation and was assumed to vanish when the crack opening displacement w reached a certain critical

91

S. P. Shah (ed.), Toughening Mechanisms in Quasi-Brittle Materials, 91–124.
© 1991 Kluwer Academic Publishers.

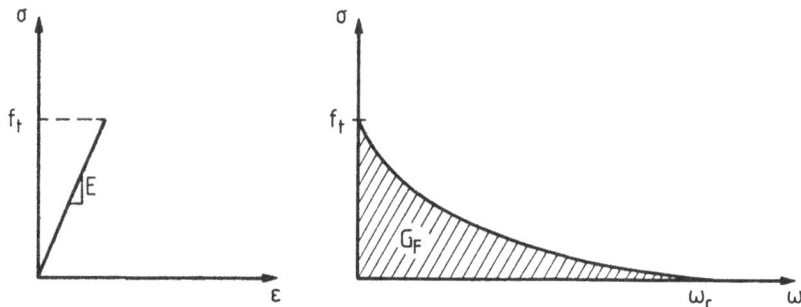

Figure 1: Pre-peak (linear) response and post-peak tension softening

value w_c. The fracture response of concrete in this method is characterized by fracture energy G_F defined as the energy required to open a unit crack fully, or in other words, as the area under the strain-softening curve between $w = 0$ and $w = w_c$. Moreover, from the material parameters E, f_t, G_F an independent parameter with the units of length $\ell_{ch} = G_F E / f_t^2$ is defined and is called the characteristic length.

An extensive round-robin testing programme organised by RILEM FMC-50 (1985) and using the same concrete mix in all participating laboratories confirmed earlier doubts that G_F varied significantly with the size of test specimens for one and the same mix. In the light of this confirmation it is somewhat surprising that this method is still widely used. It is now generally accepted that the size (scale) effect is due to curing conditions which result in cracking, to the development of microcracks at the aggregate/paste interfaces or the paste itself, to crack bridging between the aggregate particles, etc. These processes consume energy and lead to the observed non-linear load response, not only in the post-peak regime but already in the pre-peak behaviour. In the above method, the material response is assumed to be linear right up to the peak load (Fig. 1).

In order to better understand the reasons behind the observed size (scale) effect and, what is more important, to propose a reasonably size-independent test method(s) for determining the intrinsic mode I fracture toughness of concrete the RILEM TC-89 (FMT) was set up in Paris in 1986, under the chairmanship of Professor S.P. Shah. Of the several sub-committees formed at this Paris meeting, sub-committee A (Chair: B.L. Karihaloo) was charged with analysing available test data from three-point bend notched specimens (including the data from above-mentioned round-robin testing programme) according to the two-parameter model (Shah & Jenq 1985) and the size-effect law (Bažant & Pfeiffer 1986). These two proposals were before the committee at its inception. However, of the extensive body of raw test data available for analysis, only a very small fraction was suitable for analysis according to these two proposals. The reasons for this were twofold. First, very few sets of data were accompanied by load-CMOD diagrams and unloading compliance figures necessary for analysis according to the two-parameter model. Secondly, even fewer sets of data met the rather strict size requirements of the size-effect law. The available data were therefore also analysed according to a third method, namely that based on effective crack model (Karihaloo & Nallathambi 1986a). This lecture will review all three methods in relation to three-point bend notched specimens on the basis of one and the same set of test data, as well as on the basis of their maximum load prediction following the methodology proposed by Planas & Elices (1987). This latter method of comparison also serves as a link between the above three methods and the earlier method due to Hillerborg *et al.* (1976).

Sub-committee B (Chair: P. Rossi) on the other hand, was asked to consider other, more compact specimen geometries for the determination of fracture properties of plain concrete. This lecture will review the three compact specimen geometries analysed by Sub-committee B, namely the tapered double cantilever beam (TDCB), the cylindrical wedge splitting specimen and the cubical wedge splitting specimen.

Both sub-committees have now finalised their reports in anticipation of their publication in October 1990. These reports have not only eased the task of preparing this review lecture, but more importantly have provided a valuable reference source. To keep this lecture within manageable length, frequent reference will be made to these reports.

2. The Two-Parameter Model (TPM)

The two-parameter model or the effective Griffith crack model (Shah & Jenq 1985) allows for both pre-critical and post-critical crack growth. The pre-critical stable crack growth is accompanied by an increasing stress intensity factor K_I (R-curve behaviour). It is assumed that the load-CMOD (Fig. 2) is more or less linear up to about half the maximum load and the corresponding crack tip opening displacement (CTOD) is zero. Significant inelastic displacement and slow crack growth occur when the load increases from about $0.5P_{max}$ on the ascending branch of the load-CMOD plot to $0.95P_{max}$ on the descending (tension-softening) branch. The latter load level is associated with the instant of growth of the original notch, i.e. with critical K_I which is designated K_{Ic}^s and with the critical value of $CTOD$ of the original notch tip which is designated $CTOD_c$.

The load-CMOD diagram is used to calculate both E and K_{Ic}^s by the traditional compliance approach. The initial compliance C_i of the load-CMOD plot (Fig. 2)

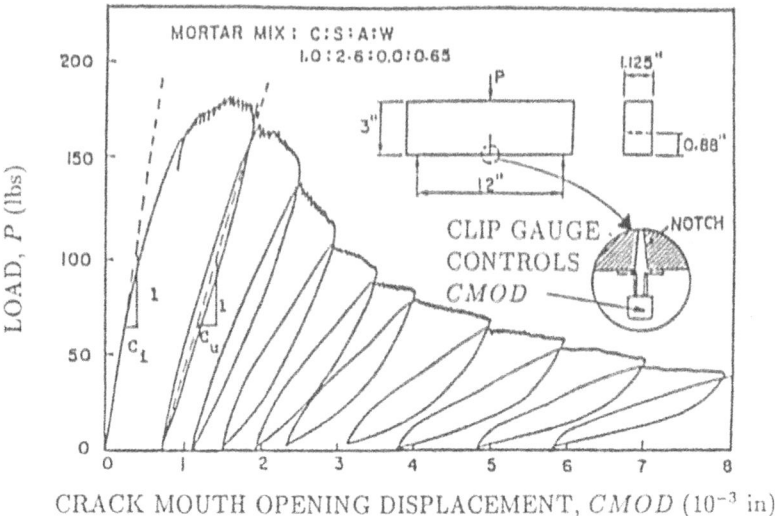

Figure 2: A typical load-CMOD plot (Jenq & Shah 1985)

is measured and used for determining the elastic modulus E of the mix.

$$E = 6S(a_0 + H_0) V_1 \left(\frac{a_0}{W} \right) / (C_i BW^2),$$ (1)

where H_0 is the thickness of the knife-edge used for holding the clip gauge and the function $V_1(a_0/W)$ established from finite element calculations is

$$V_1 \left(\frac{a_0}{W} \right) = 0.76 - 2.28 \left(\frac{a_0}{W} \right) + 3.87 \left(\frac{a_0}{W} \right)^2 - 2.04 \left(\frac{a_0}{W} \right)^3 + \frac{0.66}{\left(1 - \frac{a_0}{W} \right)^2}.$$ (2)

For a notched three-point bend specimen, B, W, S are respectively the width, depth and loaded span of the beam. It should be mentioned that the principle of two-parameter model should be applicable not only to notched beam specimens but to other geometries as well. However, recent calculations would seem to cast doubt on its applicability to geometries other than the notched beam. We shall have more to say on this subject when we discuss compact specimen geometries.

The unloading compliance C_u (Fig. 2) corresponding to $0.95 P_{max}$ on the descending branch of load-CMOD plot is used to obtain an augmented traction-free notch depth $\underline{a}$ (equal to a_0 plus the effective slow crack growth size) by solving the following equation

$$E = 6S(\underline{a} + H_0) V_1 \left(\frac{\underline{a}}{H} \right) / (C_u BW^2).$$ (3)

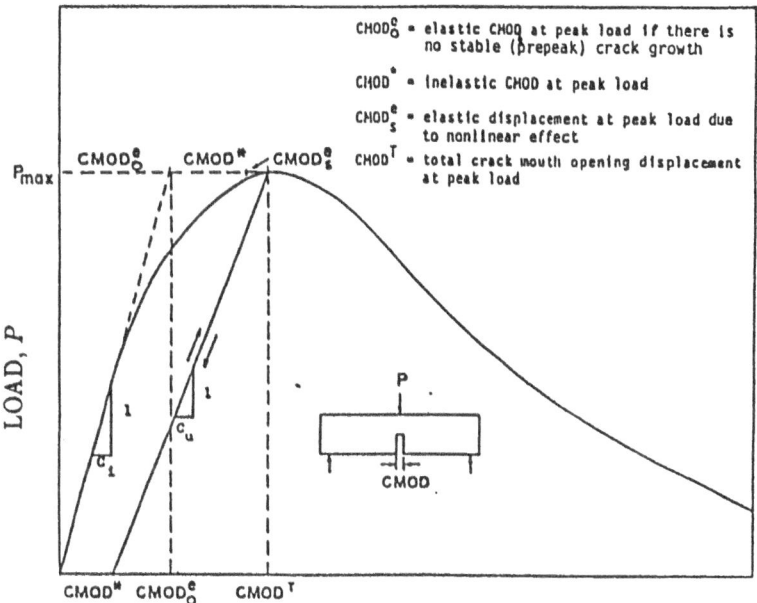

Figure 3: Decomposition of CMOD due to non-linear effect

$\underline{a}$ is determined from (3) by a trial and error procedure or from nomograms provided by John, Shah & Jenq (1987). Having calculated $\underline{a}$, the critical stress intensity factor K^s_{Ic} is determined from the LEFM formula (Srawley 1976) after replacing in it a_0 by $\underline{a}$

$$K^s_{Ic} = \frac{3P_{max} S}{2BW^2} \sqrt{\underline{a}} F(\alpha), \qquad (4)$$

where $\alpha = \underline{a}/W$ and

$$F(\alpha) = \frac{1.99 - \alpha(1 - \alpha)(2.15 - 3.93\alpha + 2.7\alpha^2)}{(1 + 2\alpha)(1 - \alpha)^{3/2}}. \qquad (5)$$

The effective notch depth $\underline{a}$ may also be used to calculate the critical crack tip (original notch) opening displacement $CTOD_c$

$$CTOD_c = \frac{6P_{max}S\underline{a}}{W^2BE} V_1(\alpha)\{(1 - \beta)^2 + (-1.149\alpha + 1.081)(\beta - \beta^2)\}^{1/2}, \qquad (6)$$

where $\alpha = \underline{a}/W$ and $\beta = a_0/\underline{a}$.

Practical difficulties are faced in unloading the specimen at precisely $0.95P_{max}$ on the descending branch of load-CMOD plot. Even in laboratories which are equipped to perform a stable bend-test, unloading the specimen at $0.95P_{max}$ is not always successful, so that inaccuracies in the determination of C_u are unavoidable. For those laboratories which cannot perform a stable three-point bend test

Shah & Jenq recommend that C_u be approximated by its value corresponding to $CMOD^* = 0$ (Fig. 3). They suggest that K_{Ic}^s and $CTOD_c$ determined under this approximation are about 10% to 25% higher than the values calculated using the C_u corresponding to $0.95P_{max}$ on the descending branch (Fig. 2). A summary of results according to TPM based on test data from several laboratories around the world (Karihaloo & Nallathambi 1990a) is given in Table 1.

TABLE 1. Summary of Results According to TPM

Sl No	g	n	f_c'	E	Range a_0/W	Range $\underline{a}/W$	K_{Ic}^s Mean(sd)	$CTOD_c(mm)$ Mean(sd)
1	19	6	55.8	36.8	0.29-0.67	0.55-0.81	0.931(0.263)	0.0148(0.0076)
2	19	4	53.1	38.4	0.29	0.41	1.054(-)	0.0153(-)
3	19	3	54.4	39.3	0.30-0.67	0.42-0.73	1.128(0.269)	0.0200(0.0085)
4	19	25	53.1	38.4	0.25-0.61	0.44-0.83	1.146(0.014)	0.0316(0.0067)
5	19	22	54.4	39.3	0.16-0.57	0.54-0.69	1.220(0.102)	0.0312(0.0087)
6	3	12	-	36.8	0.13-0.51	0.15-0.56	0.894(0.068)	0.0042(0.0013)
7	6	2	60.7	33.5	0.20-0.21	0.25-0.28	1.141(0.095)	0.0145(0.0057)
8	13	3	45.5	31.0	0.20-0.21	0.25-0.35	1.475(0.191)	0.0220(0.0086)
9	13	3	43.4	31.0	0.20-0.21	0.24-0.28	1.530(0.022)	0.0169(0.0024)
10	19	8	25.2	27.2	0.29-0.33	0.35-0.55	0.976(0.103)	0.0170(0.0068)
11	32	17	31.0	32.3	0.50	-	1.211(0.121)	-
12	2	11	35.0	25.7	0.50	-	0.790(0.090)	-
13	8	2	110.	56.5	0.33	-	2.130(-)	0.0338(0.0039)

Notes:

1. g = Max. aggregate size (mm); n = No. of specimens tested.

2. f_c' in MPa, E in GPa, K_{Ic}^s in $MPa\sqrt{m}$

3. The entries have been grouped according to mix variables only because they do not vary with the size of test specimens. Thus entries differ only by the maximum size of coarse aggregate (g) used in the mix and other mix parameters, e.g. water/cement ratio, texture of coarse aggregate. That K_{Ic}^s (but not $CTOD_c$) is relatively insensitive to the specimen size is best judged from Fig. 4 which shows the relative K_{Ic}^s of a mix calculated by dividing the K_{Ic}^s of a particular specimen group from this mix with the K_{Ic}^s for the specimen of least depth from this group and mix. The various plots on Fig. 4 refer to different mixes.

The two fracture parameters K_{Ic}^s and $CTOD_c$, together with an appropriate approximation of the tension-softening behaviour may be used to calculate f_t and any other parameters required for the full description and finite element modelling of the fracture process. We shall demonstrate this calculation later in this lecture.

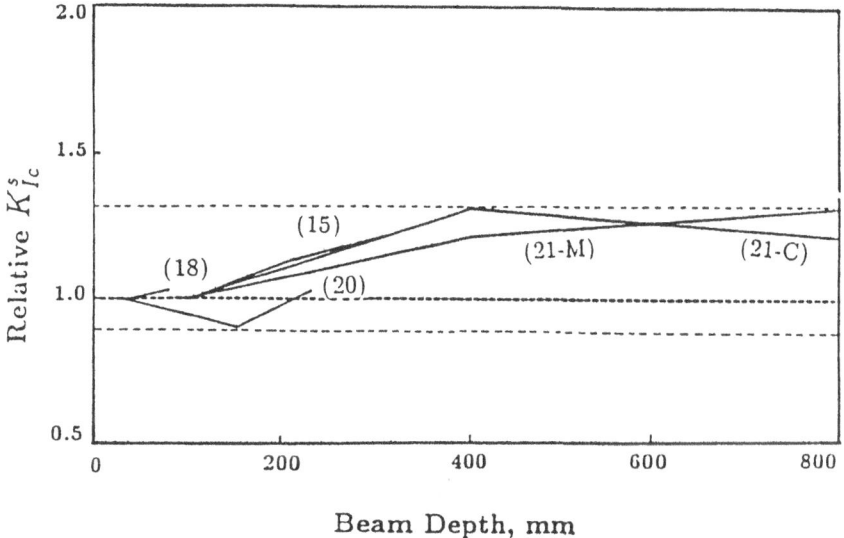

Figure 4: Variation of relative K_{Ic}^s with specimen depth

3. The Size-Effect Law (SEL)

In this model (Bažant & Pfeiffer 1986) which is applicable to materials whose fracture front is blunted by a non-linear zone of distributed cracking and damage (process zone), the fracture energy G_f (not to be confused with G_F used previously) is defined as the specific energy required for crack growth in an infinitely large structure, for which the LEFM is strictly valid. This definition is obviously independent of specimen size and geometry, provided the law for extrapolating the results of geometrically similar specimens of finite size to infinite specimen size is known and is unaffected by other size effects, such as those due to hydration heat or shrinkage.

An exact form of the scaling law for blunt fracture is not known, although an approximate form which appears to be sufficient for practical purposes was proposed by Bažant (1984)

$$\sigma_N = \beta f_t \left(1 + \frac{W}{\lambda_0 g}\right)^{-1/2}, \qquad (7)$$

in which σ_N = nominal stress at failure, g = maximum size of aggregate in the mix, and β, λ_0 = empirical constants. It appears that Eqn (7) is adequate for a size range 1:30 and is not affected by alterations to the specimen geometry provided W is appropriately reinterpreted.

As applied to the notched three-point bend specimen, it is recommended that the span-to-depth ratio S/W be at least 2.5. Moreover, it is recommended that a_0/W be in the range $0.15 < a_0/W \le 0.4$, B and W not be less than $3g$, and that the notch width be as small as possible and not exceed $0.5g$. For the scaling law to retain its validity, tests have to be performed on specimens of at least three different sizes, characterized by depths $W_1, \ldots, W_n$ and loaded spans $S_1, \ldots, S_n$. It is necessary that the smallest depth W_1 not be larger than $5g$ and the largest depth W_n not be smaller than $15g$, and that the maximum W to minimum W be at least 4. These size requirements may necessitate fabrication of very bulky specimens if large size aggregate is used in the mix. It is also recommended that the ratios of the adjacent sizes be roughly constant, with as broad a size range as feasible. Ideally, the specimens of all sizes should be geometrically similar in two dimensions, with the third dimension (width B) the same for all specimens to avoid introduction of undesirable size effects of thickness.

In actual testing, one needs to record only the maximum loads $P_1, \ldots, P_n$. This does not require the use of closed-loop systems and is thus a great advantage of this method of testing. A linear regression of the depths $X_j = W_j (j = 1, \ldots, n)$ against the inverse square of the nominal stress at failure ($= P_j^*/BW_j$) is carried out and the slope A of the regression line is determined

$$A = \frac{\sum_j (X_j - \bar{X})(Y_j - \bar{Y})}{\sum_j (X_j - \bar{X})^2}, \tag{8}$$

where

$$\bar{X} = \frac{1}{n} \sum_j X_j, \qquad \bar{Y} = \frac{1}{n} \sum_j Y_j. \tag{9}$$

$Y_j = (BW_j/P_j^*)$, and $(\bar{X}, \bar{Y})$ defines the centroid of all data points. P_j^* which is related to P_j takes into account the self-weight of the specimen, the effect of any overhang, i.e. if L_j is much larger than S_j, and of any geometrical dissimilarity in the specimens (Karihaloo & Nallathambi, 1990a). Having determined the slope A of the regression line (Fig. 5), the fracture energy G_f is calculated from

$$G_f = g\left(\frac{a_0}{W}\right)/(EA), \tag{10}$$

where the non-dimensional energy release rate is

$$g\left(\frac{a_0}{W}\right) = \left(\frac{S_m}{W_m}\right)^2 \left(\frac{a_0}{W}\right)[1.5F\left(\frac{a_0}{W}\right)]^2, \tag{11}$$

and $(\alpha_0 = a_0/W)$

$$F(\alpha_0) = F_4(\alpha_0) + \frac{(S_m/W_m) - 4}{4}[F_8(\alpha_0) - F_4(\alpha_0)], \tag{12}$$

$$F_4(\alpha_0) = 1.090 - 1.735\alpha_0 + 8.20\alpha_0^2 - 14.18\alpha_0^3 + 14.57\alpha_0^4, \tag{13}$$

$$F_8(\alpha_0) = 1.107 - 1.552\alpha_0 + 7.71\alpha_0^2 - 13.55\alpha_0^3 + 14.25\alpha_0^4 \tag{14}$$

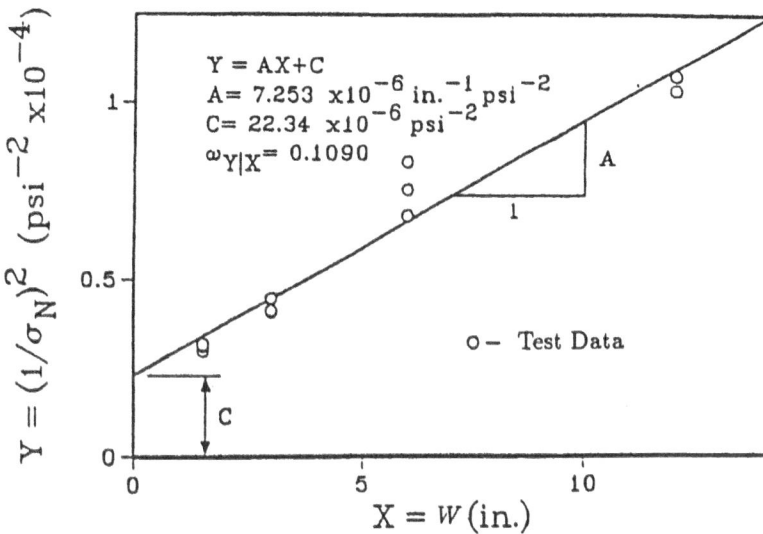

Figure 5: Linear regression plot constructed from measured maximum load values

In Eqn (11), S_m and W_m are respectively the median span and median depth of the group of geometrically similar specimens. The interpolation formula (12) is not recommended for use outside of the range $3 < S/W < 10$. Since LEFM is applicable at the limit of extrapolation, the fracture toughness may be calculated from fracture energy G_f using the plane stress relationship

$$K_{Ic}^b = \sqrt{G_f E}. \tag{15}$$

Finally, in order to establish the confidence level of calculated G_f it is necessary to perform statistical analysis of the test data. In particular, it is necessary to calculate the coefficient of variation w_A of the slope of regression line A and the relative width of scatterband m. It is suggested that w_A not exceed about 0.08 and the value of m be around 0.15. These statistical measures prevent situations in which the size range used is insufficient compared to the scatter of results. Three such situations are illustrated in Fig. 6. Fig. 6a shows the situation in which A is uncertain, while Fig. 6b illustrates the case of large scatter necessitating the use of a very broad range of sizes. Fig. 6c on the other hand illustrates the case of small scatter permitting the use of a narrow range of sizes.

It should however be borne in mind that since the value of w_A can be reduced by increasing the number of test specimens even with a wide scatterband, it is necessary also to limit the value of m, in addition to limiting the value of w_A.

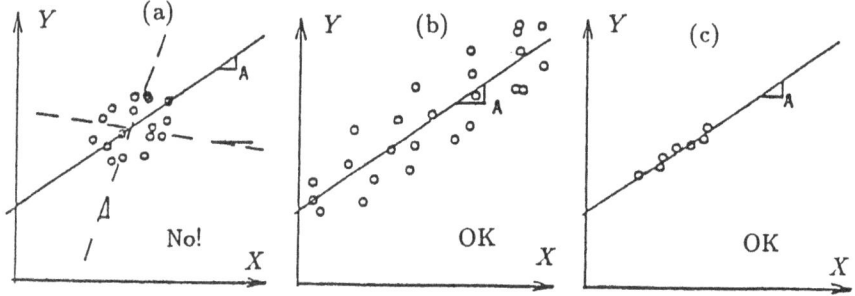

Figure 6: Examples of correct and incorrect regression plots

A summary of results according to SEL based on test data from several laboratories (Karihaloo & Nallathambi 1990a) is given in Table 2.

4. The Effective Crack Model (ECM)

The effective crack model (Karihaloo & Nallathambi 1986a, 1989a) is based on the assumption that the effect of various (non-linear) energy consuming processes taking place in the fracture process zone can be represented by a supplementary traction-free crack (Fig. 7). The latter, when added to the pre-existing notch depth a_0 gives the size of the effective notch a_e. It is evident that this assumption is also made in the TPM. However, as will become clear in the sequel, the method of determining a_e differs from that of $\underline{a}$ in TPM. What is more important, it is much easier to determine a_e accurately than it is to determine $\underline{a}$ without the use of a servo-hydraulic testing machine. That is the major advantage of ECM over TPM, otherwise in principle they are very similar.

It is also obvious from Fig. 7 that the size of the traction-free supplementary crack $\Delta a_e = a_e - a_0$ cannot be equal to the size of the fracture process zone which has residual stress carrying capacity.

The application of ECM to notched three-point bend specimen (Fig. 8) does not require any load/deflection information past the peak load. However, if a closed-loop testing machine is available, then CMOD or load-point displacement may be used as a feed-back signal to achieve stable failure.

The Young modulus (E) of the mix is calculated from the initial, linear segment of the continous load-deflection plot. However, if such a plot cannot be obtained, E may be determined by testing cylindrical specimens, preferably using two electrical strain gauges with gauge length at least $3g$ glued opposite to each other at mid-height.

TABLE 2. Summary of Results According to SEL

Sl. No.	g	n	Size, mm S B W	a_0/W	A	G_f (J/m)	K^b_{Ic}	ω_A	m
1		6	0600 80 076						
2		4	1000 80 140						
3	20	4	1200 80 200	0.200	0.759E-01	22.986	0.874	0.144	0.161
4		4	1500 80 240						
5		4	1800 80 300						
6		6	0600 80 076						
7		4	1000 80 140						
8	20	4	1200 80 200	0.300	0.123E-00	23.989	0.892	0.185	0.211
9		4	1500 80 240						
10		4	1800 80 300						
11		6	0600 80 076						
12		4	1000 80 140						
13	20	4	1200 80 200	0.400	0.231E-00	21.600	0.847	0.102	0.121
14		4	1500 80 240						
15		4	1800 80 300						
16		3	095 38 038						
17		3	191 38 076						
18	13	3	381 38 152	0.167	0.578E-02	42.531	1.084	0.133	0.146
19		3	762 38 305						
20		3	095 38 038						
21		3	191 38 076						
22	5	3	381 38 152	0.167	0.873E-02	23.670	0.883	0.031	0.048
23		3	762 38 305						
24		3	0400 100 100						
25	19	2	0800 100 200	0.400	0.270E-01	71.059	1.520	0.537	0.585
26		2	2000 100 500						
27		8	0400 100 100						
28		2	0800 100 200						
29	19	2	1200 100 300	0.200	0.709E-02	94.058	1.751	0.331	0.435
30		1	2000 100 500						
31		1	3200 100 800						

1. g = Max. aggregate size in mm; n = Number of specimens tested.

2. Of the hundreds of specimen groups for which peak load values were available only the above few groups satisfied the rather strict size requirements of SEL. Even for these few groups, statistical measures (ω_A should be less than 0.08 and m should be about 0.15) point towards poor quality of results.

3. Extreme care should be exercised in determining slope of regression line A. A slight error in its determination can significantly alter G_f.

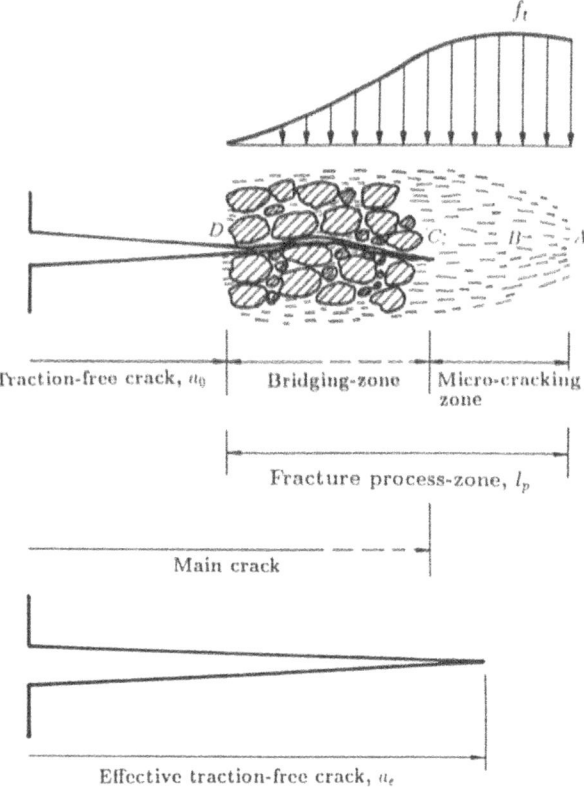

Figure 7: Schematic illustration of process zone and effective notch depth

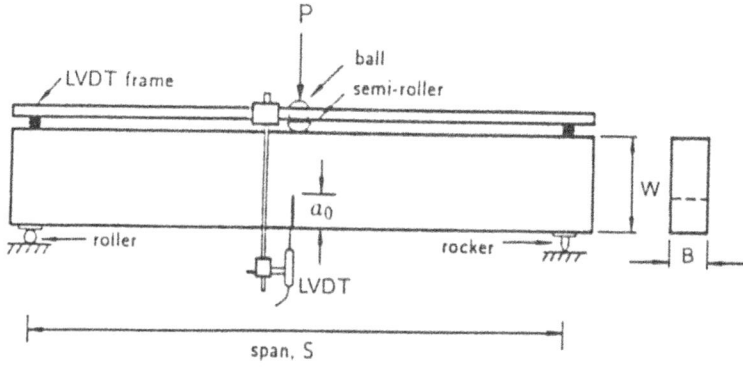

Figure 8: Loading apparatus and LVDT fixing arrangement

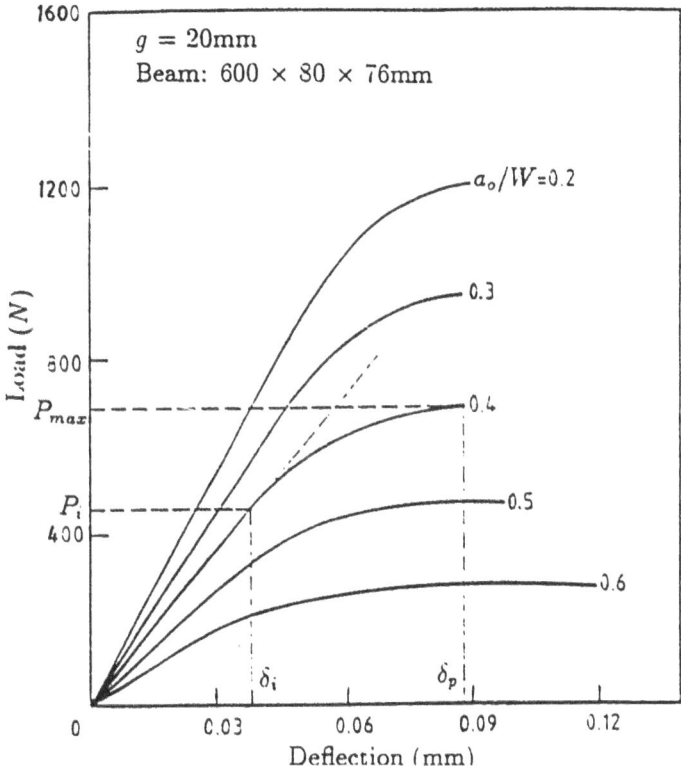

Figure 9: Typical load-deflection plots up to peak load for various a_0/W ratios

Typical load-deflection plots up to the peak load are shown in Fig. 9, for various a_0/W ratios. It should be emphasized that these plots need not be drawn continuously; it is sufficient to measure mid-span load-point deflection at several load levels up to the peak load. From these plots, P_i, P_{max} (P_i in linear range) and the corresponding mid-span deflections δ_i, δ_p are read, and E calculated from the following equation (Karihaloo & Nallathambi 1989a)

$$E = \frac{P_i}{4B\delta_i}\left(\frac{S}{W}\right)^3\left[1 + \frac{5qS}{8P_i} + \left(\frac{W}{S}\right)^2\left\{2.70 + 1.35\frac{qS}{P_i}\right\} - 0.84\left(\frac{W}{S}\right)^3\right]$$
$$+\frac{9}{2}\frac{P_i}{B\delta_i}\left(1 + \frac{qS}{2P_i}\right)\left(\frac{S}{W}\right)^2 F_2(\alpha_0), \tag{16}$$

where q is the self-weight of the beam per unit length, and

$$F_2(\alpha_0) = \int_0^{\alpha_0} \beta\, F_1^2(\beta)\, d\beta, \tag{17}$$

with $\alpha_0 = a_0/W$, and for $S/W = 4$, $F_1(\beta)$ is given by Eqn (5). A slightly less accurate expression (error $< 1\%$) for $F_1(\beta)$ is available in the range $0.1 < \beta < 0.6$

for two span to depth ratios ($S/W = 4$ and 8) with linear interpolation permitted within, and outside of, these ratios

$$F_1(\beta) = A_0 + A_1\beta + A_2\beta^2 + A_3\beta^3 + A_4\beta^4 \tag{18}$$

where

$$
\begin{aligned}
A_0 &= +0.0075\frac{S}{W} + 1.90 \\
A_1 &= +0.0800\frac{S}{W} - 3.39 \\
A_2 &= -0.2175\frac{S}{W} + 15.40 \\
A_3 &= +0.2825\frac{S}{W} - 26.24 \\
A_4 &= -0.1450\frac{S}{W} + 26.38
\end{aligned}
\tag{19}
$$

The coefficients $A_i(i = 0, 1, ..., 4)$ have been obtained by linear interpolation from the coefficients given by Brown & Srawley (1966) for $S/W = 4$ and 8.

The reduction in the stiffness of the beam (Fig. 8) is a result of both the stable crack growth and the formation of the discontinuous process zone ahead of the visible crack. It is however difficult to separate these two causes. Therefore it is assumed that the critical notch depth a_e may be calculated by introducing a fictitious beam containing a notch a_e whose unchanged stiffness (proportional to E) would be equal to the reduced stiffness of the real beam containing a notch of depth a_0, i.e.

$$
\begin{aligned}
\delta_p &= \frac{P_{max}}{4BE}\left(\frac{S}{W}\right)^3\left[1 + \frac{5qS}{8P_{max}} + \left(\frac{W}{S}\right)^2\left\{2.70 + 1.35\frac{qS}{P_{max}}\right\} - 0.84\left(\frac{W}{S}\right)^3\right] \\
&\quad + \frac{9}{2}\frac{P_{max}}{BE}\left(1 + \frac{qS}{2P_{max}}\right)\left(\frac{S}{W}\right)^2 F_2(\alpha_e),
\end{aligned}
\tag{20}
$$

where

$$F_2(\alpha_e) = \int_0^{\alpha_e} \beta F_1^2(\beta)\, d\beta \tag{21}$$

Here $\alpha_e = a_e/W$ and $F_1(\beta)$ is again given by Eqn (5) or Eqn (18). It will be noticed that Eqn (20) equates the stiffnesses and not the energies, as should be done in the ECM. Later in this lecture, we shall use the actual comparison of energies and shall show that a_e thus calculated is indeed very close to that calculated from Eqn (20).

$\alpha_e = a_e/W$ is calculated from Eqn (20) by a trial and error procedure described by Karihaloo & Nallathambi (1989a). This procedure was applied to all notched three-point bend test data available to the authors from various laboratories (Karihaloo & Nallathambi 1990a) and the corresponding a_e/W was calculated. The following regression equation gave the best fit

$$\frac{a_e}{W} = \gamma_1 \left(\frac{\sigma_n}{E}\right)^{\gamma_2} \left(\frac{a_0}{W}\right)^{\gamma_3} \left(1 + \frac{g}{W}\right)^{\gamma_4}, \tag{22}$$

where, as before, g is the maximum size of the aggregate used in the mix, $\sigma_n = 6M/(BW^2)$, $M = (P_{max} + qS/2)S/4$, and $\gamma_1 = 0.088 \pm 0.004$, $\gamma_2 = -0.208 \pm 0.010$, $\gamma_3 = 0.451 \pm 0.013$ and $\gamma_4 = 1.653 \pm 0.109$. The elastic modulus E in Eqn (22) was determined from the initial, linear segment of the load-deflection plot using Eqn (16). This method of determining E is more consistent with the ECM. Having determined the effective crack length a_e, the critical stress intensity factor according to the ECM is calculated using

$$K_{Ic}^e = \sigma_n \sqrt{a_e} F_1(a_e/W), \tag{23}$$

where $F_1(a_e/W)$ is given by Eqn (5) or Eqn (18).

It is possible to improve the above expression K_{Ic}^e by considering the true state of stress at the front of a pre-crack in a three-point bend specimen. Elastic finite element calculations show (Nallathambi & Karihaloo 1986b) that this stress field consists not only of a tensile stress normal to the crack plane (as assumed in the derivation of Eqn (23)) but also of a significant (tensile) stress in the crack plane and of a shear stress. By making an allowance for the true stress state ahead of the existing crack front, Eqn (23) becomes

$$\bar{K}_{Ic}^e = \sigma_n \sqrt{a_e} Y_1(\alpha) Y_2(\alpha, \beta), \tag{24}$$

where as before $\alpha = a_e/W$, $\beta = S/W$, and

$$Y_1(\alpha) = A_0 + A_1\alpha + A_2\alpha^2 + A_3\alpha^3 + A_4\alpha^4 \tag{25}$$
$$Y_2(\alpha, \beta) = B_0 + B_1\beta + B_2\beta^2 + B_3\beta^3 + B_4\alpha\beta + B_5\alpha\beta^2 \tag{26}$$

Regression coefficients A_i, B_j, $(i = 0, ..., 4; j = 0, 1, ..., 5)$ are given in Table 3.

TABLE 3. Regression coefficients A_i, $B_j(i = 0, ..., 4; j = 0, ..., 5)$

i/j	0	1	2	3	4	5
A_i	3.6460	-6.7890	39.2400	-76.8200	74.3300	-
B_j	0.4607	0.0484	-0.0063	0.0003	-0.0059	0.0033

Fig. 10 shows the relative variation of K_{Ic}^e with specimen depth. The plots refer to different mixes. It is clear that K_{Ic}^e is reasonably independent of the size of the specimen. Further decrease in variation is expected when the tests are conducted in strict accordance with the requirements of the ECM.

Tables 4 and 5 compare the fracture parameters K_{Ic}^e, Δa_e predicted by the ECM with the predictions from the TPM and SEL (Nallathambi & Karihaloo, 1990a)

TABLE 4. Summary of Results According to (ECM) & (TPM)

Sl No	n	g	f'_c	E	Range a_0/W	Range a_e/W	Range $\underline{a}/W$	K^e_{Ic} Mean(sd)	K^s_{Ic} Mean(sd)	$CTOD_c$ Mean(sd)
1	90	2	44.3	26.3	.20-.60	.27-.63	-	0.633(.065)	-	-
2	84	5	42.1	29.3	.20-.60	.27-.64	-	0.641(.057)	-	-
3	60	10	40.3	32.0	.20-.50	.28-.56	-	0.706(.046)	-	-
4	30	14	40.9	32.6	.20-.40	.28-.47	-	0.728(.018)	-	-
5	30	20	37.6	33.0	.20-.40	.29-.49	-	0.776(.028)	-	-
6	76	20	38.0	33.2	.20-.60	.29-.69	-	0.884(.057)	-	-
7	6	19	55.8	36.8	.29-.67	-	.55-.81	-	0.931(.263)	.0148(.0076)
8	3	19	53.1	38.4	.29-.50	.36-.55	.41-	0.965(.045)	1.054(-)	.0153(-)
9	3	19	54.4	39.3	.30-.67	.37-.73	.42-.73	1.145(.098)	1.128(.269)	.0200(.0085)
10	20	19	53.1	38.4	.15-.61	.23-.68	.44-.83	0.797(.075)	1.146(.014)	.0316(.0067)
11	20	19	54.4	39.3	.19-.57	.27-.63	.40-.71	0.971(.078)	1.220(.102)	.0312(.0087)
12	12	10	29.0	21.7	.50-.50	.54-.55	-	0.760(.039)	-	-
13	8	10	58.9	24.5	.30-.50	.37-.55	-	0.908(.068)	-	-
14	8	10	33.1	19.7	.50-.50	.54-.55	-	0.859(.091)	-	-
15	15	10	55.5	29.8	.20-.50	.27-.55	-	1.023(.052)	-	-
16	14	20	36.2	24.0	.50-.50	.53-.55	-	1.031(.129)	-	-
17	15	16	38.3	34.1	.33-.33	.40-.41	-	1.120(.217)	-	-
18	8	19	29.0	32.5	.20-.40	.27-.47	-	1.413(.272)	-	-
19	4	19	34.3	33.2	.20-.40	.28-.48	-	1.759(.172)	-	-
20	16	19	26.3	32.0	.20-.40	.28-.48	-	1.232(.147)	-	-
21	12	3	-	36.8	.13-.50	.18-.56	.15-.56	0.926(.061)	0.894(.068)	.0042(.0013)
22	2	6	60.7	33.5	.19-.21	.28-.29	.25-.28	1.221(.061)	1.141(.095)	.0145(.0057)
23	3	13	45.5	31.0	.19-.21	.27-.29	.25-.35	1.429(.046)	1.475(.191)	.0220(.0086)
24	3	13	43.4	31.0	.20-.21	.28-.29	.24-.28	1.610(.026)	1.530(.022)	.0169(.0024)
25	12	13	34.1	27.7	-	-	-	0.975(.150)	-	-
26	12	5	48.4	32.9	-	-	-	1.004(.068)	-	-
27	8	19	25.2	27.2	.29-.33	.36-.40	.35-.55	0.982(.169)	0.976(.103)	.0170(.0068)
28	6	13	48.5	33.3	.50-.51	.62-.73	-	0.421(.091)	-	-
29	11	8	93.0	32.0	.50-.50	.53-.54	-	1.198(.199)	-	-
30	11	8	28.0	31.0	.50-.50	.55-.57	-	0.830(.117)	-	-
31	6	12	68.0	39.0	.50-.50	.55-.57	-	1.356(.101)	-	-
32	6	12	21.0	26.0	.50-.50	.57-.58	-	0.727(.038)	-	-
33	17	32	31.0	32.3	.50-.50	.56-.57	-	1.585(.210)	1.211(.121)	-
34	11	2	35.0	25.7	.50-.50	.53-.58	-	1.102(.098)	0.790(.090)	-
35	2	8	110.	56.6	.33-.33	.37-.38	-	1.896(.064)	2.130(-)	.0338(.0039)

Notes:

1. g = Max. aggregate size in mm; n = No. of specimens tested.

2. f'_c in MPa, E in GPa, K_{Ic} in $MPa\sqrt{m}$, $CTOD_c$ in mm.

3. Absence of an entry in K^s_{Ic} and $CTOD_c$ columns means that load-CMOD plot was not available.

TABLE 5. Summary of Results According to ECM & SEL

Sl No	g	n	Size, mm S B W	a_0/W	A	G_f (J/m)	K^b_{Ic}	ω_A	m	K^e_{Ic} Mean(sd)
1		6	0600 80 076							
2		4	1000 80 140							
3	20	4	1200 80 200	0.200	0.07590	22.986	0.874	0.144	0.161	
4		4	1500 80 240							
5		4	1800 80 300							
6		6	0600 80 076							
7		4	1000 80 140							
8	20	4	1200 80 200	0.300	0.12300	23.989	0.892	0.185	0.211	0.867(.063)
9		4	1500 80 240							
10		4	1800 80 300							
11		6	0600 80 076							
12		4	1000 80 140							
13	20	4	1200 80 200	0.400	0.23100	21.600	0.847	0.102	0.121	
14		4	1500 80 240							
15		4	1800 80 300							
16		3	095 38 038							
17		3	191 38 076							
18	13	3	381 38 152	0.167	0.00578	42.531	1.084	0.133	0.146	0.975(.150)
19		3	762 38 305							
20		3	095 38 038							
21		3	191 38 076							
22	5	3	381 38 152	0.167	0.00873	23.670	0.883	0.031	0.048	1.004(.068)
23		3	762 38 305							
24		3	0400 100 100							
25	19	2	0800 100 200	0.400	0.02700	71.059	1.520	0.537	0.585	1.264(.293)
26		2	2000 100 500							
27		8	0400 100 100							
28		2	0800 100 200							
29	19	2	1200 100 300	0.200	0.00709	94.058	1.751	0.331	0.435	1.208(.260)
30		1	2000 100 500							
31		1	3200 100 800							

Notes:

1. g = Max. aggregate size in mm; n = Number of specimens tested.

2. All K_{Ic} values are given in $MPa\sqrt{m}$.

3. For the purposes of comparison G_f has been converted to an equivalent fracture toughness value K^b_{Ic} using the LEFM plane stress relationship $K^b_{Ic} = \sqrt{G_f E}$.

4. See also notes 2, 3 appearing after Table 2.

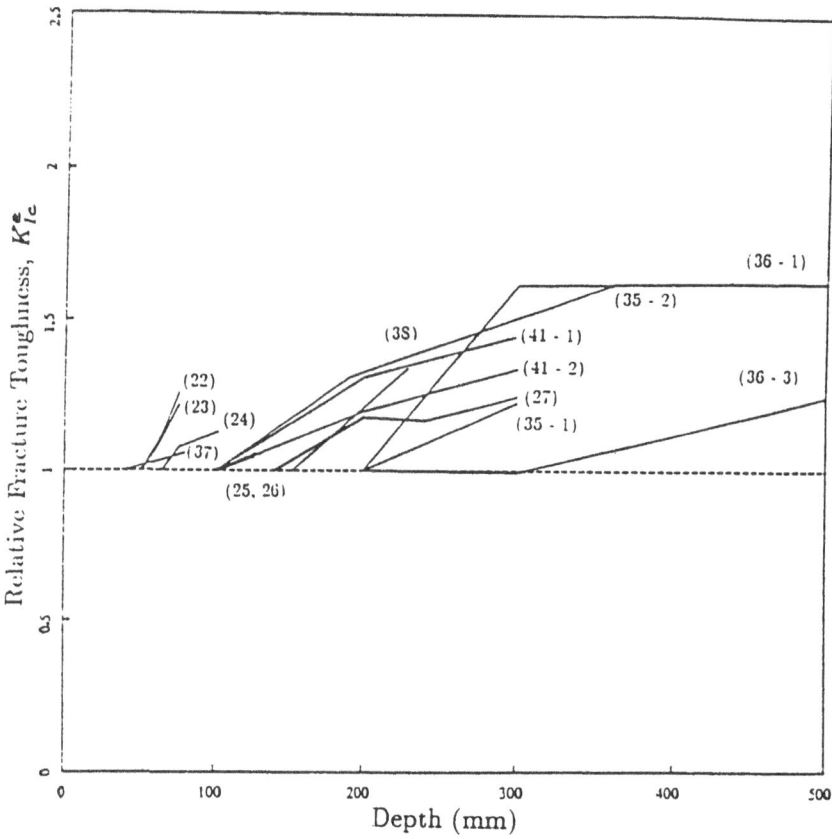

Figure 10: Variation of relative K_{Ic}^e with specimen depth

5. Comparison of ECM with TPM

As seen from Table 4 the fracture parameters of plain concrete calculated using the effective crack model(ECM) and the two parameter model(TPM) are in good agreement despite the fact that they were determined from essentially separate series of test data. This was because of the paucity of simultaneous load-displacement and load-CMOD measurements from one and the same three-point bend specimen.

An investigation was conducted (Nallathambi & Karihaloo, 1990a) to compare the results of the two models using the necessary data from the same test specimens. To achieve this aim, both load-displacement and load-CMOD plots are simultaneously recorded for all notched specimens tested in three-point bending. A further aim of this investigation was to make this comparison for a wide variety of concrete mixes, ranging in (cylinder) compressive strength from about $25MPa$ to nearly $80MPa$.

It is found that irrespective of the concrete strength, the fracture parameters calculated using the effective crack model (namely, the effective fracture toughness K_{Ic}^e and the effective traction-free notch length a_e) are practically indistinguishable from the corresponding parameters calculated using the two parameter model (namely, the fracture toughness K_{Ic}^s and the notch length $\underline{a}$; the latter being also used to determine the critical crack tip opening displacement$(CTOD_c)$).

Table 6 shows selected mix properties and the elastic modulus E calculated in four different ways, both direct and indirect. It will be appreciated that all four values of E for each of the mixes are in very good agreement with one another. This proves conclusively that any variations in fracture parameters between various models (Hillerborg's cohesive crack, ECM, TPM and SEL) cannot be attributed to differences in the measurement of E.

TABLE 6. Mix Properties

Mix	Compressive Strength $f_c(MPa)$	Elastic Modulus, (GPa)				Tensile Strength $f_t^\odot(MPa)$
		E^*	E_c^{**}	$E^\dagger$	$E^\ddagger$	
C1	26.8	24.62	24.51	25.56(.35)	25.04(.29)	2.58
C2	39.0	33.80	29.56	29.87(.21)	31.56(.64)	3.11
C3	49.4	34.65	33.27	33.28(.22)	32.96(.24)	3.50
C4	67.5	37.20	38.89	37.13(.23)	38.39(.82)	4.09
C5	78.2	40.30	41.86	40.99(.60)	40.26(.99)	4.41

*   Determined from separate cylinder tests (using strain gauges)
**   Estimated from the relationship, $E_c = 4734\sqrt{f_c}MPa$ $(=57000\sqrt{f_c}psi)$
†   Calculated from $P - \delta$ plot
‡   Calculated from $P - CMOD$ plot
⊙   Estimated from the relationship, $f_t = 0.4983\sqrt{f_c}\ MPa(=6\sqrt{f_c}psi)$

Table 7 shows the fracture parameters calculated according to ECM and TPM. As mentioned above, the agreement between the two models is excellent. It should however be mentioned that the regression formula (Eqn 22) overestimates slightly the values of a_e/W (see Table 7). It is therefore recommended that a_e be calculated from Eqn (20) using the trial and error procedure given by Karihaloo & Nallathambi (1989a).

Another approach to comparing ECM with TPM is by calculating the tensile strength f_t, the total work of fracture W_d, and/or the critical crack opening w_c using the pair of fracture parameters defined by these models. These additional (but dependent) parameters are in any case required for a full description and finite element implementation of the fracture process. As an additional benefit this comparison approach allows us to put Δa_e on a sound physical foundation. It will be recalled that Δa_e was calculated above from a comparison of stiffnesses, whereas the basic assumption in ECM is that it be determined by equating the actual work of fracture W_d with the energy required to create a hypothetical supplementary traction-free crack $\Delta a_e = a_e - a_0$. This will be done in the sequel.

TABLE 7. Fracture Parameters for Various Mixes

Mix/	C1	C2	C3	C4	C5
Data	Mean(sd)	Mean(sd)	Mean(sd)	Mean(sd)	Mean(sd)
a_0/W	0.295(.001)	0.296(.000)	0.295(.001)	0.293(.003)	0.293(.003)
$a_e/W^{\dagger}$	0.443(.005)	0.441(.001)	0.435(.004)	0.428(.002)	0.419(.006)
$a_e/W^{\ddagger}$	0.447(.003)	0.454(.001)	0.446(.003)	0.442(.004)	0.426(.008)
$\underline{a}/W^{\odot}$	0.443(.015)	0.442(.006)	0.436(.001)	0.430(.002)	0.413(.006)
K_{Ic}^e	0.992(.015)	1.265(.013)	1.376(.020)	1.502(.046)	1.881(.095)
K_{Ic}^s	0.993(.054)	1.269(.028)	1.381(.031)	1.509(.040)	1.847(.098)
$CTOD_c$	0.033(.010)	0.026(.001)	0.026(.001)	0.024(.001)	0.026(.001)

$\dagger$ ECM
$\ddagger$ Eqn 22
$\odot$ TPM
K_{Ic} in $MPa\sqrt{m}$, $CTOD_c$ in mm

For the above comparison approach to work, it is necessary to judiciously approximate the post-peak tension softening diagram. For simplicity this diagram is usually approximated by a linear segment, although the present authors have recently given an analytical solution (Karihaloo & Nallathambi, 1989b)[1] for a highly non-linear law that better approximates the observed post-peak behaviour (Fig. 11) particularly at P_{max} (horizontal tangent at $\sigma = f_t$).

The transmitted stress-displacement law in the post-peak region is approximated as follows:

$$\frac{\sigma}{f_t} = 1 - \frac{w}{w_c} \quad (Linear) \tag{27}$$

$$\frac{\sigma}{f_t} = \left[1 - 9.2431\beta^2 + 33.8259\beta^3 - 59.4248\beta^4 + 49.3000\beta^5 - 15.4722\beta^6\right] \quad (Non\text{-}lin), \tag{28}$$

where $\beta = w/w_c$.

For the assumed $\sigma - w$ laws it was shown by Nallathambi & Karihaloo, (1990b) that

$$K_{Ic}^e = 0.7071\sqrt{E'w_c f_t}$$

---

[1]Several unfortunate errors in this paper have been corrected in (Nallathambi & Karihaloo 1990b).

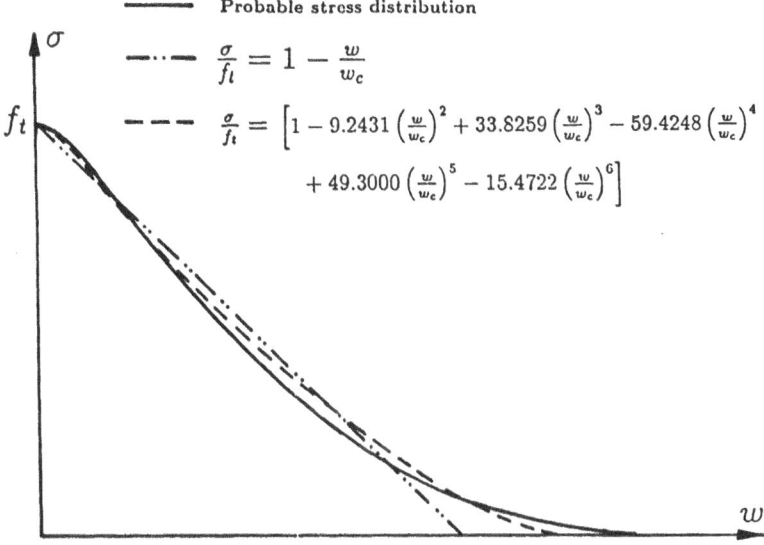

Figure 11: Approximations to tension-softening diagram

$$W_d = \int_0^{\ell_{pc}} \int_0^{w_s} f^{-1}(w)\, dw\, ds = 0.1050 E' w_c^2. \quad (Linear\ law) \qquad (29)$$

and

$$K_{Ic}^e = 0.7043 \sqrt{E' w_c f_t}$$

$$W_d = \int_0^{\ell_{pc}} \int_0^{w_s} f^{-1}(w)\, dw\, ds = 0.0521 E' w_c^2. \quad (Non\text{-}linear\ law) \qquad (30)$$

where $f^{-1}(w)$ represents the right hand side of Eqns. (27, 28). Equating W_d to $(K_{Ic}^e{}^2/E)\Delta a_e$ gives

$$\Delta a_e = 0.210 E w_c / f_t \quad (Linear\ law)$$
$$\Delta a_e = 0.105 E w_c / f_t \quad (Non\text{-}linear\ law)$$

Finally, one may also calculate the critical process zone size ℓ_{pc} corresponding to the two assumed $\sigma - w$ laws:

$$\ell_{pc} = 0.366 E w_c / f_t \quad (Linear\ law)$$
$$= 0.359 E w_c / f_t \quad (Non\text{-}linear\ law) \qquad (31)$$

TABLE 8. Additional Material Parameters Calculated From ECM and TPM

	$CTOD_c$	$w_c(mm)$		$f_t^\dagger$	$f_t(MPa)$		$\ell_{pc}(mm)$	
Mix	(mm)	Lin	Non-lin		Lin	Non-lin	Lin	Non-lin
C1	0.0332*	0.0208	0.0296	2.58	3.70	2.63	52.6	103.1
C2	0.0263	0.0213	0.0318	3.11	4.78	3.39	51.2	100.4
C3	0.0261	0.0219	0.0305	3.50	5.29	3.76	49.5	97.0
C4	0.0242	0.0202	0.0293	4.09	5.88	4.18	47.7	93.6
C5	0.0261**	0.0221	0.0322	4.41	7.61	5.40	44.8	87.8

Notes:

† Calculated from the empirical relationship $f_t = 0.4983\sqrt{f_c}(MPa)$

* Possible inaccuracy of measurement

** Could be influenced by the presence of superplasticizer

Using the pair of fracture parameters $(K_{Ic}^e, \Delta a_e)$ or $(K_{Ic}^s, \Delta \underline{a})$ from Table 7 in Eqs. (29)-(31) one can calculate w_c without knowing f_t. The results are given in Table 8, which also shows $CTOD_c$ according to the TPM. If one now uses the calculated values of w_c to determine f_t and ℓ_{pc} (Eqn 31) and compares the resulting f_t with that given by the well-known empirical formula (Table 8), then it transpires that the non-linear approximation gives the better fit. It is worth noting that f_t is the only material parameter for which an independent formula, albeit an empirical one, exists. There is at present no similar independent way of checking the accuracy of $CTOD_c$ or ℓ_{pc}. It would seem though that the linear $\sigma - w$ approximation underestimates both w_c and ℓ_{pc}. Finally, it is rather surprising that $CTOD_c$ (Table 8) exceeds w_c predicted by the linear approximation and is very close to that predicted by the non-linear approximation, whereas one would have expected it to be smaller than both.

6. Comparison of FCM, ECM, TPM and SEL

The above comparison was restricted to two methods, namely TPM and ECM. Moreover, the comparison was based on the fracture parameters determined using these methods. To the extent that both these methods approximate the actual pre-peak behaviour and use LEFM, albeit after accounting for the stable slow crack growth, it is advisable to compare them at the asymptotic limit of large size structures when LEFM is strictly applicable. The comparison will not be restricted to TPM and ECM but will also include SEL and fictitious cohesive crack model (FCM) of Hillerborg et al. (1976).

The methodology for such a comparison was proposed by Planas & Elices (1987) and will not be repeated here. Readers will find a succinct account in the report

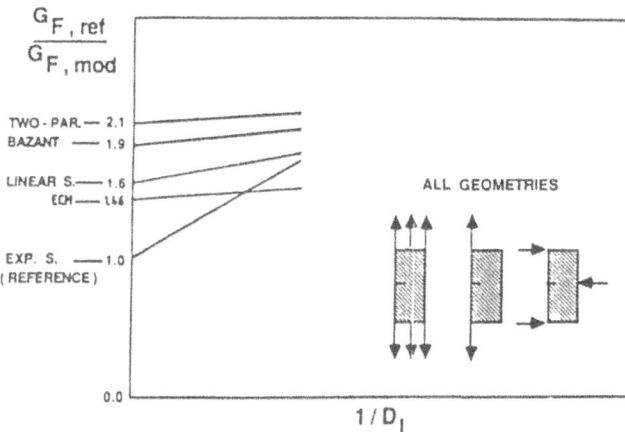

Figure 12: Ratio of $G_{F,reference}$ (FCM with quasi-exponential softening) to $G_{F,model}$ predicted for large sizes by ECM, TPM, SEL and FCM with linear softening

by Karihaloo & Nallathambi (1990a). It will suffice here to summarize the major conclusion reached by Planas & Elices. It should be mentioned though that Planas & Elices did not include ECM in their comparison. However, their methodology was recently followed by Karihaloo & Nallathambi (1990b) and applied also to ECM.

The major conclusion flowing from this extensive comparative study is that all four models are able to predict accurately the maximum load carrying capacity of concrete specimens in the practical range of sizes used in the laboratory but that the predictions for large size structures are less accurate. The latter part of this conclusion is illustrated on Fig. 12 which compares the predictions of the various models in the limit as the representative size of the structure (D_i) tends to infinity. It should be mentioned that the comparison for ECM is only applicable to notched beam geometry and not to the other two geometries shown in Fig. 12.

Planas & Elices argue that the reason for the discrepancy among the models at large sizes is due to the inconsistency in the definition of size effect in the various models and therefore in that of the fracture energy based on this effect. It should however be mentioned that the prediction of ECM is the least conservative of the models studied.

7. Mode I Fracture Toughness from Compact Specimen Geometries

In the foregoing we analysed and compared four methods (models) for determining mode I fracture toughness of concrete from notched three-point bend specimens. Only three of these models (TPM, SEL, ECM) were considered by Sub-Committee A. FCM was included in the above discussion for completeness of presentation. Except for ECM which is applicable only to notched three-point bend geometry, it is claimed that TPM, SEL and FCM can be applied to any geometry. We shall however see in the sequel that their applicability to geometries other than the notched three-point bend is not fully validated. In fact, it would seem that $CTOD_c$ of TPM is particularly susceptible to geometry and cannot therefore be regarded as a material parameter without further validation. K_{Ic}^s on the other hand, would seem to be relatively insensitive to geometry changes.

Sub-committee B which analysed three compact geometries did not choose any of the above models *per se* as a starting point. Nevertheless, it is clear that its approach can be regarded as a judicious mix of the effective Griffith crack model (which forms the basis of TPM and ECM) and the scaling law (extrapolation to infinite size specimens) implied in SEL (Rossi *et al.* 1990). They started from the same premise as SEL, namely that LEFM is strictly applicable in the asymptotic limit of very large concrete structure (representative size $D \to \infty$). They claim that it is possible to obtain the intrinsic K_{Ic} for ordinary concrete using a DCB test specimen 3.5m long and 1.1m wide. They recognized however the difficulty associated with handling and testing such a specimen in a *normal* laboratory and proposed to carry out parallel tests on small specimens. The latter will yield *non-objective* K_{Ic} (i.e. size-dependent K_{Ic}), whilst the large specimens would yield the *objective* K_{Ic}. Their aim is to find an analytical relation between these two K_{Ic} with a view to establishing the scaling law (scale factor). If this aim is attained, then it would be possible to determine the intrinsic fracture toughness of 'in-situ' concrete from core samples.

With these two broad objectives in mind Rossi *et al.* (1990) chose the tapered DCB (Fig. 13) as the *infinitely* large specimen for determining the *objective* K_{Ic} and cylindrical and cubical wedge splitting geometries (Fig. 14) as the small specimens for determining the *non-objective* K_{Ic}. The choice of the last two geometries was dictated by the desire to accommodate on the one hand core samples drilled from existing concrete structures and dams and on the other fresh concrete specimens prepared using large aggregate (Jenq & Shah 1988).

The crack opening displacement was used as the feedback signal to obtain stable failure. Loading and unloading were performed during the test to determine the compliance function that was approximated by only four data points (crack lengths) resulting in a surprising inflexion point in the compliance function. Rossi *et al.*, (1990) are aware of the inaccuracies so caused and have also used the program FRANC to improve the accuracy. A sample calibration curve is shown in Fig 15 for the cubical WS specimen.

Typical load-COD plots for the three geometries are shown in Figs. 16-18, respectively. Concrete A refers to a mix with $f_c = 51\ MPa$, E=36.6 GPa and indirect

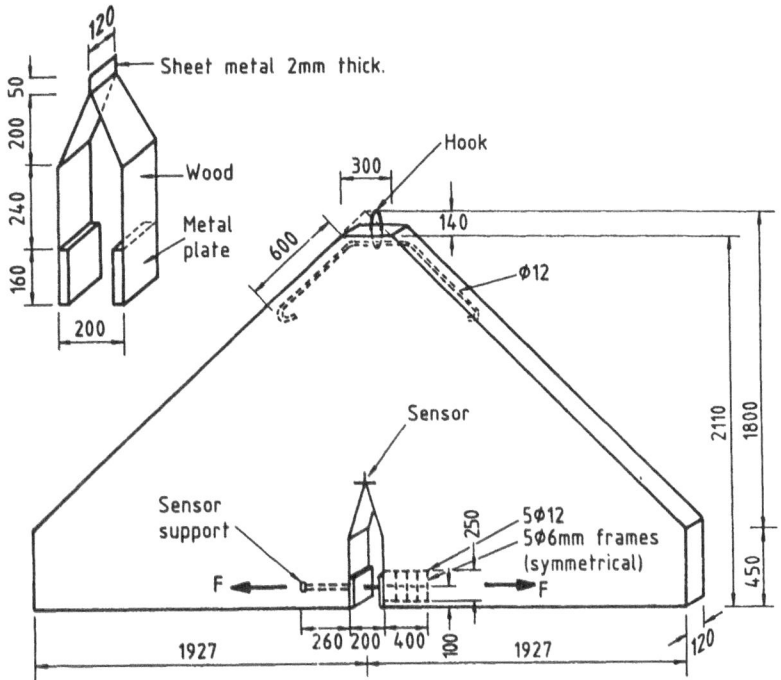

Figure 13: The tapered double cantilever beam (TDCB) specimen.

tensile strength $= 3.7\ MPa$, whereas concrete B refers to a mix with the corresponding properties 55 MPa, 36 GPa and 4.5 MPa, respectively.

The maximum size of aggregate used in concrete A was $20mm$ and that in concrete B was 12.5mm. Finite element calculations were performed to establish the compliance function $C(a)$ for each geometry. However, the above remark regarding the accuracy should be borne in mind in the practical use of these functions.

TDCB Specimen ($0.55 \leq a \leq 1.55m$):

$$C(a) = \frac{1}{E}(3410.75a^3 - 7470.05a^2 + 5700.1a - 1281)\ (m/N) \tag{32}$$

Cubical WS Specimen ($0.059 \leq a \leq 0.15m$):

$$C(a) = \frac{10^4}{E}(268.1a^3 - 58.7a^2 + 4.4a - 0.1)\ (m/N) \tag{33}$$

Cylindrical WS Specimen ($0.047 \leq a \leq 0.09m$):

$$C(a) = \frac{10^6}{E}(33.03a^3 - 5.19a^2 + 0.28a - 0.005)\ (m/N) \tag{34}$$

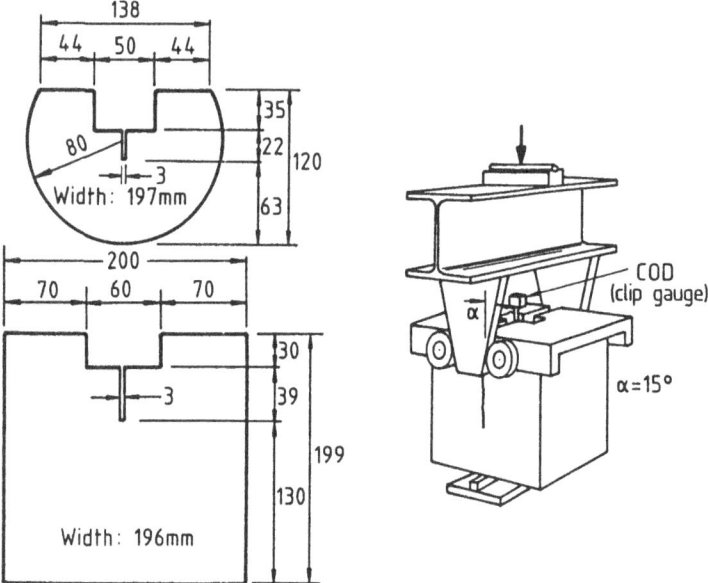

Figure 14: The cylindrical and cubical wedge splitting specimens, showing the load application device.

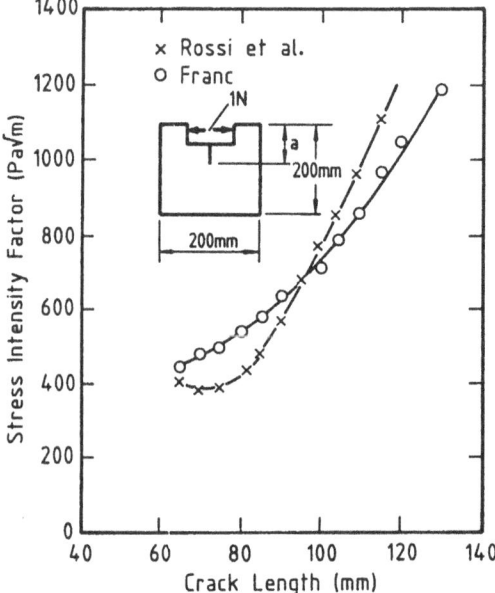

Figure 15: Stress intensity factor per unit applied force *vs* the crack length for the cubical WS specimen

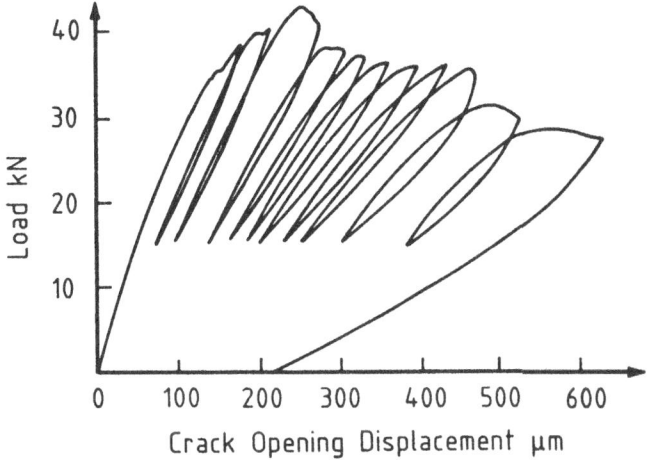

Figure 16: Load-COD plot for TDCB specimen from concrete A

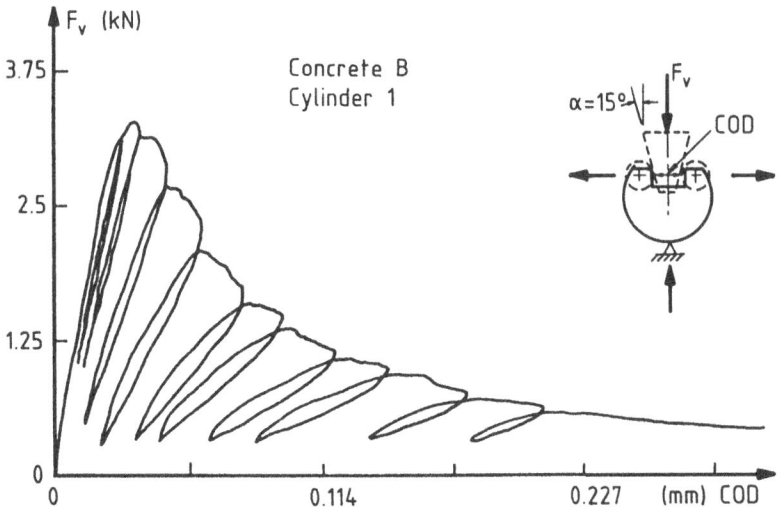

Figure 17: Load-COD plot for cylindrical WS specimen from concrete B

Then from the standard plane stress relationship between the stress intensity factor and rate of change of compliance with crack growth the following expressions for the R-curve result:

TDCB Specimen ($0.55 \leq a \leq 1.55m$):

$$K_I(a) = \frac{1}{\sqrt{2}\,B}(10232.25a^2 - 14940.1a + 5700.1)^{1/2}\,F\ (MPa\sqrt{m}), \qquad (35)$$

Cubical WS Specimen ($0.059 \leq a \leq 0.15m$):

$$K_I(a) = \frac{100}{\sqrt{2}\,B}(804.3a^2 - 117.4a + 4.4)^{1/2}\,F_s\ (MPa\sqrt{m}), \qquad (36)$$

Cylindrical WS Specimen ($0.047 \leq a \leq 0.09m$):

$$K_I(a) = \frac{1000}{\sqrt{2}\,B}(99.09a^2 - 10.38a + 0.28)^{1/2}\,F_s\ (MPa\sqrt{m}), \qquad (37)$$

where B = thickness of the specimen.

To avoid having to use unreliable surface observations of crack advance, Rossi et al. (1990) use the notion of effective crack. The length of this crack a_{eff} is determined by equating the experimental compliance of the real crack having an irregular front preceded by a microcracked zone with that of an effective crack with a regular front (Fig. 19).

They also calculate an effective elastic modulus E_{eff} for use in compliance expressions (32)-(34). It is calculated by matching the theoretical compliance corresponding to a_0 with the experimental compliance during the first loading (i.e. in the elastic domain before the formation of any microcracks at the notch tip). The critical stress intensity factor K_{Ic} for each geometry is then calculated from (35)-(37) with a replaced a_{eff} and E by E_{eff}.

Typical $K_{Ic} - a_{eff}$ curves for the three geometries are shown in Fig. 20-22 respectively.

It is claimed that the mean value of the *objective* K_{Ic} following from the TDCB specimen is 2.21 and 2.11 $MPa\sqrt{m}$ for mixes A and B respectively, whereas the mean value and variation = $(K_{Ic})_{max} - (K_{Ic})_{min}/(K_{Ic})_{max}$ of the *non-objective* K_{Ic} determined from cylindrical WS specimen is 1.44 (33.7%) and 1.52 (14.3%) for mixes A and B, and from cubical WS specimen is 1.82 (17.4%) and 1.97 (11.8%), respectively.

Rossi et al. (1990) have shown that as the TDCB specimen is tested in a vertical position, the influence of the self-weight upon K_{Ic} is negligible. They also claim that the influence of the reinforcing bar (required for suspending the specimen) on the energy consumption is negligible. This is a rather doubtful claim in the light of the stiffening effect of the reinforcing bar on the TDCB system as a whole. It would be helpful if Rossi et al. could provide some quantitative estimate of this stiffening

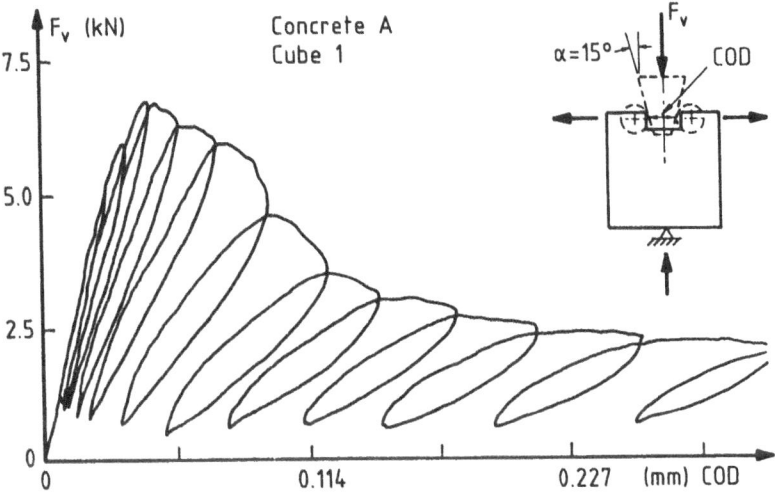

Figure 18: Load-COD plot for cubical WS specimen from concrete A

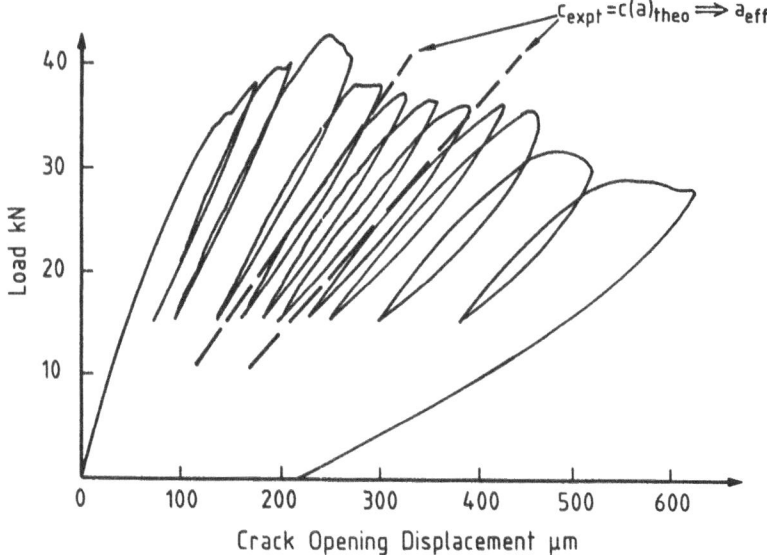

Figure 19: Determination of a_{eff}

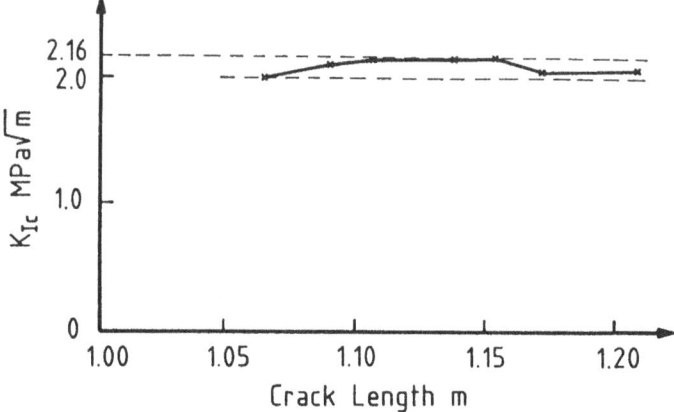

Figure 20: K_{Ic} *vs* a_{eff} for TDCB specimen from concrete A

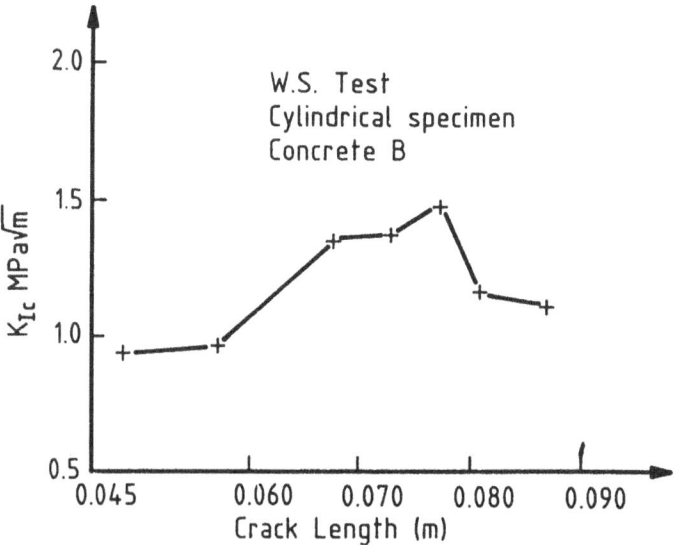

Figure 21: K_{Ic} *vs* a_{eff} for cylindrical WS specimen from concrete B

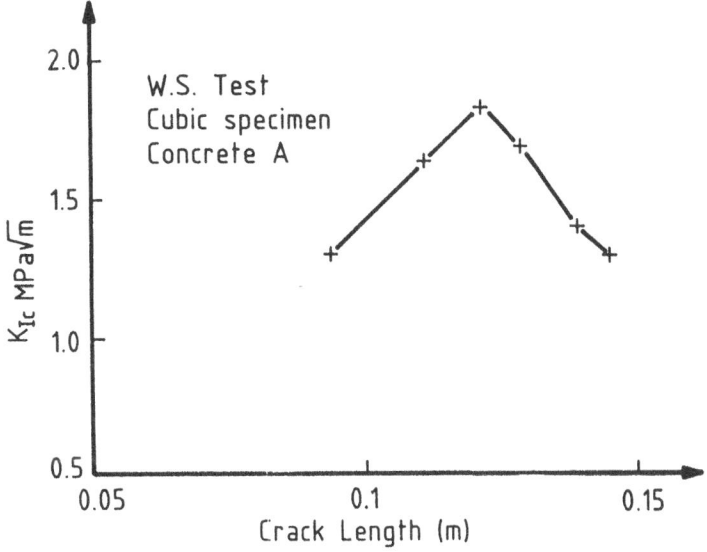

Figure 22: K_{Ic} *vs* a_{eff} for cubical WS specimen from concrete A

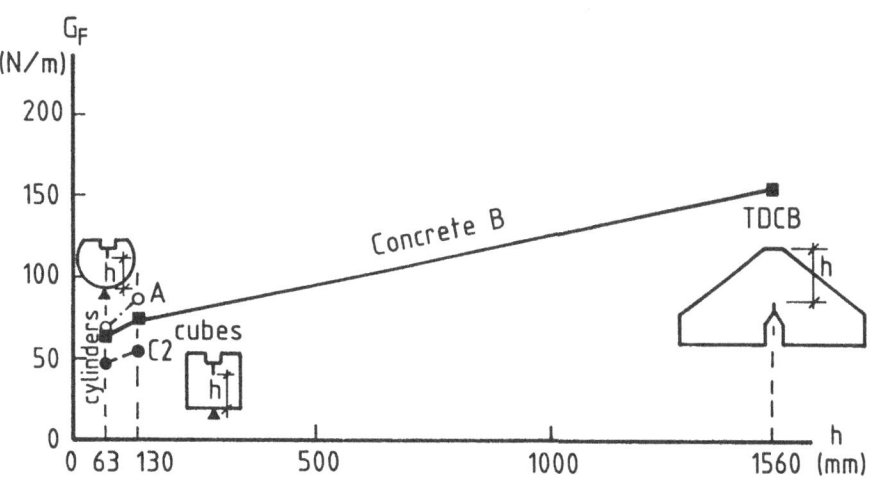

Figure 23: Specific fracture energy G_F *vs* ligament length

effect. Until such further evidence is forthcoming, the *objective* K_{Ic} determined from the TDCB must be regarded with caution.

They also propose the following regression equations for calculating the *objective* K_{Ic} of plain concrete from the results of small specimens

$$K_{Ic}(\text{Cubical WS}) = K_{Ic}(\text{TDCB})(1.075 - 0.075\frac{g}{g^*}) \qquad (38)$$

$$K_{Ic}(\text{Cylindrical WS}) = K_{Ic}(\text{TDCB})(0.802 - 0.042\frac{g}{g^*}) \qquad (39)$$

where $g^* = 6mm$ and $6 \leq g(mm) \leq 20$.

They have also calculated the specific fracture energy G_F as a function of the ligament length and found that it increases with increasing h, besides the maximum aggregate size, g. It is reasoned that the increase in G_F with h (Fig. 23) is because of the changes in the width and length of the fracture process zone as it propagates along the ligament, so that the longer the ligament, the more energy is expended outside of the localized fracture process zone.

Based on G_F and K_{Ic}, the Sub-committee report ends with the finite element evaluation of the maximum load based on a bilinear approximation to the tension-softening diagram, in much the same manner as it has been done by Petersson (1981).

8. Concluding Remark

The aim of this lecture was to review the various test methods and models considered by sub-committees A and B of RILEM TC-89 (FMT) and to present the results as they appear in the final reports of these sub-committees, with very occasional remarks on the drawbacks of one method or the other. In a general review of this nature it is very tempting to criticise methods or models which may be competing with one's own. It is to be hoped that this reviewer has avoided that temptation, and that the presentation will result in a lively debate and discussion.

9. References

Bažant, Z.P., (1984) Size effect in blunt fracture: concrete, rock, metal, *J. of Eng. Mech., ASCE*, **110**, pp. 518-535.

Bažant, Z.P., Kim, J.K. and Pfeiffer, P.A., (1986) Determination of fracture properties from size effect tests, *J. of Struct. Eng., ASCE*, **112**, No.2, pp. 289-307.

Brown, W.F. and Srawley, J.E., (1966) Crack toughness testing of high strength matallic matyerials, *ASTM-STP No. 410*, p. 130.

Hillerborg, A., Modeer, M. and Petersson, P.E. (1976), Analysis of Crack Formation and Crack Growth in Concrete by Means of Fracture Mechanics and Finite Elements, *Cement Concr. Res.*, **6**, pp. 773-782.

Jenq, Y.-S. and Shah, S.P., (1985) Two parameter fracture model for concrete, *ASCE, J. Eng. Mech.*, **111**, pp. 1227-1241.

Jenq, Y.-S. and Shah, S.P., (1988) On the concrete fracture testing methods, *Proceedings of the International Workshop on Fracture Toughness & Fracture Energy*, Sendai, pp. 443-463.

John, R., Shah, S.P. and Jenq, Y.-S., (1987) A fracture mechanics model to predict the rate sensitivity of mode I fracture for concrete, *Cem. and Conc. Research*, **17**, pp. 249-262.

Karihaloo, B.L. and Nallathambi, P., (1989 a) An improved effective crack model for the determination of Fracture toughness of concrete, *Cem. and Conc. Research*, **19**, pp. 603-610.

Karihaloo, B.L. and Nallathambi, P., (1989 b) Effective crack model and tension-softening models, *Fracture of Concrete and Rock: Recent Developments* (Eds: S. P. Shah, S. E. Swartz and B. Barr), Elsevier Science., pp. 701-710.

Karihaloo, B.L. and Nallathambi, P., (1990 a) Final report of Sub-committee *A* (Chairman B. L. Karihaloo) notched beam test: mode I fracture toughness of *RILEM TC89-FMT*, (Chairman: S. P. Shah), p.86. To be published by Chapman & Hall, Oct. 1990.

Karihaloo, B.L. and Nallathambi, P., (1990 b) Size effect prediction from effective crack model for plain concrete, *Matériaux et Constructions*, (in press).

Nallathambi, P. and Karihaloo, B.L., (1986 a) Determination of specimen-size independent fracture toughness of plain concrete, *Magazine of Concrete Research*, **38**, pp. 67-76.

Nallathambi, P. and Karihaloo, B.L., (1986 b) Stress intensity factor and energy release rate for three-point bend specimens, *Engineering Fracture Mechanics*, **25**, pp. 315-321.

Nallathambi, P. and Karihaloo, B.L., (1990a) Performance of effective crack and two-parameter models, *ASCE, J. of Eng. Mech.*, (submitted for publication).

Nallathambi, P. and Karihaloo, B.L., (1990b) Fracture of concrete: application of effective crack model, *Proc. of the 9th Int. Conf. on Expt. Mech.*, Lyngby, Denmark.

Petersson, P.E. (1981), Crack Growth and Development of Fracture Zones in Plain Concrete and Similar Materails, *Report TVBM 1006*, University of Lund, Sweden.

Planas, J. and Elices, M. (1987), Asymptotic Analysis of the Development of a Cohesive Crack Zone in Mode I Loading for Arbitrary Softening Curves, *Proceedings SEM-RILEM International Conference*, Houston, Eds: S. P. Shah and S. E. Swartz, pp. 384.

RILEM FMC-50 (1985), Determination of the fracture energy of mortar and concrete by means of three-point bending tests on notched beams, *RILEM, Matériux et Constructions*, **18**, pp. 287-290.

Rossi, P, Brühwiler, E, and Chhuy, S., (1990) Fracture mechanics of concrete: Test methods, *Sub-committee B, RILEM TC89-FMT*, Compact Specimen Testing, p. 33. To be published by Chapman & Hall, Oct. 1990.

Srawley, J. E., (1976) Wide-range stress intensity factor expressions for ASTM E399 standard fracture toughness specimens, *Int. J. of Fracture*, **12**, pp. 475-476.

Report on Session 2

FRACTURE IN CONCRETE AND ROCK

J. MAZARS
ENS de Cachan/C.N.R.S./Université Paris 6
Laboratoire de Mécanique et Technologie
61, avenue du Président Wilson
94235 CACHAN Cedex (France)

ABSTRACT. Two main aspects were treated in this session. The first one is about   concrete and rock fracture and concerns the relationships between microphenomena and macroscopic behavior as well for the experimental analysis as for the theoretical aspects. The second one is   related to the experimental procedure to determine toughness properties of concrete. The discussion included both a debate on same specific aspects and complementary contributions on undeveloped points   during presentations.

Introduction

This report concerns the 3 papers of Session 2. The papers on "Microcracking and Damage in concrete" by D. François, "Cracking Damage and Fracture in stressed rock : a holistic approach" by P.G. Meredith and "Test methods for determining mode I Fracture Toughness of concrete" by B.L. Karihaloo and P. Nallathambi. As the title shows, we can consider that there were two main points aborded in this paper : one concerns the experimental point of view and theoretical aspects of the relationships between the micromechanisms of damage and microcracking and their consequences in terms of macroscopic behavior. The other point treated by the third paper, focuses on the experimental procedure to determine toughness properties of concrete. Thus, the discussions on this session can be related with the same schedule. These discussions pointed out both the difficulties to realize and analyze precise and fiable experiments and also the importance to use the good tools to perform good predictive calculations. For these two reasons they were a useful contribution to this topic in this workshop.

Microphenomenon-Macroconsequences

Meredith presented in his paper an integrated discussion on the fracture and failure of rocks that are different from other quasi brittle material, due to the presence of pores fluids (the properties of which are different than those in concrete) which introduce strong time dependent properties and scaling relations. Nevertheless the trend for mode I Fracture is the same than others materials, but it seems that for rocks it is the maximum   grain size, that is the controlling influence. Like in concrete, rocks under compressive loading exhibits strain softening and volume expansion due to nucleation propagation and interaction of a distibuted array of microcracks. These phenomena are pressure and time dependent. To keep out the particularity of this process, different experimental analysis were presented : the Elastic waves velocity changes with   the state of damage, acoustic

S. P. Shah (ed.), Toughening Mechanisms in Quasi-Brittle Materials, 125–127.
© 1991 *Kluwer Academic Publishers.*

emission which can provide informations about the rate of damage evolution. Measurement of changes in transport properties, such as fluid permeability or electrolytic conductivity, provides information about the interaction and linkage of dilatant microcracks leading to macroscopic fracture and failure.

Some precisions were given during discussions in particular on the evolution of acoustic emission during a test on a sandstone sample under confining pressure which failed by localized shear faulting after a dynamic stress drop. Before the peak the A.E. rate increases rapidly with the onset of dilatancy due to the growth of new microcracks. In the post peak, strain softening evolutions are dominated by the interaction and coalescence of these microcracks to form a throughgoing fault. Around the peak strange phenomenon of A.E. falling down to an apparent quiescence, is in fact caused by saturation of the monitoring system caused by a cascade of events due to dynamic failure. The analogy of such an instability with earthquake after shock sequence, followed by the A.E. rate recover before decaying, is very interesting. In this field and in spite of the scaling effect some specific characteristics from geological structures to rock laboratory sample are scale invariant, it is the case of shear fault systems with one dominant throughgoing fault, over a range of scale from fractions of a mm to hundreds of km, see figure 13 of the Meredith's paper which superposes the size distribution of four fault systems, with each normalized to the length of the dominant throughgoing fault. In the same part of the discussion a question was on the sensivity of acoustic techniques to observe microruptures ; the answer of the speaker was that generally the sensivity corresponds more or less to the grain size, in fact it depends on the energy dissipated during the process and on the damping of the medium, and in some case, limestone for example, the results of A.E. are very insatisfactory.

François, in his presentation discussed cracking of brittle medium at the microscale considering that microdefects can be considered as microcracks, the propagation of which respects crack propagation criteria involving the two modes of fracture, I and II. In the discussion, it was pointed out that mode II can't exist alone at the local scale and could be considered only at the global scale. Perhaps at the begining of an evolution or in the case of fiber reinforced composites which in certain cases fibers stops mode I and forces mode II. François suggested that in such a case, of special local phenomena, it would be interesting to introduce a characteristic length, linked to the size of heterogeneities, and then to define the main phenomena in a global manner at this level.

Modelling at the macrolevel

François presented some recent results obtained mainly in France, using damage mechanics. The goal of which is to represent with a homogeneous continuum the effects of discontinuous phenomena which take place in a heterogeneous multiphasic media, including opening and closure of cracks under multiaxial states, fatigue, failure, ...

Kachanov began the debate considering that the dependence stiffness-damage is not realistic because in certain specific cases (as it was be shown in his presentation session 6) an increase of damage, such as an array of microcracks, does not act on the stiffness and even can increase the strength of the media. In the same part of the discussion Krajcinovic expressed his disagreament with the partition $\sigma = \sigma^+ + \sigma^-$, used to separate tensile and compression effects, and shows a contreversial example where superposition leads to instability. The debate on these two points was intense and François concluded saying that even if in a few specific cases results are unrealistics, internal variables such as damage are very useful to represent complex and multiple phenomena.

About the fatigue model presented, some questions were asked to François, among this the following asked by Shah : What about the effects of the frequencies and the ratio between upper and lower value of the load on the fatigue process ? The author indicates that these points are implicitely included in the model due to the introduction of the characteristics strain response of the material in the steady part of the process, but the change of the characteristics of the loading during the process is not yet introduced.

Experimental procedure to determine K_{IC}

After an exhaustive presentation by Karihaloo of the work done by the RILEM committee 89 FMCT, the discussion was intense and focused first on the experimental procedures.

The effects of friction on the wedge splitting test are important and therefore acted on the response of the specimen, increasing the ultimate load and the size of the loop in the unloading-reloading curves, for these reasons , is it really a good test ?

Rossi, who was in charge of this part in the committee, says that the tests were performed with precautions and the analysis, to establish the compliance function, took into account the real loading conditions using F.E.M. code.

A debate on the best way to measure compliance increases the exchange of points of view. Is the best way on the unloading curve ? Certainly not, the closure of cracks being difficult at reloading is preferable and it was choosen a secant value.

An other interesting question was : what means objective ?
Rossi answered this question and said K_{IC} is the constant value obtained with a large specimen in a steady state of the evolution of the process zone. The results obtained at L.C.P.C. on DCB specimen larger than 2 m exhibit a quasi constant value which is the objective K_{IC} value. When specimens are smaller this objective value cannot be reached directly and must be deduced by a specific law including the size of the specimen. Therefore the size effect is an important aspect in the determination of the fracture toughness parameters. This notion is explicitly introduced, in the case of the size effect law proposed by Bazant and implicitly in the case of other models (two parameter model and effective crack model).

Conclusions

This session was definitely interesting and this for two principal reasons : One is the extent of the subjects considered from micro to macro analysis and for concrete and rocks. The other is that the discussion pointed out about the difficulties of the experimental analysis and the necessity to obtain characteristics values of fracture toughness properties whatever are the size and the type of the specimens used. In fact, it is the only way to obtain toughness properties useful in the design.

Theoretical Fracture
Mechanics Considerations

RATE EFFECT, SIZE EFFECT AND NONLOCAL CONCEPTS FOR FRACTURE OF CONCRETE AND OTHER QUASI-BRITTLE MATERIALS

Zdeněk P. Bažant
Walter P. Murphy Professor of Civil Engineering
Northwestern University, Evanston, Illinois 60208, USA

ABSTRACT. The lecture reviews several recent results achieved at Northwestern University in the problem of size effects and nonlocal concepts for concrete and other brittle hetero-geneous materials, and presents a new method for calculating the load-deflection curves of fracture specimens or structures with time-dependent crack growth and viscoelastic material behavior. The results reviewed deal with the size effect law in fracture and its exploitation for determining material fracture characteristics, statistical generalization of the size effect law with a nonlinear reformulation of Weibull's weakest-link theory, determination of the size dependence of the fracture energy determined by work-of-fracture method, nonlocal models for smeared cracking and damage, microstructural determination of the nonlocal material properties and fracture process zone behavior, size effect in fatigue fracture of concrete, and use of the size effect for determining the fracture properties of high-strength concrete.

1. Introduction

Fracture analysis of concrete structures has to deal with two important complicating characteristics: the distributive nature of cracking and damage in concrete, which causes the fracture process zone to be relatively large and engenders a size effect, and time dependence of both the crack growth and the material behavior. The nonlinear behavior caused by the existence of a large fracture process zone has been in the focus of attention for some time and its treatment is becoming quite well understood [1-3, etc.]. Attention to the size effect is more recent [8] but it has already led to some useful extensions of fracture theory and a new method for determining material fracture properties [4-7, 9-12]. The existence of the rate effect has been known for a long time and has been studied extensively with regard to dynamic fracture. However, the nonlinear fracture aspects of the rate effect, which are manifested in interaction with the size effect, have not received attention until recently, although they are no doubt very important for predicting the response of structures.

The present lecture intends: (1) to present a new effective and relatively simple method for calculating the load-deflection response of a structure with a large fracture process zone, time-dependent fracture growth, and viscoelastic material properties; and (2) to review several recent results achieved at Northwestern University. No claims for exhaustive or even balanced coverage of the latest developments are made. Due to exploding research

S. P. Shah (ed.), Toughening Mechanisms in Quasi-Brittle Materials, 131–153.

activity, this would be beyond the scope of the present paper.

2. Review of Some Recent Results

2.1 A REVIEW OF SIZE-EFFECT - THE SALIENT CHARACTERISTIC OF FRACTURE

The structural size-effect is the most important characteristic and the easiest measurable consequence of nonlinear fracture behavior. Considerable attention has been devoted at Northwestern University to the phenomenologic description of the size effect [8-12] as well as its physical mechanism. We will begin by a brief overview of the latest phenomenologic characterization of fracture in terms of the size effect and the consequences for the measurement of fracture properties.

The size effect may be defined in terms of the nominal strength $\sigma_N = c_n P_u/bd$ (for 2D) or $c_n P_u/d^2$ (for 3D) in which P_u = maximum load of geometrically similar specimens of size d (dimension) and, in case of two dimensions, thickness b; c_n = factor chosen for convenience. Under the assumption that there is a large fracture process zone that is not negligible compared to d, and that the crack at failure of geometrically similar structures of different sizes is also geometrically similar, the nominal strength approximately obeys the size effect law [8]:

$$\sigma_N = B f_u (1 + \beta)^{-1/2}, \qquad \beta = d/d_0 \tag{1}$$

in which f_u = any measure of material strength, e.g., the tensile strength, and B, d_0 = empirical constants. This law describes a smooth transition from plastic limit analysis, for which there is no size effect (σ_N = constant) to linear elastic fracture mechanics (LEFM), for which the size effect is the maximum possible, given by $\sigma_N \propto \beta^{-1/2}$. The plot of Eq.1 is shown in Fig.1. The horizontal asymptote represents the limiting case of plastic limit analysis, and the inclined asymptote of slope - 1/2 the limiting case of LEFM. Parameter d_0, called the transitional size, corresponds to the intersection of these two asymptotes. Eq.1 has originally been derived by dimensional analysis and similitude arguments, based on the hypothesis that the energy release due to fracture depends not only on the fracture length but also on a second length characteristic that is approximately a material property and characterizes either the effective length or the effective width of the fracture process zone, or the nonlocal properties of an equivalent continuum.

Under certain further simplifying assumptions based on equivalent LEFM, it has been shown [11] that the size effect law from Eq.1 can also be written in the form

$$\sigma_N = c_n \left(\frac{E\, G_f}{g'(\alpha_0)c_f + g(\alpha_0)d} \right)^{1/2} \tag{2}$$

in which G_f = fracture energy of the material, defined as the energy required for crack propagation in an infinitely large specimen, E = Young's elastic modulus, c_f = effective length of the fracture process zone in an infinitely large specimen (a material constant), α_0 = a_0/d, a_0 = length of initial notch of crack, and $g(\alpha)$ = non-dimensionalized energy release rate of the specimen of the given geometry, which is obtained by writing the LEFM solution for the energy release rate in the form $G = P^2 g(\alpha)/E\,b^2\,d$ where P is the applied load or reaction, $\alpha = a/d$, $a = a_0 + c$, c = crack extension from the notch or initial crack tip. These formulas are valid for plain stress. For plain strain or axisymetric propagation, E must be replaced by $E/(1 - \nu^2)$. Defining the so-called intrinsic strength $\tau_N = (P_u/bd)g'(\alpha_0)$ and intrinsic size $\bar{d} = g(\alpha_0)d/g'(\alpha_0)$, one can rewrite Eq.2 in the form

$$\tau_N = \left(\frac{E\,G_f}{c_f + \bar{d}}\right)^{1/2} \tag{3}$$

which involves only material constants EG_f and c_f. Comparing Eq.2 or 3 with Eq.1, one gets the following expressions for material fracture constants

$$G_f = \frac{B^2 f_u^2}{c_n^2 E} d_0 g(\alpha_0), \quad c_f = d_0 \frac{g(\alpha_0)}{g'(\alpha_0)} \tag{4}$$

Thus, after measuring the maximum loads for geometrically similar specimens of sufficiently different sizes, one can determine B and d_0 by least-square fitting all the data (Eq.1 can be rearranged to a linear regression plot), and then evaluate the fracture energy and the effective process zone length from Eq.4 (strictly on the basis of maximum load data). This method is probably the easiest to implement in the laboratory (even a soft testing machine is adequate and no measurements of displacements or crack lengths are required). The method has been verified by numerous tests on concrete and rock. The results are, by definition, size independent and they were also proven to be approximately shape independent, since very different fracture specimen geometries furnished approximately the same results, as expected theoretically.

The ratio β, which may be calculated by one of the following two expressions,

$$\beta = \frac{g(\alpha_0)}{g'(\alpha_0)}\frac{d}{c_f}, \quad \beta = \frac{B^2 g(\alpha_0)}{c_n^2}\frac{d\,f_u^2}{E\,G_f} \tag{5}$$

is called the brittleness number. For $\beta \to 0$, plasticity applies, and for $\beta \to \infty$, LEFM applies. For $\beta < 0.1$ it is possible to use plasticity as an approximation, and for $\beta > 10$ it is possible to use LEFM. For the intermediate β-values, nonlinear fracture mechanics must be used. However, if the transitional size d_0 is determined, an approximate prediction of maximum load can be obtained by interpolating between the solutions of plasticity and LEFM according to Eq.1. This should be useful for design; proposals to modify the existing design formulas for diagonal shear failure of beams (without or with stirrups, unprestressed or prestressed), punching shear failure of slabs, torsional failure of beams, pullout failures of bars and of studded anchors have been made and verified by extensive tests [14-22].

Based on size effect measurements, other basic nonlinear fracture characteristics can also be obtained. The critical crack-tip opening displacement may be determined as

$$\delta_c = \frac{8\,K_{If}}{E}\sqrt{\frac{c_f}{2\pi}}, \quad \text{with} \quad K_{If} = \sqrt{E\,G_f} \tag{6}$$

Furthermore, the R-curve (resistance curve) for the specimen or structure can be calculated as

$$R(c) = G_f \frac{g'(\alpha_1)}{g'(\alpha_0)} \frac{c}{c_f}, \quad \text{with} \quad \frac{c}{c_f} = \frac{g'(\alpha_0)}{g(\alpha_0)} \left(\frac{g(\alpha_1)}{g'(\alpha_1)} - \alpha_1 + \alpha_0 \right) \tag{7}$$

in which α_1 is a dummy parameter representing the relative crack length for the structure size for which $R(c)$ corresponds to the maximum load (Fig.2). Choosing various values of α_1, the values of R (critical G-value required for further crack growth) and c can be calculated from Eq.7, and thus the $R(c)$-curve defined parametrically. This curve is by definition size independent but depends on the geometry of the structure. For very different geometries, very different R-curves can be obtained. Eq.7 defines the master R-curve for an infinitely large specimen. For a specimen of finite size, the R-curve given by Eq.7 is followed only up to the maximum load P_u, and after that the actual R-curve is constant (horizontal), with the R-value equal to that attained at the peak load [23, 24]. The reason is that for prepeak loading the fracture process zone grows in size while remaining attached to the crack or notch tip (provided structures with $g'(\alpha_0) > 0$ are considered), whereas in post-peak softening the fracture process zone gets detached from the notch tip and travels ahead retaining approximately a constant size.

Using the equivalent LEFM approach, the curve of load or reaction P versus the load-point displacement u may be calculated from the equations

$$u = P \left(C_0 + \frac{2}{bE} \int_0^\alpha g(\alpha')d\alpha' \right), \quad P = b\sqrt{\frac{Ed}{g(\alpha)} R(c)} \tag{8}$$

in which C_0 is the compliance for a specimen without any crack. Choosing various values of c, with $\alpha = (a_0 + c)/d$, the values of P and u can be evaluated from Eq.8, defining the load-deflection curve parametrically.

Eq.8 provided a strong verification of the size effect method of determining fracture properties. The material fracture parameters were determined solely from the maximum loads measured on geometrically similar rock fracture specimens of very different sizes [23]; then the R-curve was calculated from Eq.7, and from that the load-deflection diagram shown in Fig.3 from Eq.8 was computed. The results showed excellent agreement with the measured load deflection curve (Fig.3). Similar agreement has been obtained for concrete [13].

2.2. STATISTICAL GENERALIZATION AND WEIBULL'S EFFECT

The fact that Eq.1 or Eq.3 can be algebraically rearranged to a linear regression plot of $Y = \tau_N^{-2}$ versus $X = d$ makes it possible to obtain easily the statistics of the material fracture parameters. The coefficients of variation of fracture toughness (defined for a specimen of infinite size), the effective length of the fracture process zone, and the fracture energy may be approximately obtained as

$$\omega_{K_{If}} = \frac{1}{2}\omega_A, \quad \omega_{c_f} = (\omega_A^2 = \omega_c^2)^{1/2}, \quad \omega_{G_f} = (4\omega_{K_{If}}^2 + \omega_E^2)^{1/2} \tag{9}$$

in which ω_A and ω_C are the coefficients of variation of the aforementioned slope and of the

Y-intercept of the linear regression plot, and ω_E is the coefficient of variation of the elastic modulus of concrete [11].

Eq.9 takes care only of the uncertainty of the material parameter values in the foregoing deterministic model. More realistically, one should note that the failure process in itself is stochastic, and the simplest vehicle to take that into account is Weibull's reasoning. However, the classical Weibull-type formulations do not apply to concrete structures because they exhibit stable growth of cracking with significant stress redistributions prior to maximum load. Good results, however, can be obtained with a nonlocal generalization of Weibull approach [25], in which the survival probability of the structure is calculated as the joint probability of survival of all the material elements based on the stress distribution just prior to failure, in which the material failure probability is determined from Weibull distribution using the nonlocal stress average, $\bar{\sigma}_i$;

$$- \ln(1 - P_f) = \int_V \sum_{i=1}^{n} \left(\frac{\bar{\sigma}_i}{\sigma_0}\right)^m \frac{dV}{V_0}, \quad \bar{\sigma}_i(x) = \int_V \alpha(s - x)\sigma_i(s)dV(s) \qquad (10)$$

in which P_f = failure probability of the structure, σ_i = principal stresses ($i = 1, 2, 3$), V = volume of the structure, V_0 = volume of a small representative volume of the material, x, s = coordinate vectors, $\alpha(s - x)$ = given weighting function of a nonlocal material model (based on characteristic length ℓ); and m, σ_0 = Weibull modulus and scale parameter determined by fitting Weibull distribution to direct tensile test data (assuming a zero Weibull threshold). It has recently been found [25] that the nonlocal Weibull concept leads, under certain approximations, to the following generalization of the size effect law (Fig.4).

$$\bar{\sigma}_N = B f_u (\beta^{2n/m} + \beta)^{-1/2} \qquad (11)$$

in which the overbar denotes the mean nominal strength, $\bar{\sigma}_N$; and $n = 2$ or 3 for two- or three-dimensional similarity. For concrete, typically $m = 12$. For large structure sizes, Eq.11 aproaches LEFM, same as Eq.1. For small structure sizes, $\beta \to 0$, Eq.11 asyptotically approaches the classical Weibull size effect, $\bar{\sigma}_N = \beta^{-n/m}$, which gives a rather weak size effect, $\bar{\sigma}_N = \beta^{-1/6}$ for two-dimensional similarity. Thus, Eq.11 represents a smooth transition from the classical Weibull size effect to LEFM. Eq.11 has been shown to agree with the data for concrete somewhat better than Eq.1, but the difference is rather small except when dealing with very small structure sizes. The formulation in Eq.2-7 can be generalized in accordance with Eq.11.

2.3. SIZE DEPENDENCE OF FRACTURE ENERGY OBTAINED BY CURRENT RILEM METHOD

The fact that the size effect method based on the maximum load yields excellent predictions of the load-deflection curves, in good agreement with measurements, makes it possible to exploit this formulation for examining the fracture energy determined from the area under the load-deflection curve, which represents the work-of-fracture method proposed for ceramics by Nakayama [26], and by Tattersall and Tappin [27], and introduced for concrete by [28,29]. The work of fracture has been calculated for concrete specimens on the basis of the load-deflection curve obtained from the R-curve (Eqs.7 and 8), keeping the R-value constant for the post-peak softening (area in Fig.2c). This calculation indicates a size effect on the value of the fracture energy G_f^R, due to the fact that the peak load occurs at different points of the R-curve for specimens of different sizes. The calculation results are shown in Fig.2d; note that the size dependence of G_f^R is quite strong, in fact stronger than that of the R-curve, although not as strong as that of the apparent fracture energy G_c determined by LEFM method. This agrees with the conclusions of Planas and Elices [4], who showed that the fracture energy measurements according to the RILEM standard, which is based on the work-of- fracture method, must be extrapolated to a specimen of infinite size in order to obtain consistent (size independent) results.

2.4. NONLOCAL DAMAGE MODELS

In finite element analysis of damage and cracking in concrete structures, the size effect has long been neglected. Unfortunately, most of the existing models are based on plasticity or its modifications and exhibit no size effect, which is unacceptable for concrete structures. Modeling of the size effect should be accepted as the basic criterion for correctness of a finite element code. The only way to achieve a correct size effect in agreement with Eq.1 is to either use some type of a nonlinear fracture model for a line crack with cohesive crack-bridging zone, or a nonlocal form of a finite element code for distributed damage of smeared cracking. The latter approach is more versatile and perhaps somewhat more realistic due to the diffuse nature of cracking in reinforced concrete structures. A nonlocal generalization of the classical smeared cracking formulation has been introduced in [30], and a good agreement with size effect data and with Eq.1 has been demonstrated. A more realistic constitutive flaw for the evolution of damage or cracking in the fracture process zone is the microplane model, in which the material properties are characterized separately on planes of various orientation in the material. This model has recently been generalized to a nonlocal form, and it was again demonstrated that such a generalization agrees well with size effect fracture data as well as Eq.1 (Fig.5); see [31].

2.5. MICROMECHANICS MODELING

It is very difficult to identify the strain-softening constitutive relations for the fracture process zone on the basis of measurements alone. Therefore, micromechanics modeling could be of great help. Micromechanics models need to represent systems of microcracks that are observed experimentally. Therefore, initial studies of micromechanics of fracture of concrete concentrated on the analysis of an array of cracks in a homogeneous elastic matrix. Some observed features could be reproduced with such models, particularly the strain-softening behavior. This was, for example, demonstrated for an array of parallel microcracks spaced on a cubic lattice and subjected to a microscopic uniaxial stress field. Application of the homogenization conditions to such a crack array also showed that the corresponding macroscopic smoothing continuum is nonlocal, and of the nonlocal damage type. Stability analysis of the interacting crack systems, however, indicated that such a model is unrealistic because only one of the cracks can grow in a stable manner, which is of course not what is seen in experiments. The reason for this discrepancy no doubt consists in the micro-inhomogeneity of the material, especially the presence of harder inclusions.

Interaction of cracks and inclusions in an elastic matrix has been studied in [32], using a Green's function approach (Fig.6). Approximate solutions have been obtained for a crack interacting with many inclusions, for various geometric configurations. The solution was used to obtain an apparent R-curve of a microcrack in a smoothed homogeneous matrix such that its growth is the same as the growth of the actual crack interacting with inclusions. It was found that in many situations the apparent R-curve is rising, which has a stabilizing effect on the system of cracks. The apparent rising R-curve can stabilize a system of many cracks, such that many cracks can grow simultaneously, in agreement with observations.

As a conclusion from this study, it appears that a study of crack arrays in a homogenous continuum is in general insufficient, and the presence of inhomogeneities representing the aggregate pieces must be considered simultaneously in the analysis. It should be also noted that this result is similar to that of Gao and Rice [33], who used perturbation method; however, they considered only the case when the elastic moduli of matrix and inclusions differ very little. A special problem of this type has also been solved by Mori et al. [34].

2.6. SIZE EFFECT CORRECTION TO PARIS LAW FOR FATIGUE FRACTURE

Under repeated loading, cracks tend to grow, which is described by the well-known Paris law [35, 36]. Applicability of this law to fatigue crack growth in concrete has been verified by Swartz et al. [37]. Since Paris law describes the crack growth as a function of the amplitude of the stress intensity factor, a question arises with respect to the size effect. In monotonic loading, the stress intensity factor does not provide sufficient characterization of fracture when different sizes are considered, as is known from the previously discussed size effect law. The same phenomenon must be expected for cyclic fracture, especially since fracture under monotonic loading can be regarded as a limiting case of fracture under cyclic loading. Recent fatigue fracture experiments on notched concrete beams at Northwestern University [38] have shown that the fatigue crack growth in geometrically similar specimens of different sizes can be described by the following law:

$$\frac{\Delta a}{\Delta N} = C \left(\frac{\Delta K_I}{K_{I_c}} \right)^n, \qquad K_{I_c} = K_{I_f} \left(\frac{\beta}{1 + \beta} \right)^{1/2} \tag{12}$$

in which K_{I_f} = fracture toughness for an infinitely large specimen, ΔK_I = amplitude of the stress intensity factor, K_{I_c} = apparent fracture toughness derived from Eq. 1; $\Delta a / \Delta N$ = crack length extension per cycle; and C, n = constants. For $\beta \rightarrow \infty$, this equation reduces to the well-known Paris law. For normal size concrete specimens, however, the deviations from Paris law are quite significant. This is revealed by the experimental results in Fig.7 for three different sizes in the ratio 1:2:4. In this plot, the Paris law gives one inclined straight line of slope n for all sizes, but it is seen from Fig.7 that for each size the test results allign on different straight line for each size. The three solid straight lines represent Eq.12.

2.7. FRACTURE OF HIGH STRENGTH CONCRETE

It has already been well established that high strength concrete is more brittle than normal strength concrete. This question has been investigated at Northwestern University using the size effect method of determining material fracture properties [13]. Concrete of 28-day standard compression strength 12,000 psi, typical for high-rise construction in the Chicago area, has been used.

The results are summarized in Fig.8, which shows the relative values of various material properties compared to the normal strength concrete, particularly the compression strength f_c', modulus of rupture f_r, Young's modulus E, fracture toughness K_{I_c} and fracture energy G_f (both for an infinitely large specimen), effective length of the fracture process zone c_f, and Irwin's characteristic size of the nonlinear zone ℓ_0. Whereas the compression strength is 2.6-times higher than that of normal-strength concrete, the fracture toughness is increased only by about 25%, fracture energy by about 15%, and the effective lengths of the fracture process zone is decreased 2.5 times and the characteristic size of the nonlinear zone is *decreased* approximately 5-times.

Consequently, the brittleness number of the high strength concrete structure is approximately 2.5-times *higher* than the brittleness of an identical structure made of normal-strength concrete. This aspect of high strength concrete is unfavorable for design and requires special attention.

3. Effect of Rate of Loading and Creep

Fracture of rocks as well as ceramics is known to exhibit a significant sensitivity to the rate of loading. For concrete, the influence of the rate of loading on fracture propagation is even more pronounced and is further compounded by viscoelasticity of the material in the entire structure. To calculate the response of a structure, as well as to be able to evaluate laboratory measurements, the most important is the determination of the load or reaction P as a function of the load-point displacement u and time t for a prescribed loading regime. The following simple method has been formulated for this purpose.

We begin by rewriting Eq.8 for a structure with rate-independent fracture as follows

$$u = \frac{P}{E}\bar{C}(a), \quad \bar{C}(a) = \bar{C}_0 + \frac{2}{b}\phi(\alpha), \quad \phi(\alpha) = \int_0^\alpha [k(\alpha')]^2 d\alpha' \tag{13}$$

in which $\bar{C}(a)$ is the secant compliance of the structure at crack length a calculated for a unit value of Young's modulus ($E' = 1$) and the actual Poisson ratio ν; $\bar{C}_0$ is the elastic compliance of the same structure with $E' = 1$ with no crack; $\alpha = a/d$; $[k(\alpha)]^2 = b^2 g(\alpha)$; function $k(\alpha)$ is defined by writing the known solution of the stress intensity factor in the form $K_I = P\, k(\alpha)/b\sqrt{d}$. Eq. 13 is valid for plane stress conditions.

For plane strain conditions, E needs to be replaced by $E/(1 - \nu^2)$; this replacement needs to be carried out in all the subsequent analysis.

Outside the fracture processs zone, concrete behaves as a linearly viscoelastic (aging) material described, for uniaxial stress, by the stress-strain relation

$$\varepsilon(t) = \int_{t_0}^t J(t,t')d\sigma(t') \quad \text{(Stieltjes integral)} \tag{14}$$

provided that there is no shrinkage and thermal expansion or that they have a negligible effect; $\sigma, \varepsilon =$ uniaxial stress and strain, $J(t,t') =$ given compliance function of the material that characterizes creep, representing strain at age t caused by a unit uniaxial stress applied at age t'. Eq. 14 may be written in an operator form as $\varepsilon(t) = \underline{E}^{-1}\sigma(t)$ where $\underline{E}^{-1}$ is the creep operator defined by Eq. 14.

The load-displacement relation for a structure exhibiting creep may be obtained from the corresponding elastic relation by replacing $1/E$ with the corresponding creep operation $\underline{E}^{-1}$. Doing this in Eq. 13, one gets

$$u(t) = \int_{t_0}^t J(t,t')d\{P(t')\bar{C}[a(t')]\} \tag{15}$$

The solution needs to be carried out numerically. To this end, time t is subdivided by discrete times $t_r(r = 0, 1, 2, 3 \ldots)$ into time steps $\Delta t_r = t_r - t_{r-1}$. Time t_0 represents the age at the first loading. Using the trapezoidal rule (the error of which is proportional to Δt^2), we may approximate Eq. 15 as

$$u_r = \sum_{s=1}^r J_{r,s-\frac{1}{2}}(P_s\bar{C}_s - P_{s-1}\bar{C}_{s-1}) \tag{16}$$

where subscript r refers to time t_r and $s - \frac{1}{2}$ refers to time $t_s - \Delta t_s/2$; $P_s = P(t_s); \bar{C}_s = \bar{C}[a(t_s)]; J_{r,s-\frac{1}{2}} = J(t_r, t_{s-\frac{1}{2}})$; and the initial load value is $P(t_0) = P_0 = 0$. Writing Eq. 16 for t_{r-1} instead of t_r, i.e.,

$$u_{r-1} = \sum_{s=1}^{r-1} J_{r-1,s-\frac{1}{2}}(P_s\bar{C}_s - P_{s-1}\bar{C}_{s-1}) \tag{17}$$

and subtracting this from Eq. 16 one gets

$$\Delta u_r = \frac{1}{E_r''}(P_r\bar{C}_r - P_{r-1}\bar{C}_{r-1}) + \Delta u_r'' \tag{18}$$

in which $\Delta u_r = u_r - u_{r-1}$, $1/E_r'' = J_{r,r-\frac{1}{2}}$ and

$$\Delta u_r'' = \sum_{s=1}^{r-1}(J_{r,s-\frac{1}{2}} - J_{r-1,s-\frac{1}{2}})(P_s\bar{C}_s - P_{s-1}\bar{C}_{s-1}) \quad \text{for } r > 2 \tag{19}$$

(Bažant, ed., 1988, p. 116). The accuracy, however, is somewhat improved if, instead of $J_{r,r-\frac{1}{2}}$, the effective modulus approximation

$$1/E_r'' = J_{r,r-1} = J(t_r, t_{r-1}) \tag{20}$$

is used for the last step. Eq. 19 is applicable only for $r \geq 3$. For the first two steps, the approximations of Eq. 15 are

$$\text{for} \quad r = 1: \quad \Delta u_1 = u_1 = J_{1,0} P_1 \bar{C}_1 \tag{21}$$
$$\text{for} \quad r = 2: \quad \Delta u_2 = J_{2,1}(P_2 \bar{C}_2 - P_1 \bar{C}_1) + (J_{2,\frac{1}{2}} - J_{1,\frac{1}{2}}) P_1 \bar{C}_1$$

The foregoing analysis must be supplemented by a law for the growth of crack length $a(t)$. Materials that under a constant very fast loading rate follow linear elastic fracture mechanics exhibit, at slower rates, crack growth that approximately obeys the law

$$\dot{a} = \kappa_c \left(\frac{K_I(P)}{K_{If}} \right)^n \exp\left[-\frac{U}{R_0} \left(\frac{1}{T} - \frac{1}{T_0} \right) \right] \tag{22}$$

in which K_I = stress intensity factor, U = activation energy of crack growth, R_0 = gas constant, T = absolute temperature, T_0 = reference temperature, and κ_c, n = empirical material constants. The applicability of this well-known relation to concrete has been verified in [39]. However, to take into account nonlinear fracture properties and obtain the correct transitional size effect (agreeing with the size effect law), Eq. 22 must be generalized as:

$$\dot{a} = \kappa_c \left(\frac{K_I(P)}{K_I^R(c)} \right)^n \exp\left[-\frac{U}{R_0} \left(\frac{1}{T} - \frac{1}{T_0} \right) \right] = f(P, a), \quad \text{with } K_I(P) = \frac{Pk(\alpha)}{b\sqrt{d}} \tag{23}$$

where f is a function of P and a, as defined by this equation, and $K_I^R(c)$ is the given R-curve (determined in advance for the given structure geometry). For the time step t_{r-1}, t_r), Eq. 23 yields

$$\Delta a_r = f(P_{r-\frac{1}{2}}, a_{r-\frac{1}{2}}) \Delta t_r \tag{24}$$

The following algorithm may now be used in every time step Δt_r (for $r > 2$), in which the previous values $a_0, a_1, \ldots a_{r-1}; P_0, \ldots P_{r-1}, u_0, \ldots u_{r-1}$ are already known.

1. Setting $P_{r-\frac{1}{2}} = P_{r-1}$ and $a_{r-\frac{1}{2}} = a_{r-1}$, calculate $\Delta a_r = f(P_{r-a}, a_{r-1}) \Delta t_r$ (Eq. 24) as the first estimate, and set $a_{r-\frac{1}{2}} = a_{r-1} + \frac{1}{2} \Delta a_r, a_r = a_{r-1} + \Delta a_r$. Evaluate $\bar{C}_r = \bar{C}(a_r)$ from Eq. 13.

2. Loop on iterations.

3. Calculate $\Delta u_r''$ from Eq. 19. Then, using prescribed Δu_r (or prescribed P_r), calculate P_r (or Δu_r) from Eq. 18. Then calculate Δa_r from Eq. 24 and obtain updated values $a_{r-\frac{1}{2}} = a_{r-1} + \frac{1}{2} \Delta a_r$ and $a_r = a_{r-1} + \Delta a_r$. Evaluate updated $\bar{C}_r = \bar{C}(a_r)$ from Eq. 13.

4. Check the given tolerance citerion, requiring that the absolute value of the change of P_r in the last iteration be less than $|eP_r|$ where e is a given small number (e.g., $e = 10^{-6}$). If violated, go to 2 and start the next iteration. If satisfied, go to 1 and start the first iteration of the next time step Δt_{r+1}.

The limiting case of time-dependent propagation law with elastic material behavior is obtained from the preceding algorithm if one sets $J(t, t') = 1/E$ for $t \geq t'$.

An interesting question is how to obtain the solution when the crack propagates according to the classical time-independent law of fracture mechanics (with R-curve) while the material creeps. This situation is approached when $\dot{a} \to 0$ for $K_I < K_I^R(c)$, and $\dot{a} \to \infty$ for $K_I > K_I^R(c)$, i.e., when for $K_I < K_I^R(c)$ the rate $\dot{a}$ is almost 0 and for $K_I > K_I^R(c)$ the rate $\dot{a}$ is extremely large. Such behavior is obtained from Eq. 23 when $n \to \infty$. The foregoing algorithm, however, cannot be expected to converge well for extremely large n values, and it is preferable to obtain the limiting case of time-independent crack propagation law directly. In that case, for a propagating crack we simply have the condition $K_I = K_I^R(c)$, which replaces Eqs. 22-24 while Eqs. 13-21 remain applicable. The values of $\bar{C}_r$ and $\bar{C}_s$ in Eqs. 18-21 must now be such that the relation $P_r = b\sqrt{d}\, K_I^R(c_r)/k(\alpha_r)$ be always satisfied, when $c_r = a_r - a_0$, $\alpha_r = a_r/d$. From this relation, the load increment is

$$\Delta P_r = b\sqrt{d} \left(\frac{K_I^R(c_{r-1} + \Delta a_r)}{k\left(\alpha_{r-1} + \frac{\Delta a_r}{d}\right)} - \frac{K_I^R(c_{r-1})}{k(\alpha_{r-1})} \right) \tag{25}$$

At the same time, Eq. 18 may be rewritten in the form

$$\Delta u_r = \frac{1}{E_r''} \left[(P_{r-1} + \Delta P_r)\bar{C}(a_{r-1} + \Delta a_r) - P_{r-1}\bar{C}_{r-1} \right] + \Delta u_r'' \tag{26}$$

Before stargin the solution of time step Δt_r, the values of $E_r'', \Delta u_r'', P_{r-1}, a_{r-1}, c_{r-1}, \alpha_{r-1}$ are known, and Eqs. 25-26 contain two unknowns, Δa_r and either ΔP_r or Δu_r. If ΔP_r is prescribed, then Δa_r may be solved first from Eq. 25 (e.g., by Newton iterations) and then Δu_r from Eq. 26. If Δu_r is prescribed, then Eqs. 25 and 26 represent two simultaneous nonlinear equations for Δa_r and ΔP_r, and they may again be solved iteratively.

The most important simplification in the preceding solution is the approximation of a crack with a finite fracture process zone and cohesive crack bridging zone by an equivalent sharp crack which supposedly gives about the same overall response of the specimen. This is no doubt adequate for a sufficiently large structure but inadequate for a sufficiently small structure. For smaller structures it is necessary to solve the problem taking at least into account a crack bridging zone of a finite length, which requires postulating a relationship between the crack bridging stress and the crack opening displacement as a material property. This relationship involves both instantaneous response and crack bridging creep. A solution of this type has recently been formulated.

The rate-of-loading effect on fracture has been studied experimentally by the size effect method. The most interesting result [40] is that the effective length of the fracture process zone decreases as the loading rate increases, and thus the response is getting more brittle, closer to LEFM. This is seen in Fig.9 which shows that, for specimens of 3 sizes (1:2:4), for max σ_N-points shift to the right (i.e., toward a higher brittleness β) as the time to reach the peak load increases (tests at constant displacements rates). Measurements are continuing.

4. Closing Remarks

As a final comment on the rate and creep effects, many concrete structures, (for example dams), develop large cracks over a long period of time.   Taking the rate effects in fracture growth as well as material creep (and shrinkage) farther away from the fracture process zone into account is essential for realistic predictions. To present the mathematical groundwork representing perhaps the simplest possible formulation has been one goal of the present workshop contribution. The other goal has been to review a host of recent developments which all exploit in some way a knowledge of the size effect due to fracture This effect itself is a consequence of the nonlocal character of damage in this type of materials.

Acknowledgement

Partial financial support from NSF Center for Science and Technology of Cement-Based Materials at Northwestern University is gratefully acknowledged. Ravindra Gettu of Northwestern University deserves thanks for some expert help in preparation of this manuscript.

References

1. A. Hillerborg, M. Modéer, and P.-E. Petersson, "Analysis of Crack Formation and Crack Growth in Concrete by Means of Fracture Mechanics and Finite Elements," *Cement and Concrete Research* 6(6) (1976), 773-782.

2. Y. S. Jenq and S. P. Shah, "A Fracture Toughness Criterion for Concrete," *Engineering Fracture Mechanics* 21(5) (1985), 1055-1069.

3. S. E. Swartz and T. M. E. Refai, "Influence of Size Effects on Opening Mode Fracture Parameters for Precracked Concrete Beams in Bending," in *Fracture of Concrete and Rock, Proceedings of SEM-RILEM Int. Conf.*, Houston, June 1987, edited by S. P. Shah and S. E. Swartz, Springer-Verlag, NY (1989), 242-254.

4. J. Planas and M. Elices, "Conceptual and Experimental Problems in the Determination of the Fracture Energy of Concrete," in *Fracture Toughness and Fracture Energy, Test Methods for Concrete and Rock*, Preprints of the Proceedings of an Int. Workshop Sendai, Japan, October, 1988, edited by M. Izumi, Tohoku Univ., Sendai, Japan (1988), 1-18.

5. J. Planas and M. Elices, "Size Effect in Concrete Structures: Mathematical Approximations and Experimental Validation," in *Cracking and Damage, Strain Localization and Size Effect*, Proceedings of France-U.S. Workshop, Cachan, France, 1988, edited by J. Mazars and Z. P. Bažant, Elsevier, London (1989), 462-476.

6. H. Horii, Z. Shi, and S.-X. Gong, "Models of fracture process zone in concrete, rock and ceramics," in *Cracking and Damage, Strain Localization and Size Effect*, Proceedings of France-US Workshop, Cachan, France, 1988, edited by J. Mazars and Z. P. Bažant, Elsevier, London (1989), 104-115.

7. M. Elices and J. Planas, "Material Models," in *Fracture Mechanics of Concrete Structures*, RILEM TC90-FMA, edited by L. Elfgren, Chapman and Hall, London, (1989), 16-66.

8. Z. P. Bažant, "Size Effect in Blunt Fracture: Concrete, Rock, Metal," *J. Engng. Mech.*, ASCE, Vol. 110, No. 4 (April 1984), 518-535.

9. Z. P. Bažant and P. A. Pfeiffer, "Determination of Fracture Energy from Size Effect and Brittleness Number," *ACI Mater. J.*, Vol. 84, No. 6, (Nov.-Dec. 1987), pp. 463-480.

10. Z. P. Bažant, "Fracture Energy of Heterogeneous Materials and Similitude," *Preprints, SEM/RILEM Int. Conf. on Fracture of Concrete and Rock* (Houston), edited by S. P. Shah and S. E. Swartz, Soc. for Exp. Mech., 1987; also in *Fracture of Concrete and Rock SEM-RILEM Int. Conf.*, edited by S. P. Shah and S. E. Swartz, Springer-Verlag, New York, (1989), 229-241.

11. Z. P. Bažant and M. T. Kazemi, "Size Effect in Fracture of Ceramics and Its Use for Determining Fracture Energy and Effective Process Zone Length," Report No. 89-6/498s, (1989) ACBM Center, Northwestern University; *J. of American Ceramic Soc.*, in press.

12. Z. P. Bažant, "Mechanics of distributed cracking," *Appl. Mech. Reviews ASME*, 39 (1986), 675-705.

13. R. Gettu, Z. P. Bažant, and M. E. Karr, Fracture Properties and Brittleness of High Strength Concrete, Report No. 89-10/B627f, Center for Advanced Cement-Based Materials, Northwestern University (1989); also *ACI Materials Journal*, in press.

14. Z. P. Bažant and M. T. Kazemi, "Size Effect Tests of Diagonal Shear Fracture," *ACI Materials Journal* (1990), in press.

15. Z. P. Bažant and Hsu-Huei Sun, "Size Effect in Diagonal Shear Failure: Influence of Aggregate Size and Stirrups," *ACI Materials Journal* 84(4) (1987), 259-272.

16. Z. P. Bažant and J. K. Kim, "Size Effect in Shear Failure of Longitudinally Reinforced Beams," *Am. Concrete Institute Journal* 81 (1984), 456-468; Disc. 796 & Closure 83 (1985), 579-583.

17. Z. P. Bažant and Z. Cao, "Size Effect in Shear Failure of Prestressed Concrete Beams," *Am. Concrete Inst. Journal* 83 (1986), 260-268.

18. Z. P. Bažant and Z. Cao, "Size Effect in Punching Shear Failure of Slabs," *ACI Structural Journal* (Am. Concrete Inst.) 84 (1987), 44-53.

19. Z. P. Bažant and S. Sener, "Size Effect in Pullout Tests," *ACI Materials Journal* 85 (1988), 347-351.

20. Z. P. Bažant, S. Sener, and P. C. Prat, "Size Effect Tests of Torsional Failure of Plain and Reinforced Concrete Beams," *Materials and Structures* RILEM, Paris, (1988), 425-430.

21. P. Marti, "Size Effect in Double-Punch Tests on Concrete Cylinders," *ACI Materials Journal* 86(6) (1989), 597-601.

22. R. Eligehausen and J. Ožbolt, "Size Effect in Anchorage Behavior," Proc., ECF8, "Fracture Behavior and Design of Materials and Structures," (1990), 2671-2677.

23. Z. P. Bažant, R. Gettu, and M. T. Kazemi, Identification of Nonlinear Fracture Properties from Size Effect Tests and Structural Analysis Based on Geometry-Dependent R-Curve, Report No. 89-3/498p, Center for Advanced Cement-Based Materials, Northwestern Univ., Evanston, IL., 1989.

24. Z. P. Bažant and M. T. Kazemi, "Size Dependence of Concrete Fracture Energy Determined by RILEM Work-of-Fracture Method," Report No. 89-12/B623s, (1989), NSF Center for Science and Technology of Advanced Cement-Based Materials, Northwestern University, Evanston, IL; also *Int. J. of Fracture*, in press.

25. Z. P. Bažant and Y. Xi, "Statistical Size Effect in Concrete Structures: Nonlocal Theory," Report No. 90-5/616s, (1990), Center for advanced Cement-Based Materials, Northwestern University, Evanston, IL.

26. J. Nakayama, "Direct Measurement of Fracture Energies of Brittle Heterogeneous Materials," *Journal of American Ceramics Society* 48(11) (1965), 583-87.

27. H. G. Tattersall and G. Tappin, "The Work of Fracture and Its Measurement in Metals, Ceramics and Other Materials," *Journal of Material Science* 1(3) (1966), 296-301.

28. A. Hillerborg, "The Theoretical Basis of a Method to Determine the Fracture Energy G_F of Concrete," *Materials and Structures* 18(106) (1986), 291-96.

29. A. Hillerborg, "Results of Three Comparative Test Series for Determining the Fracture Energy G_F of Concrete," *Materials and Structures* 18(107) (1986), 407-13.

30. Z. P. Bažant and F.-B. Lin, "Nonlocal Smeared Cracking Model for Concrete Fracture," *J. of Struct. Eng. ASCE* 114(11) (1988), 2493-2510.

31. Z. P. Bažant and J. Ožbolt, "Nonlocal Microplane Model for Fracture, Damage and Size Effect in Structures," *J. of Eng. Mech. ASCE*, (1990), in press.

32. J. Pijaudier-Cabot, Z. P. Bažant and Y. Berthaud, "Interacting Crack Systems in Particulate or Fiber-Reinforced Composites," Proc., 6th Int. Conf. on Numerical Methods on Fracture Mechanics (1990), held in Freiburg, West Germany, April 1990.

33. H. Gao and J. R. Rice, "A First order Perturbation Analysis on Crack Trapping by Arrays of Obstacles," Report Division of Appl. Sciences, Harvard Univ., Cambridge MA, Nov. 1988.

34. T. Mori, K. Saito and T. Mura, "An Inclusion Model for Crack Arrest in a Composite Reinforced by Sliding Fibers," *Mech. of Materials* Vol. 7 (1988), 49-58.

35. P. C. Paris, M. P. Gomez, and W. E. Anderson, "A Rational Analytic Theory of Fatigue," *The Trend in Engineering* 13(1) (Jan. 1961).

36. P. C. Paris and F. Erdogan, "A Critical Analysis of Propagation Laws," *Transactions of ASME, J. of Basic Engrg.* 85 (1963), 528-534.

37. S. E. Swartz and C. G. Go, "Validity of Compliance Calibration to Cracked Beams in Bonding," J. of Experimental Mech. 24(2) (June 1984), 129-134.

38. Z. P. Bažant and K. Xu, "Size Effect in Fatigue Fracture of Concrete," Report No. 89-12/623s, (1989), ACBM Center, Northwestern University; submitted to *ACI Mat. J.*.

39. Z. P. Bažant and P. C. Prat, "Effect of Temperature and humidity on fracture energy of concrete," *ACI Materials J.* 84 (July 1988), 262-271.

40. Z. P. Bažant and R. Gettu, "Determination of Nonlinear Fracture Characteristics and Time Dependence from Size Effect," in *Fracture of Concrete and Rock: Recent Developments*, Proc. Int. Conf. in Cardiff, UK, edited by S. P. Shah, S. E. Swartz and B. Barr, Elsevier Applied Science, London, (1989), 549-565.

41. Z. P. Bažant, Editor (1988), "Mathematical Modeling of Creep and Shrinkage of Concrete," John Wiley and Sons, Chichester and New York.

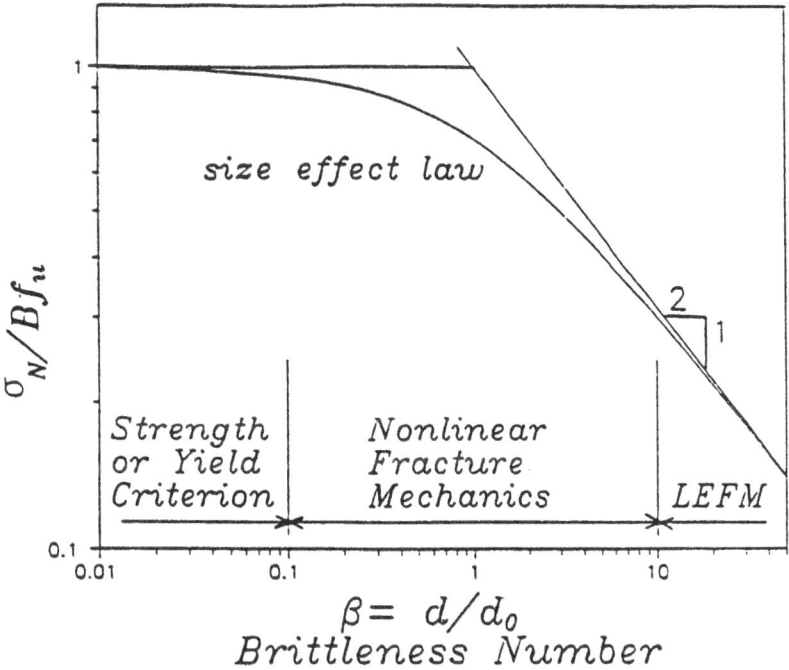

Fig.1. Size effect law for quasi-brittle structures.

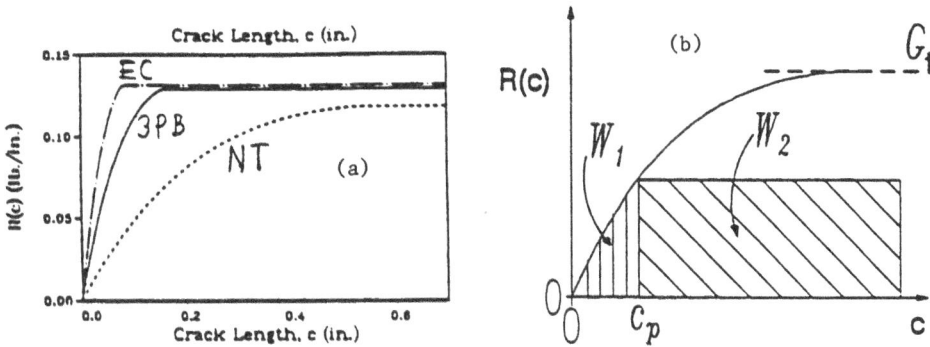

Fig.2. R-curves and effect of size on fracture energy, along with an idealized picture of the evolution and movement of fracture process zone.

148

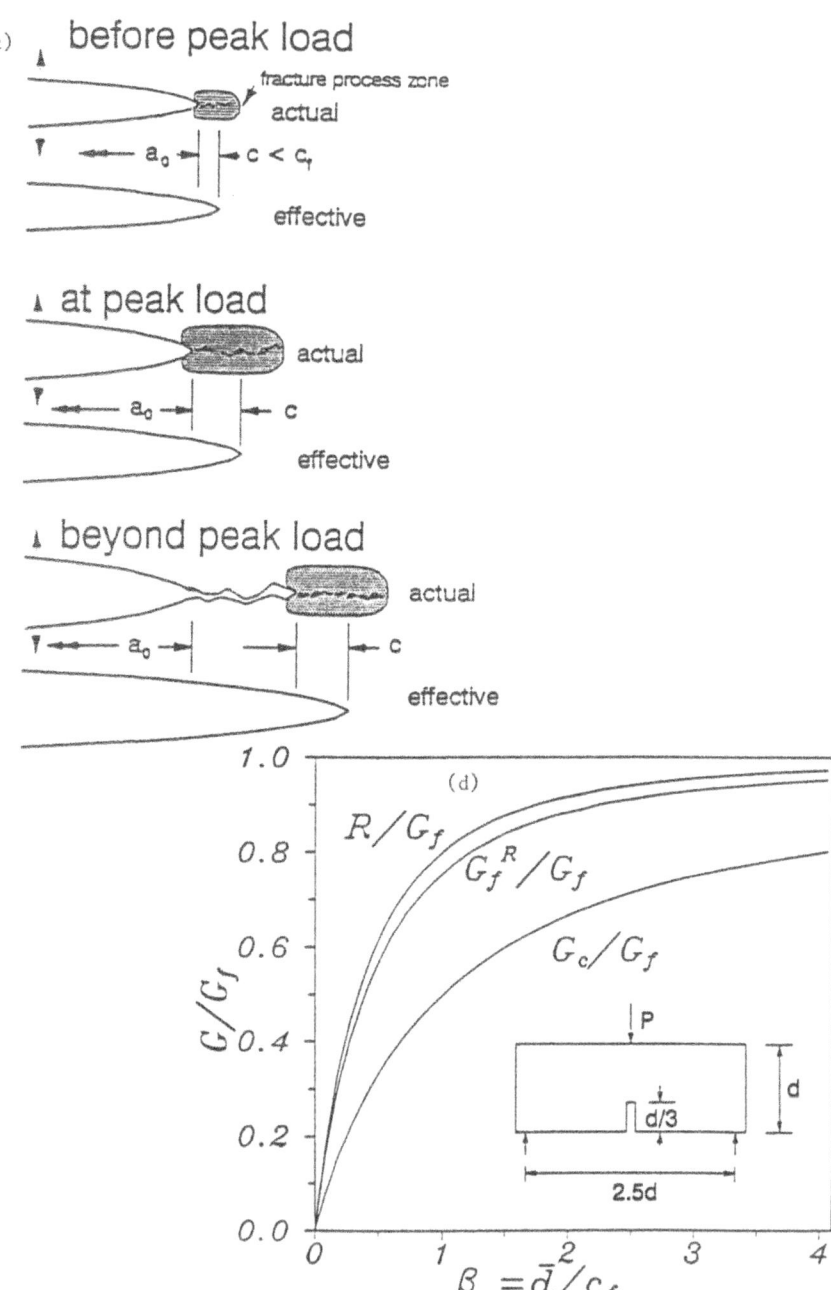

Fig. 2 continued

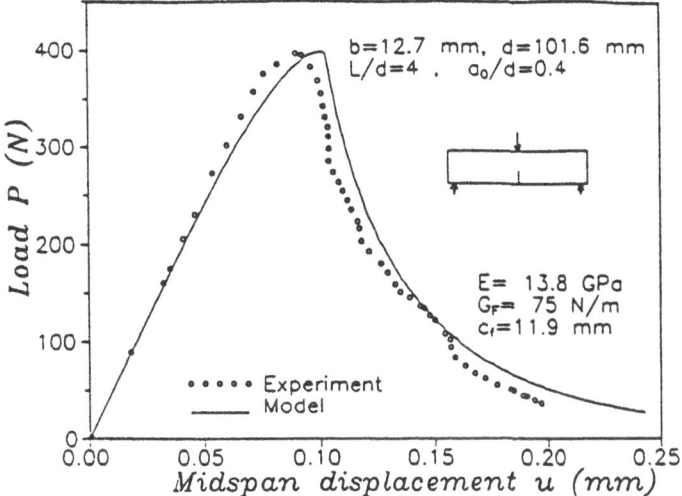

Fig.3. Prediction of the load-deflection curve from measured maximum loads of rock fracture specimens of different sizes, in comparison with the load- deflection curve measured by Bažant, Gettu, and Kazemi [23].

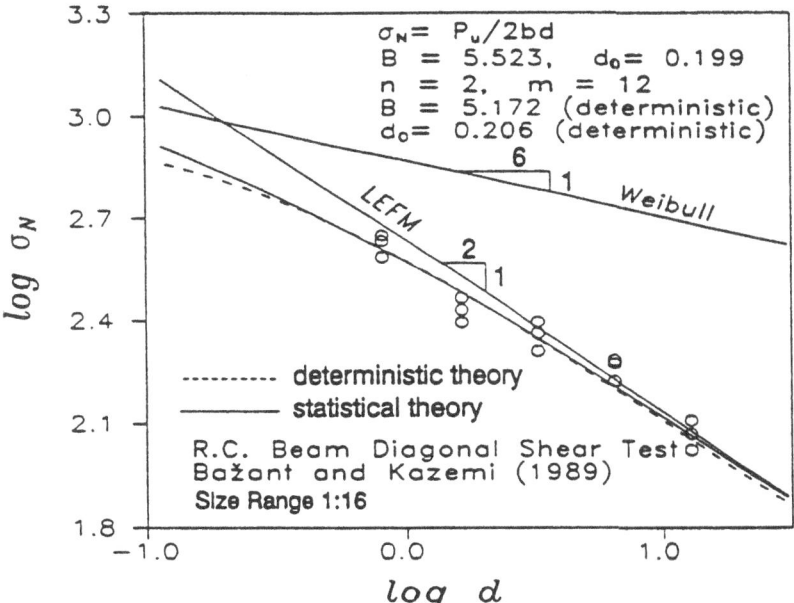

Fig.4. Statistical generalization of the size effect law.

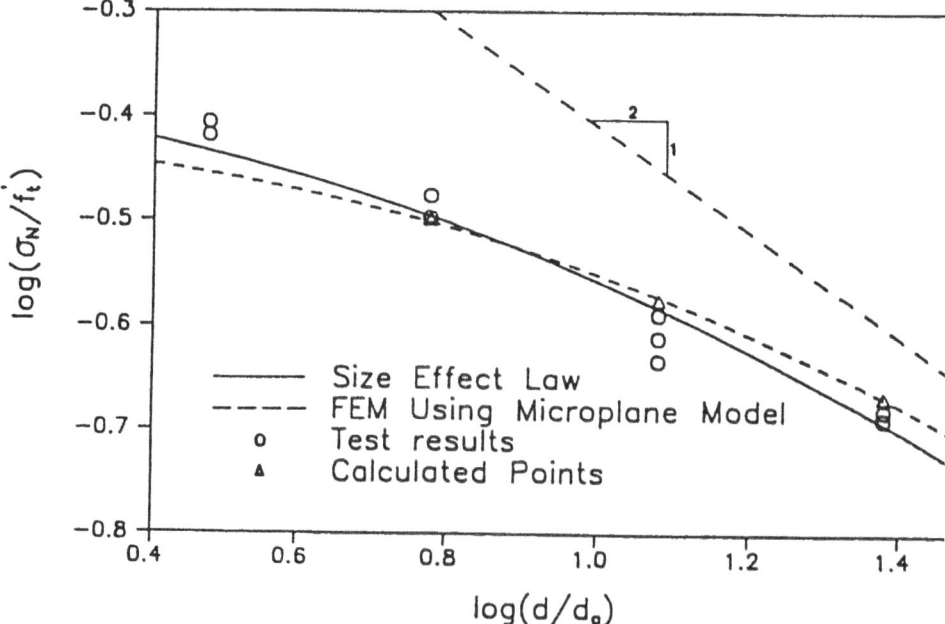

Fig.5. Nominal strengths of fracture specimens of different sizes calculated by a finit element program based on nonlocal microplane material model, compared with the size effec law and with experimental measurements of Bažant and Pfeiffer [9].

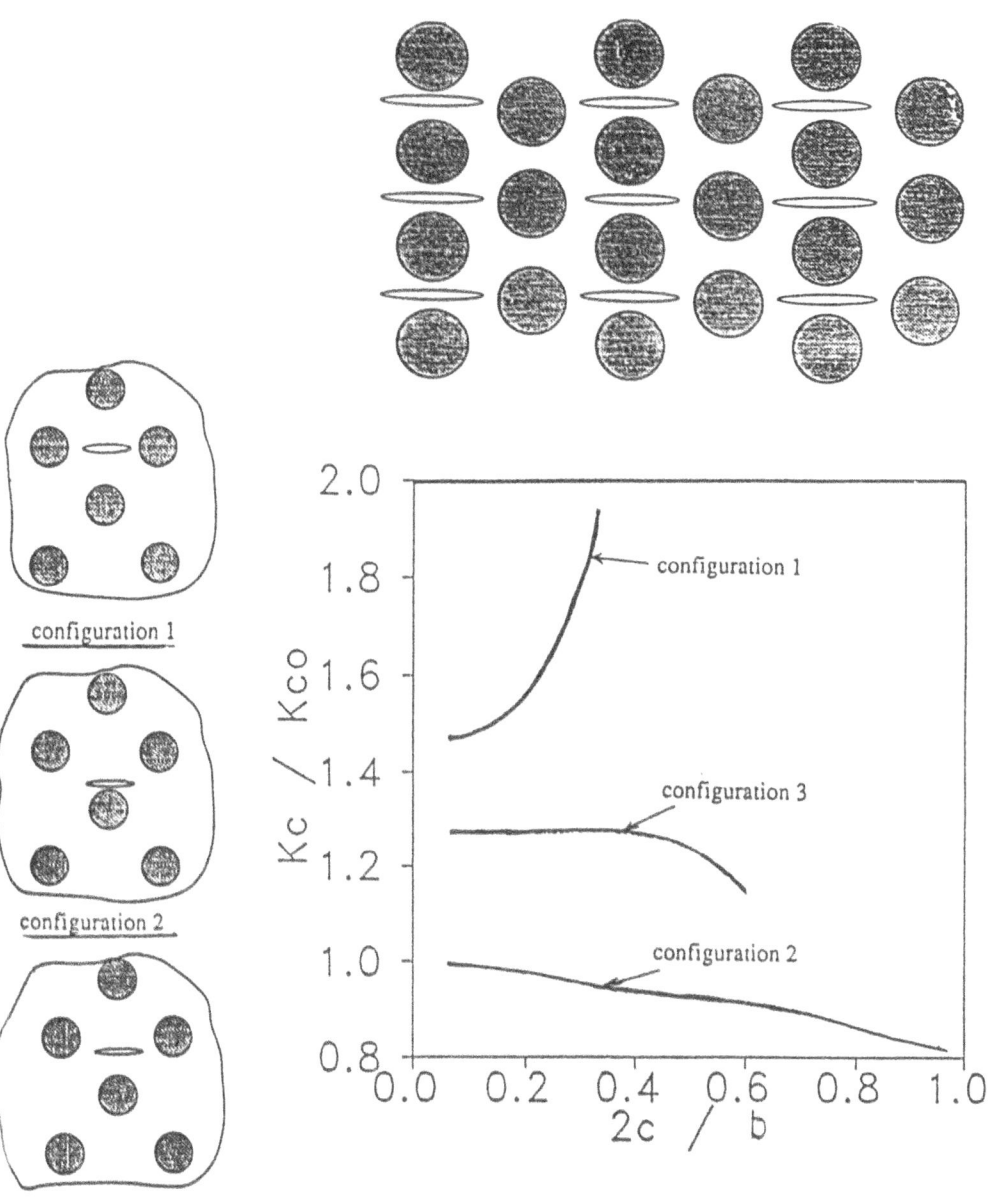

configuration 1

configuration 2

configuration 3

Fig.6. Approximation of fracture process zone by an elastic matrix with microcracks and inclusions, and apparent R-curves for the microcracks calculated by Pijaudier Cabot, Bažant and Berthaud [32].

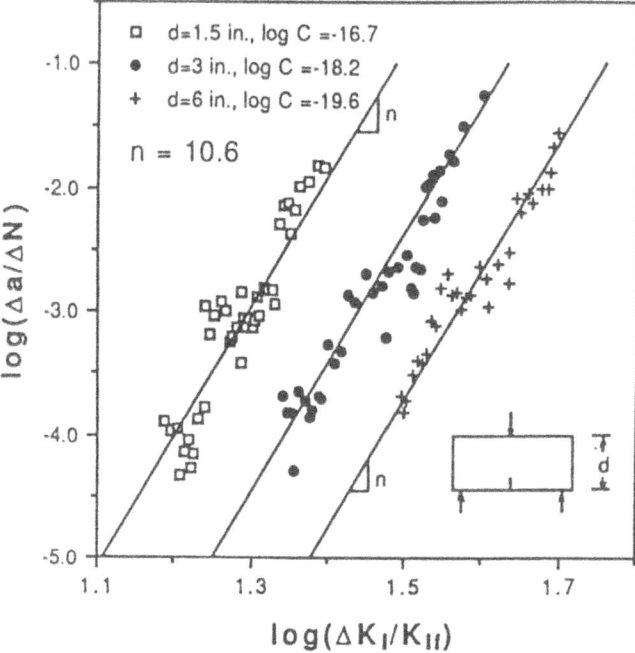

Fig.7. Test results of Bažant and Xu [38] on fatigue fracture of concrete specimens of three different sizes and their comparison with size-adjusted Paris law.

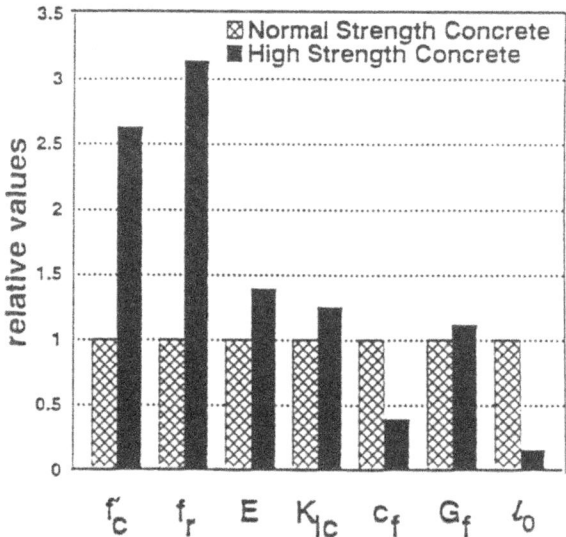

Fig.8. Comparison of nonlinear fracture characteristics of high strength concrete and normal concrete [13].

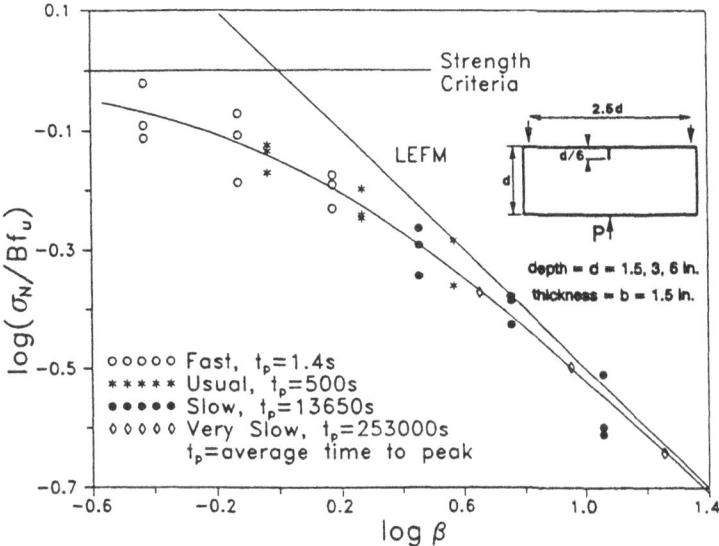

Fig.9. Size effect on nominal strengths at different loading rates, reported by Bažant and Gettu [40].

MICROMECHANICS OF DEFORMATION IN ROCKS

John M. Kemeny
Department of Mining and Geological Engineering
University of Arizona
Tucson, AZ 85721

Neville G. W. Cook
Earth Sciences Division
Lawrence Berkeley Laboratory and
Department of Materials Science and Mineral Engineering
University of California, Berkeley, CA 94720

ABSTRACT. Laboratory testing of rocks subjected to differential compression have revealed many different mechanisms for extensile crack growth, including pore crushing, sliding along pre-existing cracks, elastic mismatch between grains, dislocation movement, and hertzian contact. Micromechanical models based on fracture mechanics have been developed for these different mechanisms by many different researchers. In this paper, the K_I solutions for these micromechanical models are reviewed. Because of the similarity in rock behavior under compression in a wide range of rock types, it is not surprising that these micromechanical models have many similarities. This may explain the success of models based on certain micromechanisms in spite of the lack of evidence for these mechanisms in microscopic studies. Based on these similarities, a generic micromechanical model is proposed that in some way takes into account all of the above phenomena. It is demonstrated how the K_I solutions from the micromechanical models can be used to derive nonlinear stress-strain curves that exhibit strain-hardening and strain-softening, dilatation, σ_2 sensitivity, and rate dependence. By using subcritical crack growth, transient and tertiary creep behavior can also be predicted. Also, it is shown how these micromechanical models can form the basis for continuum damage models using the finite element method.

1 Introduction

Rock is a very heterogeneous material, containing different and often anisotropic crystals, as well as structural weaknesses at all scales. These weaknesses include grain boundaries, pores, and cracks on the small scale, and joints, faults, and bedding planes on the larger scale. When rock is subjected to differential compressive stresses, extensile, opening-mode microcracks grow from these flaws and degrade the properties of the rock. Our understanding of the fundamental mechanisms involved in the micromechanics of brittle and semi-brittle rock deformation has increased greatly in the past decade, due to the many microscopic studies that have been carried out on rocks subjected to compressive stresses in the laboratory. It has been found, for instance, that microcrack growth occurs

155

S. P. Shah (ed.), Toughening Mechanisms in Quasi-Brittle Materials, 155–188.
© 1991 *Kluwer Academic Publishers.*

preferentially in the direction of the maximum principal stress, and this results in stress-induced anisotropic rock properties (Wong, 1985; Fredrich et al., 1989; Zheng, 1989). As more cracks grow and the cracks increase in length, crack interaction becomes important, and cracks can coalesce to form large scale splitting or shear fractures. The growth of cracks has been shown to be closely associated with the macroscopic constitutive behavior of the rock. For instance, the initial, stable growth of microcracks is associated with strain-hardening stress-strain behavior, and the interaction and coalescence of cracks is associated with unstable strain-softening stress-strain behavior (Hallbauer et al., 1973). Both strain-hardening and strain-softening behavior exhibit dilatation as a result of extensile crack growth. Also, experiments have shown that for tests conducted at higher confining pressures, the average length of the cracks is reduced, promoting the transition from brittle to semi-brittle behavior (Fredrich et al., 1989; Zheng, 1989). Microstructural parameters such as grain size, porosity, and initial crack density play an important role in rock deformation, and recent experimental studies have helped elucidate their effects (Zhang et al., 1989; Fredrich et al., 1989). An example of microcrack growth in Indiana limestone is presented in Figure 1, revealing several of the different mechanisms that can cause crack growth under compression (from Zheng, 1989).

Another aspect of rock deformation and failure of great importance is the time and rate-dependence of rock deformation. Recent experimental studies have shown that both time and rate dependent behavior in brittle rocks can be the result of rate-controlled processes acting at the tips of cracks where stress concentrations exist (Sano et al., 1981; Carter et al., 1981; Atkinson, 1984). This time-dependent crack growth occurs at values of the stress intensity factor below the fracture toughness of the material, and is referred to as sub-critical crack growth. Subcritical crack growth is the result of several mechanisms that can occur simultaneously, including stress corrosion, diffusion, dissolution, ion exchange, and microplasticity (Atkinson, 1984).

Fracture mechanics theory (i.e., Lawn and Wilshaw, 1975) is now routinely being applied to many different aspects of the micromechanics of rock deformation and failure. Fracture mechanics models have been developed to analyse crack growth under compressive stresses due to grain crushing (Zhang et al., 1989), sliding along pre-existing cracks (Nemat-Nasser and Horii, 1982; Steif, 1984; Ashby and Hallum, 1986; Horii and Nemat-Nasser, 1986; Kemeny and Cook, 1987a,b), stress concentrations around pores (Sammis and Ashby, 1986), elastic mismatch between grains (Dey and Wang, 1981), dislocation movement (Krajcinovic, 1989; Wong, 1990), and combinations of these mechanisms (Costin, 1985). One of the important aspects of these models is that even though they are based on linear elastic fracture mechanics (LEFM), nonlinear stress-strain behavior can be predicted due to the growth of the cracks in an otherwise linear-elastic solid. For instance, the models predict that the initial growth of cracks results in strain-hardening stress-strain behavior. Also, the models that include crack interaction show that crack interaction can cause a transition from strain-hardening to strain-softening stress-strain behavior. Several fracture mechanics models have been able to predict the transition from axial splitting to shear faulting as the confining pressure is increased (Horii and Nemat-Nasser, 1986; Kemeny and Cook, 1987a). Also, a few of the models have implemented a crack growth criterion based on subcritical crack growth, and these models are able to predict the creep and rate dependence in rock (Costin, 1985; Kemeny, 1990).

The experiments and theoretical models described above are in general based on the standard triaxial test consisting of an axial stress and a confining pressure. Ideally, under these boundary conditions the stress state is constant throughout the body, and a rotation of the principal stresses does not occur during loading. In actual field situation where rock deformation and failure are occurring in a rock mass, stress gradients occur due to the complicated boundary conditions imposed. These stress gradients play an important role in

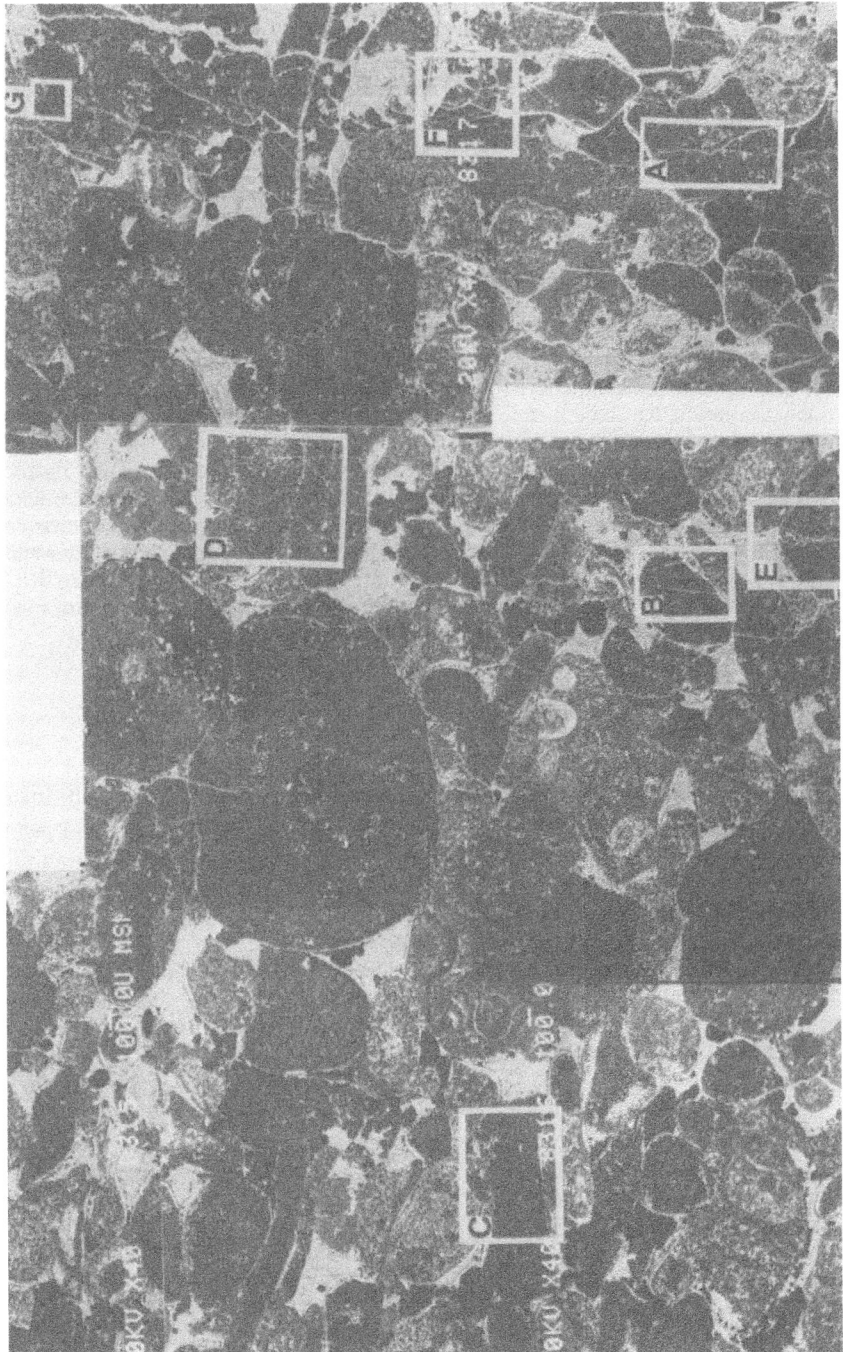

Figure 1. SEM micrograph of Berea sandstone subjected to uniaxial compression (σ_1 vertical) showing stress-induced microcracks formed by several different micro-mechanisms, from Zheng (1989).

the fracture systems that develop, and in rock mass stability. Experimental studies have been conducted looking at the effects of stress gradients on rock deformation and fracture formation (Guenot, 1989; Gough and Bell, 1982; Haimson and Herrick, 1985; Santerelli, 1987; Ewy, 1989). Fracture mechanics models based on the micromechanics of crack growth, interaction, and coalescence are now being developed that can take into account complicated boundary value problems with stress gradients. These models are formulated by implementing micromechanical models into an elastic finite element code. In each of the elements, the micromechanical models are utilized to calculate the effects of crack growth, interaction, and coalescence, and the stress gradient effects are accounted for by the finite element calculations. Examples of finite element damage models that have been developed in this way are given in Costin and Stone (1987), Krajcinovic (1989), and Kemeny and Tang (1990).

This paper discusses the use of fracture mechanics in modelling the deformation and failure of rocks subjected to differential compressive stresses. In section 2 that follows, linear elastic fracture mechanics theory is briefly reviewed. Then, in section 3, some of the micromechanical models that have been developed to model the different mechanisms for crack growth under compression are reviewed. In section 4, it is shown how these models can be used to develop nonlinear constitutive relations for rock. This includes micromechanical models for nonlinear constitutive relations based on the rate-independent fracture toughness, rate dependent models based on subcritical crack growth, and creep due to subcritical crack growth. In section 5, these micromechanical models are implemented into an elastic finite element code, to look at the effects of microcrack growth on complicated boundary value problems. As an example, we look at borehole breakout using both the rate-independent and subcritical crack growth models.

2 Fracture Mechanics Preliminaries

The analyses in this paper are limited to two-dimensional, linear-elastic bodies subjected to compressive principal stresses σ_1 and σ_2, and containing simple configurations of cracks. In this section, some of the important aspects of linear elastic fracture mechanics are briefly discussed. The discussion is limited to those topics that will be used in later sections of this paper. For a complete discussion of linear elastic fracture mechanics theory, see Liebowitz (1968), Lawn and Wilshaw (1975), or others.

Following the usual notation in linear elastic fracture mechanics, three types of stress intensity factors, K_I, K_{II}, K_{III}, are distinguished, which relate to the three types of crack displacements: mode I (opening), mode II (sliding), and mode III (tearing). The stresses near the crack tip then have the following form (Lawn and Wilshaw, 1975):

$$\left\{ \begin{array}{c} \sigma_1 \\ \sigma_2 \\ \sigma_{12} \end{array} \right\} = \frac{K_Q}{\sqrt{2\pi r}} \left\{ \begin{array}{c} f_1{}^Q \\ f_2{}^Q \\ f_{12}{}^Q \end{array} \right\} \tag{1}$$

where r is the radial distance from the crack tip, θ is the angle measured from the plane of the crack, Q takes on the values I, II, and III for the three cracking modes, and $f_{ij}{}^Q$ are functions of θ. The stress intensity factors take into account the crack geometry and boundary conditions, and can be extremely simple for simple crack configurations (e.g., Rooke and Cartwright, 1976). For instance, for a single flat crack of length $2l$ oriented along the x axis in an infinite body subjected to normal crack face tractions $\sigma(x)$, the mode I stress intensity factor at the tip $x=l$ is given by (Cherepanov, 1979):

$$K_1 = \frac{1}{\sqrt{\pi l}} \int_{-l}^{l} \sigma(x) \sqrt{\frac{l+x}{l-x}} \, dx \tag{2}$$

Much of the usefulness of the stress intensity factors for rock deformation lies in their relationship to the strain energy of the body that contains the cracks. The energy release rate, G, is defined as:

$$2G = \frac{\partial U_e}{\partial l} \tag{3}$$

where U_e is the elastic strain energy of the solid that contains the crack, and $2l$ is the crack length. The relationship between G and the stress intensity factors is given by:

$$G = \frac{K_I^2}{E'} + \frac{K_{II}^2}{E'} + \frac{K_{III}^2}{E}(1+\nu) \tag{4}$$

where $E' = E$ for plane stress and $E' = E/(1 - \nu^2)$ for plane strain. By integrating 2G from zero to the given crack length, the additional strain energy due to the crack, U_e, can be calculated.

A convenient method for calculating the displacements of an elastic body containing cracks is Castigliano's theorem (Sokolnikoff, 1956). From Castigliano's theorem the displacement x_i that occurs under a load P_i when a linear elastic body is subjected to loads $P_1, P_2, \dots P_n$ is given by:

$$x_i = x_i \text{ (uncracked body)} + \frac{\partial U_e}{\partial P_i} \tag{5}$$

The displacements due to the uncracked body are already known for the simple configurations considered in this paper. Equations (3) to (5) give the procedure for calculating the additional displacements due to the cracks once the stress intensity factor solutions are known.

Two criteria for crack growth will be considered in this paper. The first crack growth criterion (rate-independent) is based on the work of Barrenblatt (1962) and others, and states that crack growth occurs when:

$$K_I = K_{IC} \tag{6}$$

where K_{IC} is the fracture toughness. Assuming that K_{IC} does not change with crack growth, the criterion for unstable crack growth is:

$$K_I \geq K_{IC} \text{ and } \frac{\partial K_I}{\partial l} > 0 \tag{7}$$

A second crack growth criterion (rate-dependent) is based on crack growth that occurs below K_{IC} and is referred as subcritical crack growth. A common empirical equation used to describe subcritical crack growth is based on the power law formulation of Charles(1958), and is given by:

$$\frac{\partial l}{\partial t} = A \, K_I^n \tag{8}$$

where A and n are material constants.

3 Micromechanical Models for Crack Growth Under Compression

As discussed in the introduction, microcrack growth under compression occurs by many different mechanisms, and in the past few years, mechanical models have been developed for many of these mechanisms. In this section we review the stress intensity factor solutions for several of these micromechanical models. The discussion is limited to the axial growth of extensile cracks, i.e., the formation of shear bands or faults is not considered. It is found that the micromechanical models for crack growth under compression have many similarities, and at the end of this section we introduce a generic model for crack growth under compression that encompasses all the models. The stress intensity factor solutions discussed here will be used in section 3 to derive nonlinear constitutive relations.

3.1 CYLINDRICAL PORE MODEL

A two-dimensional cylindrical pore is subjected to maximum and minimum principal stresses σ_1 and σ_2, respectively, as shown in Figure 2. For $\sigma_1 > 3\sigma_2$, tension will occur at the boundary of the pore in the direction of the maximum principal stress. As $\sigma_1 - 3\sigma_2$ increases, eventually a pair of tensile stresses will initiate and grow in the direction of σ_1.

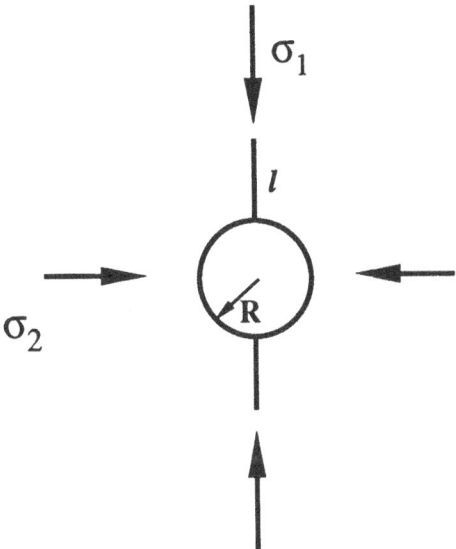

Figure 2. Cylindrical pore model for crack growth under compression.

When the length of the tensile crack, l, is small compared with the radius of the pore, R, i.e., $l<<R$, a small-crack approximation to the stress intensity factor is appropriate, given by:

$$K_I = 1.12\ (\sigma_1 - 3\sigma_2)\ \sqrt{\pi l} \qquad (9)$$

This solution is based on the edge crack subjected to a uniform tensile far field stress (Rooke and Cartwright, 1976). Notice that K_I for this configuration increases with increasing crack length, and for a fixed σ_1, equation (7) predicts that the crack will grow in an unstable manner.

When the tensile cracks are long in relation to the size of the pore, i.e., $R<<l$, the stress intensity factor can be approximated by a straight crack oriented in the direction of σ_1 with a set of point forces at the center of the crack. The stress intensity factor for this configuration is given by:

$$K_I = \frac{C\ R\ (\sigma_1 - 3\sigma_2)}{\sqrt{\pi l}} - \sigma_2 \sqrt{\pi l} \qquad (10)$$

162

where C is a constant. Note that this solution takes into account the opening force due to σ_1 and the closing effect that σ_2 has on the long crack of length $2l$. Also note that from equation (7), this solution predicts stable crack growth (K_I decreases for increasing l for fixed σ_1 and σ_2). Thus equations (9) and (10) together predict that the tensile cracks will initially grow in an unstable fashion, and stabilize at some crack length.

In order to determine the length of crack at which the crack stabilizes, a solution is needed that is valid at all crack lengths. A stress intensity factor solution that encompasses both the small and large crack behavior is given by Sammis and Ashby (1986):

$$K_I = \left[\frac{1.1(\sigma_1 - 2.1\sigma_2)}{(1+l/R)^{3.3}} - \sigma_2 \right] \sqrt{\pi l} \tag{11}$$

The K_I vs. l behavior for this solution is shown in Figure 3, for various values of σ_2/σ_1. K_I initially increases with increasing l, reaches a maximum, and thereafter decreases with increasing l. The maximum represents the transition from stable to unstable crack growth. Figure 3 shows that the maximum occurs at a length $l<R$, and decreases with increasing σ_2/σ_1.

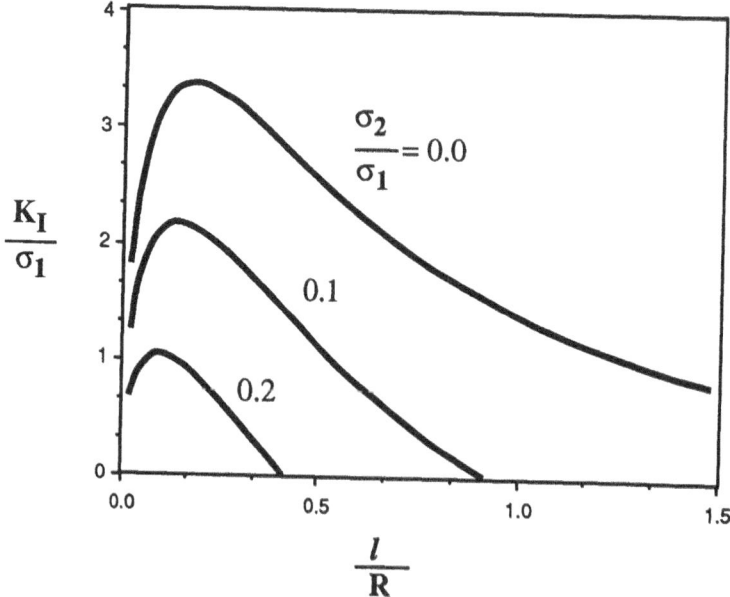

Figure 3. Results from the pore model of Sammis and Ashby for different values of σ_2/σ_1.

3.2 SLIDING CRACK MODEL

Consider an initial straight crack of length $2l_0$, at an angle β to the principal stresses σ_1 and σ_2, as shown in Figure 4. The shear stress along the crack over and above that due to friction (assuming a linear coefficient of friction, μ) is given by:

$$\tau^* = \frac{1}{2}\left[(\sigma_1\text{-}\sigma_2)\sin2\beta - \mu(\sigma_1+\sigma_2+(\sigma_1\text{-}\sigma_2)\cos2\beta)\right] \tag{12}$$

The stress intensity factors for the initial configuration are given by:

$$K_I = 0 \tag{13}$$

$$K_{II} = \tau^* \sqrt{\pi l_0}$$

Assuming that crack growth will occur in the direction such that $\sigma_{\theta\theta}$ at the crack tip is maximized, crack growth will occur at an angle of about 70° from the plane of the crack. These out of plane cracks are referred to as wing cracks.

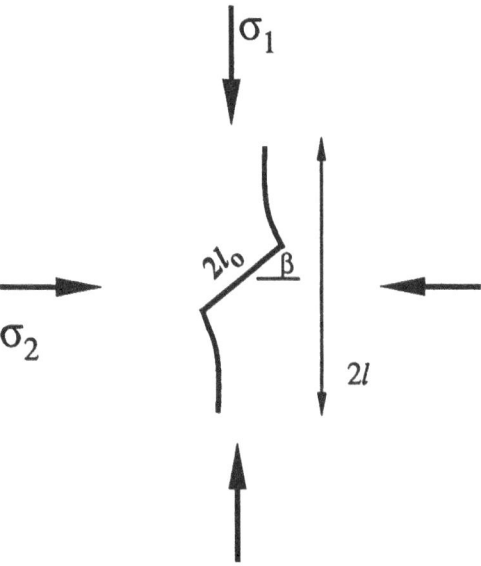

Figure 4. Sliding crack model for crack growth under compression.

When the wing cracks are small, i.e., $l << l_0$, the stress intensity factors are given by Cotterell and Rice (1980):

$$K_I = \frac{3}{4} (\sin\beta/2 + \sin3\beta/2) \, \tau^* \sqrt{\pi l_0}$$

$$(14)$$

$$K_{II} = \frac{1}{4} (\cos\beta/2 + 3\cos3\beta/2) \, \tau^* \sqrt{\pi l_0}$$

This solution cannot be used for stability analysis, since $\partial K_I/\partial l$ is identically equal to zero.

As the crack grows longer, it grows in the direction of σ_1, and the stress intensity factors at the tip of the wing cracks becomes predominantly K_I. A long crack approximation for the sliding crack for $l >> l_0$ is given by Kemeny and Cook (1987a):

$$K_I = \frac{2l_0 \, \tau^* \cos\beta}{\sqrt{\pi l}} - \sigma_2 \sqrt{\pi l} \qquad (15)$$

The long crack approximation to the sliding crack is similar to the long crack approximation for the cylindrical pore model, when $l_0 \approx R$, where R is the radius of the pore. Compared with the Sammis and Ashby approximation to the pore model (equation 11), K_I in equation (15) decreases much less rapidly.

One approximation to the sliding crack valid at all crack lengths is given by Horii and Nemat-Nasser (1986). This model considers an initial crack length $2l_0$ at angle β with straight wing cracks of length l which are at an angle α from the direction of the initial crack, where α varies with the length of the wing cracks:

$$K_I = \frac{2l_0 \, \tau^* \sin\alpha}{\sqrt{\pi(l+l^*)}} - \frac{\sqrt{\pi l}}{2} \left[\sigma_1 + \sigma_2 + (\sigma_1 - \sigma_2)\cos2(\alpha+\beta)\right] \qquad (16)$$

where $l^* = 0.27l_0$. The stress intensity factor for a given wing crack length is given by the value of α that maximizes K_I. This model predicts stable crack growth for σ_2 compressive, but shows a transition to unstable crack growth when σ_2 is only slightly tensile (Horii and Nemat-Nasser, 1986).

3.3 ELASTIC MISMATCH MODEL

Consider a body containing two materials (or one material with a change in anisotropy orientation) and subjected to principal stresses σ_1 and σ_2, as shown in Figure 5. In two dimensions, each material can be described by a 3x3 compressibility matrix, as given by:

$$\left\{ \begin{array}{c} \varepsilon_1 \\ \varepsilon_2 \\ \varepsilon_{12} \end{array} \right\} = S \left\{ \begin{array}{c} \sigma_1 \\ \sigma_2 \\ \sigma_{12} \end{array} \right\} \tag{17}$$

where, for instance, S_{21} gives the strain in the 2 direction due to a stress σ_1. The S_{ij} can, for instance, take into account the strong anisotropy that occurs within individual grains.

For $\sigma_1 > \sigma_2$, differential expansion will occur in the 2 direction, and a tensile stress will develop at the interface in the material with the smaller lateral expansion. For a small crack in this material at the interface, the stress intensity factor can be approximated by the edge crack with point forces P at the edge (Rooke and Cartwright, 1976):

$$K_I = 2.6 \frac{P}{\sqrt{\pi l}} \tag{18}$$

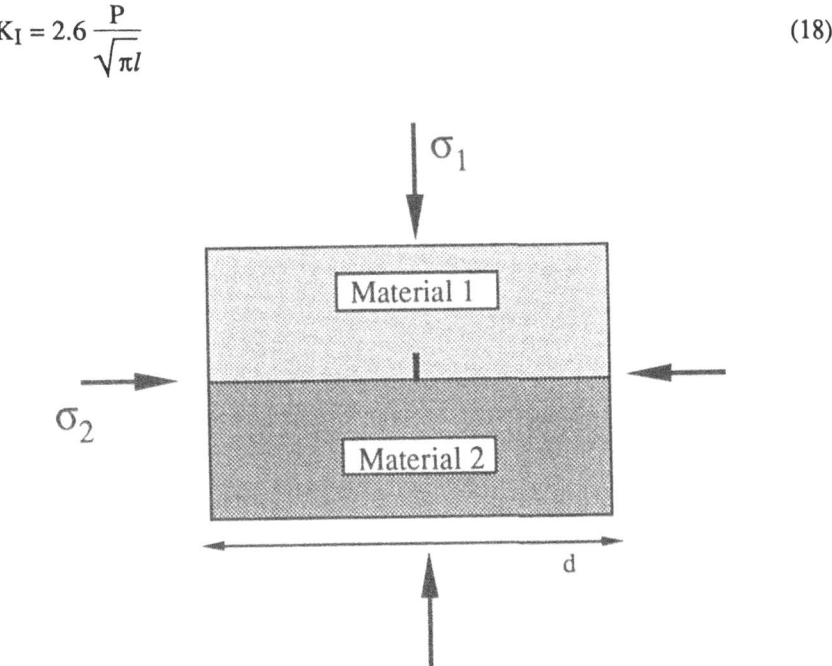

Figure 5. Elastic mismatch model.

Dey and Wang (1981) give an approximate solution for the point force P in terms of the S matrix and the width of the body, d:

$$P = \frac{0.45\sigma_1 d}{\pi} \left[\frac{\sigma_1(S_{12}{}^2 - S_{12}{}^{1)} - \sigma_2(S_{22}{}^2 - S_{22}{}^1)}{\sigma_1 \, S_{12}{}^2} \right] \tag{19}$$

where the superscript indicates material 1 or 2. Under a uniaxial stress ($\sigma_2 = 0$) and for a large contrast in compressibility between materials 1 and 2, this reduces to:

$$K_I = \frac{1.04 \; \sigma_1 \; d}{\pi \, \sqrt{\pi l}} \tag{20}$$

If we take d to be the grain size, then this result is very similar to the results for the long-crack approximations to the pore and sliding crack models when $d/2 \approx R \approx l_0$.

3.4 DISLOCATION PILE-UP MODEL

When a solid is stressed beyond its yield point, dislocations will be created, giving rise to plastic slip. The yield point for different minerals can differ by orders of magnitude, causing grain boundaries to become barriers to dislocation movement. At such a barrier, a dislocation pile-up can occur. As the strength of the pile-up increases, a tensile crack can develop at the edge of the pile-up, referred to as a Zener-Stroh crack.

Consider a slip plane subjected to a resolved shear stress τ_r, as shown in Figure 6. This stress has to exceed the lattice friction stress τ_f before the dislocations can move. If there are n edge dislocations of Burgers vector magnitude $\bar{b}$ in a pile-up of length L, the driving stress τ^* is given by (Wong, 1990):

$$\tau^* = \frac{n \, \bar{b} \, G}{L \, \pi \, (1-v)} \tag{21}$$

The stress distribution due to the pile-up at a distance x and angle θ is given by (Krajcinovic, 1989):

$$\sigma_r = \frac{n \, G \, \bar{b} \, \sin\theta}{2\pi \, (1-v) \, x} \tag{22}$$

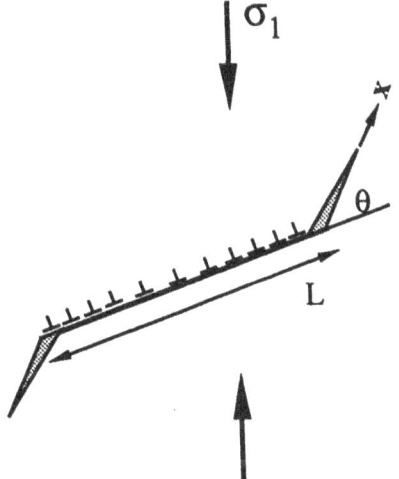

Figure 6. Dislocation pile-up model.

Using the stress distribution in equation (22), and using the formula for K_I for a variable stress as given in (2), this gives (Wong, 1990):

$$K_I = \tau^* \sin\theta \, L \sqrt{\frac{\pi}{2l}} \qquad (23)$$

There is also a term for the reduction in K_I due to the stress normal to the crack (Wong, 1990). Note that this solution has the same form as the stress intensity factor for the sliding crack model given in equation (15).

3.5 HERZTIAN CRACK MODEL

Consider the simple model of two spheres of radius R in contact and under a compressive force P. Using Herztian contact theory (e.g., Johnson, 1985), the pressure distribution along the contacting portions of the spheres is given by:

$$P(r) = P_0 \left[1 - \left(\frac{r}{a}\right)^2 \right]^{\frac{1}{2}} \qquad (24)$$

where r is radial distance, and a is the radius of contact given by:

$$a = \frac{\pi P_0 R (1-v^2)}{E} \tag{25}$$

and P_0 is the maximum pressure at the center of the circle of contact given by:

$$P_0 = \frac{3}{2} \frac{P}{\pi a^2} \tag{26}$$

At the edge of the contact area, a tensile stress σ_r exists, the maximum value of which is given by:

$$\sigma_r = \frac{P (1-2v)}{2 \pi a^2} \tag{27}$$

Based on this tension, the stress intensity factor for a small crack in this tension region is given by (Zhang et al., 1989):

$$\begin{aligned} K_I &= 1.12 \, \sigma_r \sqrt{\pi l} \\ &= \left[\frac{1.12 \, (1-2v) \, P}{2\pi \left[\frac{3PR(1-v^2)}{2E} \right]^{2/3}} \right] \sqrt{\pi l} \end{aligned} \tag{28}$$

This model predicts unstable crack growth and is only valid for $l \ll a$. A tensile stress will also develop in the sphere similar to that in a Brazilian test (Sternberg and Rosenthal, 1952). A small flaw in the center of the sphere will be subjected to a mode I stress intensity factor which for a two dimensional disk of radius R can be approximated by (Rooke and Cartwright, 1976):

$$K_I = \frac{P}{R} \sqrt{\frac{l}{\pi}} \left[1 + \frac{3}{2} \left(\frac{a}{R} \right)^2 \right] \tag{29}$$

Again this solution predicts unstable crack growth.

3.6 GENERIC MODEL

Because of the similarity in rock behavior under compression in a wide range of rock types, it is not surprising that the micromechanical models discussed in the previous sections have many similarities. This may explain the success of certain models such as the sliding crack and pore models, in spite of the lack of evidence for these models in microscopic studies. The similarities in behavior between the different models include:

1. Crack growth predominantly in the σ_1 direction.
2. K_I proportional to a distance parameter such as pore size, grain size, initial crack length, etc.
3. Crack growth unstable when the crack length on the order of the small parameter.
4. Crack growth stable when the crack length is large compared with the small parameter.
5. K_I very sensitive to σ_2.
6. K_I linearly proportional to $\sigma_1 - C\sigma_2$, where C is a constant.

Because of these similarities, is seems appropriate to develop a 'generic' micromechanical model that in some way takes into account all of the above phenomena. Here a generic model is proposed, which consists of a crack of length $2l$ oriented in the direction of σ_1, and subjected to a tensile stress σ_0 over a region of length 2a, as shown in Figure 7. The length of this tensile region remains fixed as the crack grows. Initially the crack length $2l$ can be smaller than the length 2a, and for this case the behavior should be described by the small-crack approximations. As the crack grows and becomes long compared with a, the model should behave like the long-crack approximations. Based on item 6 above, we take σ_0 to be linearly proportional to $\sigma_1 - C\sigma_2$, i.e.:

$$\sigma_0 = C_1 (\sigma_1 - C_2\sigma_2) \tag{30}$$

where C_1 and C_2 are constants. A closed form solution for K_I for this configuration can be derived from the Green's function solution given in equation (2). The stress distribution is given by:

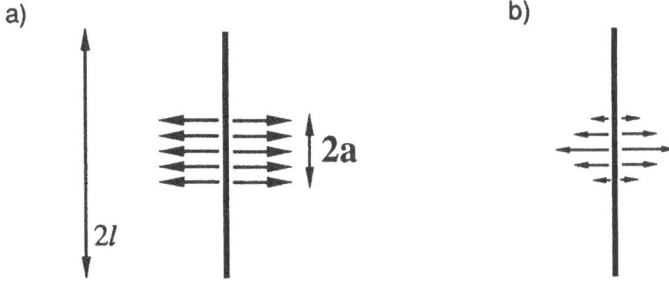

Figure 7. Generic model for crack growth under compression. a) constant stress σ_0 over a distance a at the crack center. b) linearly decreasing stress at the crack center.

$$\sigma(x) = \begin{cases} C_1(\sigma_1 - C_2\sigma_2) & |x| \le a \\ 0 & a < |x| \le l \end{cases} \tag{31}$$

Thus K_I for this configuration is given by:

$$K_1 = \frac{C_1(\sigma_1 - C_2\sigma_2)}{\sqrt{\pi l}} \int_{-a}^{a} \sqrt{\frac{l+x}{l-x}}\, dx - \sigma_2 \sqrt{\pi l}$$

$$= \frac{2C_1(\sigma_1 - C_2\sigma_2)\sqrt{l}}{\sqrt{\pi}} \sin^{-1}(a/l) - \sigma_2 \sqrt{\pi l} \tag{32}$$

Note we have added the effect of σ_2 closing the crack. The results of this equation with $C_2 = 3$ are shown in Figure 8, at different values of σ_2/σ_1. Figure 8 shows that initially K_I increases with increasing crack length, and starts to decrease as soon as $l > a$. Thus initial unstable behavior is predicted, followed by stable crack growth for $l > a$. The model has three free parameters, a, C_1 and C_2.

Consider a slightly more realistic configuration where the stress σ_0 is not constant over the region 2a but decreases linearly as shown in Figure 7. K_I for this case can be calculated in the same manner and gives:

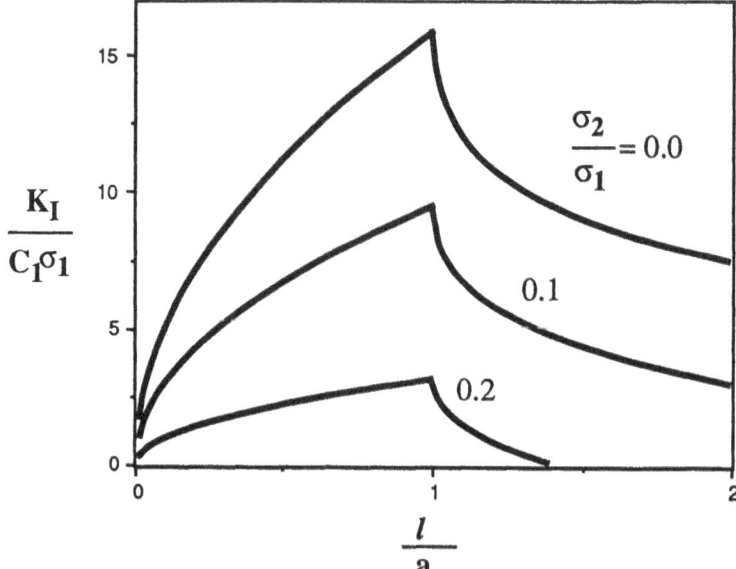

Figure 8. Results of the generic model in Figure 7a for different values of σ_2/σ_1, and taking $C_2 = 3$.

$$K_I = \frac{2C_1(\sigma_1-C_2\sigma_2)\sqrt{l}}{\sqrt{\pi}}\left[(2-l/a)\sin^{-1}(a/l) + \sqrt{1-a^2/l^2}\right] - \sigma_2\sqrt{\pi l} \qquad (33)$$

The results of this equation with $C_2 = 3$ are shown in Figure 9, at different values of σ_2/σ_1. In contrast with the results in Figure 8, the peak in the K_I vs. l curves in Figure 9 decreases with increasing σ_2/σ_1. This matches closely with the Sammis and Ashby (1986) model shown in Figure 3.

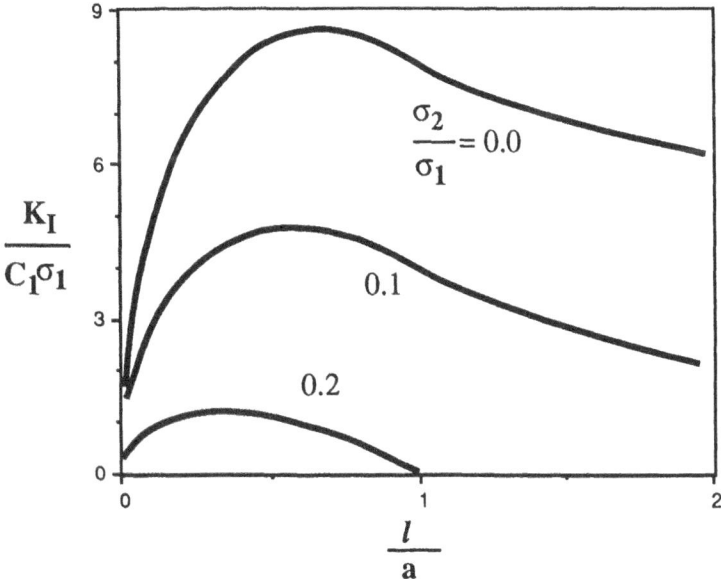

Figure 9. Results of the generic model in Figure 7b for different values of σ_2/σ_1, and taking $C_2 = 3$.

3.7 CRACK INTERACTION

The effects of crack interaction were not considered in any of the models described above. In general, the two extremes of crack interaction effects (in two dimensions) are collinear interaction and parallel interaction. Simple models for the effects of collinear and parallel crack interaction are considered here. Recall that many of the long-crack approximations are based on a crack oriented in the direction of σ_1 and subjected to point normal forces P at the center of the crack. Collinear crack interaction can be included by considering a collinear array of cracks, each containing center point forces P. The stress intensity solution for this is given by (Rooke and Cartwright, 1976):

$$K_I = \frac{P}{\sqrt{bsin(\pi l/b)}} - \sigma_2\sqrt{2btan(\pi l/2b)} \tag{34}$$

where b is the center-to-center distance between the cracks. This solution can form the basis for interaction effects in the models described above. For instance, for the long-crack approximation to the sliding crack as given in equation (15), the inclusion of crack interaction gives:

$$K_I = \frac{2l_0\tau^*cos\beta}{\sqrt{bsin(\pi l/b)}} - \sigma_2\sqrt{2btan(\pi l/2b)} \tag{35}$$

The intensity of crack interaction is a function of l/b, which varies from l_0/b initially to 1 when the cracks have coalesced into a splitting fracture. Similarly, the effects of parallel interaction can be included by considering a doubly periodic array of cracks, each containing center point forces P. For the sliding crack model, the K_I solution becomes (Kemeny and Cook, 1987a):

$$K_I = \frac{2l_0\tau^*cos\beta}{\sqrt{bsin(\pi l/b)}} \left[1-2sin(\pi l/2b)\frac{\ln\left[\frac{tan(\pi l/2b)}{tan(\pi l_0/2b)}\right]}{\frac{\pi f}{4b} + \ln\left[\frac{cos(\pi l_0/2b)}{cos(\pi l/2b)}\right]} \right] \tag{36}$$

where 2f is the spacing between parallel columns of cracks.

4 Derivation of Nonlinear Stress-Strain Curves

The K_I solutions as given in the previous section can be used to derive nonlinear stress-strain curves due to the growth of cracks under compressive stresses. Experiments have indicated a close relationship between nonlinear constitutive rock behavior and the growth, interaction, and coalescence of microcracks (e.g., Hallbauer et al., 1973). In general, the initial growth of cracks is associated with strain-hardening behavior, and the interaction and coalescence of cracks is associated with strain-softening behavior. These results also agree with experiments on plexiglass sheets containing slits and subjected to compressive stresses (Horii and Nemat-Nasser, 1986). In addition to the K_I solutions given in last section, we have the choice of the crack growth criterion to use, namely:

$K_I = K_{IC}$ crack growth criterion

or

Subcritical crack growth criterion - $\partial l/\partial t = A(K_I)^n$

If the $K_I = K_{IC}$ criterion is used, then rate-independent stress-strain curves can be derived. If the subcritical crack growth criterion is used, then rate-dependent stress-strain curves and creep behavior can be derived. In this section the derivation of nonlinear stress-strain relations and creep are demonstrated, using first the K_I solution for the long-crack approximation to the sliding crack without crack interaction (equation 15), and secondly, using the long-crack approximation to the sliding crack with crack interaction (equation 35). The results show that without crack interaction, the following characteristics of rock behavior can be modelled:

1. Strain-hardening
2. Sensitivity of stress-strain curves to σ_2.
3. Rate-dependence of stress-strain curve
4. Transient creep behavior

With crack interaction, the following two additional characteristics of rock behavior can be modelled:

5. Strain-softening
6. Creep Rupture

Also, it is found that using realistic material parameters, the models give an excellent match with laboratory data.

4.1 NONLINEAR STRESS-STRAIN CURVES - $K_I = K_{IC}$ CRITERION

Consider a body of width $2w$, height $2h$, and unit thickness, containing a single sliding crack, and subjected to principal stresses σ_1 and σ_2. The displacement at the boundary of the body will consist of the displacement of the body if the crack was not present plus the additional displacement due to the crack. The displacement due to the body with no crack is simply given by:

$$\delta_e = \frac{\sigma_1 \, 2 \, h}{E'} \tag{37}$$

The additional displacement due to the crack can be calculated using Castigliano's theorem, as given in equation (5), and using the K_I solution for the non-interacting sliding crack as given in equation (15). This gives:

$$\delta_c = \frac{\partial U_e}{\partial P} = \frac{\partial}{\partial P}\left\{ 2\int_{l_0}^{l} \frac{K_I^2}{E} \, dl \right\} \tag{38}$$

174

$$= \frac{8l_0^2 c(sc-\mu c^2)}{2wE}\left[\frac{2\tau^* c}{\pi} \ln\frac{l}{l_0} - \sigma_2(\frac{l}{l_0}-1) \right]$$

where throughout the rest of this section c and s are used for $\cos\beta$ and $\sin\beta$, respectively. At this point the displacement due to N non-interacting sliding cracks is considered, which will be N times the displacement given above. The total strain is calculated from the total displacement (elastic plus crack displacements) and gives:

$$\varepsilon_1 = \frac{\sigma_1}{E}\left[1 + \frac{8\chi c(sc-\mu c^2)}{\sigma_1}\left(\frac{2\tau^* c}{\pi}\ln\frac{l}{l_0} - \sigma_2(\frac{l}{l_0}-1) \right) \right] \tag{39}$$

where we define χ as the initial crack density, given by:

$$\chi = \frac{N\, l_0^2}{V} \tag{40}$$

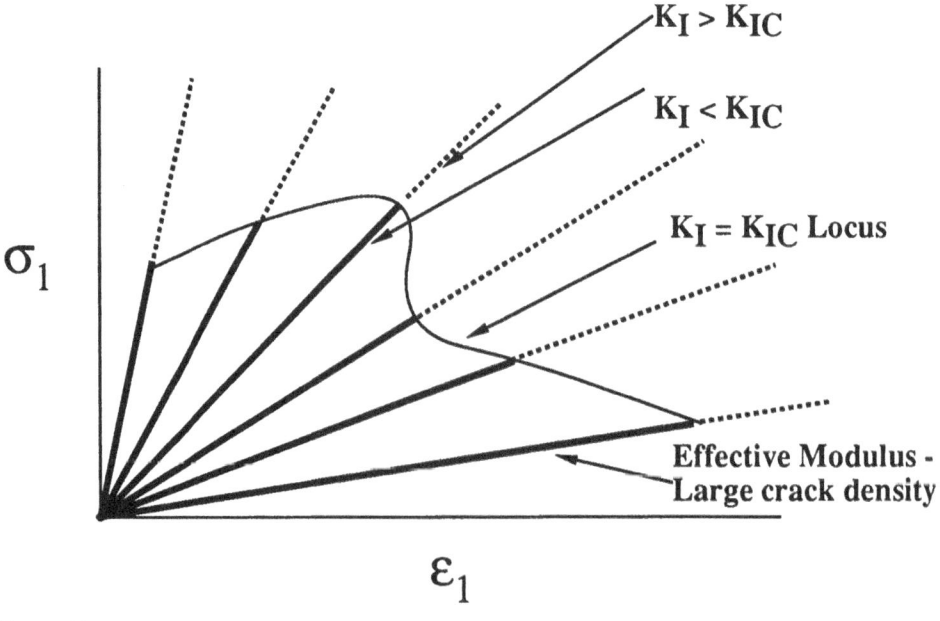

Figure 10. Illustration of the derivation of nonlinear stress-strain curves using the effective moduli for the body containing cracks of different lengths (straight lines), along with the K_I = K_{IC} criterion for crack growth for each of these crack lengths (solid points). Nonlinear stress-strain relation is the curve connecting these points.

V is the volume which is equal to 2w x 2h for a two dimensional body of unit thickness. For a fixed crack length, equation (39) represents a linear relationship between stress and strain. On a plot of stress vs. strain, this represents a series of straight lines with different slopes for different values of l, as shown on Figure 10. These lines would be the effective moduli for proportional loading.

The crack growth criterion $K_I = K_{IC}$ is now introduced. On each of the straight lines (i.e., for a given crack length $2l$), there will be a point on the line where the $K_I = K_{IC}$ criterion is met, which is given by:

$$K_{IC} = \frac{2l_0 \, \tau^* \, \cos\beta}{\sqrt{\pi l}} - \sigma_2 \sqrt{\pi l} \qquad (41)$$

This equation can be solved for σ_1 (τ^* is a function of σ_1, see equation 12), which gives the stress at which cracking will occur along any of the straight lines on Figure 10. Initially, the crack has a length l_0, and as the stress is increased loading will initially follow this line. When equation (41) above is satisfied for $l = l_0$, the crack will start to grow, and the stress and strain values will follow the nonlinear curve defined by the locus of points calculated from the stresses in equation (41) and the strains from equation (39). The results of this procedure are plotted in Figure 11 for material parameters representing Westerly granite (material properties given in Table 1), for different values of σ_2. Note that this model predicts initial linear loading, followed by strain hardening. Also, the results are very sensitive to small increases in σ_2.

Table 1. Material property values for Westerly granite and Oshima granite.

Property	Westerly granite	Oshima granite
β	45°	45°
μ	0.4	0.3
l_0/b	0.25	0.15
E	60 GPa	50 GPa
ν	0.25	0.2
b	600 μm	600μm
K_{IC}	1 MPa√m	-
n	-	31
A	-	6x10⁻⁸

For the case of uniaxial loading ($\sigma_2 = 0$), a closed form solution for this nonlinear curve can be derived by eliminating the crack length l from the two equations, which gives:

176

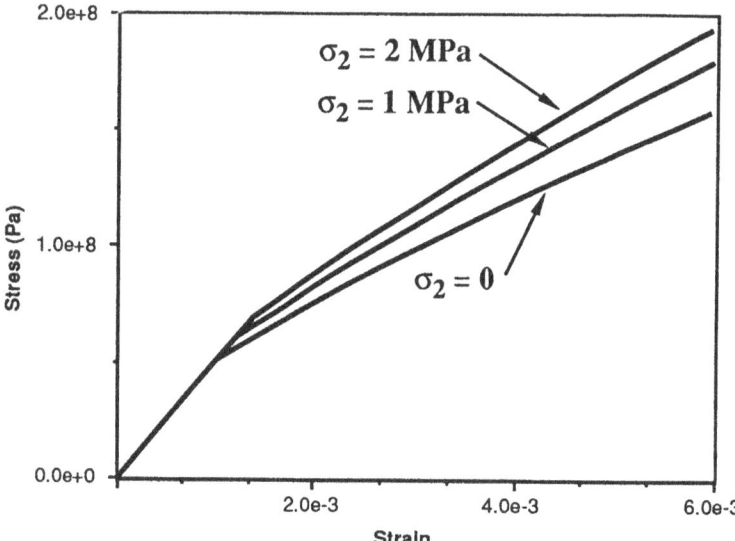

Figure 11. Nonlinear stress-strain curves calculated from the sliding crack model without crack interaction using the $K_I = K_{IC}$ crack growth criterion. Parameter values used for Westerly granite (see Table 1).

$$\varepsilon_1 = \frac{\sigma_1}{E}\left[1 + \frac{8\chi c(sc - \mu c^2)}{\sigma_1}\left(\frac{2\tau^* c}{\pi}\ln\frac{4l_0\tau^{*2}c^2}{K_{IC}^2\pi}\right)\right] \tag{42}$$

This analysis is now repeated, using the K_I solution for the model with crack interaction as given in equation (35). First the linear stress-strain behavior is calculated for a body containing N cracks, where collinear crack interaction is included. Using Castigliano's theorem, the total strain is given by:

$$\varepsilon_1 = \varepsilon_1^{elastic} + \frac{16\cos\beta\chi(\sin\beta\cos\beta - \mu\cos^2\beta)}{\pi\ E}\left[\tau^* \cos\beta \ \ln\frac{\tan(\pi l/b)}{\tan(\pi l_0/b)}\right.$$

$$\left. - \sigma_2\frac{b}{l_0}\ln\frac{\tan(\pi/4(1 + l/b))}{\tan(\pi/4(1 + l_0/b))}\right] \tag{43}$$

Again the crack density parameter χ has been used as given in equation (40). The second equation that is needed is the crack growth criterion, given by:

$$K_{IC} = \frac{2l_0\tau^*\cos\beta}{\sqrt{b\sin(\pi l/b)}} - \sigma_2\sqrt{2b\tan(\pi l/2b)} \qquad (44)$$

Taken together, equations (43) and (44) represent the nonlinear stress-strain behavior, i.e., the stress-strain curve is the locus of stresses calculated from equation (44) and strains calculated from equation (43) for crack lengths that vary from l_0/b to 1. Nonlinear stress-strain curves calculated for different values of σ_2 are presented in Figure 12. The material properties used are that of Westerly Granite, as given in Table 1. Also in Figure 12, these results are compared with experimental results from Wawersik and Brace (1971). The model results are able to reproduce many of the features of the stress-strain behavior of Westerly granite. This includes the initial strain-hardening due to the initial stable growth of the wing cracks before crack interaction, and the strain-softening behavior due to crack interaction. Also, the model predicts the large increase in strength with very small increases in confining stress. The model predicts that ultimate failure occurs by the wing cracks coalescing to form a single, macroscopic axial-splitting crack. The model does not predict the transition to shear faulting that occurs at higher values of confining stress. More sophisticated results could be produced by considering more complex crack interaction effects.

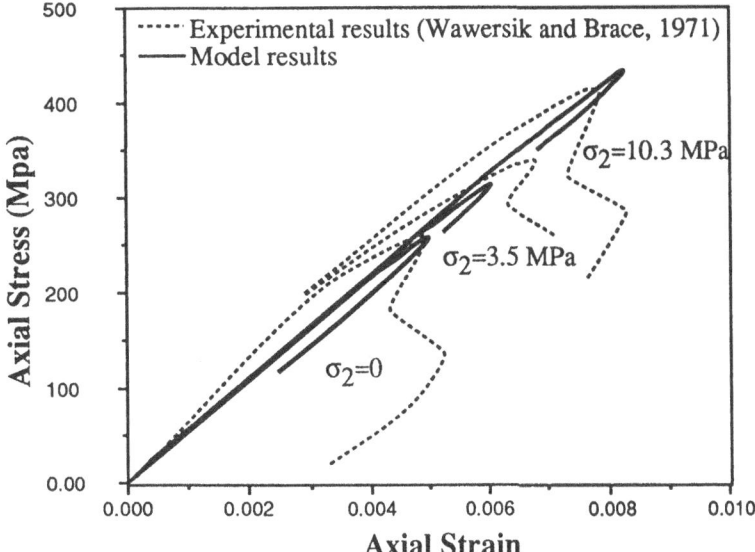

Figure 12. Nonlinear stress-strain curves calculated from the sliding crack model with crack interaction using the $K_I = K_{IC}$ crack growth criterion. Parameter values used for Westerly granite (see Table 1). Experimental results by Wawersik and Brace (1971) at the same confining stresses (dashed lines).

4.2 NONLINEAR STRESS-STRAIN CURVES - SUBCRITICAL CRACK GROWTH

Consider a body of width 2w, height 2h, and unit thickness containing N noninteracting sliding cracks, and subjected to uniaxial loading ($\sigma_2 = 0$). On the boundary of the body, a fixed uniaxial strain rate $\dot{\varepsilon}_0$ is applied. Assuming the strain at t=0 is equal to zero, this gives:

$$\varepsilon_1 = \dot{\varepsilon}_0 t \tag{45}$$

Using the value for the stress intensity factor from equation (15), the subcritical crack growth equation becomes:

$$\partial l/\partial t = A \left[\frac{2l_0\tau^*\cos\beta}{\sqrt{b\sin(\pi l/b)}} \right]^n \tag{46}$$

The linear stress-strain relation for this configuration is the same as equation (39) that was derived in the previous section (and setting $\sigma_2 = 0$). Equation (39) can be solved for σ_1 and along with equations (45) and (46), this gives:

$$\partial l/\partial t = A \left[\frac{2l_0\dot{\varepsilon}_0 t E(sc-\mu c^2)c}{\sqrt{\pi l}\left[1+\frac{16\chi c^2(sc-\mu c^2)^2}{\pi}\ln\frac{l}{l_0}\right]} \right]^n \tag{47}$$

This nonlinear ordinary differential equation, along with the initial condition that $l = l_0$ at t=0, can be solved numerically to give crack length as a function of time ($l(t)$). This, along with equation (39), gives σ_1 as a function of time:

$$\sigma_1(t) = \frac{\dot{\varepsilon}_0 t E}{1+\frac{16\chi c^2(sc-\mu c^2)^2}{\pi}\ln\frac{l(t)}{l_0}} \tag{48}$$

The strain as a function of time is given by equation (45), and the nonlinear stress-strain curve is the plot of these $\sigma(t)$, $\varepsilon(t)$ pairs for different times. Some numerical results are shown in Figure 13, using the properties of Oshima granite as given in Table 1. The results are shown for two values of the loading rate that differ by an order of magnitude.

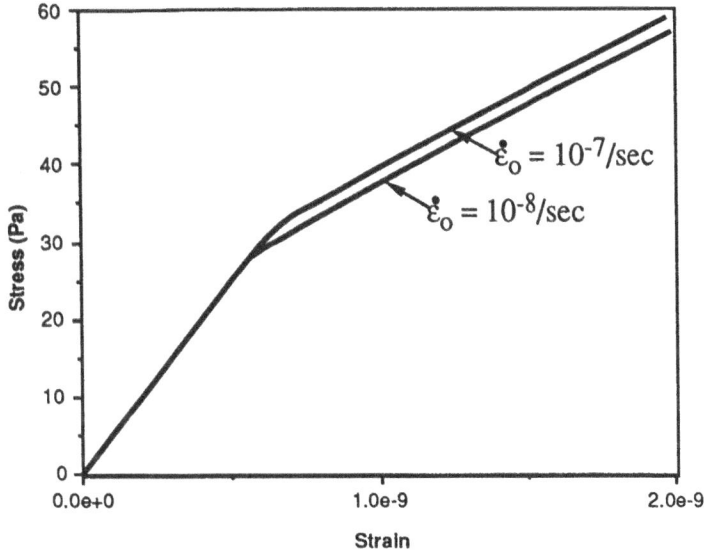

Figure 13. Nonlinear stress-strain curves from the sliding crack model without crack interaction using the subcritical crack growth criterion. Parameter values for Oshima granite (see Table 1).

As shown in Figure 13, the stress-strain curve is sensitive to the loading rate. Initially, the cracks grow at extremely small rates (due to the low K_I and the n value of 30), and in this region the results for the two loading rates follow the same elastic slope. As the cracks start to grow at higher rates (K_I gets closer to K_{IC}), the case with the higher loading rate results in higher stresses, which agrees with experimental data (Jaeger and Cook, 1979).

The analysis is now repeated, using the stress intensity factor solution that includes collinear crack interaction (equation 35). This solution is used in the subcritical crack growth equation, along with the linear stress-strain relation for the case of collinear crack interaction as given in equation (43). As before, this results in a nonlinear ordinary differential equation. Using the initial condition that $l/b = l_0/b$ at t=0, this equation can be integrated to give $l/b(t)$. This along with equation (43) can be used to give σ_1 as a function of time. These results are presented for the material properties of Oshima granite (Table 1) in Figure 14a for four values of applied strain rate. The results are compared with the experimental data of Sano et al. (1981) at the same strain rates in Figure 14b. The results show several interesting features. Initially, the behavior is linear, and this is due to the region when the cracks are growing at a very low rate. As the crack velocity increases, strain-hardening behavior is initially predicted, followed by strain-softening behavior. The volume strain is initially compressional, dominated by the solid matrix before the cracks begin to grow. As the cracks grow, the volume strain becomes dilatational. The stresses at which this occurs, and the amount of dilatation, match closely with the results of Sano et al. (1981), as shown in Figure 14b. The results show an increase in strength with increasing loading rate, which is in agreement with the experimental results.

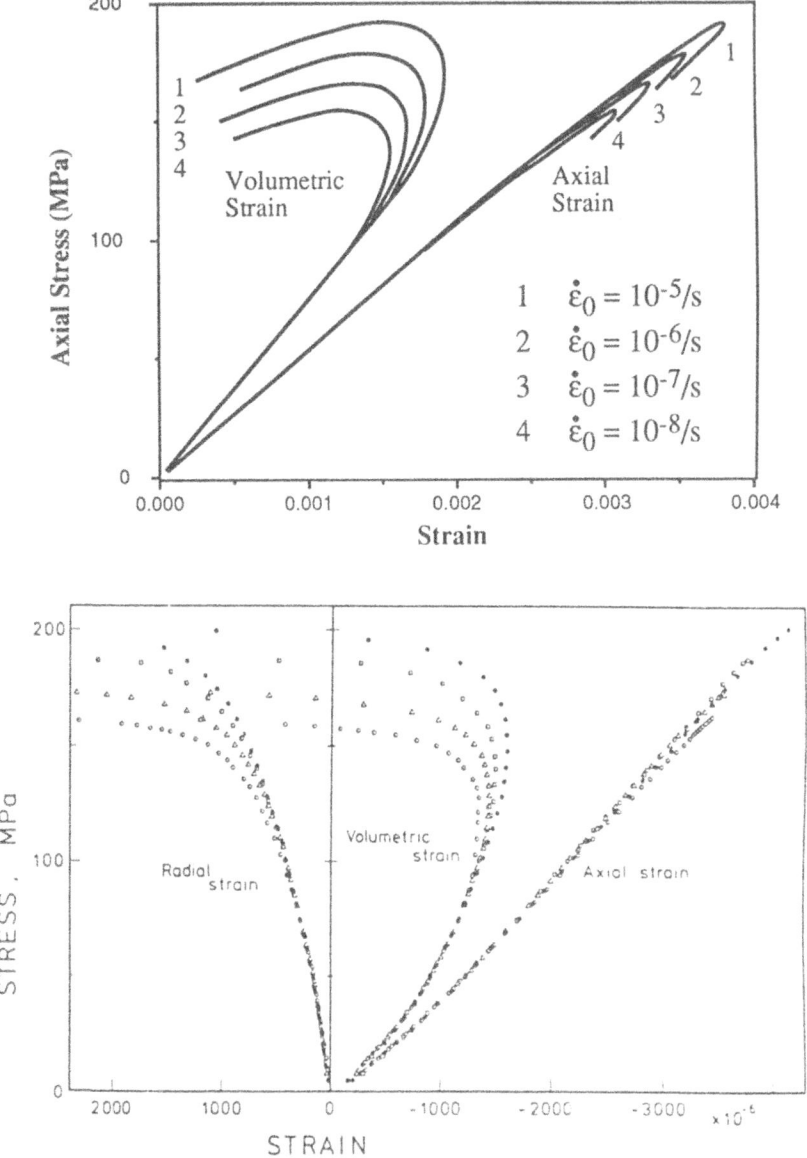

Figure 14. a) Nonlinear stress-strain curves from the sliding crack model with crack interaction using the subcritical crack growth criterion, at four values of applied strain rate. Parameter values for Oshima granite (see Table 1). b) Experimental results by Sano et al. (1981) at the same strain rates.

4.3 CREEP

For the case with no crack interaction, consider the same two equations as before, i.e., the linear stress-strain equation as given in equation (39), and the crack velocity equation as given in equation (46). To model the creep behavior of crack growth under compression, we now consider the boundary conditions of a constant uniaxial stress σ_1. In this case, using the initial condition that $l = l_0$ at t=0, equation (46) can be integrated analytically to give:

$$l = \left[A\frac{(2l_0\tau^*\cos\beta)^n}{\pi^{n/2}}(1+n/2) + l_0^{1+n/2} \right]^{\frac{1}{1+n/2}} \tag{49}$$

and the strain as a function of time becomes:

$$\varepsilon_1 = \frac{\sigma_1}{E}\left[1+\frac{16\chi c^2(sc-\mu c^2)^2}{\pi(1+n/2)}\ln\left(1+\frac{A(2l_0\tau^*\cos\beta)^n t(1+n/2)}{\pi^{n/2}} \right) \right] \tag{50}$$

This can be rewritten into the form:

$$\varepsilon_1 = \varepsilon_0 + C_1\ln(1+C_2t) \tag{51}$$

where

$$\varepsilon_0 = \frac{\sigma_1}{E}$$

$$C_1 = \frac{\sigma_1 16\chi c^2(sc-\mu c^2)^2}{E\pi(1+n/2)}$$

$$C_2 = \frac{A(2l_0\tau^*\cos\beta)^n(1+n/2)}{\pi^{n/2}}$$

Equation (51) predicts transient creep and is of the exact form that is often used to describe the transient creep in brittle rocks from experimental studies (Jaeger and Cook, 1979).

The above analysis is now repeated using the stress intensity factor solution that includes collinear crack interaction, equation (35). When this solution is used in the subcritical crack growth equation, it can be integrated to give $l/b(t)$, and using equation (43), $\varepsilon(t)$. The

182

results for different values of creep stress are presented in Figure 15. For small times for all values of σ_1 and for all times for low values of σ_1, the results agree with the results with no crack interaction, i.e., transient creep. For values of σ_1 close to the failure stress, a transition from transient creep to creep rupture is seen.

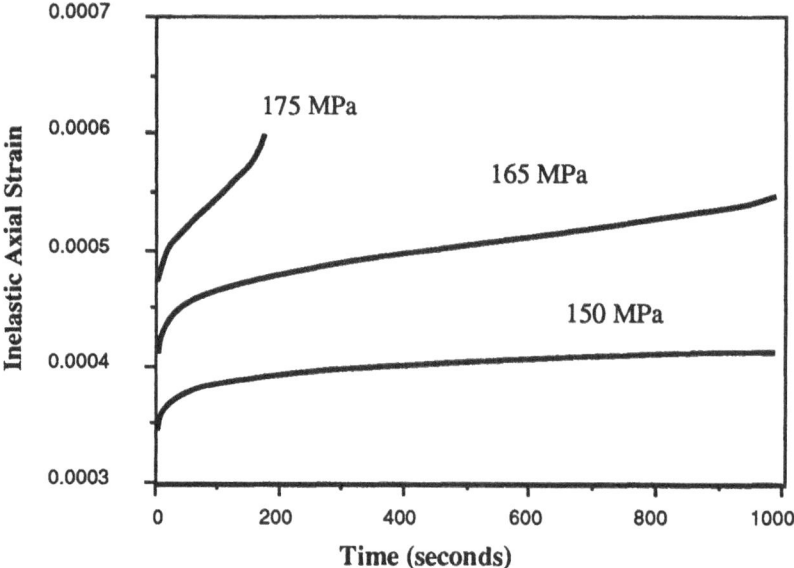

Figure 15. Creep vs. time using the sliding crack model with crack interaction and using the subcritical crack growth criterion, at three values of creep stress. Parameter values for Oshima granite (see Table 1).

5 Continuum Damage Modelling

The nonlinear results derived in the previous section were made under the assumption that the principal stresses σ_1 and σ_2 are uniform throughout the body (no end effects or stress gradients). Using the finite element method, these results can be extended to complicated boundary conditions that contain stress gradients. This is accomplished by implementing the procedures described in the previous section into an elastic finite element code on an element by element basis, where in each element it is assumed that the principal stresses are uniform. Like the results in the previous section, nonlinear behavior is predicted due to crack growth in an otherwise linearly elastic body. Also as in the previous section, two relationships are needed, the first being the linear stress-strain relationship for each element as a function of its crack density, and the second being a crack growth criterion. If the K_I $=K_{IC}$ crack growth criterion is used, then rate-independent nonlinear behavior will be predicted. If the subcritical crack growth criterion is used, then rate-dependent nonlinear behavior and creep will predicted. Results using both the K_I $=K_{IC}$ and subcritical crack growth criteria are presented in the following sections. First the issue of stress-induced anisotropy is discussed.

5.1 STRESS-INDUCED ANISOTROPY

The micromechanical models reviewed in section 2 predict that crack growth under compression occurs primarily in the direction of σ_1. For a body that is initially isotropic, crack growth in the direction of σ_1 will result in stress-induced anisotropic rock properties, and this must be taken into account in the finite element calculations. In the case of an initially isotropic body, crack growth in the direction of the maximum principal stress will render the material transversely isotropic. The linear stress-strain relations under the assumption of transverse isotropy are given by:

$$\varepsilon_x = \frac{\sigma_x}{E_1} - \frac{v_2 \sigma_y}{E_2} - \frac{v_1 \sigma_z}{E_1}$$

$$\varepsilon_y = -\frac{v_2 \sigma_x}{E_2} + \frac{\sigma_y}{E_2} - \frac{v_2 \sigma_z}{E_2}$$

$$\varepsilon_z = -\frac{v_1 \sigma_x}{E_1} - \frac{v_2 \sigma_y}{E_2} + \frac{\sigma_z}{E_1} \tag{52}$$

$$\varepsilon_{xy} = \frac{2(1+v_1)}{E_1} \sigma_{xy}$$

$$\varepsilon_{xz} = \frac{1}{G_2} \sigma_{xy}$$

$$\varepsilon_{yz} = \frac{1}{G_2} \sigma_{yz}$$

Note that there are five independent elastic constants, E_1, E_2, v_1, v_2, and G_2. In two dimensions, the linear stress-strain relationship can be put into the form (plain strain):

$$\left\{ \begin{array}{c} \sigma_x \\ \sigma_y \\ \sigma_{xy} \end{array} \right\} = D \left\{ \begin{array}{c} \varepsilon_x \\ \varepsilon_y \\ \varepsilon_{xy} \end{array} \right\} \tag{53}$$

where the D matrix is given by:

$$D = \frac{E_2}{(1+v_1)(1-v_1-2nv_2^2)} \begin{bmatrix} n(1-v_2^2) & nv_2(1+v_1) & 0 \\ n(1+v_1)v_2 & (1-v_1^2) & 0 \\ 0 & 0 & m(1+v_1)(1-v_1-2nv_2^2) \end{bmatrix} \tag{54}$$

where $n=E_1/E_2$ and $m=G_2/E_2$. The above formula assume that the cracks grow parallel to the x axis. If crack growth is inclined relative to the x axis, then the D matrix can be transformed by a matrix containing the direction cosines (Zienkiewicz and Taylor, 1989). Using equation (52), Castigliano's theorem, and one of the K_I solutions for crack growth under compression from section 2, the five independent elastic constants for stress-induced anisotropy can be determined. For instance, for the sliding crack model with collinear crack interaction (equation 35), this gives:

$$E_1 = \frac{E}{1 + \frac{16\chi}{\pi}\left[(sc-\mu c^2)-\frac{\sigma_2}{\sigma_1}(sc+\mu s^2)\right]c^2(sc-\mu c^2)\ln\left[\frac{\tan(\pi l/2b)}{\tan(\pi l_0/2b)}\right]}$$

$$E_2 = \frac{E}{1 + \frac{16\chi}{\pi}(b/l_0)^2\ln\sec(\pi l/2b)}$$

$$v_1 = \frac{E_1 v}{E} \tag{55}$$

$$v_2 = \frac{E_2}{E}\left[v + \frac{16\chi}{\pi}(sc-\mu c^2)\left[c(cs+\mu s^2)\ln\frac{\tan(\pi l/2b)}{\tan(\pi l_0/2b)}\right] + \frac{b}{l_0}\ln\frac{\tan(\pi/4(1+l/b))}{\tan(\pi/4(1+l_0/b))}\right]$$

$$G_2 = \frac{E}{2(1+v) + \frac{16\chi}{\pi}(b/l_0)^2\ln\sec(\pi l/2b)}$$

5.2 FINITE ELEMENT DAMAGE MODEL - $K_I = K_{IC}$ CRITERION

The D matrix for stress-induced anisotropy discussed above is a measure of the damage that occurs in each of the elements due to crack growth and interaction. This damage measure, along with the $K_I = K_{IC}$ crack growth criterion, has been implemented into an elastic finite element code to produce a nonlinear damage model. An addition to the model was made to account for the presence of isotropic damage in addition to the anisotropic damage due to the growth of axial extensile cracks. This allows for the many additional sources of damage such as the linking of axial cracks via a shear crack. Results of this model for the geometry of a thick walled cylinder are presented in Figure 16. The thick walled cylinder is subjected to a vertical external stress of 260 MPa and a horizontal stress of 130 MPa. The properties assumed are that of Westerly granite given in Table 1, and the sliding crack model with crack interaction was used. As shown in Figure 16, an elliptically shaped breakout region occurs, which agrees with the numerical results of Zheng et al. (1989) and experimental results of Ewy (1989) and others. It is interesting to note that

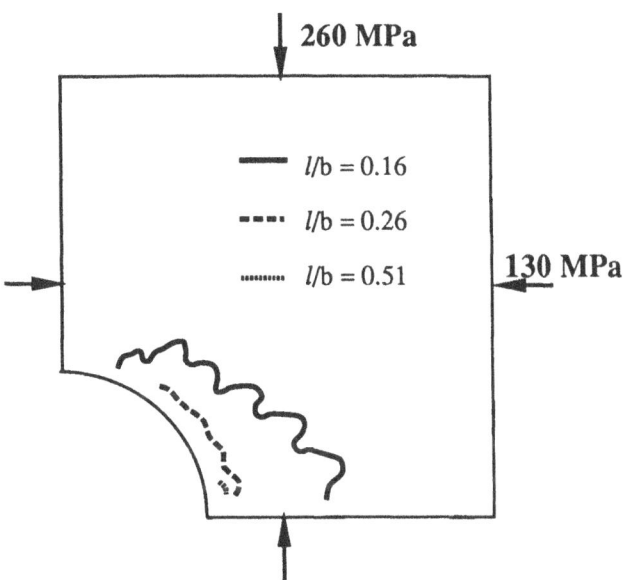

Figure 16. Contours of damage for a thick-walled cylinder subjected to horizontal and vertical stresses of 260 and 130 MPa, respectively. Damage model based on $K_I = K_{IC}$ crack growth criterion and sliding crack model with crack interaction.

damage progresses into the rock without completely failing the rock in the damage zone. For instance, at the boundary of the hole, the damage variable, l/b, reaches a value of 0.508 ($l/b=1$ is complete failure). This agrees with experimental results (e.g., Ewy, 1989).

5.3 FINITE ELEMENT DAMAGE MODEL - SUBCRITICAL CRACK GROWTH

Here the subcritical crack growth criterion is used as the basis for a finite element damage model. Implementing the subcritical crack growth criterion rather than the $K_I = K_{IC}$ criterion results in several improvements in the damage model. One improvement is that the increase in crack length in each element at each time increment is calculated explicitly from the subcritical crack growth equation. Another improvement is the sensitivity of the results to the applied rate of loading. This rate dependence plays an important role in actual problems in geomechanics. In drilling, for instance, stress concentrations around the borehole can occur almost simultaneously with the drilling. In laboratory experiments, however, rock samples are usually pre-drilled, followed by the application of load at slower loading rates. These differences can be examined using the rate-dependent damage model. Some results of the rate-dependent damage model are presented in Figure 17. A thick walled cylinder is subjected to a fast and a slow loading rate. In the fast loading rate, σ_1 and σ_2 are 0.25 and 0.125 MPa/sec, respectively. In the slow loading rate, σ_1 and σ_2 are 0.125 and 0.0625 MPa/sec, respectively. Figure 17 compares the $l/b=0.3$ contour

186

between the fast and slow loading rate results. The results show that the breakout is more developed for the slow loading rate, which agrees with the results in section 4.2.

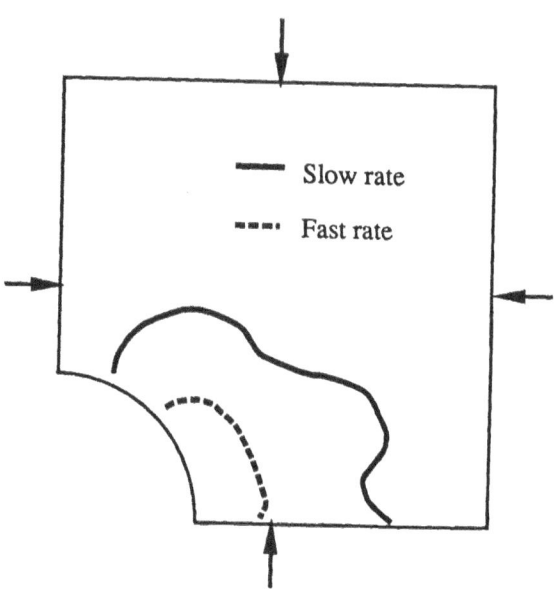

Figure 17. Comparison of the $l/b=0.3$ contour between a fast and a slow loading rate applied to the thick-walled cylinder. Damage model based on subcritical crack growth criterion.

6 Conclusions

Linear elastic fracture mechanics is an important tool in understanding the nonlinear deformation of rocks. There are many different mechanisms for extensile crack growth in rocks due to differential compression, including pore crushing, sliding along pre-existing cracks, elastic mismatch between grains, dislocation movement, and hertzian contact. We have reviewed several fracture mechanics models for these different mechanisms, and many similarities between the models are evident. Based on these similarities, a generic model is proposed that encompasses the different models. Models to take into account the effects of crack interaction are also discussed. From the micromechanical models, nonlinear stress-strain curves are derived that exhibit strain-hardening and strain-softening, dilatation, σ_2 sensitivity, and rate dependence. Also, it is shown how these models can form the basis for continuum damage models using the finite element method.

Acknowledgements

This work was supported, in part, by the Director, Office of Basic Energy Sciences, Division of Engineering, Mathematics, and Geosciences, of the U.S, Department of Energy under contract DE-AC03-76SF00098. Special thanks are given to Dr. Ziqiong Zheng for allowing us to use the photograph shown in Figure 1, and to F. F. Tang for providing the rate-dependent finite element damage results.

References

Ashby, M.F. and Hallam, S.D. 1986. The failure of brittle solids containing small cracks under compressive stress. Acta Metall. 34:497-510.

Atkinson, B. K. 1984. Subcritical crack growth in geologic materials. J. Geophys. Res., 89:4077-4114.

Atkinson, B. K., and Meredith, P. G. 1987. The theory of subcritical crack growth with applications to minerals and rocks, In: Fracture Mechanics of Rock, edited by B. K. Atkinson, Academic Press, London, pp 111-166.

Barrenblatt, G. I. 1962. Mathematical theory of equilibrium cracks in brittle fracture, Adv. Appl. Mech., vol. 7., Academic Press, NY.

Carter, N.L., Anderson, D.A., Hansen, F.D., and Kranz, R.L. 1981. Creep and creep rupture of granitic rocks, In The Mechanical Behavior of Crustal Rocks, Geophysical Monograph 24, AGU, Washington, D.C.

Charles, R.J. 1958. Static fatigue of glass, J. Appl. Phys. 29:1549-1560.

Cherepanov, G.P. 1979. Mechanics of Brittle Fracture, McGraw-Hill, NY.

Costin, L.S. 1985. Damage mechanics in the post-failure regime. Mech. Mat. 4:149-160.

Costin, L.S. and Stone, C.M. 1987. Implementation of a finite element damage model for rock. Proc. 2nd Int. Conf. Constitutive Laws for Eng. Mat. 2:829-840.

Dey, T.N. and Wang, C.Y. 1981. Some mechanisms of microcrack growth and interaction in compressive rock failure, Int. J. Rock Mech. & Min. Sci., 18:199-209.

Ewy, R. 1989. Ph.D. thesis, University of California, Berkeley.

Fredrich, J.T., Evans, B. and Wong, T.-f. 1989. Micromechanics of the brittle to plastic transition in Carrara marble. J. Geophys. Res. 94:4129-4145.

Fredrich, J.T.. and Wong, T.-f. 1986. Micromechanics of thermally induced cracking in three crustal rocks. J. Geophys. Res. 91:12743-12764.

Gough, D.I., and Bell, J.S. 1982. Stress orientations from borehole wall fractures with examples from Colorado, east Texas, and northern Canada, Can. J. Earth Sci., 19:1358-1370.

Guenot, A. 1989. Borehole breakouts and stress fields. Int. J. Rock Mech. 26:185-195.

Haimson, B.C. and Herrick, C.G. 1985. In situ stress evaluation form borehole breakouts experimental studies, Proc. U.S. Symp. Rock Mech. 26th, 1207-1218.

Haimson, B.C. and Herrick, C.G. 1989. Borehole Breakouts and In Situ Stress. Proceedings, Energy-Source Technology Conference, Houston, TX.

Hallbauer, D.K., Wagner, H., and Cook, N.G.W. 1973. Some observations concerning the microscopic and mechanical behavior of quartzite specimens in stiff, triaxial compression tests. Int. J. Rock Mech. Min. Sci. 10:713-726.

Horii, H. and Nemat-Nasser, S. 1986. Brittle failure in compression: splitting, faulting, and brittle-ductile transition. Phil. Trans. Royal Soc. London. 319:337-374.

Jaeger, J.C., and Cook, N.G.W. 1979. Fundamentals of Rock Mechanics, 3rd Edition, Chapman and Hall, London.

Johnson, K.L. 1985. Contact Mechanics. Cambridge University Press, Cambridge.

Kemeny, J. M., and Cook, N. G. W. 1987a. Crack models for the failure of rock under compression. Proc. 2nd Int. Conf. Constitutive Laws for Eng. Mat. 2:879-887.

Kemeny, J.M. and Cook, N.G.W. 1987b. Determination of rock fracture parameters from crack models for failure in compression. Proc. 28th US Symposium on Rock Mechanics, 1:367-374.

Kemeny, J. M. 1990. A model for nonlinear rock deformation under compression due to subcritical crack growth, To appear.

Kemeny, J.M. and Tang, F.F. 1990. A Numerical Damage Model for Rock Based on Microcrack Growth, Interaction, and Coalescence, To appear, Symposium on Damage Mechanics in Engineering Materials at the ASME Winter Annual Meeting, Nov. 25-30, 1990, Dallas, Tex.

Krajcinovic, D. 1989. Damage mechanics, Mech. Mats., 8:117-197.

Kranz, R. L. 1983. Microcracks in rocks: a review. Tectonophysics. 100:449-480.

Lawn, B.R. and Wilshaw, T.R. 1975. Fracture of Brittle Solids. Cambridge University Press, Cambridge.

Liebowitz, H. 1968. Fracture: An Advanced Treatise, Academic Press, NY.

Nemat-Nasser, S. and Horii, H. 1982. Compression-induced nonplanar crack extension with application to splitting, exfoliation, and rock burst. J. Geophys. Res. 87:6805-6822.

Rooke, D.P., and Cartwright, D.J. 1976. Compendium of Stress Intensity Factors, The Hillingdon Press, Middx.

Sammis, C.G. and Ashby, M.F. 1986. The failure of brittle porous solids under compressive stress states. Acta Metall. 34:511-526.

Sano, O., Ito, I., and Terada, M. 1981. Influence of strain rate on dilatancy and strength of Oshima granite under uniaxial compression, J. Geophys. Res., 86:9299-9311.

Santarelli, F.J., and Brown, E.T. 1987. Performance of deep wellbores in rock with a confining pressure-dependent elastic modulus, Proc. 6th Int. Congr. Rock Mech. Montreal, pp. 1217-1222, Balkema, Rotterdam.

Sokolnikoff, I.S. 1956. Mathematical Theory of Elasticity, McGraw-Hill, NY.

Steif, P.S. 1984. Crack extension under compressive loading. Eng. Fract. Mech. 20:463-473.

Sternberg, E. and Rosenthal, F. 1952. The elastic sphere under concentrated loads, J. Appl. Mech. 19:413-21.

Wawersik, W. R. and Brace, W. F. 1971. Post-failure behavior of a granite and diabase. Rock Mech. 3:61-85.

Wong, T.-f. 1990. A note on the propagation behavior of a crack nucleated by a dislocation pileup, J. Geophys. Res., 95:8639-8646.

Wong, T.-f. 1982. Micromechanics of faulting in Westerly granite. Int. J. Rock Mech. Min. Sci. 19:49-64.

Wong, T.-f. 1985. Geometric probability approach to the characterization and analysis of microcracking in rocks. Mech. of Mat. 4:261-276.

Zhang, J., Wong, T.-f., and Davis, D.M. 1989. Micromechanics of pressure-induced grain crushing in porous rocks. J. Geophys. Res., in press.

Zheng, Z., Kemeny, J. M., and Cook, N. G. W. 1989. Analysis of wellbore breakouts. J. Geophys. Res. 94:7171-7182.

Zheng, Z. 1989. Ph.D. thesis, University of California, Berkeley.

Zienkiewicz, O.C. and Taylor, R.L. 1989. The Finite Element Method. McGraw-Hill, London.

ASYMPTOTIC ANALYSIS OF COHESIVE CRACKS AND ITS RELATION WITH EFFECTIVE ELASTIC CRACKS

J. PLANAS and M. ELICES
Department of Materials Science. Escuela de Ingenieros de Caminos
Universidad Politécnica de Madrid. Ciudad Universitaria
28040-Madrid
Spain

ABSTRACT. Replacement of a cohesive crack problem by an approximate linear elastic problem was proven useful in some instances. This contribution provides a unified treatment of a wide class of equivalences between such problems and sets limiting conditions for the effective crack models. After reviewing the cohesive crack model, the concept of *equivalent crack* is introduced and some examples of equivalences are discussed. Finally, the equivalences are analyzed in the limit of very large sizes, and it is shown that some equivalences, different for small sizes, merge when the specimen size is increased. Some equivalences, however, remain different, particularly the equivalence which is at the root of the *R-CTOD-curve* model. Equations to improve the predictions of this model are given for large sizes.

1. Introduction

Modelling crack initiation and propagation in cohesive materials leads to solving a non linear fracture mechanics problem and, with a few exceptions, one has to resort to numerical analysis. Under these circumstances it is quite natural to look for approximate solutions based on Linear Elastic Fracture Mechanics (LEFM), a field where computation is easier and more experience is available.

The substitution of an actual fracture process —with a plastic or non-linear zone surrounding the crack tip— by an effective or equivalent crack was probably the first approximation to non-linear fracture problems. In the early approaches the effective crack extension was mildly related to the size of the plastic zone in quite an intuitive way, but the ability of the effective crack to represent the actual behaviour was not deeply investigated at a theoretical level, although it was extensively used in experiment interpretation and in design.

The *equivalent* linear elastic problem has to be solved in conjunction with an associate R-curve or, otherwise stated, a crack growth rule has to be independently stated as a relationship between the effective crack growth resistance and the effective crack extension. This is not the only price one has to pay for simplifying the cohesive crack; it was proven that the R-curve is geometry and size dependent —i.e., is not a material property— and, consequently, the equivalence becomes severely restricted.

In spite of these shortcomings there is some evidence [1] showing that for usual geometries and available sizes, differences among R-curves (when adequately formulated)

189

S. P. Shah (ed.), Toughening Mechanisms in Quasi-Brittle Materials, 189–202.
© 1991 *Kluwer Academic Publishers.*

are well inside the experimental scatter and, in this respect, they may be considered a material property for practical purposes, as long as peak-load versus size prediction is concerned.

However, there is no way to know *a priori* whether this R-curve which fits very well the peak-load size effect will be equally accurate in describing other aspects of the behavior, for example the load-displacement curve, or any other curve.

The aim of this contribution is to deepen in this research line and provide limits of validity of the effective elastic crack models as approximations of cohesive crack models. In this paper the concept of effective elastic crack is discussed first and afterwards the equivalence for large sizes is explored with the help of the asymptotic method, already developed by the authors [2, 3, 4].

2. The Equivalent Elastic Crack

We shall restrict to a class of cohesive materials and loading according to the following hypothesis:

H.1.- *Loading*: Loading is assumed to consist in monotonic mode I crack growth, achieved by symmetric proportional loading of an initially symmetric specimen.

H.2.- *Bulk Behaviour*: The material displays linear elastic bulk behaviour —with Young modulus E and Poisson's ratio v— as long as the major principal stress does not reach a critical value f_t.

H.3.- *Crack Initiation*: When the maximum principal stress reaches f_t (the tensile strength) fracture is initiated and strain localization takes place in what is called the Fracture Process Zone (FPZ). The FPZ is modelled as a cohesive crack where the strain localization is idealized as a displacement jump or crack opening, while cohesive stresses simulate the softening behaviour.

H.4.- *Crack Evolution*: Once the cohesive crack has formed, the stress transferred through the crack faces is assumed to depend upon the relative displacement of the crack faces.

For these loading conditions, the stress normal to the crack plane, σ, is supposed to be given by a single-valued, non-negative function of the crack opening w, i.e.:

$$\sigma = F(w) \quad \text{where} \quad F(0) = f_t \quad \text{and} \quad F(w) \geq 0 \tag{2.1}$$

The material function $F(w)$, together with E and v, suffice to characterize the cohesive material behaviour, as far as monotonic mode I is concerned. Based on this function, usually called the softening curve, several definitions were done:

a.- The work needed to monotonically open a crack of unit surface up to w, the *specific work supply*, is a material function given by

$$W_F(w) = \int_0^w F(w')dw' \tag{2.2}$$

b.- The specific work supply needed to fully open a unit surface is also a material property called the *specific fracture energy G_F* (or, simply, fracture energy), i.e.,

$$G_F = \int_0^{w_c} F(w')dw' \tag{2.3}$$

where, it is assumed that the value of $F(w)$ is zero for crack openings exceeding w_c.

Now, let us go back to the concept of *equivalent crack*, and consider a particular equivalence —the force-displacement equivalence— to fix the idea. Two geometrically identical cracked samples, as shown in figure 1, are loaded under displacement control u. One sample is made with a cohesive material, as defined above, and the other is made with a linear elastic material. The measured response of the two samples —i.e. the loads P and P_{eq} — for every displacement u will be different, but we can force the responses to match each other, $P = P_{eq}$, by choosing a suitable *equivalent crack length a_{eq}* and a suitable *equivalent crack growth resistance R_{eq}* at each deformation level.

In doing so, we force both samples to exhibit the same P-u behaviour, but in general, the equivalence ends here; stress or displacement fields, or relevant parameters like *CMOD* or *CTOD*, are not the same. Moreover, the price paid for the equivalence is that the linear elastic material has not a constant crack growth resistance. Instead, a changing value with crack length is needed in order to keep the P-u equivalence. Moreover the R-Δa curve obtained is not a material property, it depends on the geometry and specimen size.

At first sight, the advantages of using this equivalence are not obvious since there are not simple rules for the generation of the R-curves for every geometry and size. However, in some circumstances this can be done as we shall see later.

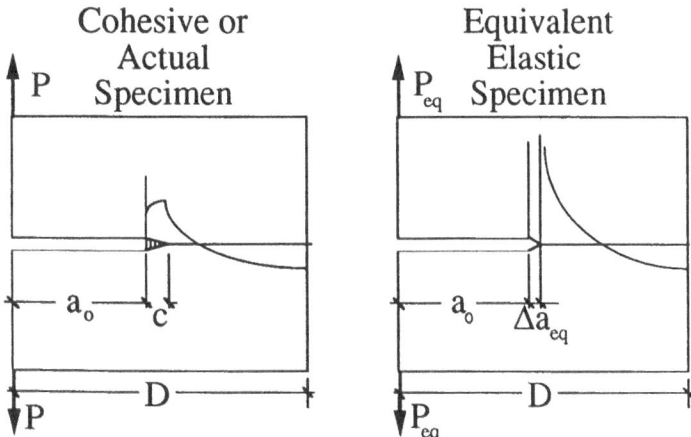

Figure 1. Definition of equivalent elastic specimen.

In a general equivalence, the equivalent resistance to crack growth may be obtained (for a monotonic process), by using the classical concepts of LEFM, because the equivalent specimen is, by construction, linear elastic. Given the equivalent load P_{eq} and the equivalent crack length a_{eq}, we may compute the stress intensity factor for the equivalent crack, which will always take the form:

$$K_{Ieq} := K_{Ieq}(P_{eq}, a_{eq}) = \frac{P_{eq}}{B\sqrt{D}} S\left(\frac{a_{eq}}{D}\right) \tag{2.4}$$

where B is the specimen thickness, D one of its characteristic in-plane dimensions, and $S(a/D)$ is the geometrical shape factor.

Owing to the postulated linearity of the material of the equivalent specimen, the resistance to crack growth may be written in any of the following equally valid forms:

$$R_{eq} := G_{Req} = J_{Req} = \frac{K_{Req}^2}{E'} \tag{2.5}$$

where G_R is the resistant-energy release rate, J_R the resistant-J integral, and K_R the resistant-stress intensity factor, and E' is the effective elastic modulus for generalized plane stress or plane strain. Since the equivalent crack is on the monotonic quasi-static loading curve, hence in a state of incipient growth, the resistant-K must be equal to the driving-K in Eq. (2.4), and then

$$R_{eq} := \frac{[K_{Ieq}(P_{eq}, a_{eq})]^2}{E'} \tag{2.6}$$

2.1. P-Y EQUIVALENCES

This kind of equivalence is shown in figure 1. The actual sample is sketched on the left, its cohesive zone has grown monotonically up to C, and the corresponding load is $P(C)$. The equivalent sample, made with an elastic non cohesive material, is sketched on the right and it is loaded with *same P* value (hence the P equivalence labeling). Notice that the crack length is not a_0 but $a^{P-Y} = a_0 + \Delta a^{P-Y}$, where P stands for the imposed load and Y for the magnitude related with the second degree of freedom. One should realize that the stress and displacement fields of the right hand sample are known when the load and the crack length are known. Since load is fixed, only one degree of freedom remains.

2.1.1. *P-u Equivalence.* When load point displacement u is chosen as a second variable, one arrives at the load-displacement equivalence. If P and u are measured in the actual sample, the *P-u equivalent elastic crack length* a^{P-u} can be computed from:

$$C_{eq}(a^{P-u}) = \frac{u}{P} \tag{2.7}$$

where $C_{eq}(a)$ is the expression for the compliance of the equivalent specimen for a crack length a, which may be obtained from linear elastic analyses.

Equation (2.7) determines the equivalent crack length at each loading step, from which, and the known load, other magnitudes may be computed for the P-u equivalent specimen.

In particular, the *P-u equivalent crack growth resistance* R^{P-u} may be obtained from equations (2.6) and (2.4) substituting $R_{eq} = R^{P-u}$, $a_{eq} = a^{P-u}$ and $P_{eq} = P$. The resultant R-curve will be geometry and size dependent because of the implicit size and geometry dependence of both members of equation (2.7)

2.1.2. *P-CMOD Equivalence.* When *CMOD* (Crack Mouth Opening Displacement) is chosen instead of the displacement associated to the load, one has a *P-CMOD* equivalence. The *P-CMOD equivalent elastic crack length* $a^{P\text{-}CMOD}$ can be computed from an equation entirely similar to (2.7),

$$C_{eq}^{CMOD}(a^{P\text{-}CMOD}) = \frac{CMOD}{P} \tag{2.8}$$

where $C_{eq}^{CMOD}(a)$ is the expression for the compliance associated to *CMOD* for a non cohesive sample with a crack of length *a*.

The *P-CMOD crack growth resistance* may be obtained —after computing the equivalent crack length from (2.8)— from (2.6) and (2.4) with the adequate change of indices as done before for the P-u equivalence. The same comments as before, regarding size and geometry dependence of R-curves, can be done.

It can not be stated, at first sight, that equations (2.7) and (2.8) are equivalent and, hence, there is no reason for $a^{P\text{-}u}$ and $a^{P\text{-}CMOD}$ to take the same value.

2.2. X-Y EQUIVALENCES

The above reasoning used to set *P-X* equivalences can be generalized to any couple of variables *X-Y*. Now, the actual specimen and the equivalent (or *virtual*) specimen are not bearing the same load, in general, and the equivalent load $P^{X\text{-}Y}$ and equivalent crack length $a^{X\text{-}Y}$ corresponding to the virtual specimen can be computed by equating *X* and *Y* in both specimens:

$$X_{eq}(P^{X\text{-}Y}, a^{X\text{-}Y}) = X \tag{2.9}$$

$$Y_{eq}(P^{X\text{-}Y}, a^{X\text{-}Y}) = Y \tag{2.10}$$

where the left hand members are the elastic expressions for magnitudes *X* and *Y* in the equivalent specimen and the right hand members the actual values of *X* and *Y* (either measured or computed using the cohesive model).

Equations (2.9) and (2.10) determine the equivalent load and crack length, from which any other magnitude can be found for the equivalent specimen. In particular, the equivalent crack growth resistance is, again, obtained from (2.4) and (2.6) with the obvious equivalence specifiers: $R_{eq} = R^{X\text{-}Y}$, $a_{eq} = a^{X\text{-}Y}$ and $P_{eq} = P^{X\text{-}Y}$.

2.2.1. *J-CTOD Equivalence.* The couple *J-CTOD* is an example of the generalized *X-Y* equivalence The variable *J* stands for the J-integral, with the essential remark that when the cohesive sample is considered, the J-integral has to be taken over a path always surrounding the cohesive zone. Under such circumstances it was shown [5] that:

$$J = W_F(CTOD) \tag{2.11}$$

where W_F, the specific work supply, was defined in (2.2). For the non-cohesive sample, *J* is equal to K_I^2/E' as stated before in Eq. (2.5). The variable *CTOD* is the crack opening at the initial crack tip.

The couple of equations (2.9) and (2.10) for this particular equivalence are, after use of (2.6) and (2.11):

$$\frac{[K_{Ieq}(P^{J\text{-}CTOD}, a^{J\text{-}CTOD})]^2}{E'} = W_F(CTOD) \qquad (2.12)$$

$$w_{eq}(P^{J\text{-}CTOD}, a^{J\text{-}CTOD}, a_0) = CTOD \qquad (2.13)$$

where $w_{eq}(P, a, a_0)$ is the crack opening at the initial crack tip location a_0 of the equivalent specimen subject to load P when the crack length is a.

These two equations may be used in different ways. We first notice that once expressions for K_I and w —depending only on geometry and loading— and for W_F —depending only on material— are given, (2.12) and (2.13) provide the parametric representation of the P-a curve, with parameter $CTOD$. We further notice that the LH member of (2.12) is the equivalent crack growth resistance (2.6) which, according to (2.12) is a unique function of $CTOD$. This equivalence results, then, in a R-$CTOD$ curve model, previously analyzed by the authors on different grounds [1].

Equation (2.13) may be made more explicit by writing the $CTOD$ of the elastic sample (LH member of 2.13) in the form derived in [6]

$$w_{eq}(P^{J\text{-}CTOD}, a^{J\text{-}CTOD}, a_0) = \frac{8}{\sqrt{2\pi E'}} K_{Ieq}(P^{J\text{-}CTOD}, a^{J\text{-}CTOD})(\Delta a^{J\text{-}CTOD})^{1/2} L\left(\frac{\Delta a^{J\text{-}CTOD}}{D}\right) \qquad (2.14)$$

where $L(\Delta a/D)$ is a dimensionless function implicitly depending on shape and initial crack length and D is a characteristic structural length (for example, beam depth). The function $L(\Delta a/D)$ satisfies the condition $L(0) = 1$ for any geometry.

The use of this equation in (2.12) and (2.13) allows the elimination of the load, and further use of (2.6) reduces the equations to:

$$R^{J\text{-}CTOD} = W_F(CTOD) \qquad (2.15)$$

$$\Delta a^{J\text{-}CTOD} L^2\left(\frac{\Delta a^{J\text{-}CTOD}}{D}\right) = \frac{\pi}{32} \frac{E' \, CTOD^2}{W_F(CTOD)} \qquad (2.16)$$

The first of this equations is no more than the R-$CTOD$ equation, linked to the material softening curve by (2.2). The second equation (2.16) gives the J-$CTOD$ equivalent crack extension. The set of the two equations is the parametric representation of the R-Δa curve which must be noted to be size and geometry dependent, because of the presence of the function $L(\Delta a/D)$. Only for $D \to \infty$ the R-Δa curve becomes geometry independent because then $\Delta a/D \to 0$ and $L \to 1$.

From these equations and the relationships (2.6) and (2.4) the load $P^{J\text{-}CTOD}$ acting on the equivalent specimen for a given $CTOD$ may be obtained. There is no reason to expect that this load coincide with the actual load P. However, it was found that the maximum load can be accurately predicted using the equivalent sample, at least for notched beams. As an example, for concrete beams (where $l_{ch} = EG_F/\phi\tau^2$ was supposed equal to 0.3 m), the error was less than 5% for beam depths larger than 8 cm [1].

2.2.2. *Bazant Size-Independent R-Δa Equivalence.* As already stated, any of the above equivalences lead, at least in principle, to size dependent R-Δa curves. Bazant has put forward a method to determine a size-independent —but geometry-dependent— R-Δa curve from the knowledge of the size-effect curve, or variation of peak load P_{peak} with size D for geometrically similar precracked structures [7, 8, 9].

Bazant's approach is a general X-Y equivalence in the sense that two conditions are imposed. The first condition is that the peak load must be the same on the actual and on the equivalent elastic specimen for every size D. The second condition is that the R-Δa curve must be size independent. The only input is the size effect curve, which may be written as a function $P_{peak}(D)$, known from experiments or by computation. The unknowns are the equivalent resistance and the equivalent crack extension at peak load for every size. When the size is eliminated, the R-Δa curve appears. The result, derived and used by Bazant in many papers, is that the R-curve is the envelope of the following uniparametric family of functions, with parameter D:

$$R^{\text{B-G}} = \frac{1}{E'} \left(\frac{P_{peak}(D)}{B\sqrt{D}} \, S \left(\alpha_0 + \frac{\Delta a^{\text{B-G}}}{D} \right) \right)^2 \qquad (2.17)$$

where $\alpha_0 = a_0/D$ is constant because of geometrical similarity, and superindex B-G stand for Bazant and General, because no special size effect curve is postulated here. The size effect may come from a cohesive model, from Bazant's size effect law, or from any other imaginable model. The result depends of course on the size effect behaviour and also, at least in principle, on α_0 and on other hidden geometrical parameters in the shape function S. It is size-independent, but geometry-dependent.

3. Asymptotic Analysis of the Equivalent Crack

Let us explore, now, the *equivalences* for large specimen sizes. The authors developed a method particularly appropriate for analyzing cohesive crack models when the specimen size is large [2, 3, 4] and some results will be briefly summarized here.

It is assumed that the size of the cohesive zone, c, remains bounded as the specimen size (characterized by D) grows and that the stress and displacement fields can be developed in series of c/D. The zeroth order asymptotic approach is obtained when terms of the order of c/D or higher are neglected. When the terms linear in c/D are also retained —neglecting terms of the order of $(c/D)^2$— the first order approach is obtained.

For zeroth order approach, the stress and displacement fields *far from the cohesive crack tip* are the same as the corresponding fields of an elastic crack of length a_0. This is a well known far field property. In this approximation, the solution is written as a weighted sum of elastic solutions, with a dimensionless density function $k^*(x)$ defined on $(0,1)$ the values of which for every cohesive crack length c are determined from the equation:

$$\int_0^x k^*(u) \, (x-u)^{-1/2} \, du - F^* \left(8c^* \int_x^1 k^*(u) \, (u-x)^{1/2} \, du \right) = 0 \qquad (3.1)$$

with

$$c^* = \frac{c}{l_{ch}} = \frac{c \; \phi \tau^2}{E' \; G_F} \qquad (3.2)$$

and

$$F^*(w^*) = \frac{1}{f_t} F\left(\frac{G_F}{f_t} w^* \right) \qquad (3.3)$$

where x and u are relative positions on the cohesive zone, so that x and $u = 0$ at the initial crack tip and x and $u = 1$ at the cohesive crack tip; c^* and $F^*(w^*)$ are dimensionless forms of the cohesive zone size c and of the softening function $F(w)$ in Eq. (2.1).

Once the density function $k^*(x)$ has been obtained for a given c, the J-integral, the load expressed as nominal intensity factor (SIF for the initial crack length and the actual load), and the *CTOD* may be found to be:

$$J = 2\pi c^* \, G_F \left(\int_0^1 k^*(u) \, du \right)^2 + O\!\left(\frac{c}{D}\right) \tag{3.4}$$

$$K_{IN} = \sqrt{2\pi c^* \, E'G_F} \int_0^1 k^*(u) \, du + O\!\left(\frac{c}{D}\right) \tag{3.5}$$

$$CTOD = \frac{G_F}{f_t} 8c^* \int_0^1 k^*(u) \, u^{1/2} \, du + O\!\left(\frac{c}{D}\right) \tag{3.6}$$

where $O(c/D)$ stands for a function of the same order of its argument, and it is explicited to remember that we are in the zeroth order approach. Since Eq. (2.11) is always valid, we may use this and Eqs. (3.4) and (3.6) to find the following essential relationship between the cohesive zone size and the *CTOD*:

$$c = \frac{\pi}{32} \frac{E' \, CTOD^2}{W_F(CTOD)} <u^{1/2}>^{-2} + O\!\left(\frac{c}{D}\right) \tag{3.7}$$

where $<f(u)>$ is the k^*-*weighted average* value of any given function $f(u)$ on the $(0,1)$ interval:

$$<f(u)> = R \frac{\displaystyle\int_0^1 f(u) \, k^*(u) \, du}{\displaystyle\int_0^1 k^*(u) \, du} \tag{3.8}$$

When the approximation is extended to first order, a new far field property —not so obvious— was deduced (restrictedly in [3], a complete proof will be published elsewhere):

For a cohesive material and a general geometry under mode I loading, every far field may be approximated, up to and including order c/D, by the corresponding elastic field of a crack of length $u_0 + \Delta a^{FF}_\infty$ where Δa^{FF}_∞ is the Full Far Field effective (or equivalent) crack extension given, for each cohesive crack size, by the following equation

$$\Delta a^{FF}_\infty = c <u> \tag{3.9}$$

where $< >$ has the same meaning as before. Notice that this equation is based only on the zero order solution, and has a meaning only for first order approximation. The existence of a full far field equivalence for higher orders of approximation.is dubious.

It is possible to express Δa^{FF}_∞ in terms of the *CTOD* and to derive a lower bound for it. Substituting c from (3.7) into (3.9) we get:

$$\Delta a^{FF}{}_{\infty} = \frac{\pi}{32} \frac{E'CTOD^2}{W_F(CTOD)} \frac{<u>}{<u^{1/2}>^2} + O\left(\frac{c}{D}\right) \qquad (3.10)$$

Use of the Bunyakovsky-Schwarz inequality in the first term of the RH member of this equation, and the fact that the second term vanishes as D grows to infinite, delivers the lower bound theorem:

$$\Delta a^{FF}{}_{\infty} > \frac{\pi E}{32} \frac{CTOD^2}{W_F(CTOD)} \qquad \text{for } c/D \text{ small enough.} \qquad (3.11)$$

3.1. FAR FIELD (FF) EQUIVALENCES

Let us consider the equivalences based on variables associated with fields far away from the cohesive zone (FF, far fields), and this limited to very large sizes for which the full far field equivalence exists to a given degree of accuracy. Several among the variables used to define equivalences in the previous section are far field variables. This is obviously the case for P, u and $CTOD$, but also for J. The later is a very special case because it comes from a path independent integral that may be performed along any contour. Hence, if two specimens have identical far fields they also have identical J- integrals, eventhough their near fields may be essentially different . This is in fact the dual of a reasoning frequently used in LEFM (equal near fields imply equal Js even if far fields are essentially different).

According to the preceding, the equivalences in which both indices correspond to far fields are then mutually equivalent, and we have for a given situation (a given cohesive zone size):

$$\Delta a^{FF}{}_{\infty} = \Delta a^{P-u}{}_{\infty} = \Delta a^{P-CMOD}{}_{\infty} \qquad (3.12)$$

where subscript ∞ means infinite (very large) size.

The R-Δa curve for zero order approximation is expressed parametrically by the first term of Eq. (3.4) for R and (3.9) for Δa. To get a first order approximation, the linear term in Eq. (3.4) has to be included which requires the solution of a further integral equation similar to (3.1).

It is interesting to comment Bazant Equivalence for a Cohesive model at this point (B-C). Bazant equivalence is based on the knowledge of the peak load over a certain range of sizes. The peak load is a far field property, but the necessity of having a range of sizes to find the R-curve limits the analysis for large sizes. Indeed, if we know the size effect for only one size, no information at all about the R-curve can be obtained. If we know the size effect in the neighborhood of a size (peak load and its derivative with respect to size) only one point of the R-curve — and the slope at this point— may be obtained, which corresponds to the peak for this particular size. From the asymptotic size effect equation previously obtained by the authors, and the definition of the B-C equivalence, it may be easily proved that the B-C equivalence is also far field at the peak:

$$\Delta a^{B-C}{}_{\infty peak} = \Delta a^{FF}{}_{\infty peak} \qquad \text{and} \qquad R^{B-C}{}_{\infty peak} = R^{FF}{}_{\infty peak} = G_F \qquad (3.13)$$

3.2. THE J-CTOD EQUIVALENCE

When the variables chosen for the equivalence are related with the cohesive zone the far field property can not be directly exploited. This happens, for example, with the variable $CTOD$ and with the equivalence J-$CTOD$. In principle, there is no reason to suspect that Δa^{J-CTOD} and Δa^{FF} coincide for very large sizes, and we will see that they do not.

Consider first the *J-CTOD* equivalence in the limit of large sizes. Putting $L = 1$ for $D \to \infty$ in Eq. (2.16) one immediately obtains

$$\Delta a_\infty^{J\text{-CTOD}} = \frac{\pi}{32} \frac{E' \, CTOD^2}{W_F(CTOD)} \tag{3.14}$$

which is obviously the approximation of order zero of the *J-CTOD* equivalent crack extension , so that in a neighborhood of $c/D = 0$ one has

$$\Delta a^{J\text{-CTOD}} = \Delta a_\infty^{J\text{-CTOD}} + O\left(\frac{\Delta a_\infty^{J\text{-CTOD}}}{D}\right) \tag{3.15}$$

It appears that the RH member of equations (3.11) and (3.14) coincide, so that the lower bound theorem above turns out to state that the the *FF* equivalent crack extension is always larger than the *J-CTOD* equivalent crack extension in the limit of large sizes:

$$\Delta a_\infty^{FF} > \Delta a_\infty^{J\text{-CTOD}} \tag{3.16}$$

One may use the Full Far Field equivalence for large sizes to find first order relationships between the actual load P and displacement u and those determined through the *J-CTOD* equivalence, $P^{J\text{-CTOD}}$ and $u^{J\text{-CTOD}}$.

To obtain the relationship between the loads we just write that, by the very definitions of FF and J-CTOD equivalences

$$J = J^{FF} = J^{J\text{-CTOD}} \tag{3.17}$$

The second equality obviously imply that the stress intensity factors at the equivalent crack tips must be equal in *FF* and *J-CTOD* equivalences. Using then the expression (2.4) one finds

$$P \, S\left(\alpha_0 + \frac{\Delta a^{FF}}{D}\right) = P^{J\text{-CTOD}}.S\left(\alpha_0 + \frac{\Delta a^{J\text{-CTOD}}}{D}\right) \tag{3.18}$$

from which one obtains, accurate to first order:

$$P^{J\text{-CTOD}} = P\left(1 + \frac{S_0'}{S_0} \frac{\Delta a_\infty^{FF} - \Delta a_\infty^{J\text{-CTOD}}}{D}\right) + o\left(\frac{\Delta a_\infty^{FF}}{D}\right) \tag{3.19}$$

where $o(\Delta a/D)$ stands for a function vanishing faster than its argument.

To obtain the relationship between the displacements, we first recall that the displacement in an equivalent elastic specimen may be always written as

$$u_{eq} = C_{eq} P_{eq} = C^*\left(\frac{a_{eq}}{D}\right)\frac{P_{eq}}{B \, E'} \tag{3.20}$$

where C_{eq} is the compliance and $C^*(a/D)$ a dimensionless compliance, related to the shape factor for the stress intensity factor $S(a/D)$ in (2.4) by the relationship

$$2 S^2\left(\frac{a_{eq}}{D}\right) = C^{*\prime}\left(\frac{a_{eq}}{D}\right) \tag{3.21}$$

where $C^{*\prime}$ indicates the first derivative of C^* with respect to its argument. This equation follows easily from the well known relationship of LEFM between G and the derivative with respect to crack length of the elastic energy.

Writing now $a_{eq} = a_0 + \Delta a_{eq}$, $a_0/D = \alpha_0 = $ constant, and using (3.21) for the first derivative of C^*, we may find the first order expansion of (3.20) as

$$u_{eq} = C^*_0 \frac{P_{eq}}{B\,E'}\left(1 + \frac{2S_0^2\,\Delta a_{eq}}{C^*_0\,D}\right) + o\left(\frac{\Delta a_{eq}}{D}\right) \tag{3.22}$$

where subindex 0 for C^* and S means values for the initial crack length

According to the far field equivalence theorem, at a given instant the load and displacement in the actual specimen are equal (to first order) than those in the FF equivalent specimen, hence:

$$u = C^*_0 \frac{P}{B\,E'}\left(1 + \frac{2S_0^2\,\Delta a_\infty^{FF}}{C^*_0\,D}\right) + o\left(\frac{\Delta a_\infty^{FF}}{D}\right) \tag{3.23}$$

For the J-CTOD equivalence the analogous result is found as

$$u^{J\text{-}CTOD} = C^*_0 \frac{P^{J\text{-}CTOD}}{B\,E'}\left(1 + \frac{2S_0^2\,\Delta a_\infty^{J\text{-}CTOD}}{C^*_0\,D}\right) + o\left(\frac{\Delta a_\infty^{J\text{-}CTOD}}{D}\right) \tag{3.24}$$

The relationship between displacements is now found by eliminating P and $P^{J\text{-}CTOD}$ from (3.23), (3.24) and (3.19):

$$u^{J\text{-}CTOD} = u\left(1 + \left(\frac{S_0'}{S_0} - \frac{2S_0^2}{C^*_0}\right)\frac{\Delta a_\infty^{FF} - \Delta a_\infty^{J\text{-}CTOD}}{D}\right) + o\left(\frac{\Delta a_\infty^{FF}}{D}\right) \tag{3.25}$$

It appears, then, that up to first order the $J\text{-}CTOD$ equivalence may be used to find estimates of load and displacement that can be further corrected using (3.19) and (3.25) if an estimate of the difference between the two effective crack extensions is available. In the next section some numerical results are presented regarding this difference.

3.3.   NUMERICAL RESULTS FOR THE FF AND J-CTOD EQUIVALENCES AT INFINITE SIZE

To have a feeling of the trend of the equivalences for large sizes, a numerical analysis was performed and three softening curves, depicted in Fig. 2, were investigated; a rectangular softening (or Dugdale softening), a linear softening, and a quasi-exponential softening. The numerical method described in [3] was used to solve the integral equation (3.1) for a number of cohesive zone sizes.

At each step, the equivalent resistance to crack growth for FF and $J\text{-}CTOD$ equivalences was obtained from equation (3.4) since in these equivalences $R_{eq} = J_{eq} = J$. The FF equivalent crack extension was obtained from Eq. (3.9) and the $J\text{-}CTOD$ equivalent crack extension from Eq. (3.14), after obtention of the $CTOD$ from (3.6).

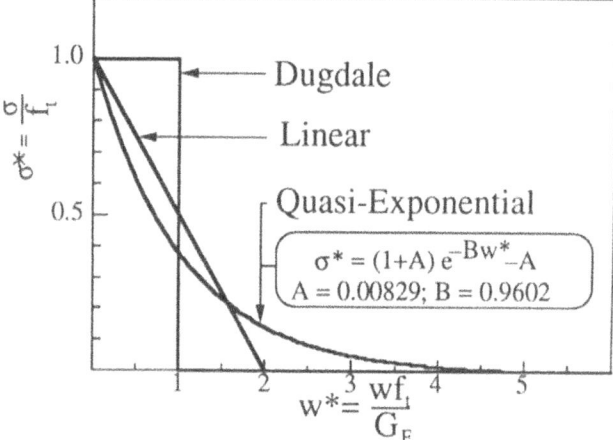

Figure 2. Softening curves analyzed in this work.

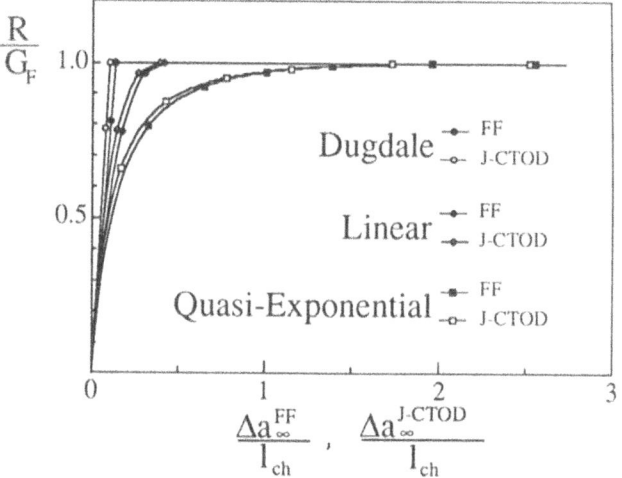

Figure 3. R-Δa curves for the FF(Far Field) equivalence and
for the J-CTOD equivalence, for infinite size.

The resulting R-Δa curves for the two equivalences and different softening curves are given in Fig. 3. For a given softening curve, we observe in this figure that although there is a neat difference between the two R-curves at a numerical level, it seems that they would be hardly distinguishable at the experimental level since both curves would fit into the usual experimental scatter band. Hence, at least for very large sizes, the *J-CTOD* equivalence provides a relatively easy-to-apply approximation of the actual overall behaviour (remember that for this equivalence the R curve is known in parametric closed form).

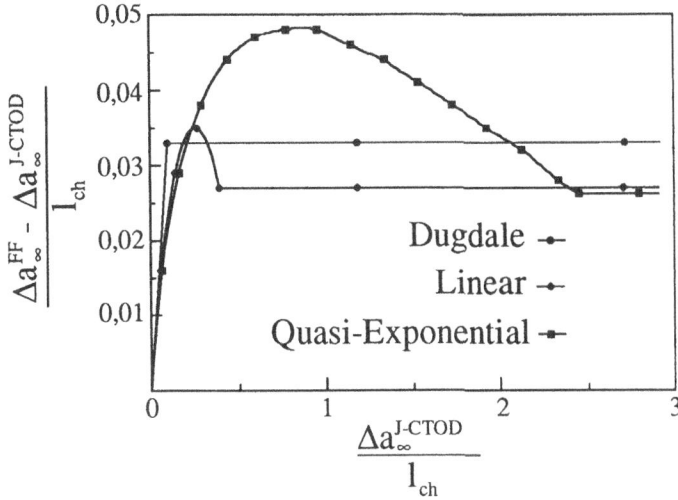

Figure 4. Difference between the equivalent crack extension
of FF and J-CTOD equivalences, for infinite size.

As far as the P-u curve (one of the basic experimental "observables") is concerned, the predictions from the J-$CTOD$ equivalence can be improved by using the relationships (3.19) and (3.25) which depend basically on the difference between the equivalent crack extensions for the FF and J-$CTOD$ approximations. Fig. 4 gives a picture of this difference for the three softening curves previously envisaged. It shows a proportional rising at the beginning of loading and ends in a plateau after the peak load, where steady-state crack growth takes place (only for infinite size, which is our case). We notice that the plateau values are very close for the progressively softening models, and may be a constant value of 2.7 % of the characteristic size l_{ch} could be a good general estimate for the difference for any reasonably smooth softening curve.

Acknowledgements.

The authors gratefully acknowledge financial support for this research provided by CICYT, Spain, under grants PB86-0494 and CE89-0012.

References

[1] Planas, J., Elices, M. and Toribio, J. (1989) 'Approximation of Cohesive Models by R-CTOD Curves', in S.P. Shah, S.E. Swartz and B. Barr (eds.), Fracture of Concrete and Rock, pp. 203-212.
[2] Planas, J. and Elices, M. (1986) 'Un Nuevo Método de Análisis del Comportamiento Asintótico de la Propagación de una Fisura Cohesiva en Modo I', Anales de Mecánica de la Fractura 3, 219-227.

[3] Planas, J. and Elices, M. (1987) 'A new method of asymptotic analysis of the development of a cohesive crack in mode I loading', Departamento de Ciencia de Materiales, ETS de Ingenieros de Caminos, Ciudad Universitaria, 28040 Madrid, Spain Report No. 87-02.

[4] Planas, J. and Elices, M. 'Nonlinear Fracture of Cohesive Materials', International Journal of Fracture, Special Issue, edited by Z. P. Bazant, in press.

[5] Rice, J. (1968) 'Mathematical Analysis in the Mechanics of Fracture', in H. Liebowitz (ed.), Fracture, Vol. 2, Academic Press, pp. 192-311.

[6] Planas, J. and Elices, M. (1990) 'Fracture Criteria for Concrete: Mathematical Approximations and Experimental Validation', Engineering Fracture Mechanics 35, 87-94.

[7] Bazant, Z. P., Kim, J-K., and Pfeiffer, P. A. (1986) 'Nonlinear Size Effect Properties from Size Effect Tests', Journal of Structural Engineering, ASCE 112, 289-307.

[8] Bazant, Z. P., and Pfeiffer, P. A. (1987) 'Determination of Fracture Energy from Size Effect and Brittleness Number', ACI Materials Journal 84, 463-480.

[9] Bazant, Z. P., Gettu, R., and Kazemi, M. T. (1989) 'Identification of Nonlinear Fracture Properties from Size Effect Tests and Structural Analysis Based on Geometry-Dependent R-Curves', Center for Advanced Cement-Based Materials, The Technological Institute, Northwestern University, Evanston, Illinois 60208 Report No. 89-3/498p.

Toughening Mechanisms in Quasi-Brittle Materials

Reporter's Summary: Session Three, Theoretical Fracture Mechanics
Considerations

by

John W. Rudnicki

This session contained three invited talks: "Fracture Mechnaics, Size
Effect and Damage Localization in Brittle Homogeneous Materials" by Zdenek
P. Bazant (Northwestern University), "Micromechanics of Deformation in Rock"
by Neville Cook (University of California, Berkeley) and "Asymptotic
Analysis of Cohesive Cracks and its Relation with Effective Elastic Cracks"
by Jaime Planas (Ciudad Universitaria, Spain).

Following Professor Bazant's talk, Professor Kobayashi noted that the
fatigue law used by Bazant (equation 12) is known as Forman's law in the
material science literature. Bazant agreed with the comment by Karihaloo
that the correction to the Paris law could be interpreted as appropriately
defining the fracture toughness. In response to a question by Wu concerning
the physical interpretation of the size effect law, Bazant noted that d_u is
proportional to a characteristic length, with the proportionality factor
depending on geometry; B is not related to the characteristic length but is
a factor that depends on the specimen geometry.

Following Professor Cook's talk, Horii commented that in analyzing the
response of rectangular arrays of cracks, shear localization occurs only if
the secondary cracks are not all the same length. Cook replied that his
considerations of crack arrays pertained to prediction of the constitutive
response; he views localization as a different phenomenon due to the highly
heterogeneous stress field. Vermeer asked whether the cases of overlapping
and joining cracks had been examined since this is often observed in
experiments. Cook replied that they had: in this case the simple analysis
based on buckling of struts created by crack growth appears to be
inappropriate; these configurations tend to fail by shear. Bazant noted
that the strain softening constitutive relation predicted in the analysis
will ultimately lead to localization and, hence, a localization limiter with
concomitant size effect will need to be introduced.

Following Professor Planas's talk, Karihaloo asked a number of
questions to clarify the range of validity of the analysis presented. In
reponse, Planas noted that the analysis is not limited to large structures
but that if the structure is large, the expansion can be truncated. The
analysis does assume that the underlying cohesive zone model is a
characteristic of the material and that the corresponding fracture energy is
independent of size. In response to a question by Kobayashi about the need
for a crack extension critierion, Planas replied that once the full
softening curve is specified a crack extension criterion is not needed.

S. P. Shah (ed.), Toughening Mechanisms in Quasi-Brittle Materials, 203–205.
© 1991 Kluwer Academic Publishers.

In addition to the invited talks, there was a short presentation by Professor Labuz (University of Minnesota) on the implications of the size effect for postpeak response, more specifically, whether the postpeak response is Class I or Class II (decreasing stress or force with increasing or decreasing displacement, respectively). Labuz noted that if the postpeak response is regarded as localized and a stress versus slip or opening cohesive zone relation is introduced, the subsequent behavior can be Class I or II depending on the size of the specimen.

The general discussion session opened with a question by Karihaloo about how geometric similiarity of the bond slip relation was ensured in the experimental results described by Bazant. Bazant replied that it was difficult to ensure the geometric similarity of bond slip but that a comparison of tests with and without hooks suggested that this was not a problem.

There was considerable discussion of the solution for a crack emanating from a circular hole, to which reference was made in Cook's presentation. Cook noted that the solution for a circular hole (without a crack) in an infinite elastic body under biaxial compression predicts tensile stresses near the hole on a line parallel to the direction of maximum applied compression. Hence, the inhomogeneity caused by the hole does make it possible to generate opening cracks. Further discussion concerned whether the value of K_I for a crack emanating from a circular hole increased for short crack lengths. Examination by this reporter of the results from Bowie ["Analysis of an Infinite Plate Containing Radial Cracks Originating at the Boundaries of an Internal Circular Hole", J. Math. and Phys., 35, p. 60, 1956], based on a conformal mapping technique and summarized in Tada [The Stress Analysis of Cracks Handbook, H. Tada, P. Paris, and G. Irwin, Del Research Corporation, Hellertown, Pennsylvania, 1973], indicates that K_I does indeed increase with cracklength for small cracklengths. More specifically, the solution approaches that for an edge crack (suitably modified to account for the stress concentration due to the hole) for short crack lengths, as noted by Kobayashi in the discussion.

There was also some discussion of results of uniaxial compression tests. Cook described some tests in which the ends of the sample were constrained by steel wires and Shah described some results obtained in his laboratory. Uniaxial compression tests are notoriously susceptible to deviations from relatively homogeneous stress caused by the end constraints. The elimination of certain features of these tests with small amounts of confining stress suggests that the results are not indicative of material behavior and should be interpretted with considerable caution.

Atkinson questioned the need for the introduction of non-local theories to eliminate the crack-tip singularity and noted that there have been some problems identified with these theories when applied to crack problems. Bazant acknowledged that these theories were not required but argued that there is a rational basis for introducing them. Furthermore, he remarked that the non-local theory currently used by himself (and Planas) differs from that introduced by Eringen for which problems have been found.

Krajcinovic asked what was the origin of nonlinear (axial) deformation in cases for which all cracks are axial. Bazant noted that finite element simulations of uniaxial compression suggest that the amount of axial cracking (splitting) depends strongly on the amount of volume change allowed. Cook remarked that nonlinear axial deformation occurs when the axial cracks are driven by local tensile stresses due to inhomogeneities.

Francois commented that for perfectly plastic materials (no strain hardening) the strain singularity ahead of the crack tip is r^{-1} for a stationary crack but decreases to ln r for a steadily moving crack. In response to his question about whether similiar behavior is predicted for cohesive zone models, Bazant replied no.

Experimental Observations

MICROSTRUCTURE, TOUGHNESS CURVES AND MECHANICAL PROPERTIES OF ALUMINA CERAMICS

STEPHEN J. BENNISON, JÜRGEN RÖDEL, SRINIVASARAO LATHABAI,
PRAPAIPAN CHANTIKUL, BRIAN R. LAWN
Ceramics Division
National Institute of Standards and Technology
Gaithersburg, MD 20899, USA

ABSTRACT. The microstructural variables that determine the toughness (T-curve) characteristics of alumina and other structural ceramics are considered. Alumina ceramics gain their toughness from shielding by grain-interlock bridging at the interface behind the crack tip. A general fracture mechanics formalism for describing the bridging is outlined in terms of desirable microstructural elements, such as weak internal boundaries, high internal stress, coarse microstructure. The T-curve imparts the quality of flaw tolerance to the strength properties. We examine this quality, under both inert and interactive environmental conditions, monotonic and cyclic loading, using indentation flaws. In situ observations of bridging sites during loading in the scanning electron microscope provide insight into the bridge degradation micromechanisms. Finally, short-crack properties, spontaneous microcracking and wear degradation, are examined in light of the bridging model. It is concluded that design with ceramics may require certain tradeoffs, long vs short cracks, high strength vs flaw tolerance, etc. The key to optimal performance in ceramics rests with microstructural processing for specific properties.

1. INTRODUCTION

It is now well established that many monophase ceramics exhibit the property of rising fracture resistance or toughness with crack extension (R-curve or T-curve) [1-14]. The magnitude of the toughness increase can be respectable, in extreme cases approaching a factor of five or so over extensions of several millimeters or hundreds of grain dimensions.

Toughness-curve characteristics dictate mechanical behavior. For example, the associated stabilizing effect on crack growth confers the

209

S. P. Shah (ed.), Toughening Mechanisms in Quasi-Brittle Materials, 209–233.
© 1991 *Kluwer Academic Publishers.*

quality of "damage tolerance" [5,9,12-14], i.e. insensitivity of strength to flaw size. This leads to the highly appealing prospect of a well-defined, flaw-insensitive stress for engineering design. However, the cost of gaining flaw tolerance is inevitably the sacrifice of toughness in the domain of "short" cracks, where properties like wear resistance are decided. Specifying the T-curve for optimal performance of ceramics involves certain tradeoffs.

A substantial body of evidence now exists to demonstrate that the principal mechanism of T-curve behavior in alumina and other ceramics is grain-localized bridging at the crack interface behind the advancing tip [1-14]. Bridging grains exert frictional closure forces across the crack walls and thereby shield the tip from the applied stress-intensity field [8,9,13,14]. The cumulation of bridging tractions over the crack interface with continued propagation leads to a rising toughness curve.

The magnitude and shape of the resultant T-curve are sensitive to the microstructure: grain size and shape [4,5,13,15]; internal residual stresses (e.g. thermal expansion anisotropy) in noncubic materials [13], especially as they may be intensified by incorporation of a second phase [16]; grain boundary energy [13]; all these are important players in the toughness. This strong influence of microstructure leads to the potential for manipulation of crack-resistance properties through controlled processing. Opportunities for the development of improved structural ceramics thereby rest with the development of novel microstructures and fabrication strategies that exploit the operative toughening micromechanisms.

In the present paper we review the toughness-curve phenomenon in ceramics, with alumina as a model system. First a microstructure-based model for grain bridging is discussed. Second, we examine the role of the T-curve in determining different mechanical properties. Primary consideration is given to **strength** properties [9,12,13,15,17,18], in both inert and interactive environments, and under quasistatic and (cyclic) fatigue loading. This is the domain of "intermediate" crack sizes (corresponding to the rising portion of the T-curve). Additional consideration is given to microfracture-induced **wear** properties [19], the domain of "short" cracks. Noting further that most conventional toughness evaluations are made in specimens with "long" cracks, we emphasize the importance of identifying any specific mechanical property with the proper crack-size domain. Then we present results of more recent, in situ observations of grain bridging in the scanning electron microscope [20]. Measurements of crack profiles demonstrate directly the nature of the closure forces exerted by the bridges on the cracks, and reveal details of fatigue processes. Finally, the prospects of innovative microstructural processing for optimizing toughness properties of ceramics, including composite systems, is discussed.

2. MICROSTRUCTURAL FRACTURE MECHANICS MODEL

2.1   GENERAL EQUILIBRIUM REQUIREMENTS

Begin by defining a net **crack-tip** stress-intensity factor condition for
a crack subject to a superposed tensile loading field, $K_a(c)$, flaw-
localized residual nucleation field, $K_r(c)$, and microstructure-
associated field, $K_\mu(c)$ [14]:

$$K_*(c) = K_a(c) + K_r(c) + K_\mu(c). \tag{1}$$

Equilibrium obtains when K_* just balances the intrinsic toughness
associated with the creation of crack surfaces:

$$K_* = T_0. \tag{2}$$

This last requirement can be restated by considering K_a and K_r in Eq. 1
as part of the net effective applied mechanical field, K_A, and $K_\mu = -T_\mu$
as part of the internal toughness, i.e.

$$\begin{aligned} K_A(c) &= K_a(c) + K_r(c) \\ &= T_0 + T_\mu(c) = T(c). \end{aligned} \tag{3}$$

The **net toughness** function T(c) constitutes the so-called T-curve, the
K-field equivalent of the R-curve.
    An alternative formulation for the crack-tip conditions may be given
in terms of the mechanical-energy-release rates [14]. Write, in
analogy to Eq. (3),

$$G_A(c) = R_0 + R_\mu(c) = R(c) \tag{4}$$

with R the crack-resistance energy. The Griffith requirement for
equilibrium is then

$$G_* = R_0 = 2\gamma_B \tag{5}$$

where γ_B is the fracture surface energy of the body in inert
atmospheres. For intergranular fracture γ_B incorporates the grain
boundary energy. By defining the connecting relations [14]

$$K_A = (G_A E')^{1/2} \tag{6a}$$
$$K_* = (G_* E')^{1/2} \tag{6b}$$

with $E' = E/(1 - \nu^2)$ in plane strain, E Young's modulus and ν
Poisson's ratio, we identify the intrinsic toughness $T_0 = (R_0 E')^{1/2}$
in Eq. 2.

2.2   MICROSTRUCTURAL STRESS-INTENSITY FACTOR

Microstructural properties are introduced into the formalism in Sect.
2.1 via $T_\mu(c)$ in Eq. 3.   For materials that toughen by bridging we need
to determine the micromechanics of grain interlock and subsequent
pullout [13].   A subsidiary element of the process is the segmentation
of the primary crack front at incipient bridging grains, driven by
conflicting tendencies to follow weak intergranular or interphase
boundaries and local thermal expansion anisotropy tensile stresses.
For the present we focus on simplistic, idealized structures in
monophase, noncubic ceramics.

Consider a **half-penny** crack, radius c, evolving in a rectangular
microstructure, Fig. 1.   The submicroscopic crack is assumed to begin
its life as a flaw in a region of most favorable internal tension, $+\sigma_R$,
and thereafter to intersect bridging grains as it expands radially
outward.   The grains which act as bridges are assumed to be those with
compressive components of the residual stress field, $-\sigma_R$, at transverse
facets.   The problem is then to determine the closure stresses exerted
by the bridges in terms of crack-wall separation.

These stresses are governed by Coulombic friction that restricts the
separation of bridging grain facets.   At initial separation the
bridging grains debond along the constrained facets and then "pull out"
until final "rupture" at some critical rupture strain.   The debond
stage consumes relatively little energy, so the constitutive relation
between closure stress p and (half-) crack-opening displacement u for a
single bridge may be written exclusively in terms of a simple tail-
dominated pullout relation [8,13]:

$$p(u) = p_M(1 - u/u_\ell) \tag{7}$$

where p_M is the maximum resistance stress (at $u = 0$) and u_ℓ is the
wall-wall half-displacement at bridge-matrix disengagement (at $p = 0$).
The essential material quantities in Eq. 7 are contained in the
parameters [13]:

$$u_\ell = \epsilon_\ell \ell/2 \tag{8a}$$
$$p_M = (\alpha_\lambda \alpha_L \epsilon_\ell \mu \sigma_R)(1 - 1/2\alpha_d^2) \tag{8b}$$

with ℓ grain size, μ friction coefficient, σ_R internal residual stress.
The α and ϵ terms are dimensionless constants for geometrically similar
microstructures:

$$\alpha_\lambda = \lambda/\ell \tag{9a}$$
$$\alpha_L = L/\ell \tag{9b}$$
$$\alpha_d = d/\ell \tag{9c}$$
$$\epsilon_\ell = 2u_\ell/\ell \tag{9d}$$

with λ the bridge cross-sectional perimeter, L the embedded grain
length, d the bridge spacing.   In Eq. 8 only the pullout distance u_ℓ
depends on the grain size;   the closure stress p_M is scale-invariant.

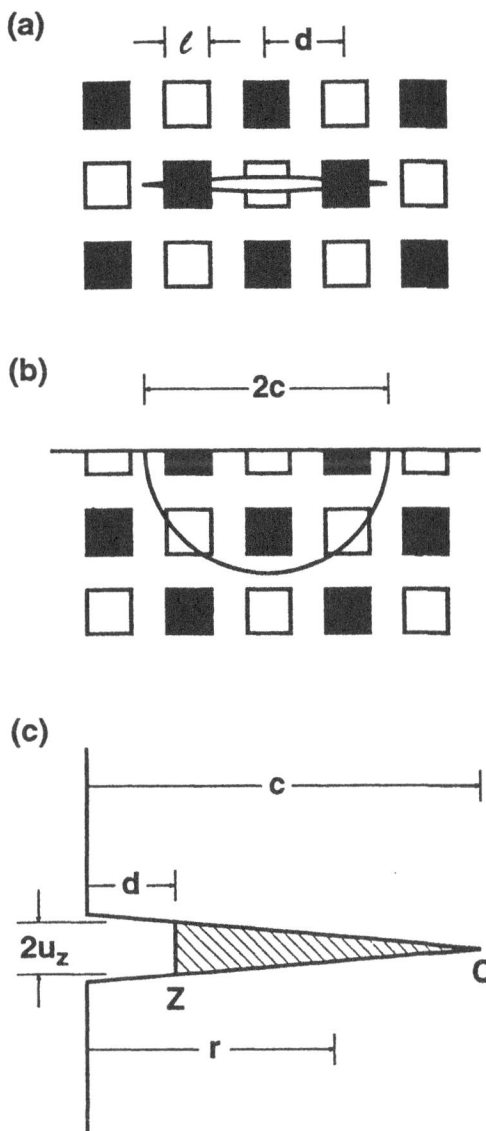

Figure 1. Microstructural model of grain bridging for penny crack in microstructure with bridging grains (squares): (a) in-plane view; (b) out-of-plane view. (c) Coordinate system for bridging zone. Crack initiates in residual tensile field at $c \leq d$, and thereafter extends into bridging field.

The toughening associated with the bridging is calculated according to crack-size domain, as follows [8,9,13,17]:

(i) **Small Cracks** ($c \leq d$), no bridge intersections. Within this region the crack experiences only the matrix tensile stress, $+\sigma_R$. Assuming this field to be uniform, we obtain

$$T_\mu(c) = -K_\mu(c) = -\psi\sigma_R c^{1/2} \tag{10}$$

where ψ is a geometry dependent coefficient. This K_μ term is not expected to be present in **straight** cracks, where tensile and compressive facets average to zero stress along the crack front.

(ii) **Intermediate Cracks** ($c \geq d$, $0 \leq u \leq u_\ell$), bridges intersected. There are two contributions to the toughness in this region, $T_\mu(c) = T'_\mu(c) + T''_\mu(c)$. The first is an opening post-intersection component from the persistent residual tensile stress field from (i) above,

$$T'_\mu(c) = -\psi\sigma_R c^{1/2}[1 - (1 - \alpha_d^2\ell^2/c^2)^{1/2}]. \tag{11}$$

The second is a countervailing closing component from the bridging tractions, Eq. 7. This contribution is most readily evaluated as a J-integral,

$$\begin{aligned} R''_\mu(u) &= 2 \int_0^{u_z} p(u)du \\ &= 2p_M u_z(1 - u_z/\epsilon_\ell\ell). \end{aligned} \tag{12}$$

The displacement $u_z = u_z(c)$ at the edge of the bridging zone (i.e. first bridge intersection at $c = d$, Fig. 1c) may be evaluated approximately from the Sneddon crack profile relation ("weak shielding" approximation)

$$u_z(c) = (\psi K_A/E')[(c^2 - \alpha_d^2\ell^2)/c]^{1/2}. \tag{13}$$

The toughness $T''_\mu(c)$ may be determined from the crack resistance $R''_\mu(c)$ by eliminating G_A and K_A from Eqs. 2-5:

$$T''_\mu(c) = E'^{1/2}\{[R''_\mu(c) + G_*]^{1/2} - G_*^{1/2}\}. \tag{14}$$

This relation is implicit in T''_μ, so Eqs. 3, 11-14 must be solved simultaneously.

(iii) **Long Cracks** ($c \gg d$, $u_z \geq u_\ell$), bridging zone now of constant size and translating with the advancing crack. In this limit, $T'_\mu \to 0$ and the steady-state toughening increment is evaluated from Eq. 14 with $R''_\mu = p_M\epsilon_\ell\ell/2$ from Eq. 12.

2.3   ENVIRONMENTALLY-ENHANCED KINETIC CRACK GROWTH

Environmentally-assisted "slow" crack growth in ceramics manifests
itself as a finite "lifetime" at sustained stress. The slow crack
growth is characterized by a crack velocity function, $v(K)$ or $v(G)$.
The fundamental form of the $v(G)$ function is [17]

$$v(G_*) = v_0 \sinh[(G_* - 2\gamma_{BE})/2\Gamma] \qquad (2\gamma_{BE} \leq G_* \leq 2\gamma_B) \qquad (15)$$

where γ_{BE} is the fracture surface energy in the presence of the
reactive environment, and v_0 and Γ are intercept and slope parameters.
This function has provision for a threshold at $G_* = 2\gamma_{BE}$, and dynamic
velocity at $G_* = 2\gamma_B$.
     Combining $v = dc/dt$ in Eq. 15 with the relations in Sects. 2.1 and
2.2 then yields a differential equation in $c(t)$ for any specified time-
dependent applied stress.

3.   STRENGTH: INDENTATION FLAWS

Now let us examine the influence of the T-curve on strength properties.
We reemphasize that strength pertains specifically to cracks in the
intermediate-crack domain, i.e. to the rising portion of the T-curve.
To place this into proper context, we again point out that most
traditional toughness evaluations are made in long-crack specimens.
Later we shall address short-crack properties, viz. spontaneous
microcracking and wear.

3.1   INERT STRENGTH

The inert strength of a ceramic material is determined by the condition
for instability under essentially equilibrium conditions: $dK_*(c)/dc \geq$
0 at $K_*(c) = T_0$ in Eq. 2 or, alternatively, $dK_A(c)/dc \geq dT(c)/dc$ at
$K_A(c) = T(c)$ in Eq. 3 [14].   The latter defines the familiar "tangency"
construction for toughness curves. The form of K_a and K_r in Eq. 1 is
determined by the relevant testing geometry and flaw type.
     In this section we consider controlled indentation radial cracks
formed at contact load P and subsequently subjected to uniform tensile
stress σ_a:

$$K_r = \chi P/c^{3/2} \qquad (16a)$$
$$K_a = \psi \sigma_a c^{1/2} \qquad (16b)$$

with χ a parameter defining the intensity of the residual contact field
[22] and ψ as defined in Eq. 10. Indentation cracks enable one to
"calibrate" T-curve parameters with maximum efficiency and, moreover,
take us close to the crack-size realm of natural flaws [12,13,23].

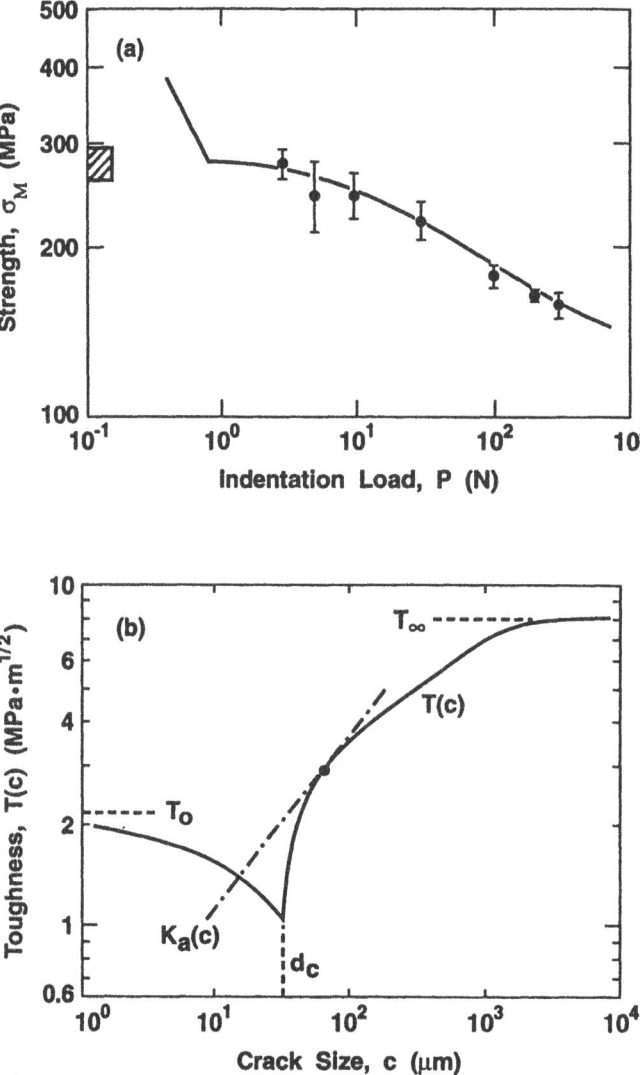

Figure 2. (a) Inert strength as function of Vickers indentation
load for a reference single-phase polycrystalline alumina, grain
size ℓ = 23 μm, tested in dry silicone oil. Shaded region at
left represents breaks from natural flaws. Solid curve is fit of
bridging T-curve formalism. (b) Deconvoluted T-curve. Dashed
line is K_a(c) tangency condition for failure from **natural** flaw (χ
= 0).

Fig. 2a shows inert strength vs indentation load data for a "reference" alumina with single phase and equiaxed microstructure, grain size $\ell = 23$ μm. We note the strong tendency to a plateau at low P. These data reflect the quality of flaw tolerance alluded to earlier. The solid curve is a data fit obtained by adjusting the microstructural parameters μ, σ_R, ϵ_ℓ and α terms in Eq. 8, using a numerical algorithm [13,17]. From this fit we deconvolute the T-curve shown in Fig. 2b, using Eqs. 10-14. Note that the intrinsic material toughness initially diminishes from $T = T_0$ with crack size, due to the action of the residual tensile field at $c \leq d$ (Fig. 1). After the first bridge intersection the toughness rises sharply as a result of dominant frictional restraining stresses from the grain pullout. This rise continues as the bridging zone expands with crack extension, until the first bridge ultimately ruptures. At this point the bridging zone translates with the crack, and a steady state obtains at $T = T_\infty$.

The T-curve strongly stabilizes the crack growth en route to failure. For simplicity, consider a natural flaw without any residual stress ($\chi = 0$). Then $K_A = K_a \propto c^{1/2}$ in Eqs. 3, 16a. We show this applied loading function at the tangency condition in Fig. 2b. Suppose the initial size of the flaw lies to the right of the dashed line somewhere on the diminishing branch of the T-curve. Then at a certain point in the loading the flaw "pops in" and arrests on the rising branch of the T-curve. Further load is needed to extend the crack in stable equilibrium up the T-curve, until the configuration at $dK_A(c)/dc \geq dT(c)/dc$ is attained, whence the specimen fails. It is this precursor stability characteristic that is responsible for the flaw tolerance characteristic in Fig. 2a: the critical failure condition depends on the final, not the initial, flaw size.

The "calibrated" T-curve contains all the necessary ingredients to predict the effects of variations in microstructural characteristics on the toughness characteristics. We return to this prospect in Sect. 5.

3.2 RATE-DEPENDENT STRENGTH

Now let us consider the deleterious, rate-dependent influence of water-containing environments on strength, in accordance with the $v(G_*)$ velocity function of Eq. 15. We again focus on indentation flaws.

Strength data for the same alumina as in Fig. 2, but now tested in water at a fixed indentation load, are shown in Fig. 3 as a function of stressing rate. The solid curve through the data is a best fit from numerical solutions of the differential equation embodied in Eq. 15 for the time to grow the crack from its initial (stable) state to final instability, the "lifetime". This is done using the T-curve calibration from Sect. 3.1 and adjusting the crack velocity parameters v_0 and Γ [17]. Note the asymptotic limits: at fast stressing rates to the inert strength; at slow rates to a "fatigue limit".

This fitting procedure allows us to deconvolute velocity functions from the strength data. We plot the results from such deconvolutions in Fig. 4 [17]. The dashed curve at left represents the fundamental

218

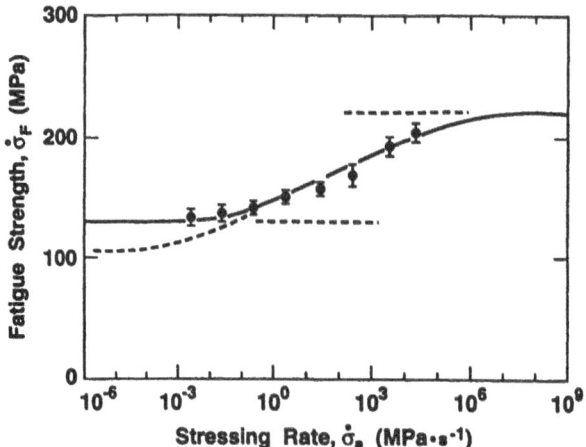

Figure 3. Strength of same alumina as Fig. 2, but as function of
stressing rate for tests in water, Vickers indentations at P = 30
N. Upper horizontal line is inert strength limit from Fig. 2,
lower line is fatigue limit. Dashed curve is equivalent response
for hypothetical material without crack velocity threshold.

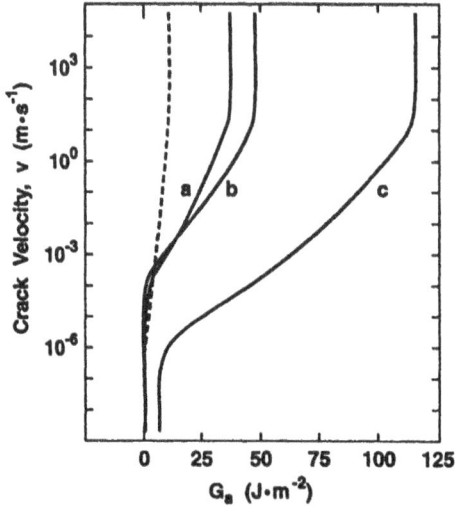

Figure 4. Velocity functions, deconvoluted from Fig. 3. Dashed
curve is intrinsic, invariant crack-tip function, $v(G_*)$. Solid
curves are history-dependent global functions $v(G_A)$, as monitored
via the applied loading: (a) P = 3 N, $\dot{\sigma}_A = 10^3$ MPa·s^{-1}; (b) P =
30 N, $\dot{\sigma}_A = 10^3$ MPa·s^{-1}; (c) P = 30 N, $\dot{\sigma}_A = 10^{-1}$ MPa·s^{-1}.

υ-G_* relation of Eq. 15, equivalent to the curve that a "global" observer at the load points would measure for a material with no shielding ($K_r = 0 = K_\mu$). The solid curves are the corresponding $\upsilon(G_A)$ curves for indentation flaws with shielding present. We see that the global velocity curve is history dependent, i.e. is a function of the residual contact field and stressing rate. The curves shift because the shielding contribution to the toughness is no longer a unique function of crack size, but depends also on the subcritical path [17]. Such curve shifts are regularly reported in the ceramics literature.

These results have a special relevance to fatigue. Many ceramics exhibit a static fatigue stress limit (analogous to the lower stress limit apparent in Fig. 3) below which the lifetime is effectively infinite. A fatigue limit is usually considered to be a direct manifestation of the crack velocity threshold, the crack-tip stress intensity below which all propagation ceases. But this is not the whole story. A strong T-curve enhances the fatigue limit, and can even generate an effective limit in materials with no detectable natural velocity threshold. To illustrate, we include as the dashed curve in Fig. 3 the computed strength vs stressing rate function for a hypothetical material with the same velocity function as our alumina but with its threshold shifted to zero G_*. A fatigue limit is still apparent, albeit at a somewhat reduced stress level. Mathematically, the existence of this limit is possible because the bridging closure term $-K_\mu$ can negate the effective applied loading term $K_A = K_a + K_r$ in Eq. 1, giving rise to a zero velocity state $K_* = 0 = G_*$ in Eq. 15.

3.3 CYCLIC FATIGUE

We have just referred to fatigue in static loading. Taking the metals literature as a guide, we might suspect some additional reduction in the fatigue limit in cyclic loading, due to some hysteresis in the crack-tip shielding. Specifically, we might anticipate mechanical degradation from deterioration of bridging ligaments in repeated loading.

Cyclic fatigue tests are most conveniently conducted with the indentation flaw configuration [18]. Results of such tests at two frequencies on alumina (this time on a coarser material than in Figs. 2 and 3) are shown in Fig. 5. Theoretical curves, obtained once more by solution of the crack velocity differential equation in Sect. 2 (but for the new grain size), exhibit a similar fatigue limit to that in Fig. 3. The data do appear to fall systematically below this predicted limit at long lifetimes, suggesting a real mechanical degradation of the microstructure.

The modest level of this degradation in Fig. 5 is not necessarily discouraging if one regards a million cycles as a reasonable service limit. Nevertheless, the potential exists for stronger effects in other configurations, e.g. long cracks and materials with more pronounced T-curves. For this reason it is proper that one should attempt to understand the underlying causes of the degradation.

220

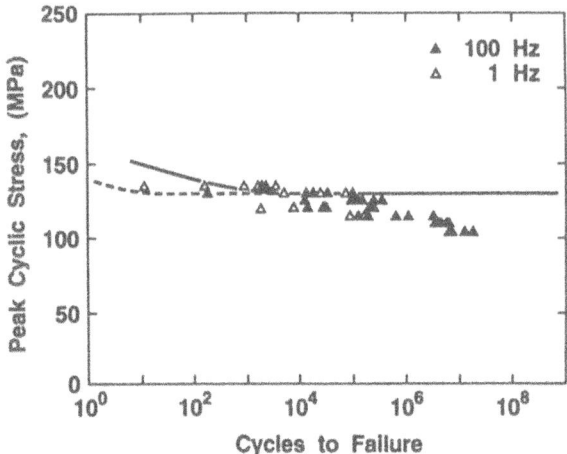

Figure 5. Cyclic fatigue plot for polycrystalline alumina, $\ell = 35$ μm, in water, Vickers indentations at P = 30 N, as function of number of cycles. Open symbols are for 1 Hz, closed symbols 100 Hz. Curves are predictions for 1 Hz (dashed) and 100 Hz (solid) assuming slow crack growth without mechanical degradation.

4. IN-SITU SCANNING ELECTRON MICROSCOPY OBSERVATIONS

4.1 OBSERVATIONS OF BRIDGE EVOLUTION

In situ observations of crack microstructure interactions using optical microscopy have played a pivotal role in identifying grain bridging as an important toughening mechanism in ceramics [7,10]. The development of miniature straining devices for operation within the scanning electron microscope (SEM) have opened the possibility for more detailed observation and quantitative analysis of bridging mechanisms [20,21].

Fig. 6 presents SEM surface views of a crack-interface bridging site in an alumina, grain size $\ell = 11$ μm, at two stages in monotonic loading [20]. The bridge has continued to evolve between the two stages, during which the crack front has undergone an incremental extension of $\approx$ 600 μm. Persistent frictional contact points P and S are indicated. In the interval between (a) and (b) secondary fractures have been initiated in the bridging grain adjoining P by the frictional tractions. During this same interval a single secondary fracture in the large grain to the right of S has closed significantly, indicating a falloff in the frictional tractions. This latter is indicative of a tail-dominated constitutive stress-separation function, as assumed in Eq. 7.

Fig. 7 presents analogous views of a frictional site in the alumina used in Fig. 5, at various stages of cyclic loading [18]. There is gradual cumulation of debris from repeated load reversal at the sliding

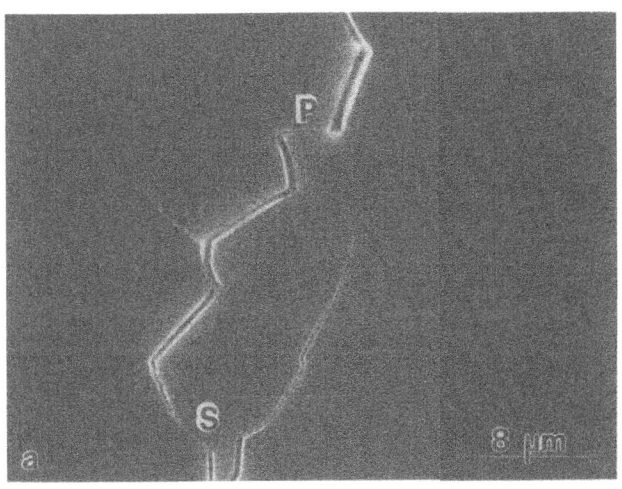

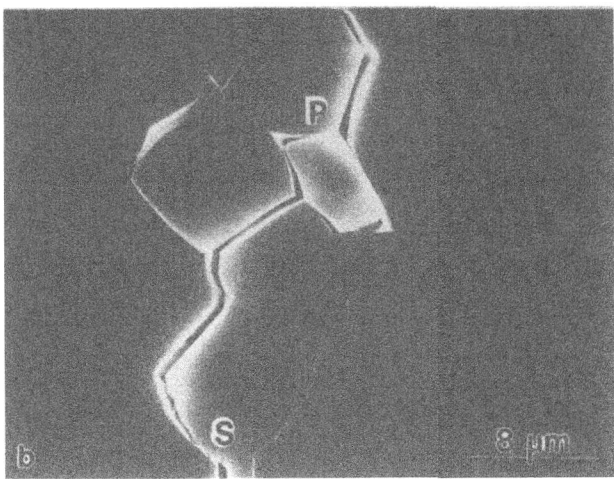

Figure 6. SEM micrographs showing evolution of grain-bridging ligament at a crack interface in alumina, $\ell = 11$ μm, at distances (a) 600 μm and (b) 1190 μm behind the advancing crack tip. P and S denote frictional contact points.

222

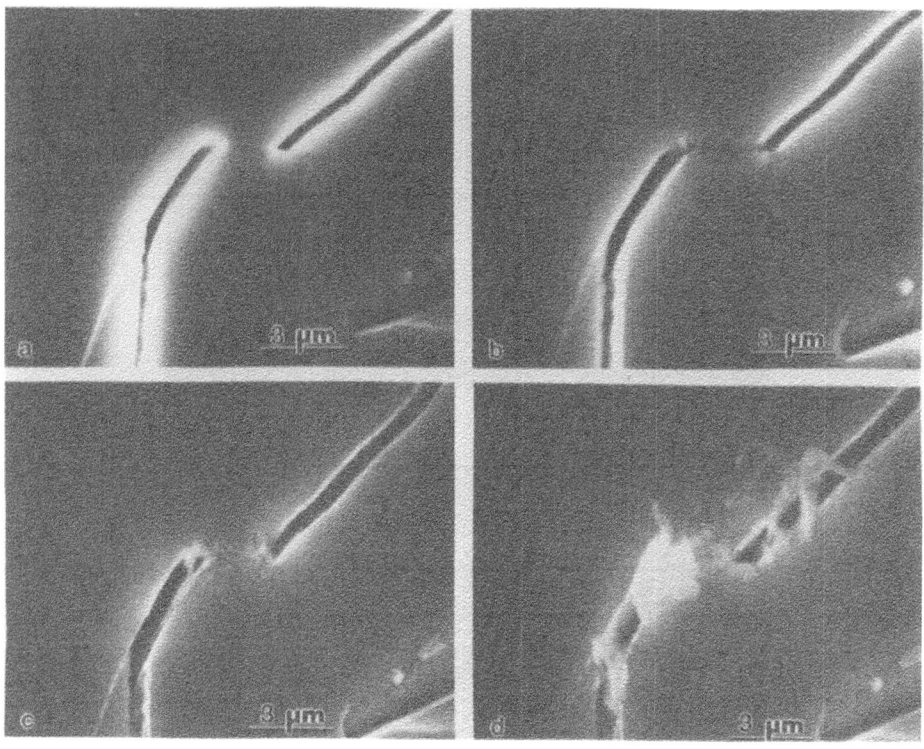

Figure 7. SEM micrographs of frictional grain facet in same alumina
as Fig. 5, at various stages of cycling loading (vacuum, 2 Hz,
unless otherwise indicated): (a) after initial propagation; (b)
after 20000 cycles, with accompanying crack propagation 90 μm;
(c) after 45000 cycles, no propagation; (d) after 50000 cycles
(air, 10 Hz), propagation 20 μm. Note debris at friction facet.

grain facet, even between (b) and (c) where no primary crack growth occurs. Stage (d) shows strongly enhanced buildup of the debris after prolonged additional cycling in air. The debris may be taken as direct evidence for degradation of the interface at the sliding facet. Analysis using the bridging formulation in Sect. 2 indicates that the reduction in friction coefficient needed to account for the mechanical fatigue effect in Fig. 5 is modest, from $\mu = 1.8$ to 1.4 [18].

4.2   QUANTITATIVE CRACK PROFILE MEASUREMENTS

The quantitative capability afforded by the SEM is demonstrated in Fig. 8 with measurements of crack-opening displacement (COD) 2u as a function of distance x behind the crack tip, for the alumina in Fig. 6. The crack profile is seen to be closer to linear than the usual Irwin-parabola for stress-free walls [20], reflecting the closure effect of bridging tractions. In the limit of small bridging zones, as pertains to small extensions Δc from a long notch, the profile may be approximated by the Barenblatt relation

$$u(x) = (8x/\pi)^{1/2}K_A/E'$$
$$-(2/\pi E')\int_0^{\Delta c} p(x')\ln[(x'^{1/2} + x^{1/2})/(x'^{1/2} - x^{1/2})]dx' \quad (16)$$

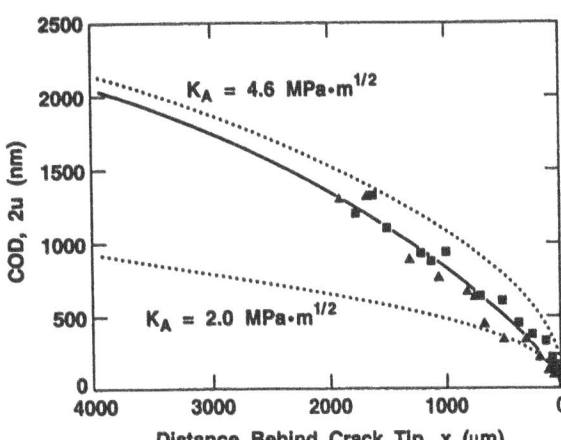

Figure 8.   Measured crack-opening displacement at crack interface in same alumina as Fig. 6, compact-tension specimen, extended crack length $\Delta c = 1.9$ mm. Solid curve is fitted profile from Eq. 16. Dashed curves are enveloping Irwin parabolas for $K_A = 4.6$ MPa·m$^{1/2}$ and $K_A = T_0 = 2.0$ MPa·m$^{1/2}$.

224

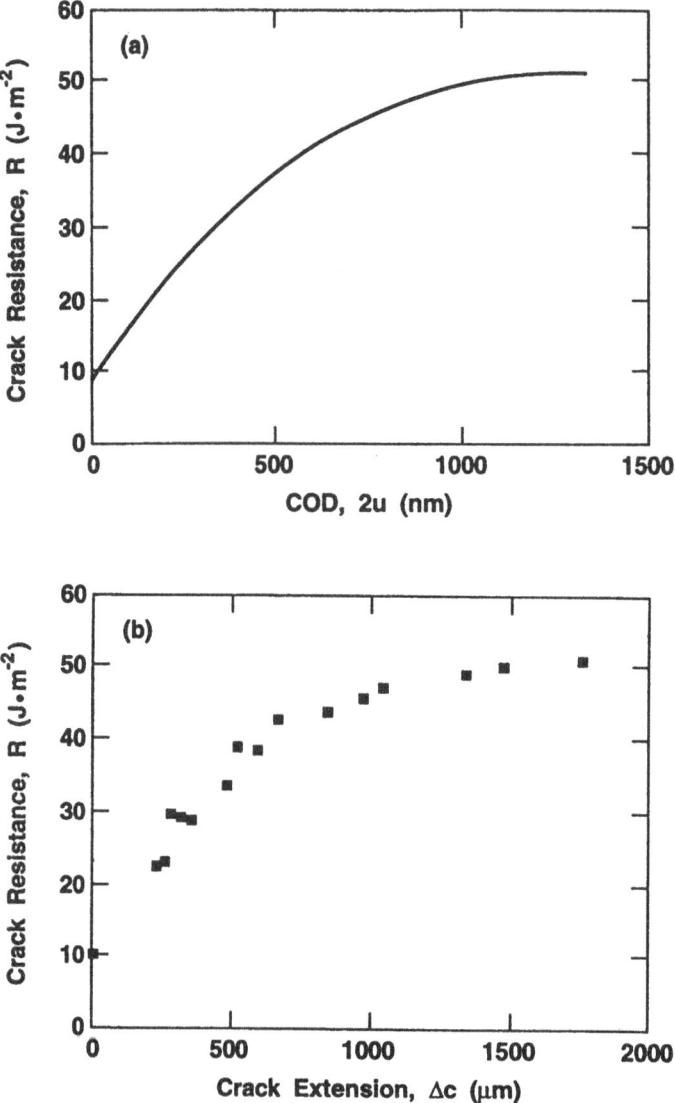

Figure 9. Crack-resistance curves corresponding to constitutive functions evaluated from crack profile in Fig. 8, (a) R(u), and (b) R(Δc).

with x now a field point at which the displacement is to be evaluated
and x' a source point for the closure stresses. The problem is to find
a solution for p(x'), thence p(u), self-consistent with the measured
u(x). Eq. 16 is a nonlinear integral equation, for which there is no
analytical solution. The solid curve in Fig. 8 is a best fit obtained
by numerical iteration.

Once p(u) is known the crack-resistance curves are readily evaluated
using the J-integral in Eq. 12 (recall that the contribution from Eq.
10 is negligible for straight cracks), identifying $2u_z$ with the notch-
tip COD. The resulting R(u) curve for the alumina is shown in Fig. 9a.
We assert that R(u) is a material property, since it is uniquely
dependent on the intrinsic function p(u). On the other hand, the
corresponding R(Δc) curve in Fig. 9b, evaluated in conjunction with the
geometry-sensitive profile u(Δc), is a function of test configuration.

5. MICROSTRUCTURAL CONTROL OF TOUGHNESS PROPERTIES

We have indicated the strong influence of microstructural variables on
toughness-curve characteristics, and the associated potential for
manipulating mechanical properties through controlled processing. Two
examples serve to illustrate the point:

5.1   GRAIN-SIZE EFFECTS

The role of grain size on the strength characteristics of alumina has
been systematically investigated in an indentation-strength study [15].
Fig. 10a plots indentation-strength data for selected grain sizes. The
solid curves correspond to predictions from the calibrated bridging
theory from Sect. 3. Fig. 10b plots the deconvoluted T-curves.
Observe the effect of coarsening the microstructure: the strengths
tend to more pronounced, low-load plateaus, hence greater flaw
tolerance; reductions of strength and toughness in the short-crack
domain are compensated by increases in the long-crack domain.

The bridging model also provides insights into the nature of flaw
states in strength-grain size characteristics. Fig. 11 is an "Orowan-
Petch" plot of strength data for failures from natural flaws ($K_r = 0$)
in our aluminas as a function of inverse square-root grain size [15].
Included as the solid and dashed curves are the predicted responses for
intrinsic (processing) grain-facet flaws ($c_f \approx d = \alpha_d \ell$) and extrinsic
(surface) flaws (c_f specified). Note that the solid and dashed curves
merge in the large grain-size (Orowan) region. The insensitivity of
the strength to any assumptions concerning type or initial size of flaw
in this region indicates an instability condition controlled by
material properties: failure is determined by a single point, the
tangency point, on the T-curve. (We point out that the solid curve is
not strictly linear in this region, as would obtain if ideal Griffith

226

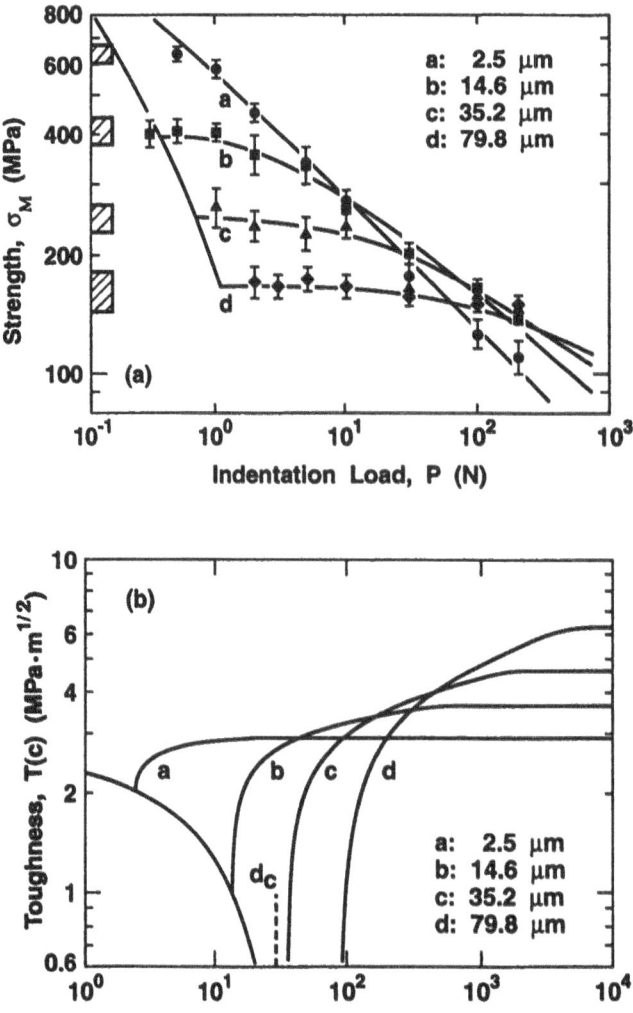

Figure 10. (a) Vickers indentation-strength data for four single-
phase aluminas of differing grain size. Solid curves are
predictions using calibrated bridging stress-intensity factor
from Sect. 3. Note enhanced plateau behavior at larger grain
sizes. Curves are theoretical fits from bridging model. (b)
Corresponding deconvoluted T-curves. Vertical dashed line
denotes intersection point of curve with crack-size axis at T = 0
for ℓ = 30 μm, corresponding to onset of spontaneous
microcracking. Note tendency for curves to cross in both (a) and
(b).

227

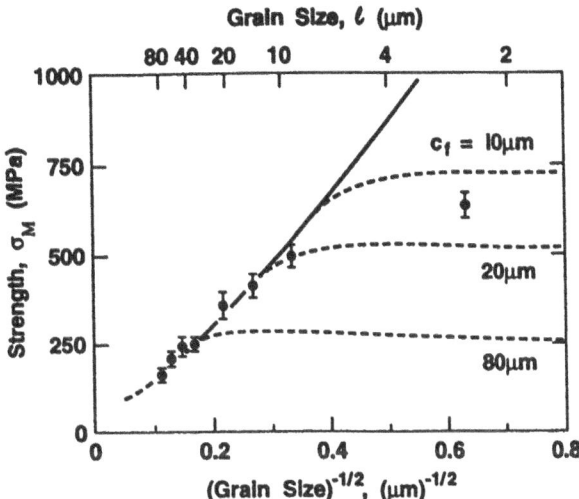

Figure 11. Strength versus inverse square-root grain size for
alumina. Data are failures for polished, unindented specimens.
Solid curve is prediction from bridging model for intrinsic
flaws. Dashed curves are predictions for extrinsic flaws of
specified initial size c_f. Note insensitivity of strength to
flaw size in large grain-size domain.

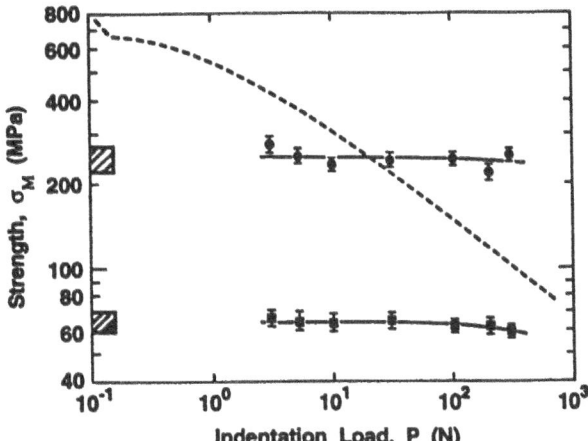

Figure 12. Inert strength as function of Vickers indentation load
for an alumina matrix composite, grain size $\ell = 6$ μm, with 20
vol.% second-phase aluminum titanate particles. Lower data are
for degraded material with $\ell = 14$ μm, above the critical grain
size for spontaneous microcracking. Included as the dashed curve
is prediction for equivalent monophase alumina ($\ell = 6$ μm).

behavior, i.e. flaw scaling with grain size, single-valued toughness T = T_0, were to prevail.) However, the curves diverge in the small grain-size (Petch) region, reflecting the sensitivity to flaw size usually associated with ceramics. On passing into this region the extrinsic flaws begin to dominate their microstructural counterparts, and traverse the tangency points on the T-curves (Fig. 10b) beyond which precursor stable crack growth prior to failure no longer occurs. Only one of the data points in Fig. 11, that of the finest grain size, appears to fall into the extrinsic region.

5.2   SECOND-PHASES

A second illustrative example demonstrates the importance of microstructural parameters other than grain size. We consider a model composite ceramic in which aluminum titanate has been added as a particulate second phase to an alumina matrix, grain size 6 μm [16]. Aluminum titanate was selected because of its unusually strong thermal expansion mismatch relative to alumina. An aging heat treatment promoting grain growth to ℓ = 14 μm caused general microcracking (see Sect. 6.1) in the composite.

Results of indentation-strength tests on the as-fabricated material are shown as the upper set of data points in Fig. 12. Included for comparison is a predicted curve for single-phase alumina of the same grain size (ℓ = 6 μm). It is immediately apparent that the composite has greatly exaggerated flaw tolerance relative to the alumina, although at the expense of short-crack strength. Also included is a data set for the heat-treated composite. The strength level for this latter material is severely reduced, as expected for an overaged microstructure.

A proper understanding of the bridging micromechanisms responsible for the striking changes in Fig. 12 is yet to be attained. Is the main role of the enhanced mismatch one of intensifying the frictional tractions at compressive bridging facets, or to enhance the formation of bridges by providing favorable tensile paths for deflection? In situ studies may help to provide answers. Once the micromechanisms are identified, the challenge will be to modify the model in Sect. 2.

These results demonstrate the capacity for improved strength properties in ceramic composites, particularly in the long-crack domain. Potential pitfalls in fabrication strategies are also apparent. The line between ultra-high toughness and unacceptable weakness may be very fine indeed.

6.   SHORT-CRACK PROPERTIES: SPONTANEOUS MICROCRACKING AND WEAR

What is good for properties that pertain to long or intermediate cracks may not be so for short cracks. The very ingredients responsible for flaw tolerance, notably internal residual stresses, can be detrimental to general microcracking and wear resistance.

6.1 SPONTANEOUS MICROCRACKING

It is well known that noncubic polycrystals tend to general microcracking above a critical grain size. Such microcracking is attributable to the action of residual thermal expansion stresses. In terms of our shielding model, the critical condition is that processing flaws at tensile grain boundary facets should be unstable in the absence of any extraneous driving force. Seen from the perspective of Fig. 10b, a minimum requirement for such spontaneous instability is that the unstable branch of the toughness curve should intersect the abscissa ($K_A = T = 0$) at the flaw size c_f. Beyond this point instability ensues, and the microcrack arrests on the stable branch. Writing $c_f = \beta\ell$, with β a dimensionless constant (geometrical scaling), the critical grain size may be determined directly from Eq. 10 at equilibrium, $K_\mu(c) = \psi\sigma_R(\beta\ell_c)^{1/2} = T_0$: thus [19]

$$\ell_c = (1/\beta\psi^2)(T_0/\sigma_R)^2. \tag{17}$$

From Fig. 10b we estimate $\ell_c \approx 30$ μm for our alumina, which accords with experimental observations of spontaneous microcracking in the same series of aluminas [15]. Fig. 13 is a micrograph of an individual microcrack in a coarse-grained alumina.

6.2 MICROFRACTURE-WEAR PROCESSES

The earliest theories of surface removal by contact-induced cracking in brittle surfaces suggested that the least severe wear rates should occur in those materials with the highest toughness [24]. Such theories presume that the toughness is single-valued. The question arises as to the behavior in ceramics with strong T-curves.

Fig. 14 presents wear data on aluminas at three grain sizes [19]. Of these aluminas, the coarser have the higher conventional (long-crack) toughness. Yet it is the coarser aluminas here that show the greater susceptibility to wear degradation. Wear is quantified in Fig. 14 by the diameter of a scar generated by a rotating hard (Si_3N_4) sphere at specified contact time. The scar diameter increases monotonically with time for each material, initially slowly and subsequently, after an "incubation" period, abruptly. Whereas the initial increase is microstructure-invariant, the transition shifts to smaller times at the larger grain sizes.

Hence the "toughest" aluminas exhibit the most rapid degradation. A qualitative resolution to the apparent anomaly is once more to be found in the T-curve construction of Fig. 10b. The coarser microstructures indeed show greater long-crack toughness. However, it is the short-crack domain that is relevant to wear; and, as we have

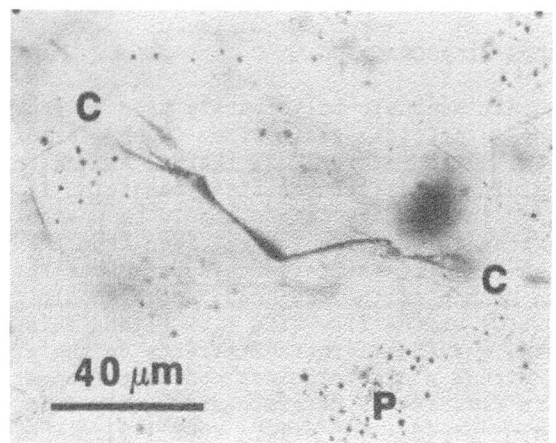

Figure 13. Optical micrograph showing extended microcrack (C-C) in alumina, $\ell = 80$ μm. Note that, contrary to "conventional wisdom", grain-size microcracks are quite visible, here in transmitted light, by virtue of residual opening associated with release of internal tensile stress. Some processing pores (P) are also visible.

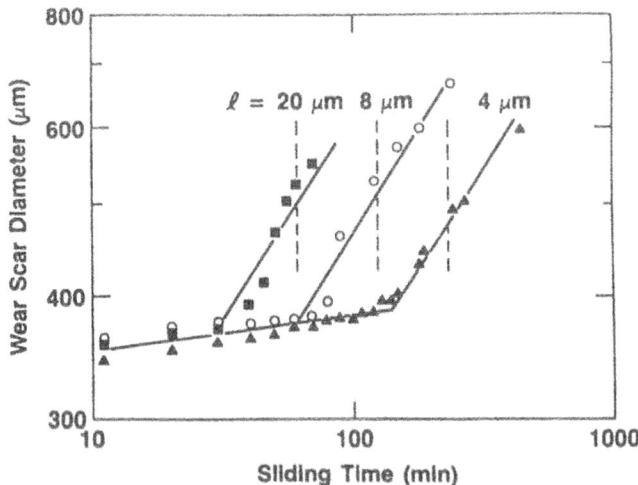

Figure 14. Wear data for aluminas at three grain sizes. Note initially slow increase of scar diameter, abrupt transition to severe microfracture-controlled wear after critical sliding time. Transition time diminishes at increasing grain size.

remarked before, the T-curves cross each other.

Microscopy reveals the initial wear to be slight, with indications of dislocation activity [19]. The abrupt increase is marked by the sudden incidence of gross chipping and grain removal. The results may therefore be interpreted in terms of a deformation-fracture transition, in which tensile stresses σ_D from the deformation augment those from the thermal expansion anisotropy, i.e. $\sigma_R' = \sigma_R + \sigma_D$. Suppose that these deformation stresses cumulate monotonically with time, i.e. $\dot{\sigma}_D = \sigma_D/t = \text{const}$. Then replacing σ_R by σ_R' in Eq. 17 allows us to solve for the time to induce microcracking [19]

$$t_* = (\sigma_R/\dot{\sigma}_D)[(\ell_C/\ell)^2 - 1] \qquad (\ell \leq \ell_C). \qquad (18)$$

Observe that $t_* = 0$ at $\ell = \ell_C$, as required. As ℓ decreases below ℓ_C, t_* diminishes, in qualitative accord with observation. The shifts in Fig. 14 are quantitatively consistent with a stress cumulation rate $\dot{\sigma}_D = 5$ MPa·s^{-1}.

The implications concerning optimization of microstructures for maximum resistance to spontaneous microcracking and microfracture-induced wear are clear in this case - refine the grain size and, if possible, avoid internal residual stresses. It is reiterated that such measures may run entirely counter to the requirements for maximum large-crack toughness and flaw tolerance.

CONCLUSIONS

We have described toughness-curve phenomena in terms of a grain-interlocking bridging model. Essential microstructural ingredients for strong T-curves are weak boundaries, high internal stresses, and coarse grains. Our model has been illustrated with data from alumina, but a wider applicability to ceramics in general is asserted.

The T-curve has positive implications in strength. These include flaw tolerance and an enhanced fatigue limit. However, such benefits could be countered to some extent by possible bridge degradation in cyclic loading.

In situ observations of bridging sites in the scanning electron microscope provide insight into the degradation micromechanisms, and allow for quantitative evaluation of the crack-interface tractions.

Short-crack properties, spontaneous microcracking and abrasive wear, may be exacerbated in materials with strong T-curves.

232

ACKNOWLEDGEMENTS

The authors wish to thank colleagues S-J. Cho, E.R. Fuller Jr., B.J.
Hockey, J.F. Kelly, Y-W. Mai and J.L. Runyan for many valuable
contributions to this work. Funding was provided by the U.S. Air Force
Office of Scientific Research and E.I. duPont de Nemours & Co. Inc.
S.J.B., J.R., S.L. are Guest Scientists on leave from the Department of
Materials Science and Engineering, Lehigh University, Bethlehem, PA,
USA; P.C. is Guest Scientist on leave from Department of Physics,
Chulalongkorn University, Bangkok, Thailand. S.L. is now at Division
of Materials Science, CSIRO, Clayton, VIC, Australia.

REFERENCES

[1] H. Hübner and W. Jillek (1977) "Sub-Critical Crack Extension and
     Crack Resistance in Polycrystalline Alumina", J. Mater. Sci. 12
     117-25.
[2] R. Knehans and R.W. Steinbrech (1982) "Memory Effect of Crack
     Resistance During Slow Crack Growth in Notched Al_2O_3 Bend
     Specimens", J. Mater. Sci. Lett. 1 327-29.
[3] R.W. Steinbrech, R. Knehans and W. Schaarwächter (1983) "Increase
     of Crack Resistance During Slow Crack Growth in Al_2O_3 Bend
     Specimens", J. Mater. Sci. 18 265-70.
[4] R. Knehans and R.W. Steinbrech (1984) "Effect of Grain Size on the
     Crack Resistance Curves of Al_2O_3 Bend Specimens"; in Science of
     Ceramics, Vol. 12, pp. 613-19, ed. P. Vincenzini. Ceramurgia,
     Imola, Italy.
[5] R.F. Cook, B.R. Lawn and C.J. Fairbanks (1985) "Microstructure-
     Strength Properties in Ceramics: I. Effect of Crack Size on
     Toughness", J. Am. Ceram. Soc. 68 604-15.
[6] M.V. Swain (1986) "R-Curve Behavior in a Polycrystalline Alumina
     Material", J. Mater. Sci. Let. 5 1313-15.
[7] P.L. Swanson, C.J. Fairbanks, B.R. Lawn, Y-W Mai and B.J. Hockey
     (1987) "Crack-Interface Grain Bridging as a Fracture Resistance
     Mechanism in Ceramics: I. Experimental Study on Alumina", J. Am
     Ceram. Soc. 70 279-89.
[8] Y-W. Mai and B.R. Lawn (1987) "Crack-Interface Grain Bridging as a
     Fracture Resistance Mechanism in Ceramics: II. Theoretical
     Fracture Mechanics Model", J. Am. Ceram. Soc. 70 289-94.
[9] R.F. Cook, C.J. Fairbanks, B.R. Lawn and Y-W. Mai (1987) "Crack
     Resistance by Interfacial Bridging: Its Role in Determining
     Strength Characteristics", J. Mater. Research 2 345-56.
[10] P.L. Swanson (1988) "Crack-Interface Traction: A Fracture-
     Resistance Mechanism in Brittle Polycrystals"; in Advances in
     Ceramics, Vol. 22, pp. 135-55, Fractography of Glasses and
     Ceramics. American Ceramic Society, Columbus, OH.

[11] R.W. Steinbrech and O. Schmenkel (1988) "Crack-Resistance Curves of Surface Cracks in Alumina", J. Am. Ceram. Soc. 71 C271-73.

[12] S.J. Bennison and B.R. Lawn (1989) "Flaw Tolerance in Ceramics With Rising Crack-Resistance Characteristics" J. Mater. Sci. 24 3169-75.

[13] S.J. Bennison and B.R. Lawn (1989) "Role of Interfacial Grain-Bridging Sliding Friction in the Crack-Resistance and Strength Properties of Nontransforming Ceramics", Acta Metall. 37 2659-71.

[14] Y-W. Mai and B.R. Lawn (1986) "Crack Stability and Toughness Characteristics in Brittle Materials", Ann. Rev. Mat. Sci. 16 415-39.

[15] P. Chantikul, S.J. Bennison and B.R. Lawn (1990) "Role of Grain Size in the Strength and R-Curve Properties of Alumina," J. Am. Ceram. Soc., 73 2419-27.

[16] J.L. Runyan and S.J. Bennison, "Fabrication of Flaw-Tolerant Aluminum-Titanate-Reinforced Alumina", J. Europ. Ceram. Soc., in press.

[17] S. Lathabai and B.R. Lawn (1989) "Fatigue Limits in Noncyclic Loading of Ceramics with Crack-Resistance Curves", J. Mat. Sci. 24 4298-4306.

[18] S. Lathabai, J. Rödel and B.R. Lawn, "Cyclic Fatigue from Frictional Degradation at Bridging Grains in Alumina", J. Mater. Sci., in press.

[19] S-J. Cho, B.J. Hockey, B.R. Lawn and S.J. Bennison (1989) "Grain-Size and R-Curve Effects in the Abrasive Wear of Alumina", J. Am. Ceram. Soc. 72 1249-52.

[20] J.Rödel, J.F. Kelly and B.R. Lawn, "In Situ Measurements of Bridged Crack Interfaces in the SEM", J. Am. Ceram. Soc., in press.

[21] J. Rödel, J.F. Kelly, M.R. Stoudt and S.J. Bennison, "A Loading Device for Fracture Testing of Compact Tension Specimens in the SEM," Scanning Microscopy, in press.

[22] B.R. Lawn, A.G. Evans and D.B. Marshall (1980) "Elastic/Plastic Indentation Damage in Ceramics: The Median/Radial Crack System", J. Am. Ceram. Soc. 63 574-81.

[23] B.R. Lawn, D.B. Marshall, P. Chantikul and G.R. Anstis (1980) "Indentation Fracture: Applications in the Assessment of Strength of Ceramics", J. Aust. Ceram. Soc. 16 4-9.

[24] A.W. Ruff and S.M. Wiederhorn (1979) "Erosion by Solid Particle Impact", in Treatise on Materials Science and Technology, Vol. 16, pp. 69-126, ed. C.M. Preece. Academic, New York.

CREEP DAMAGE MECHANISMS IN HOT-PRESSED ALUMINA

D.S. Wilkinson
Department of Materials Science and Engineering
McMaster University
Hamilton, Ontario
L8S 4L7

ABSTRACT. The creep fracture behaviour of two commercially available hot-pressed aluminas has been extensively studied, both in flexure and tension. Three distinct regions are observed. At high stress, failure is by slow crack growth, dominated by a single flaw. At intermediate stresses, widely distributed microcracking is observed. Microcrack linkage differs for flexure and tension, resulting in large increases in lifteime for the flexure geometry. At the lowest stresses, widely distributed cavities are observed. Nonetheless, failure still occurs by the growth of flaw related cracks. The most damaging flaws appear to be large-grained regions.

1. Introduction

The fracture strength of ceramics is known to degrade at elevated temperatures. This is typically measured by means of short term testing, usually in bending (Fig. 1). Over a wide range the strength is constant, independent of temperature. Above a critical temperature however, the short term strength begins to decrease fairly rapidly. The reasons for this can be seen in general terms by considering the Griffith equation for the fracture strength σ_f, of a brittle solid:

$$\sigma_f = \frac{K_{IC}}{\sqrt{\pi a}}$$

where K_{IC} is material toughness and a is the size of the critical flaw at fracture. From this, we see that a decrease in fracture stength can result either from changes in flaw size or toughness. At elevated temperatures slow crack growth of sub-critical flaws is possible. Thus, for a fixed value of K_{IC}, the fracture strength is decreased. Alternatively, microstructural changes may decrease K_{IC} and thus σ_f. Such microstructural changes may involve

S. P. Shah (ed.), Toughening Mechanisms in Quasi-Brittle Materials, 235–247.

236

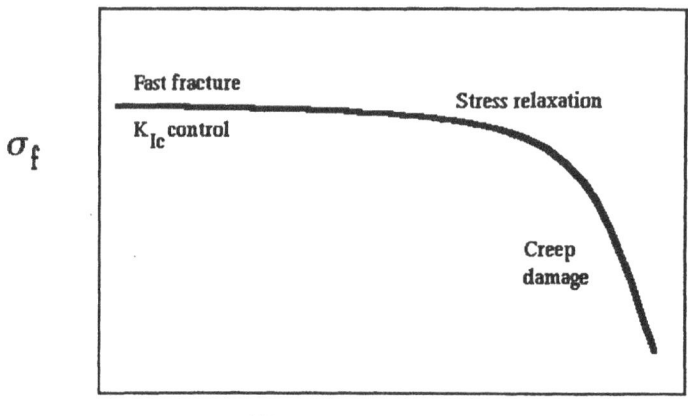

σ_f

Figure 1. The effect of temperature on the short term strength of ceramics is shown here schematically. Strength is relatively constant until creep effects begins to dominate the failure process.

microcracking and cavitation, as well as other changes (e.g. grain growth, devitrification of grain boundary glass, etc.).

It is of interest to ask whether the reduction of strength observed at high temperature is due to the slow growth of subcritical cracks or due to a reduction of toughness. In the current paper, we attempt to answer this question with respect to a microstructurally simple ceramic -- namely, hot-pressed Al_2O_3. In such a material, the microstructure is stable, except for the development of dilational damage (cavities and microcracks). Thus, the issue in question simplifies to a distinction between the growth of pre-existing flaws and the development of more general damage.

It is also important to question the relevance of data such as that shown in Fig. 1, to understanding the elevated temperature strength of ceramics over *long times* associated with service conditions. Under these conditions ceramics undergo creep. For materials such as pure Al_2O_3, which do not contain grain boundary glass, creep is due to grain boundary sliding. Such sliding needs to be accommodated at the ends of each grain facet, and it is the accommodation process which controls the creep rate. There are two main classes of accommodation process. The first involves deformation, either by diffusion or dislocation motion. The second involves dilational processes, either microcracking or cavitation. These two classes represent competing processes. If deformation dominates, then the material will be ductile. If however, grain boundary sliding is accommodated predominantly by dilitation, then premature fracture will result, and the material will be brittle.

In fine-grained alumina at high temperatures, deformation is dominated by diffusion for which the expected constitutive equation is

$$\dot{\varepsilon} = C \frac{D_b \delta}{kTd^3} \sigma$$

where d is the grain size. Therefore, the role of grain size, and especially grain size distribution, need to be considered.

Figure 2 shows a large grain surrounded by a fine-grained matrix. As this material creeps by diffusion-accommodated grain boundary sliding load is transferred to the large grain. In the limiting case in which all grains undergo the same strain, the stress concentration on the large grain is

$$k = \left(\frac{d_\ell}{d_m} \right)^3$$

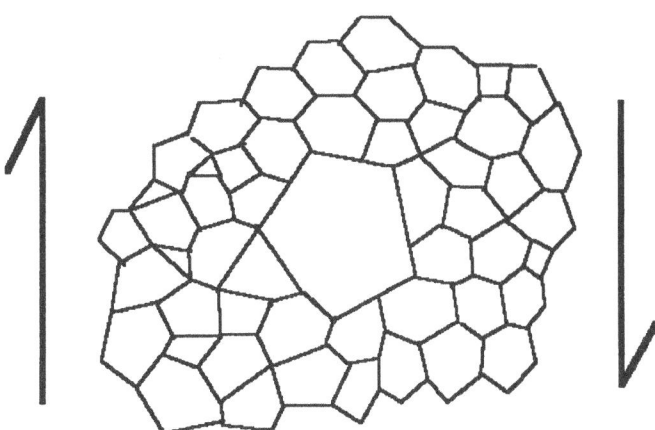

Figure 2. When deformation involves grain boundary sliding, large grains act as sites for concentrating stress.

where d_ℓ is the size of the large grain and d_m is the matrix grain size. Thus significant stress concentrations may arise due to grain size distributions.

In this paper, we present the results of an extensive study of creep damage and fracture in hot-pressed alumina. A detailed exposition of this work is published elsewhere [Robertson, et al., 1991; Wilkinson, et al., 1991]. The results are merely summarized here. However, a more extensive treatment is presented related to the analysis of damage processes and the relative importance of microcracks and general cavitation.

2. Experimental

The material used in this study was hot-pressed Al_2O_3. Two commercial materials were used. Both have low initial porosity ($<.01\%$) and a grain size of about 1 μm. The chemical composition of these materials is given in Table I. Although the composition differences do affect the creep behaviour, the fracture behaviour of the two materials are broadly similar. Creep fracture studies have been performed in both flexure and tension, over a range of temperature and stresses, and with failure times up to 1000 hours.

TABLE I. Chemical Analyses

	C	Fe_2O_3	K_2O	MgO	MnO	S	SiO_2	TiO_2	Y	Zn
ARCO	250	90	4	200	3	16	200	25	800	19
AVCO	140	130	6	3000	2	17	140	25	10	17

3. Stress Rupture Results

The creep behaviour of both materials exhibits a power-law stress dependence, $\dot{\varepsilon} \sim \sigma^n$, with a stress exponent close to 2 for both materials and over all conditions studied. Moreover, no difference is found for the creep behaviour in flexure and tension.

Creep fracture data for tests in flexure are shown in Fig. 3. There are two distinct regimes. Above a stress of 175 MPa (for AVCO alumina at 1150°C), failure occurs at short times (less than 1 hour), while failure strains are less then 1% (Fig. 3b). Below this stress the failure time increases by about 100 times, and the failure strain increases to about 12%. The ARCO material undergoes a similar transition at 180 MPa. However, below the transition stress, failure is never achieved out to the maximum available flexure strain of 18%.

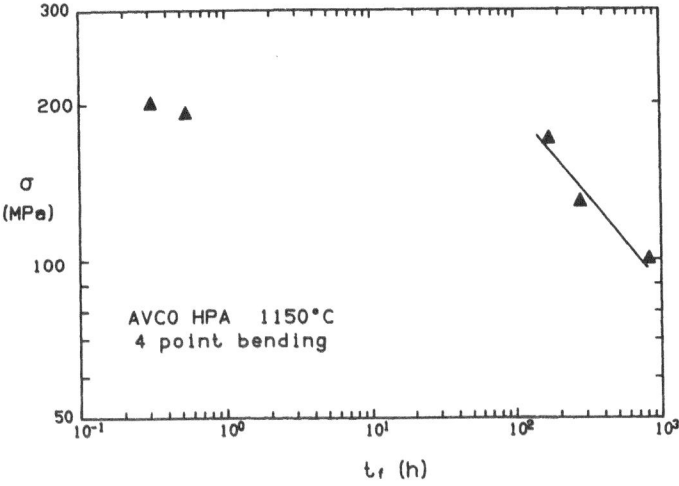

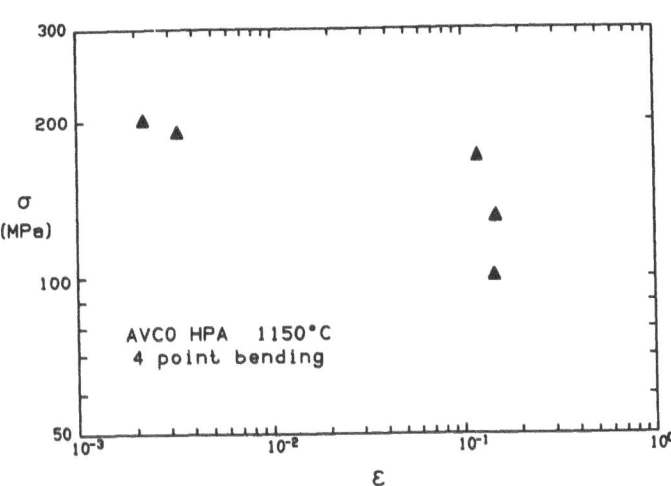

Figure 3. Stress rupture data in flexure for AVCO alumina at 1150°C, showing (a) time to failure and (b) strain to failure, as a function of applied stress.

Further testing has been performed in tension, at lower stresses, for the ARCO material. These tests indicate a second transition between 80 and 55 MPa (Fig. 4). Above 80 MPa, failure occurs after a strain of about 8%. This regime corresponds to the lower stress regime (below 180 MPa) covered in the flexure tests. However, for a given stress, lifetimes are much shorter in tension, by more than a factor of 10.

Between 80 and 55 MPa in tension there is a sharp increase in failure life (by about a factor of 3), and the ductility increases dramatically. This appears to indicate an onset of superplastic behaviour.

4. Damage Observations

The three failure regions outlined in the previous section correspond to different modes of damage accumulation and fracture behaviour. The short time/high stress failures result from the growth of a single, dominant microcrack. These nucleate and grow so rapidly that no other cracks are seen. Typical failure origins are shown in Fig. 5. They involve either large-grained regions (Fig. 5a) processing-related microcracks (Fig. 5b), or amorphous regions (not shown). Typically these regions are of 100µm dimensions and grow to several times this size prior to failure.

Below the first transition stress (at about 180 MPa) many microcracks are observed, both in flexure and tension. These cracks emanate from two different types of origin. The first is due to a chemical deposit that appears, primarily on the surface of the flexure bars, during the initial heating of the creep rig. The second consists of either a large isolated grain or a large grained region (see for example, Fig. 6).

Microcracks in tensile specimens do not interact extensively and failure is due to the propagation of a single flaw, as illustrated in Fig. 7. In flexure however, cracks do interact. Non-planar cracks link up by a "shear-band" process, first noted by Dalgleish and Evans [1985] (Fig. 8). The reason for this difference is not entirely clear. However, it is no doubt related to the constraint on crack growth which results in flexure. These cracks see a large stress gradient. In addition, at the relatively high failure strains observed here, there is a significant hinge effect. This increases the crack opening displacement necessary for crack propagation. Thus crack growth is inhibited. Moreover, a high density of microcracks helps to accommodate the strain. Microcracks in tensile specimens however are less stable. These tend to propagate without interacting with other out-of-plane cracks.

As the stress is decreased below about 80 MPa in tension, two differences in the damage behaviour are noted. First, the cavity morphology changes, as shown in Fig. 9. Above this stress, cavities are crack-like in shape. They nucleate at grain boundary triple junctions and grow across a facet. Almost no cavities larger than a single facet are observed. These cavities appear to have grown by a surface diffusion controlled process [Cocks

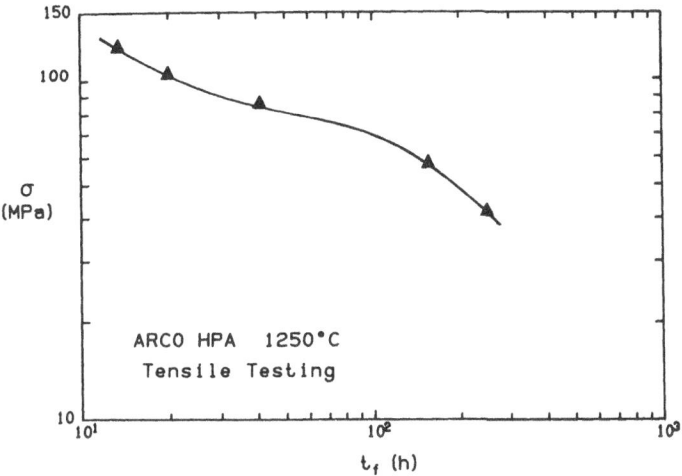

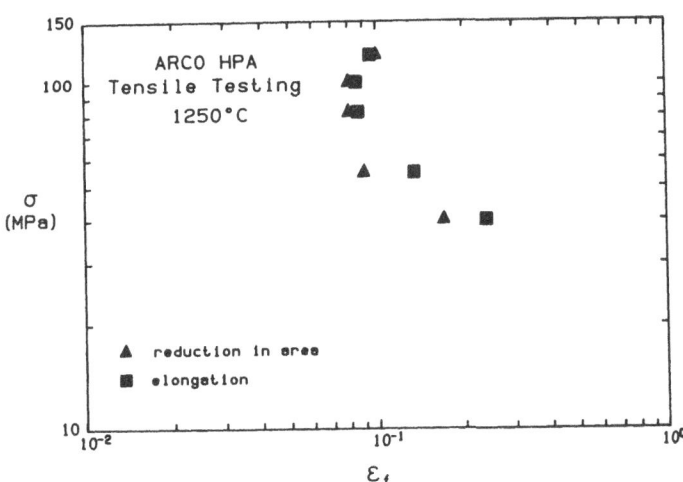

Figure 4. Stress rupture data in tension for ARCO alumina at 1250°C, showing (a) time to failure and (b) strain to failure, as a function of applied stress.

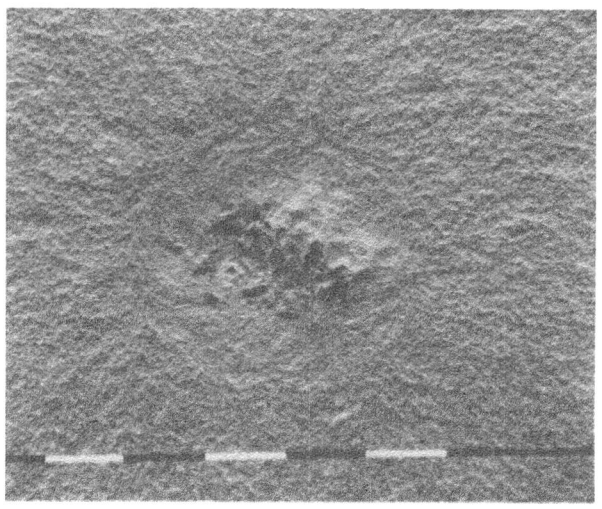

Figure 5. Two typical high stress fracture origins; (a) a large-grained region, and (b) a processing related flaw (scale markers = 0.1 mm).

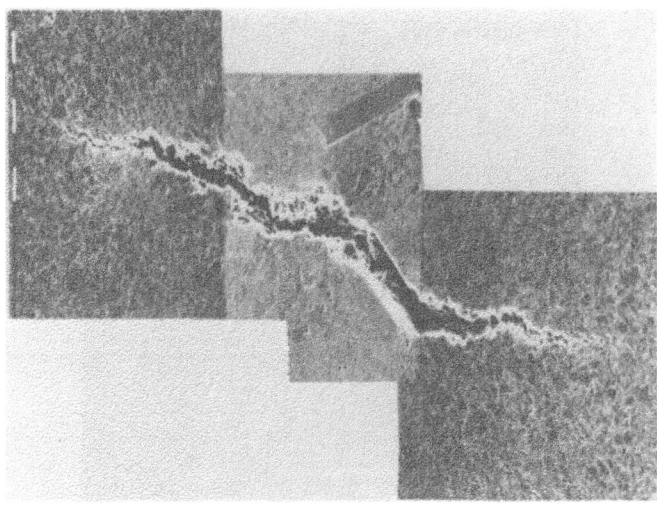

Figure 6. A microcrack emanating from a large grain (ARCO alumina, tested at 125 MPa, 1200°C, in flexure).

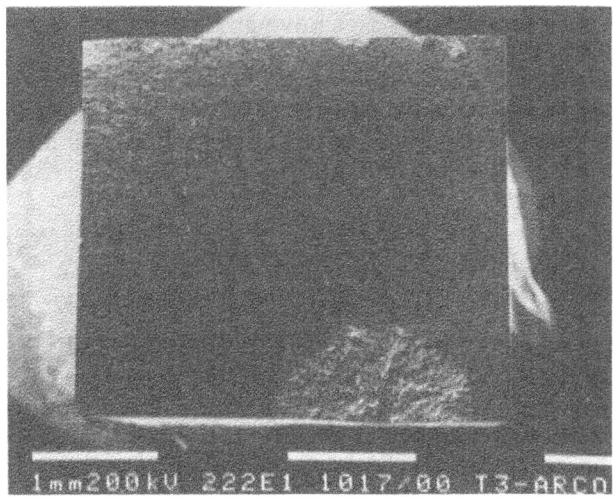

Figure 7. Fracture surface of a tensile specimen tested at 82 MPa and 1250°C.

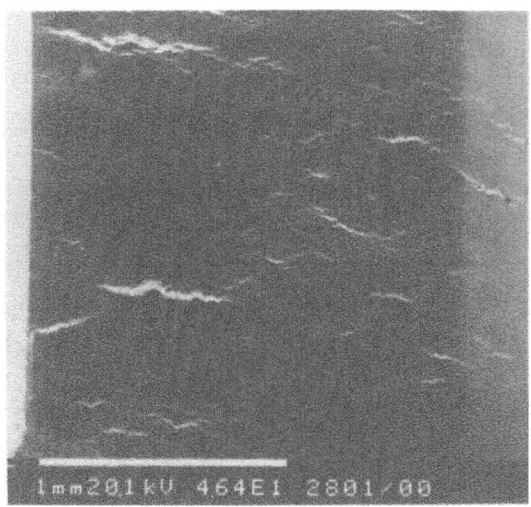

Figure 8. The tensile surfaces of a bend bar tested at 170 MPa and 1150°C. The stress axis is vertical (Scale marker = 1 mm).

and Ashby, 1982]. Below about 80 MPa, cavities are angular in shape (Fig. 9b). This suggests that cavities have grown across a facet, and have the blunted by grain boundary sliding.

A second effect of decreasing stress is seen from Fig. 10. The crack opening displacement decreases significantly as the stress is decreased. This appears to be a direct consequence of the change in cavity morphology. This slows the rate of crack extension and explains the increased lifetime on crossing this threshold.

5. Summary

In summary, the creep fracture of behaviour of hot-pressed alumina exhibits three distinct regimes. The transitions between these are characterized both by sharp increases in creep life, and by significant changes in damage morphology. Although damage tolerance increases with decreasing stress and damage becomes more widely distributed, fracture in all regions emanates from flaws. In this material the most damaging flaws are large-grained regions. At the lowest stresses only the largest of these are able to nucleate microcracks. Thus there is a maximum in the microcrack density at intermediate stresses.

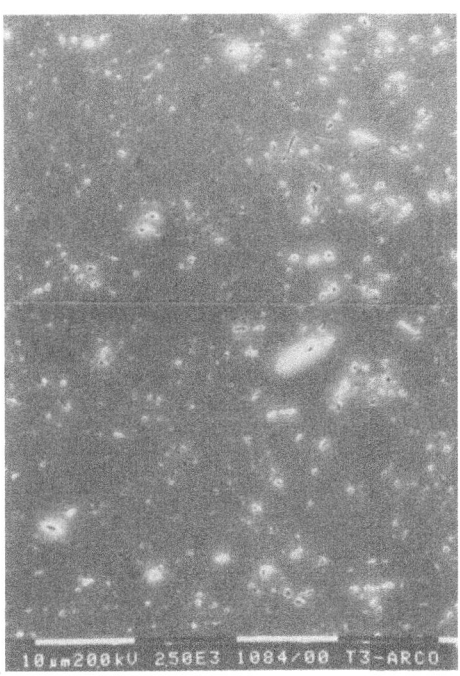

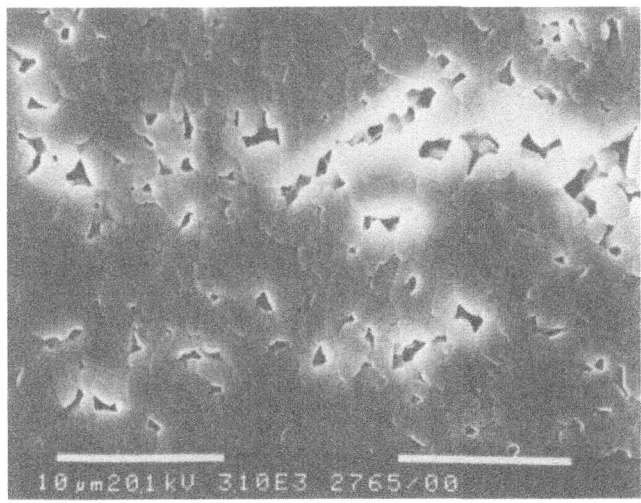

Figure 9. The cavity morphology changes on going from (a) 82 MPa to (b) 40 MPa in ARCO alumina tested at 1250°C. The tensile axis is vertical in both cases.

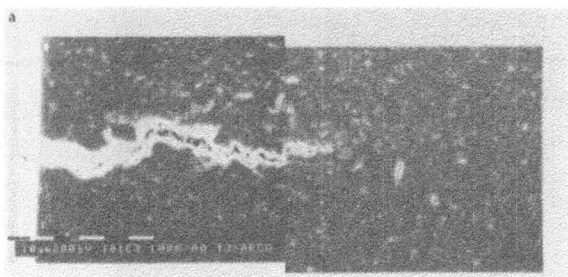

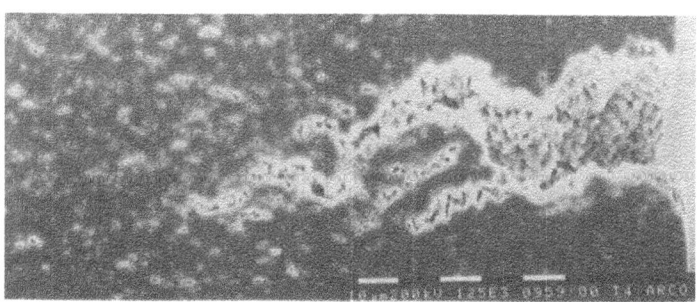

Figure 10. The crack morphology also changes with decreasing stress:
(a) 82 MPa, and (b) 55 MPa in ARCO alumina at 1250°C

We have also shown that stress rupture life is considerably shorter in tension than in flexure. Thus, flexure tests cannot be used to provide data on creep life even for microstructurally simple ceramics such as hot-pressed alumina.

6.   References

Cocks, A.C.F. and Ashby, M.F. (1982), "On creep fracture by void growth", Prog. Mater. Sci., **27**, 190-244.

Dalgleish, B.J. and Evans, A.G. (1985), "Influence of shear bands on creep rupture in ceramics", J. Amer. Ceram. Soc., **68**, 44-48.

Robertson, A.G., Wilkinson, D.S. and Cáceres, C.H. (1991), "Creep and creep fractures in hot-pressed Al_2O_3", J. Amer. Ceram. Soc., submitted.

Wilkinson, D.S. Cáceres, C.H. and Robertson, A.G. (1991), "Damage and fracture in mechanisms during high temperature creep rupture in hot-pressed Al_2O_3", J. Amer. Ceram. Soc., submitted.

STUDY OF THE FRACTURE PROCESS IN MORTAR WITH LASER HOLOGRAPHIC MEASUREMENTS

by A. Castro-Montero[1], R. A. Miller[2] and S. P. Shah[3]

ABSTRACT.   Center notched plate mortar specimens were loaded in tension. A multiple sensitivity vector holographic setup was developed to record several deformation stages during the stable crack propagation range. The three sensitivity vector setup enabled the calculation of both crack opening displacements and strain fields around the crack trajectories. An image analysis system was used to isolate the interferometric effect from the sandwich holograms resulting in fringe patterns with perfect contrast. Image analysis was also used as a faster, more accurate and more consistent method for fringe count.

After evaluation of the holograms, the existence of tensile forces transmitted through the crack faces was associated to the presence of tensile strain behind the crack tip. Existing cohesive crack models were evaluated based on experimental crack profile measurements. A crack length dependent cohesive crack model with a bilinear closing pressure was proposed. A definition of the fracture process zone (FPZ) is proposed based on the difference between experimentally observed and linear elastic fracture mechanics (LEFM) strain fields.

1. Introduction

The fracture of concrete is characterized by formation and propagation of fracture process zone. Several phenomena have been associated with fracture process zone including microcracking around the crack tip, localization and strain softening and crack bridging (often also termed aggregate interlock). A key to a better understanding of fracture mechanics of concrete is the accurate observation of fracture process zone on a microscopic scale.

This paper will examine the cohesive crack model which proposes that fracture in concrete can be modeled using linear elastic fracture mechanics (LEFM) and applying a closing pressure to the crack faces. Using experimentally measured crack opening displacements, four closing pressures proposed in literature were tested as were the basic underlying assumptions of the cohesive crack model.

In the technique described in this paper, three simultaneous holograms were taken by illuminating the specimen from three different directions.   In addition, digital image analysis techniques were developed to accurately evaluate wide field in plane deformation.   The results of the strain field calculation obtained using this newly developed technique for mortar center notched plate specimens are discussed in this paper.

2. Holographic Interferometry Test

Each interference pattern represents the component of the displacements onto the sensitivity vector k (Abramson, 1981). In order to obtain a two dimensional displacement field (in plane and out of plane), in principle, two interferograms are required. However, since there is no sign associated to the fringe counts a third interferogram is necessary. These requirements are discussed in more detail by Castro-Montero, Shah and Miller[1]. Figure 1 is a schematic view of the optical set-up with three illumination directions (R,L and C) used for this study.

---

[1] Graduate Research Assistant, Department of Civil Engineering, Northwestern University, Evanston, Illinois 60208
[2] Assistant Professor of Civil and Environmental Engineering, University of Cincinnati, Cincinnati, Ohio 45221.
[3] Professor of Civil Engineering and Director of the National Science Foundation Center for Advanced Cement Based Materials, Northwestern University, Evanston, Illinois 60208

S. P. Shah (ed.), Toughening Mechanisms in Quasi-Brittle Materials, 249–265.

For each load step a different hologram is recorded from each illumination direction (holograms R,L and C).

3. Fringe Count

The holographic interferograms were acquired into an image analysis system using a slow scan high resolution camera. The image is represented by a 512x512 2-d array of integers, each corresponding to the average light intensity across an element area referred to as pixel (Picture Element). The size of a pixel determines the spatial resolution.

The bright and dark bands of the holographic fringes can be represented as a binary image. In a binary image the pixel intensities assume only two values, ON (or 1) for perfect white and OFF (or 0) for perfect black.

Figure 2 is the enhanced interferometric image in which the intensity value is zero for dark bands and one for bright bands. Further image processing allows to assign intensity values to each fringe corresponding to the fringe order.

A pixel P is selected on the binary image (Fig. 2). A flooding procedure which assigns an arbitrarily selected intensity value (Y) to all the pixels connected to pixel P is performed. Thus, all the pixels on the band containing P are assigned the intensity value Y. The same procedure is repeated for consecutive bands. The intensity value Y is incremented by one for each band.

Figure 3 shows a fully processed interferogram in which same intensity values correspond to same fringe orders. Note that there is a different intensity level for each half a fringe (i.e., different intensity for dark and bright bands). The fringe count between any two points can now be easily obtained by computing half the difference of the corresponding intensity levels (e.g., fringe count between A and B in Fig. 3 is $(90 - 65)/2 = 12.5$).

Even though the use of the image analysis system results in very fast fringe counts, speed is not the main advantage of this technique. The isolation of the interferometric effect on the holographic images and the binarization process result in accurate and objective fringe counts. The gradual change in intensity from a dark to a bright band make it very difficult to obtain consistent fringe counts when conventional (by hand) methods are used.

4. Evaluation of Holograms

4.1. CRACK PROFILES.

For the computation of the crack opening displacements (COD) the modified Nelson and McCrickerd (Nelson and McCrickerd, 1986, Miller, Shah and Bjelkhagen, 1988) method was used. This method assumes the out of plane motion to be negligible (i.e. COD is the main component of the displacement) and that displacements are small. Since the direction of motion is assumed, one hologram is sufficient. Equation 1 gives the magnitude of the COD in terms the fringe count and the geometry of the optical set-up.

$$COD = \frac{n\lambda}{4 \cos\alpha \, \cos\psi}$$ (1)

where:

COD	=	Crack Opening Displacement. Displacement of one point on the crack face with respect to the point across the crack.
n	=	fringe count from a point on the crack face to the point across the cross the crack.
α	=	angle between illumination direction and sensitivity vector (**k**)
ψ	=	angle between direction of motion and sensitivity vector (**k**)

Typical crack profiles are shown in Fig. 4. Miller, Shah and Bjelkhagen[2] showed that short cracks have profiles which are basically the same as LEFM profiles. Longer cracks have profiles which are much thinner than LEFM crack, suggesting that applying a closing pressure to the LEFM profile may provide the correct final profile.

4.2. STRAIN FIELDS.

Fringe counts relative to the left face of the crack, or the symmetry line for points beyond the crack tip, were taken at every point on a 3 in x 3 in (76 mm x 76 mm) grid (1.5 in on each side of the crack line and 3 in in front of the notch tip) every 1/8 in (3 mm)for all three holographic images (R,L and C). A computer program was used to calculate the displacement of every point relative to the left face of the crack by locating the intersection point of the holodiagrams corresponding to each illumination direction. A third order best fit polynomial was obtained for the displacements along lines perpendicular to the crack line. For the sections crossing the crack a different polynomial was obtained for the left and right side of the crack. The strain field was obtained by differentiation and was evaluated from x=-1 in to x= +1 in every 1/8 in (3 mm) at every cross section, where x is the distance from the crack line and is positive to the right. Figure 5 shows the strain field (ϵ_{xx}) around the bottom crack for crack lengths of 1.375 in (35 mm), 1.875 in (48 mm) and 2.25 in (57 mm), corresponding to 1324 lb (6.0 KN), 1461 lb (6.6 KN) and 1601 lb (7.2 KN) of applied load. The strain fields shown in Fig. 5 include only the measured additional strain after the initial preload state.

In general, a high strain ($>100\mu\epsilon$) region can be found around and in front of the crack tip with a maximum value of approximately 300 $\mu\epsilon$. Behind the crack tip there is a zone of negative strain which suggests relaxation of tensile strain with respect to the initial preload stage. Also, note that there is a gradual descent of the strain level behind the crack tip as opposed to the sharp drop corresponding to the LEFM solution.

5. Analysis of the Closing Pressure Models with $K_I = 0$

Cohesive zone models based on the Dugdale-Barrenblatt approach of placing a closing pressure on the faces of a discrete crack could be used to account for the non-linear effects of the fracture process zone (FPZ) and aggregate interlock. Hillerborg, et. al.[3] proposed a cohesive crack model to account for the complex fracture mechanisms in concrete. Figure 6 is a summary of the closing pressure relationships studied. The following four closing pressures were chosen for this study: An exponential closing pressure proposed by Gopalaratanam and Shah[4]; a bilinear curve proposed by Roelfstra and Wittmann[5]; a trilinear closing pressure proposed by Liaw, Jeang, Hawkins and Kobayashi[6]; and a linear relationship proposed by Cedolin, Iori, and DeiPoli[7].

The applied load and the criterion $K_I = 0$ were specified and then the necessary crack length was computed as indicated by Castro-Montero, Shah and Miller[8]. This was done applying the experimental load to the FEM mesh and then propagating the crack length 3 mm (1/8 in) at a time.

Many previous studies calibrate the closing pressure vs. w relationship by comparing the theoretically predicted load-displacement relationship vs. the experimental values. Such a comparison for one of the specimens tested during the holographic study is shown in Fig. 7. The holographically measured notch tip opening displacements are compared with those obtained from the finite element analysis and with four different closing pressure vs. w relationships. It is observed that all four closing pressures give essentially the same load vs. NTOD curves. This means that a unique closing pressure vs. w relationship can not be asserted when a discrete displacement measurement is used for calibration. A similar observation was also made when the computed crack length ($K_I = 0$) is compared with the holographically measured crack length.

The authors believe that a better evaluation of the different closing pressures results from comparing the experimental crack profiles vs. predicted crack profiles. In order to provide an objective comparison, an error definition was proposed. A positive error means the closing pressure was not large enough to close the LEFM crack profile down to the holographic profile while a negative error means the closing pressure was too strong and closed the LEFM crack profile down too much.

In Fig. 8 it can be seen that there is a substantial difference between the observed and predicted crack profiles regardless of the closing pressure vs. w relationship used.

6. A Crack Length Dependent Cohesive Crack Model

A more objective method of determining closing pressure vs. crack opening relationship is not to assume any a priori shape. This was done using a Green's function type of approach. The necessary closing pressure distribution to match the holographically measured crack profiles was calculated. Three observations are pertinent from this analysis: (1) the closing pressure vs w relationship may not be invariant with respect to crack length, (2) a bilinear curve may be sufficient to approximate the closing pressure vs. w relationship and (3) as noted before, closing pressure vs w relationship appears to be a function of compressive strength.

Based on these observations a bilinear relationship was determined using the crack length dependent model proposed by Cook[9]. It was assumed that for crack lengths shorter than a given length d there is no closing pressure. For longer crack lengths a bilinear function was used. A finite difference Levenberg-Marquardt algorithm was applied to optimize the closing pressure function for each specimen. A proposed bilinear closing pressure, based on the optimization procedure is shown in Fig. 9.

For the proposed bilinear closing pressure there is a good prediction of the crack opening displacements although the crack lengths are over estimated including a segment close to the crack tip with either small or null predicted crack opening.

Jenq and Shah and Cook, et. al. suggested that K_I did not need to be equal to 0. The above bilinear closing pressure was applied and K_I corresponding to the experimentally observed crack length was calculated. When the calculated K_I values include closing pressure, an essentially constant value for the specimens with the same compressive strength is obtained regardless of crack length. In contrast, when cracks are assumed to be traction free (LEFM) the value of K_I increases with crack length. Thus, it seems that the cohesive type of model can explain the R-curve type of response reported for quasi-brittle materials. Figure 10 shows the relative error plots and typical profiles corresponding to the calculated value of K_I. Note that the predicted profiles match well with the measured ones.

7. Definition of Fracture Process Zone

In Fig. 11, the differences in strain fields between the LEFM solution and the experimentally measured values are shown. It is arbitrarily assumed that the differences are significant when they exceed the strain value of 60 $\mu\epsilon$. Positive differences mean that the tensile strain predicted by LEFM was higher while the negative values mean that the LEFM predicted lower tensile strain than the holographically measured values. The zones marked A in Fig. 11 are the regions where the LEFM solution predicts tensile strain values of 60 $\mu\epsilon$ or more higher than those measured holographically. Note that zone A remains essentially constant regardless of crack length. Since this zone is relatively small and does not change with crack length a model based on modified LEFM may be a possible approach. Zones B, with negative values, typically found behind the crack tip show that the relaxation of the observed strain is 60 $\mu\epsilon$ less than that in the elastic solution. Zones B can be defined as the wake of the fracture process zone (WFPZ) and they enclose the area where extensive microcracking has been developed and tensile forces are still transmitted through the crack. Behind this zone, the fracture process approaches a traction free condition.

Figures 12 shows the difference between the predicted response using the bilinear closing pressure and the holographic measurements. Note that there is a good correlation of the results and the zones with significant deviation from the experimental results are practically eliminated.

8. Conclusions

a. Sandwich hologram interferometry with multiple sensitivity vectors can be used to measure crack opening displacements and strain fields in mortar specimens under tensile stress. When holographic interferometry is used to measure Mode I crack opening displacements the effect of out of plane motion should be taken into account.
b. Digital image analysis facility provided a faster, more consistent and more accurate method of fringe count than the manual method.
c. There is a region of high tensile strain that moves with the tip of the propagating crack. The material behind the crack tip experiences strain relaxation. However, the gradual nature of the strain relaxation demonstrates the existence of tensile force transmitted through the crack faces.
d. Zones of nonlinear behavior can be located by computing strain field deviations form the linear elastic solution.
e. A bilinear closing pressure vs. COD cohesive crack model has been proposed to predict both crack profiles and strain field, fully characterizing the material behavior.

9. Acknowledgments

Professor Hans Bjelkhegan's advice in the development of the holographic setup is greatly appreciated. The authors thank Mr. John Schmidt for his valuable contributions in the development of the experiments. This study was made possible through the support of the Air Force Office of Scientific Research (AFOSR) (Grant No. 88-C-018P00001) Program Manager Dr. Spencer Wu). Partial support from the National Science Foundation (Grant No. NSF-DMR-8808432 Program Manager: Dr. Lance Haworth) is also appreciated.

10. References

1. Castro-Montero, A., Shah, S.P. and Miller, R.A. (1990) "Strain field measurement in fracture process zone," to be published in Engineering Mechanics, ASCE.
2. Miller, R.A., Shah, S.P., and Bjelkhagen, H. (1988) "Measurement of crack profiles in mortar using laser holographic interferometry," Experimental Mechanics, Vol 28(4), 388-394.
3. Hillerborg, A., Modeer, M. and Petersson, P. E., (1977) "Analysis of a crack formation and growth in concrete by means of fracture mechanics and finite elements," Cement and Concrete Research, Vol. 6, No. 6, pp 773-782.
4. Gopalaratanam, V. S. and Shah, S. P., (1987) "Softening response of plain concrete in direct tension," ACI Journal, Proceedings, Vol. 82, No. 3, pp. 345-356.
5. Roelfstra, R. E. and Wittmann, F. H., (1986) "A numerical method to link strain softening with fracture in concrete," in F. H. Wittmann (ed.), Fracture Toughness and Fracture Energy in Concrete, Elsevier Science Publishers, B. V., Amsterdam.
6. Jeang, F. L. and Hawkins, N. M., (1985) "Nonlinear analysis of concrete fracture," Structures and Mechanics Report, Department of Civil Engineering, University of Washington, Seattle, WA.
7. Cedolin, L., DeiPoli, S. and Iori, I.,(1987) "Tensile behavior of concrete," Journal of the Engineering Mechanics Division, ASCE, Vol. 113, No. 3, p. 431.
8. Miller, R.A., Castro-Montero, A. and Shah, S.P. (1990) "Cohesive crack models examined with laser holographic measurements," accepted for publication in the Journal of the American Ceramic Society.
9. Cook, R.F., Fairbanks, C.J., Lawn, B.R. and Mai, Y-W. (1987) "Crack resistance by interfacial bridging: its role in determining strength characteristics," Journal of Materials Research, Vol. 2(3), 345-356.

254

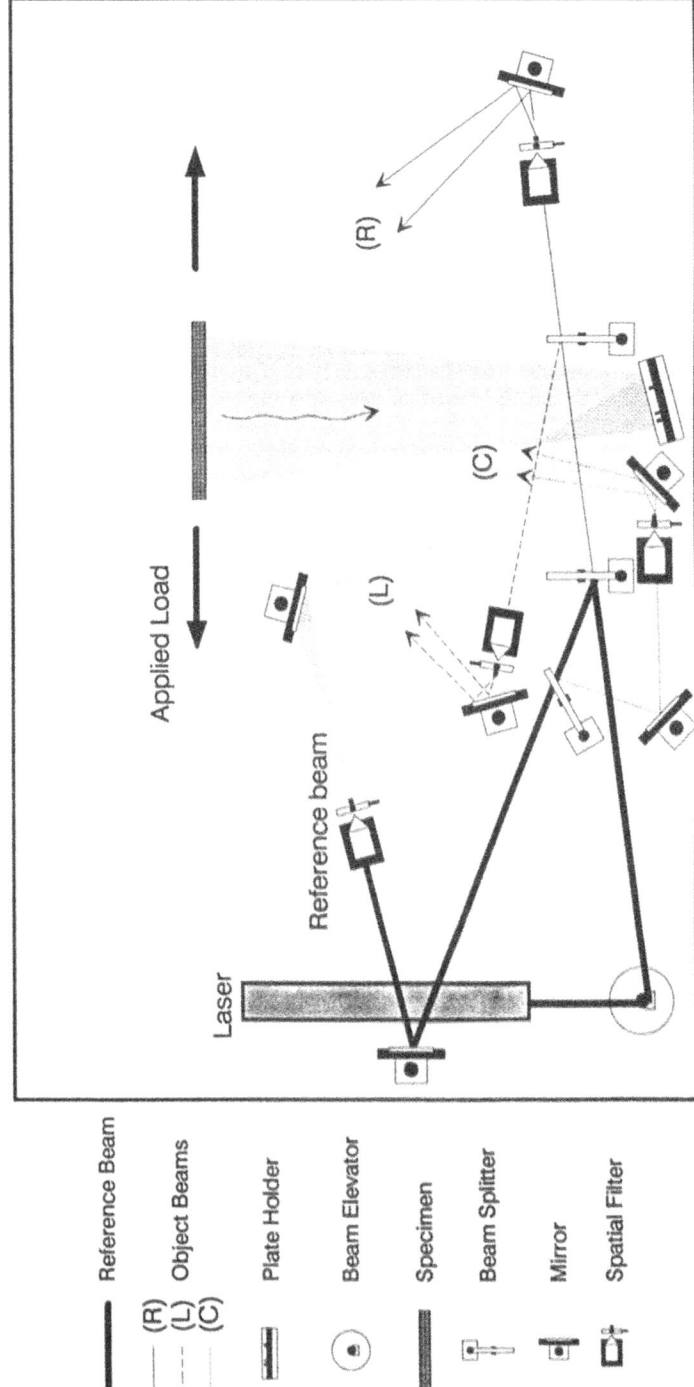

Reference Beam
(R) (L) (C) Object Beams
Plate Holder
Beam Elevator
Specimen
Beam Splitter
Mirror
Spatial Filter

Figure 1. Holographic set up with three sensitivity directions.

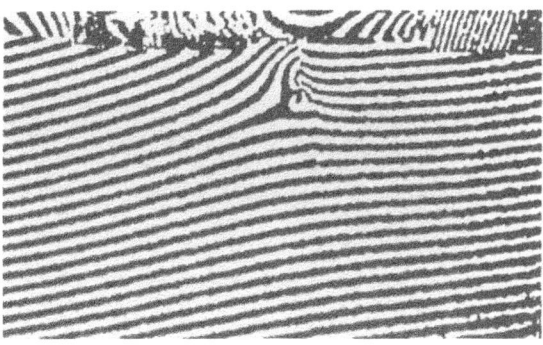

Figure 2.   Typical holographic interferogram after isolation of interferometric effect.

256

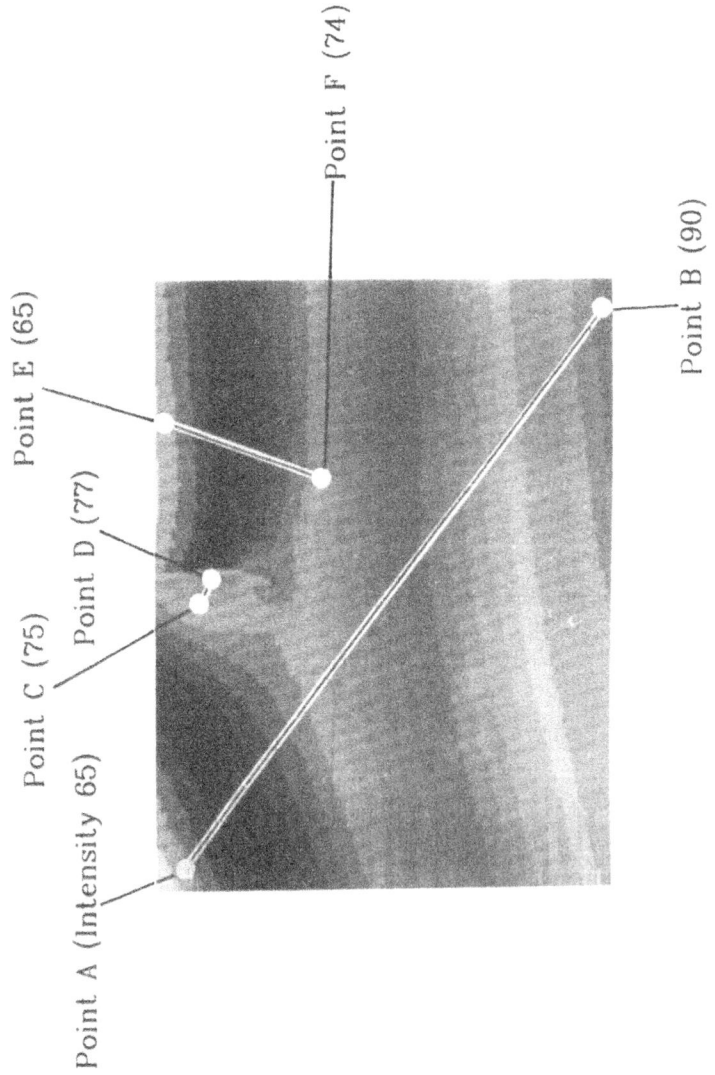

Figure 3. Fully processed interferogram for automatic fringe count.

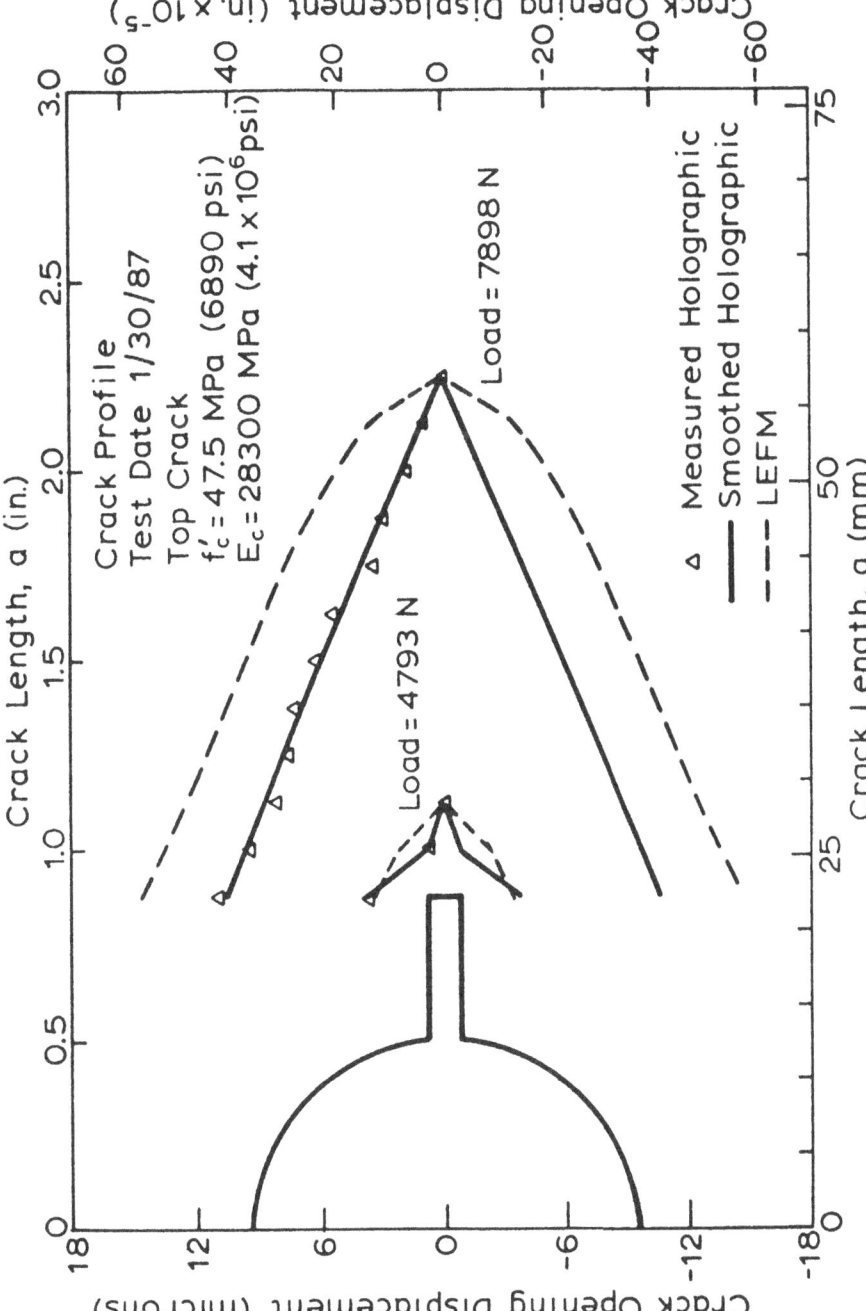

Figure 4. Typical crack profiles.

258

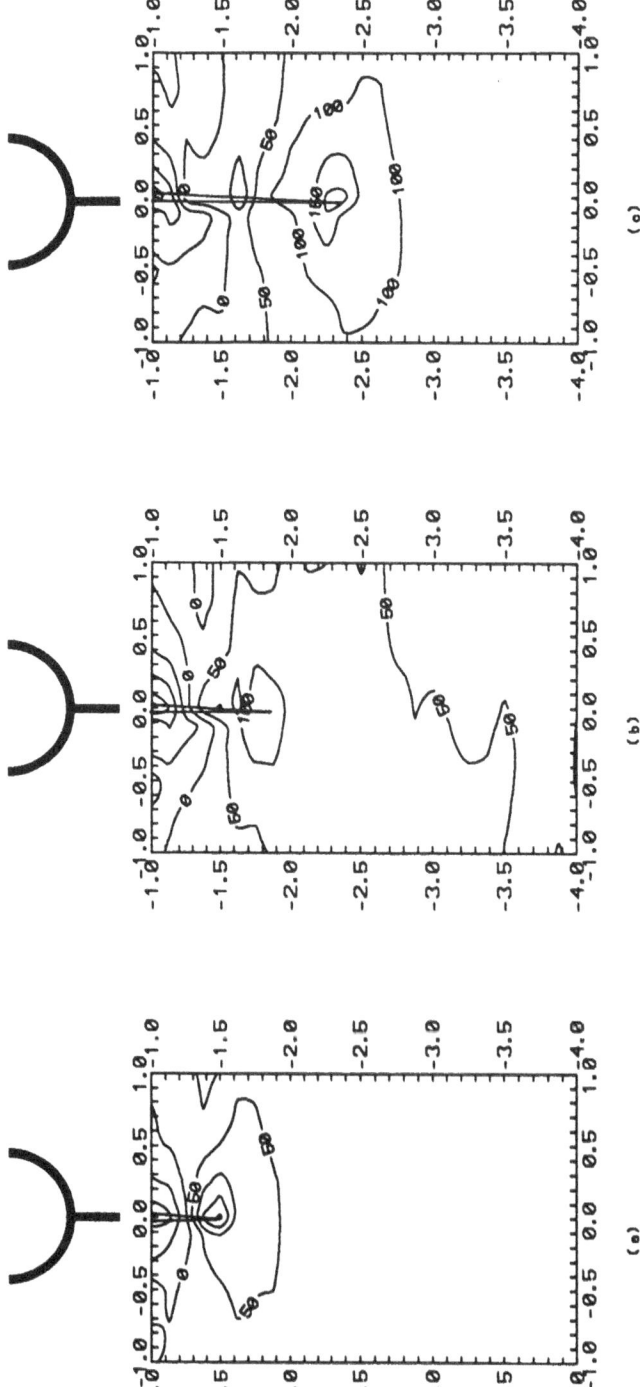

Figure 5.   Holographic strain fields.

259

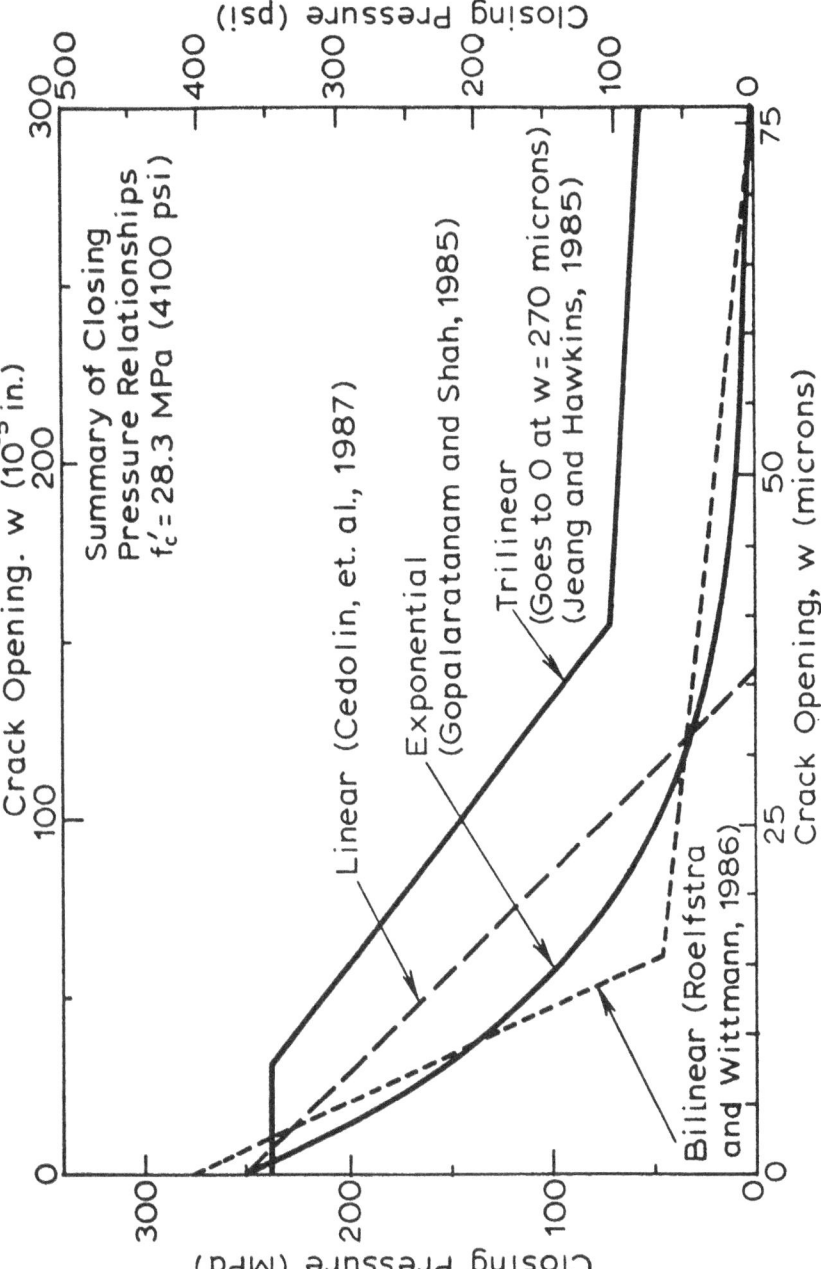

Figure 6.   Comparison of closing pressures.

260

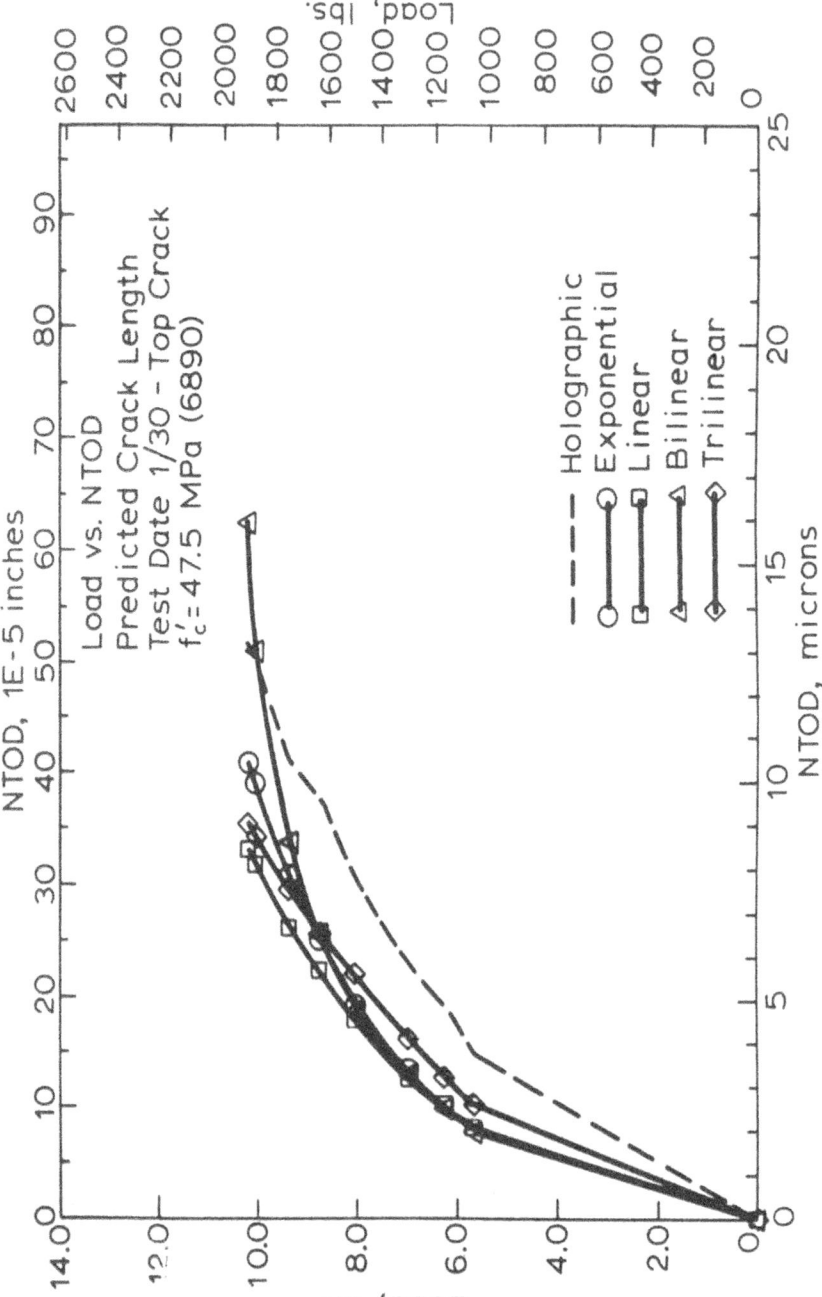

Figure 7. Load vs. Notch Tip Opening Displacement (NTOD).

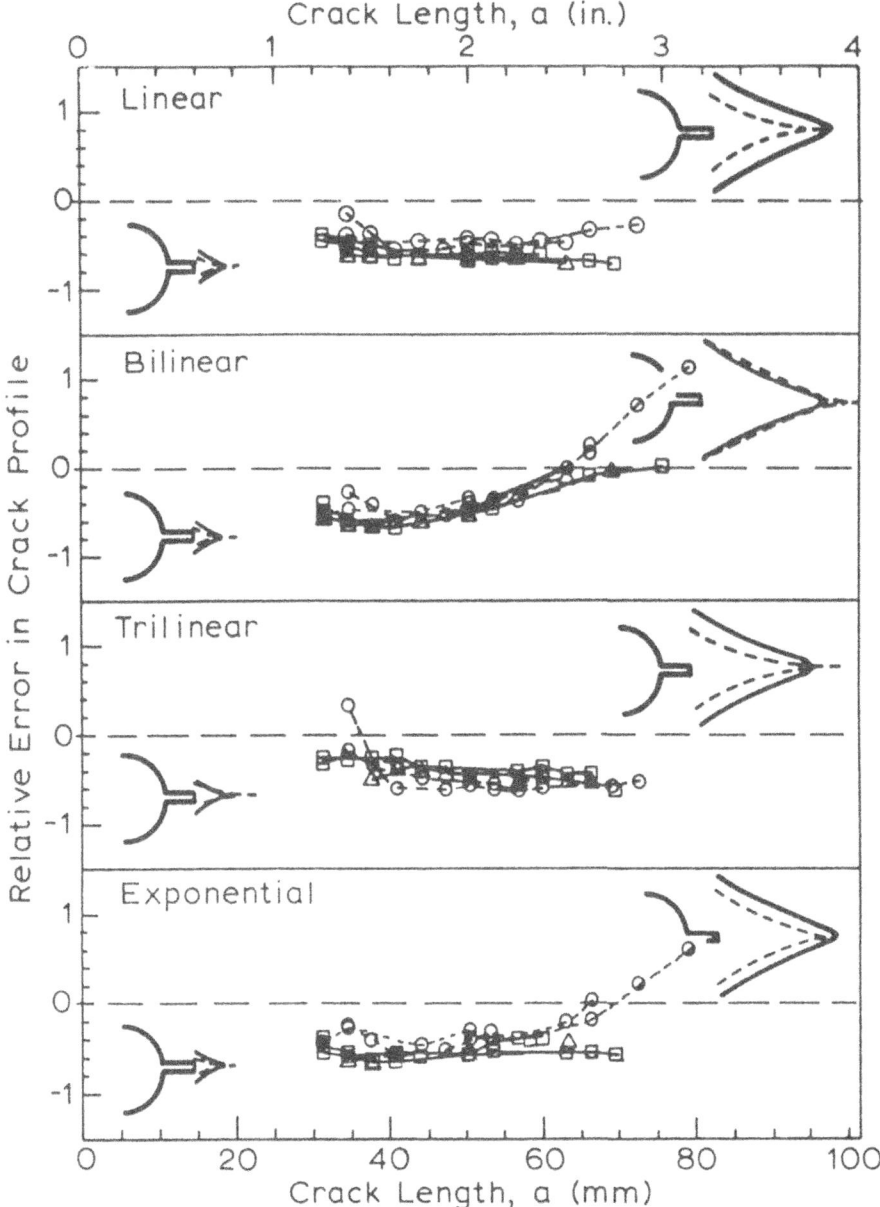

Figure 8. Crack profile comparison.

262

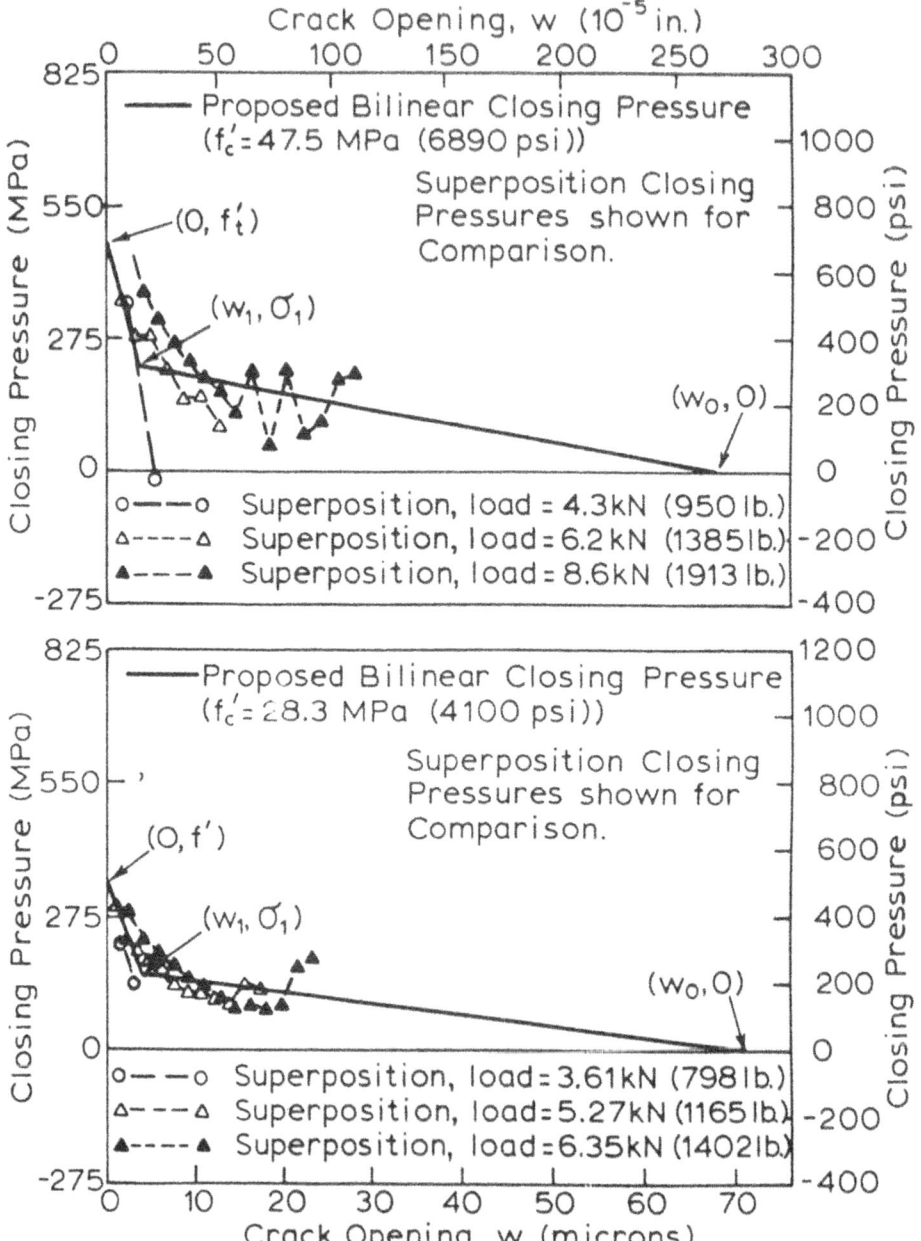

Figure 9.   Proposed bilinear closing pressure.

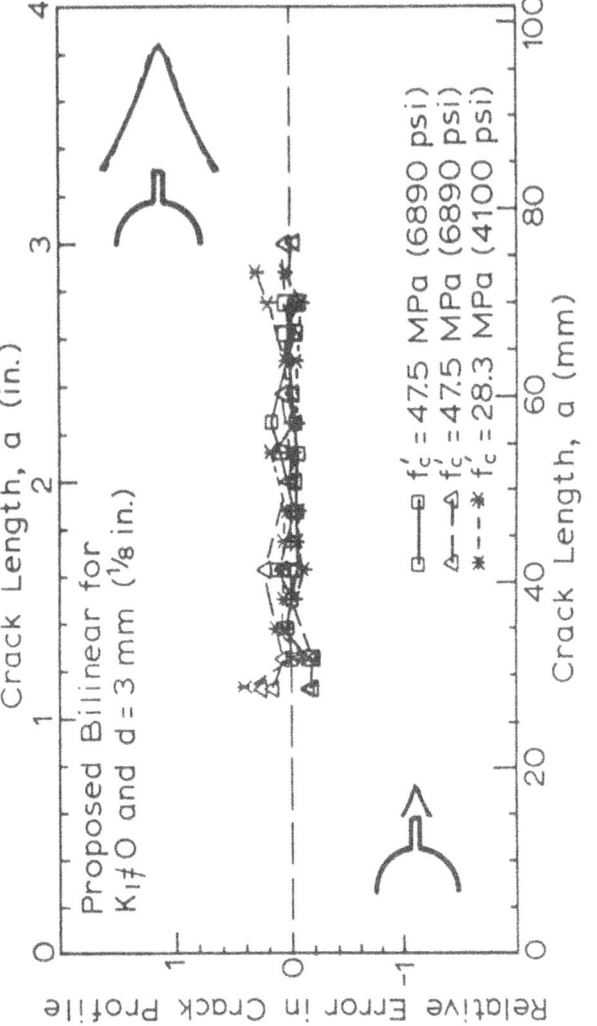

Figure 10.   Crack profile comparison for proposed bilinear closing pressure.

264

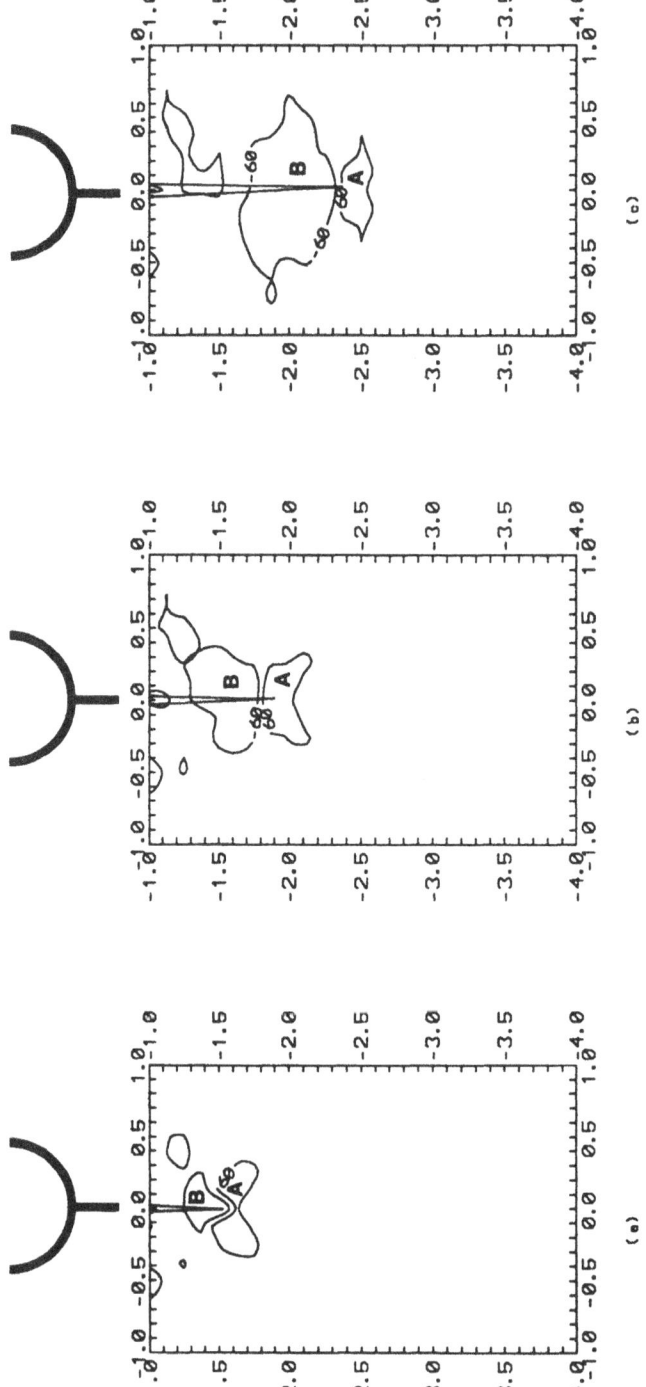

Figure 11.   Difference between experimental and LEFM strain fields.

265

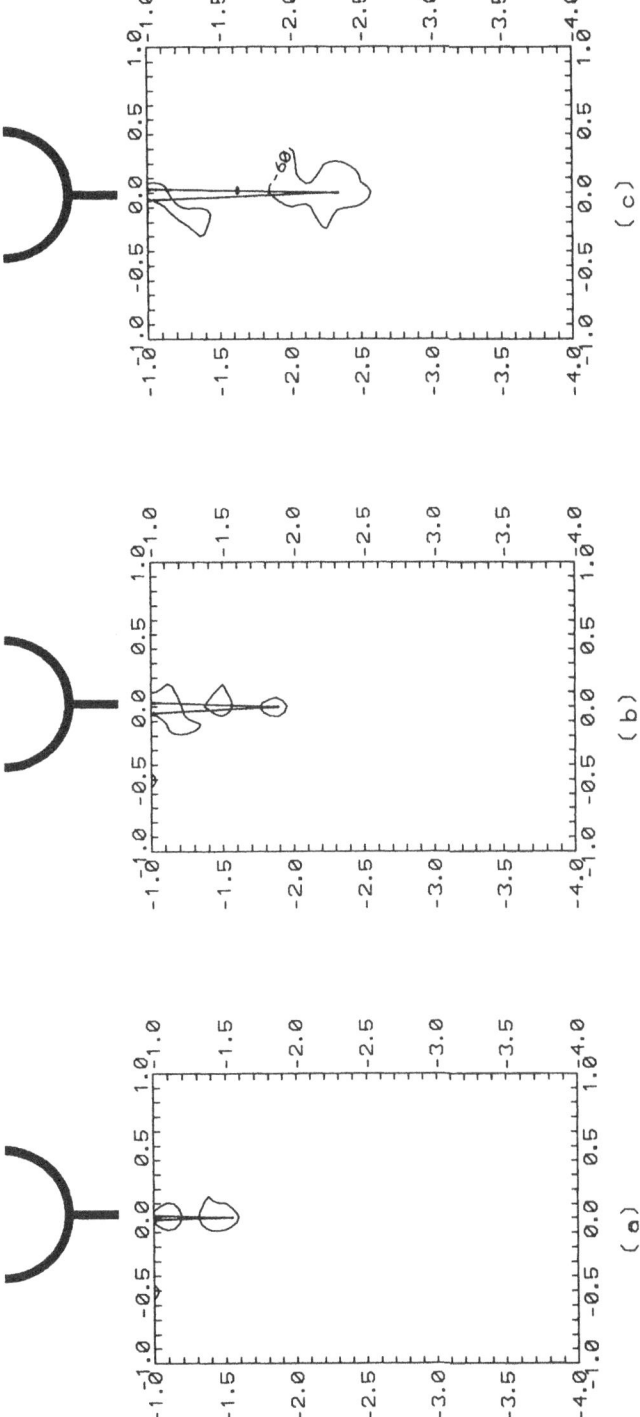

Figure 12.   Difference between experimental and LEFM with bilinear closing pressure strain fields.

.

REPORTER COMMENTS ON SESSION 4 - EXPERIMENTAL OBSERVATIONS

Joseph Homeny
Department of Materials Science and Engineering-Ceramics Division
College of Engineering
University of Illinois
105 S. Goodwin Ave.
Urbana, Illinois 61801

Session 4 dealt with experimental observations on the fracture behavior of ceramic and cement based materials. Summaries of the three presentations in this session are as follows:

(1) "R-Curves and Strength of Ceramics," S. J. Bennison, S. Lathabai, J. Rodel, and B. R. Lawn, National Institute of Standards and Technology - The researchers generated R-curves in ceramic materials by a strength-indentation method. In this manner, the short, intermediate, and long crack domains were observed experimentally. The primary mechanism responsible for rising R-curves was determined to be crack bridging by grains in the crack wake. Grain size and residual stresses were important microstructural features related to crack bridging. Increasing grain size in Al_2O_3 and increasing thermal expansion anisotropy in Al_2TiO_5 resulted in enhanced R-curve behavior. A model for grain bridging in the crack wake, involving the application of closure forces was proposed.

(2) "Damage and Fracture Mechanisms During High Temperature Creep in Hot-Pressed Alumina," D. S. Wilkinson, A. G. Robertson, and C. H. Caceres, McMaster University - The rersearchers determined the mechansms of creep and fracture in hot pressed Al_2O_3. Three stress dependent damage regimes were evident, depending on the stress level. In the slow crack growth regime (high stresses), nucleation of a single crack at a dominant processing flaw resulted in failure. In the microcrack growth and linkage regime (intermediate stresses), microcracks nucleated at processing flaws, however, crack tip stress relaxation processes were also observed. In the creep fracture regime (low stresses), cavitation, by surface diffusion and grain boundary sliding, was observed. Microcrack development was also limited in the creep fracture regime. For all three of the above stress levels, processing flaws ultimately controlled the creep strength.

(3) "Study of the Fracture Process in Mortar with Laser Holographic Measurements," A. Castro-Montero, S. P. Shah, Northwestern University, and R. A. Miller, Univesity of Cincinnati - The researchers described a method to measure crack opening displacements and strain fields in mortars using holographic techniques. A cohesive crack model, based on linear elastic fracture mechanics and closure pressures on crack faces, was used to describe the fracture behavior. For short cracks, the holographic measurements were similar to the linear elastic fracture mechanics profiles. For long cracks, the holographic measurements produced smaller crack

S. P. Shah (ed.), Toughening Mechanisms in Quasi-Brittle Materials, 267–268.
© 1991 *Kluwer Academic Publishers.*

opening displacements than linear elastic fracture mechanics profiles. This difference was attributed to aggregate bridging and microcracking in the crack wake. It was also shown that the cohesive crack model adequately explained R-curve behavior.

Comments on the above three presentations dealt with comparing R-curve behavior in ceramics and concretes. In non-transforming ceramics, R-curve behavior can be attributed primarily to crack wake processes, such as grain, whisker, and fiber bridging. In concretes, R-curve behavior can also be primarily attributed to crack wake processes, such as aggregate bridging and microcracking. It was clear that more experimental work is needed to clearly define the mechanisms responsible for R-curve behavior in these systems. It was also clear that the relative importance of crack wake versus frontal process zone effects on R-curve behavior must be more thoroughly explored.

Experimental Methods to
Assess Damage

THE FRACTURE PROCESS ZONE IN CONCRETE

SIDNEY MINDESS
Department of Civil Engineering
University of British Columbia
2324 Main Mall
Vancouver, British Columbia V6T 1W5
Canada

ABSTRACT. A number of different experimental techniques have been used to try to determine: (1) whether a process zone exists in concrete; (2) if it does, what its dimensions might be; and (3) whether it is a fundamental material property. From an extensive review of the literature, it would seem that not only the apparent size of the process zone, but its very existence, are strongly dependent on both the specimen geometry and on the methods of measurement. It is difficult to avoid the conclusion that the process zone is not a fundamental material property for cementitious composites.

1. Introduction

When a brittle material containing a crack is subjected to stress, the stress distribution ahead of the crack has the form shown in Fig. 1. That is, for a sharp crack, there are very high stress concentrations in the immediate vicinity of the crack tip. Because of the heterogeneity of concrete at both the macrostructural and microstructural levels, including discontinuities and pre-existing microcracks, there will be many highly localized areas of relative weakness in this highly stressed zone. Therefore, as a crack propagates in concrete, one would expect a great deal of microcracking to occur, largely (but not exclusively) in this highly stressed region. This region of discontinuous microcracking ahead of the continuous (visible) crack is generally referred to as the fracture process zone.

It must be emphasized, as will be seen below, that not all investigators have found such a process zone. However, if there is a true fracture process zone, or even simply random microcracking associated with crack propagation, it is important to try to quantify the amount of damage that occurs during the fracture of concrete.

There are a number of conceptual problems to be dealt with in any discussion of fracture process zones in concrete. First, as Thouless [1] has pointed out (Fig. 2), instead of looking for a process zone ahead of the crack tip, one can as easily consider a bridging zone

271

S. P. Shah (ed.), Toughening Mechanisms in Quasi-Brittle Materials, 271–286.
© 1991 Kluwer Academic Publishers.

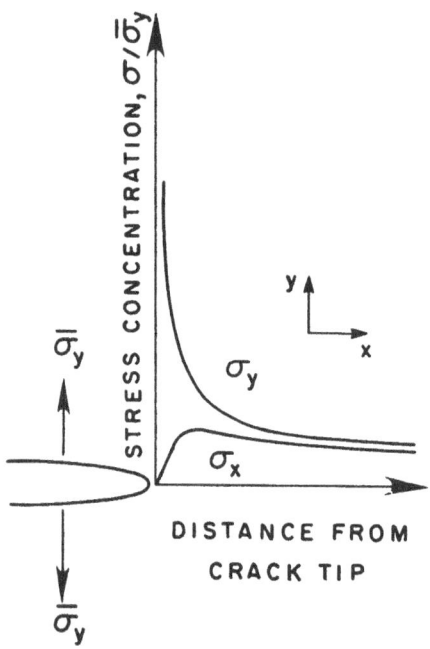

Figure 1.   Stress field ahead of a crack tip.

behind the crack tip.   Either assumption leads to similar results when applied to the problem of crack tip propagation.   Second, it appears to be impossible to define, in an unequivocal fashion, the "true" crack length in concrete [2]; the position of the tip of a propagating crack is uncertain. Since, as mentioned above, there are already pre-existing microcracks in virgin concrete, it is necessary to distinguish between these and the additional microcracks caused by the imposed stresses. At the moment, it is not possible to do this.   Third, investigators such as Bazant [3] are not really concerned with the size of the fracture process zone as determined by direct observation.   They are simply interested in an elastically equivalent crack system, which yields the correct energy dissipation and correct stress vs. displacement relationships in a concrete specimen.   Finally, it has been shown [4,5] that the crack length appears to vary across the width or thickness of the specimen.   Thus, surface crack measurements cannot accurately indicate the extent of the process zone.   In view of these complexities, it is then not surprising that there is a great deal of controversy about the process zone in concrete.

In this paper, some of the experimental techniques used to identify the process zone, and the results of these studies, will be discussed.

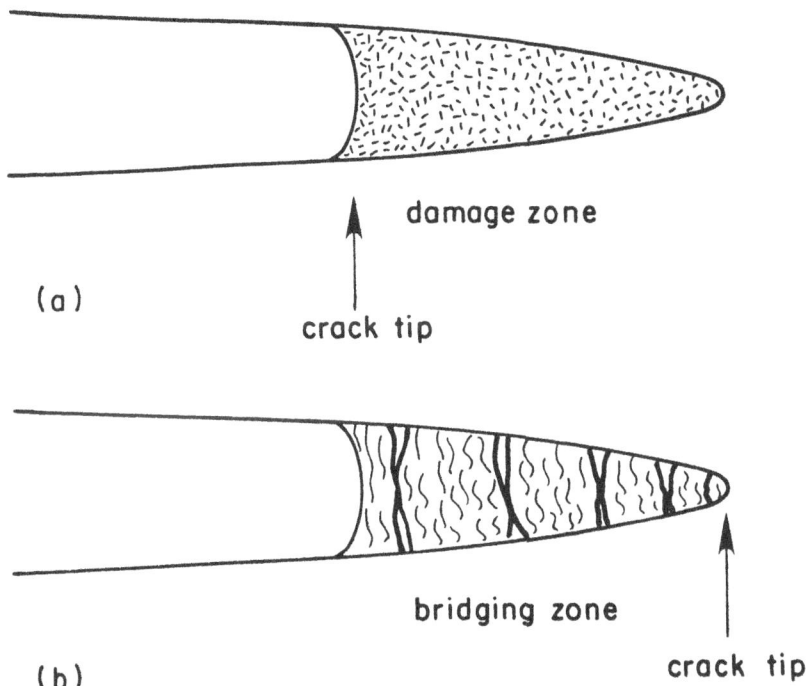

Figure 2. Schematic illustration of a crack tip, and either the (a) damage zone, or (b) bridging zone, after Thouless [1].

2. Experimental Techniques

Over the years, a great many different experimental techniques have been employed to try to find the extent of the process zone. These may be divided roughly into three categories:

Surface techniques:
- optical microscopy
- scanning electron microscopy
- electric resistance strain gauges
- photoelastic methods
- interferometry techniques

Measurements involving the specimen interior:
- X-ray techniques
- mercury penetration measurements
- dye penetrants
- infrared vibrothermography
- ultrasonic pulse velocity
- acoustic emission

Indirect methods:
- compliance measurements
- Demec gauges
- numerical methods

As indicated above, these techniques tend to measure rather different properties of the material. Some involve only the surface character-istics of the specimen, and some include an indication of what is happening in the interior of the specimen. Still others consider the system as a kind of "black box", where the physical details of the cracking are not considered directly, as long as an assumed crack system yields the correct stress vs strain and/or energy relationships. Thus, it is sometimes difficuilt to reconcile the very different conclusions that various researchers have come to with regard to the process zone.

It must also be noted that some techniques involve first loading the specimen to some stress or strain, but then unloading the specimen before the cracks are examined. Others, however, permit observations of the cracking while the specimens are still under load. Since the fine cracks tend to close and become virtually invisible on unloading, it is clear that valid measurements can only be made with those tech-niques which do not involve unloading the specimens before measurements are made.

3. Surface Techniques

The earliest attempts to examine cracking in concrete, such as the very extensive studies carried out at Cornell University using both optical microscopy [6,7] and X-ray techniques [6,8,9] were not really sensitive enough to reveal the presence of a process zone.

However, more sophisticated techniques have also not always revealed a process zone. On the one hand, Tait and Garrett [10], using in situ observations of cracking in pastes in the sample chamber of a scanning electron microscope (SEM), did identify a process zone in the vicinity of the crack tip, 1-4 mm in extent. Baldie and Pratt [11] used backscattered electron imaging on polished sections of cement paste. They found that crack growth seemed to occur by first the formation, and then the coalescence of microcracks ahead of the crack tip, but they identified "only a limited formation of an actual process zone".

On the other hand, extensive in situ observation in an SEM by Mindess et al. [12-14] showed no evidence of a process zone. Using similar techniques, Diamond and Bentur [15] concluded that while there was crack subdivision and branching near the visible crack tip, "there is no physical distinction corresponding to separate lengths of 'straight, open crack' behind a crack tip and 'process zone microcrack-ing' ahead of a crack tip".

Using a replica technique in conjunction with an SEM, Bascoul and his co-workers [16,17] also found extensive microcracking near the crack tip, but not a well-defined process zone. They concluded [17] that "Nothing allows... (us)... to think that there is a damaged zone ahead of the macrocrack except at the beginning of crack branching".

Other, more sophisticated optical microscopy techniques have also been contradictory. Eden and Bailey [18] used **diffuse illumination** with a reflected light microscope, and concluded that stable crack growth involves the formation of a process zone which grows to some characteristic size before the crack becomes unstable. Stroeven [19, 20] however, used **fluorescent oils** to help delineate the cracks in concrete. Typical results are shown in Fig. 3 [19], which represents the crack pattern of an axial section of a grooved concrete specimen subjected to direct tension, slightly beyond the ultimate load. While there is very extensive cracking, much of this was already present in the virgin specimen as a result of shrinkage. Stroeven found that the cracks developed, and grew together, in a stochastic way. Certainly, there is no "process zone" in evidence.

Still other techniques have yielded different results. **Electric resistance strain gauges** have been used to identify relatively large process zones [21,22], of the order of 10 to 15 mm wide and lengths of up to 100 mm. However, these "process zones" were defined in terms of the surface strain exceeding, typically, 3×10^{-4} rather than by direct observation of cracks. In spite of these rather large dimensions, the process zone could not, surprisingly, be identified by optical methods. **Photoelastic techniques** were also ambiguous; Van Mier [23] and Van Mier and Nooru-Mohamed [24] could not define a fracture process zone, while Stys [25] did find one.

Perhaps the most sensitive of the surface techniques for identifying cracks in concrete are the interferometry techniques: **holographic interferometry** and **speckle interferometry**. These technqiues can achieve a sensitivity down to about 1 μm. Even with this sensitivity in measuring surface displacements, however, the results are uncertain. Most studies, as with electric strain gauges, define the process zone in terms of some limiting strain [26-28]. Thus, a fairly large process zone can be identified ahead of the crack tip, as in Fig. 4 [29], even though it cannot be defined optically. On the other hand, Ferrara and Morabito [30] could not with certainty specify a well- defined process zone, while Regnault and Bruhwiler [28] could not locate, or even define, the tip of the tension-free crack.

Similarly, white light **Moiré interferometry** was used to identify rather large (20-100 mm) process zones [31-33]. Again, one wonders why zones this large could nto be observed directly by microscopic techniques.

4. Measurements Involving the Specimen Interior

In principle, measurements involving the interior of the specimen, i.e. the bulk of the material, should provide a better indication of the extent of the process zone. Unfortunately, since we cannot directly

276

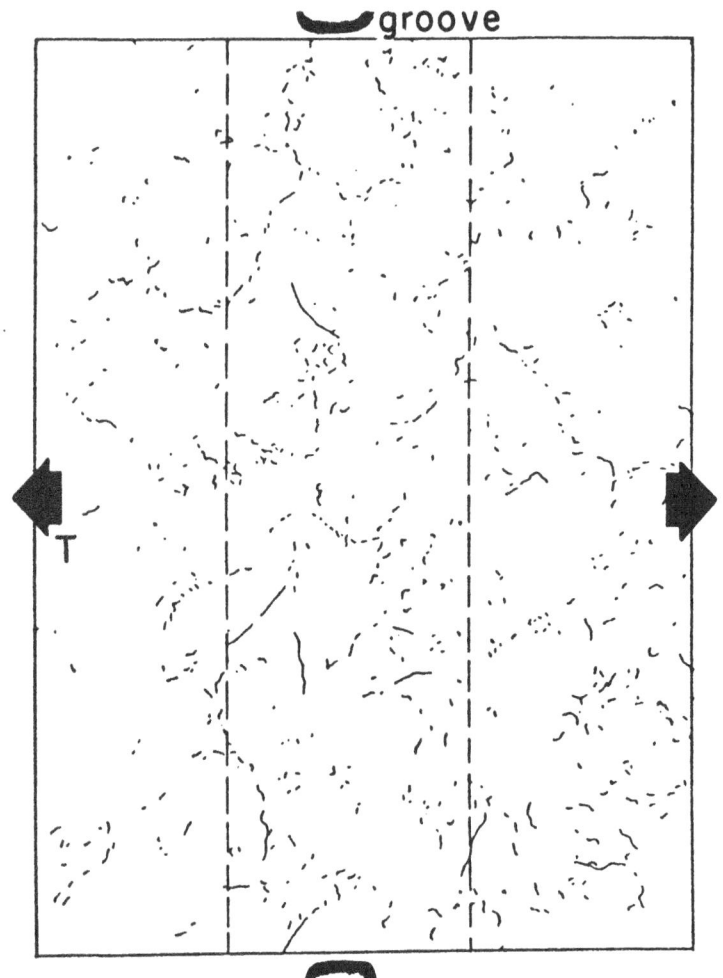

Figure 3. Manually copied crack pattern of axial section of two-sided grooved specimen subjected to direct tension slightly beyond the ultimate, after Stroeven [19].

see the interior of a specimen indirect means of locating cracks must be used for this purpose.

As indicated above, X-ray techniques [6,8,9] are not really sensitive enough for this purpose. **Mercury penetration** measurements [34] can only indicate crack widths and volumes, but cannot define the location of the microcracked region. Similarly, **dye penetrants** [35-37] can reveal the shape of the crack front, but cannot reveal the extent of the process zone.

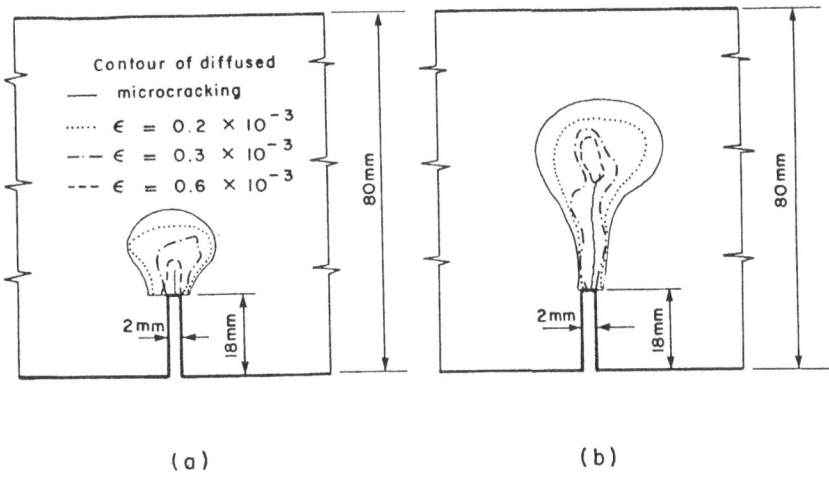

Figure 4.   Equal strain contours and the developing crack, as measured
            using optical interferometry with laser light: (a) inter-
            mediate load level; (b) load level close to failure, after
            Cedolin, Dei Poli and Iori [27].

   Infrared vibrothermography is based on the principle that heat
is generated when there is energy dissipation in a material subjected
to vibratory excitation beyond its stable reversible limit. This technique has been used [38] to observe progressive damage in a concrete
specimen under compressive loading. This is an averaging technique,
which shows the global damage within the specimen; whether this can be
equated with the microcrack patterns that occur during the growth of a
single crack under monotonically increasing loading is not certain.
   Ultrasonic pulse velocity measurements have long been used to
try to assess damage in concrete. This too is an averaging technique,
which is affected by the total damage (or cracking) between the transducers. Extensive work using this technique carried out by Alexander
and his co-workers [39-43] showed that the main crack would not propagate until a microcracked zone had developed in front of it, typically
about 1/2 of the residual beam depth in size. Similarly, Chhuy et al.
[21] found a damage zone about 100 mm long ahead of the crack tip. On
the other hand, Berthaud [44,45] found it difficult to deduce the size
of the process zone using ultrasonic pulse velocities, and Reinhardt

and Hordijk [22] concluded that such measurements do not lead to a geometrical description of the process zone.

Acoustic emission (AE) techniques have been used very extensively to try to assess the nature of the process zone. AE events occur when transient elastic waves are generated by the rapid release of energy that occurs upon cracking, and can be measured using piezoelectric transducers [46]. Maji and Shah [47-51], in a very detailed series of tests, found that beyond the peak load most of the AE events occurred near the crack tip; they deduced a process zone extending about 25 mm ahead of the crack tip, and a longer distance behind it (indicating ligament connections behind the visible crack tip). AE source locations for mortar specimens, compared to the crack tip location as determined by holographic techniques are shown in Fig. 5 [52]. Others have obtained similar results using AE techniques [21,53,54].

Berthelot and Robert [55-57] also carried out very extensive AE tests. They found that a damage zone appeared to grow in size as the crack progressed, reaching lengths of up to 160 mm and widths of up to

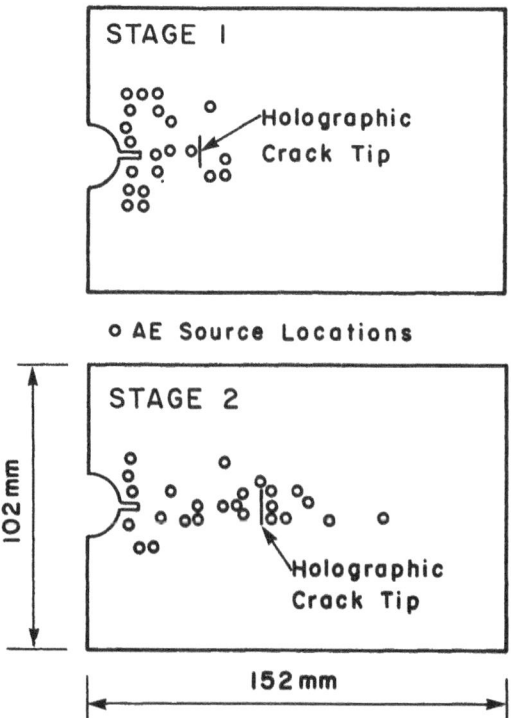

Figure 5.   AE source locations for mortar specimen, compared to the location of the crack tip found using holographic techniques, after Maji, Ouyang and Shah [52].

120 mm. Similar results were obtained by Bensouda [58]. Again, however, it is surprising that such extensive damage zones are not readily apparent by microscopic observation.

Results by Rossi [59] and Rossi et al. [60] indicated that there was not a clear transition from the uncracked concrete to the process zone. Overall, a review of the AE literature suggests that the results depend upon the specimen geometry, the sophistication of the instrumentation, and the method of analysis.

5. Indirect Methods

A number of indirect methods have also been devised to estimate the size of the process zone. The earliest of these is the use of **compliance measurements**. That is, from measurements of specimen stiffness (and the location of the visible crack tip) the extent of the process zone can be "guesstimated". Using such techniques, Karihaloo and Nallathambi [61,62] found process zones in the range of 20-40 mm, depending on the specimen size. Compliance techniques were also used by Kobayashi et al. [63], who concluded that the process zone continued to grow as the crack extended, without ever reaching a constant value. In the view of this author, however, because of the impossibility of separating the development of a process zone from slow crack growth, compliance measurements cannot provide a good estimate of the size of the process zone.

An interesting variant on compliance measurements is the use of **multi-cutting techniques**. One such technique [64] involves the cutting of thin strips of the specimen normal to the crack path and measuring their bending stiffness as a function of distance from the visible crack tip. In another version of this technique [65-67], the bridging stresses transferred within the process zone are removed by careful cutting along the plane of the original crack, and determining the compliance of the specimen at each step as the cutting progresses through the process zone. This permits an estimate of the process zone size. Hu and Wittman [65-67] found process zones of up to 43 mm long, but concluded [65] that "the length of the fracture process zone is not a material constant but depends on the actual stress gradient due to the limited specimen geometry".

Demec gauges can be used to determine the average strains in specimens over fairly large gauge lengths. Clearly, this too can only average whatever is taking place within the gauge length. However, it has been found that the use of these gauges can provide similar results to those found from ultrasonic pulse velocity measurements [39,42].

Finally, **numerical methods** (i.e., studies on 'numerical' rather than 'real' concrete) have also been used to try to deduce the size of the process zone [e.g., 68-70]. Clearly, however, the results of such methods depend entirely on the a priori assumptions made in setting up the numerical models, and this cannot provide any definitive answers.

6. Conclusions

From the above brief review, it is clear that the different test techniques provide different information. Some investigators have found no process zone at all; others have found that the process zone size varies from only a few mm to over 500 mm, as summarized in Table I. It would appear, then, that this issue is still far from being resolved. Each different techniques, it seems, is capable of examining only a limited part of the problem. (Indeed, it reminds the author of the old story of a group of blind men trying to describe an elephant.) At

Table I. Fracture process zone dimensions determined using different techniques

Ref. No.	Technique	Process Zone Dimensions (mm)	
		Length	Width
61	Compliance measurements	20-40	
63	Compliance + replica	114	
64	Multi-cutting	30-40	
65-67	"	12-43	
12	SEM	no process zone found	
15	"	no process zone found	
10	"	1-4	
21	Strain gauges	100	10
22	"		14
25	Photoelasticity	5	
39-43	Ultrasonic pulse velocity	20-160	
21	"	100	10
47,49	Acoustic emission	25 ahead of tip >25 behind tip	
53	"	105	
21	"	100	
54	"	500	
56	"	160	120
58	"	160	120
59,60	"		10
27	Interferometry	20-40	
29	"	50	
30	"	no process zone detected	
28	"	zone exists, but dimensions not definable	1/2 aggregate size
32	Moiré interferometry	100	
33	"	20-30	10
70	Numerical techniques	40-80	

present, one can only conclude that the process zone is <u>not</u> a fundamental material property, but depends on the specimen geometry, and the method of measurement.

7. Acknowledgements

This work was supported in part by a grant from the Natural Science and Engineering Research Council of Canada. The authors would also like to thank Dr. Z.P. Bazant, Dr. M.G. Alexander, Dr. A.K. Maji and Dr. S.E. Swartz for their helpful discussions of earlier drafts of this work.

8. References

1. Thouless, M.D. (1988) Bridging and Damage Zones in Crack Growth, Journal of the American Ceramic Society, <u>71</u> (6) 408-413.
2. Kasperkiewicz, J. (1988) Letter to the author, November 22.
3. Bazant, Z.P. (1988) Letter to the author, October 26.
4. Swartz, S.E. and Go, C-G. (1984) Validity of Compliance Calibration to Cracked Concrete Beams in Bending, **Experimental Mechanics**, <u>24</u> (2) 129-134.
5. Bascoul, A., Kharchi, F. and Maso, J.C. (1987) Concerning the Measurement of the Fracture Energy of a Microconcrete According to the Crack Growth in a Three Points Bending Test on Notched Beams, in S.P. Shah and S.E. Swartz (eds.), Proceedings of **SEM-RILEM International Conference on Fracture of Concrete and Rock, Houston;** Society of Experimental Mechanics, Bethal CT.
6. Slate, F.O. and Hover, K.C. (1984) Microcracking in Concrete, in A. Carpinteri and A.R. Ingraffea (eds.), **Fracture Mechanics of Concrete: Material Characterization and Testing,** Martinus Nijhoff Publishers, The Hague, pp. 137-159.
7. Slate, F.O. (1983) Microscopic Observation of Cracks in Concrete, with Emphasis on Techniques Developed and Used at Cornell University, in F.H. Wittmann (ed.) **Fracture Mechanics of Concrete,** Elsevier Science Publishers B.V., Amsterdam, pp. 75-83.
8. Slate, F.O. (1983) X-Ray Technique for Studying Cracks in Concrete, with Emphasis on Methods Developed and Used at Cornell University, in F.H. Wittmann (ed.) **Fracture Mechanics of Concrete,** Elsevier Science Publishers B.V., Amsterdam, pp. 85-93.
9. Najjar, W.D. and Hover, K.C. (1988) Modification of the X-Radiography Technique to Include a Contrast Agent for Identifying and Studying Microcracking in Concrete, **Cement, Concrete and Aggregate,** <u>10</u> (1) 15-19.
10. Tait, R.B. and Garrett, G.G. (1986) In Situ Double Torsion Fracture Studies of Cement Mortar and Cement Paste Inside a Scanning Electron Microscope, **Cem. Concrete Res.,** 16 (2) 143-155.

11. Baldie, K.D. and Pratt, P.L. (1986) Crack Growth in Hardened Cement Paste, in S. Mindess and S.P. Shah, (eds.), **Cement-Based Composites: Strain Rate Effects on Fracture**, Materials Research Society Symposia Proceedings Vol. 64, Materials Research Society, Pittsburgh, pp. 47-61.

12. Mindess, S. and Diamond, S. (1982) The Cracking and Fracture of Mortar, **Materiaux et Constructions**, 15 (86) 107-113.

13. Mindess, S. and Diamond, S. (1982) A Device for Direct Observation of Cement Paste or Mortar Under Compressive Loading Within a Scanning Electron Microscope, **Cement and Concrete Research**, 12 (569-576).

14. Diamond, S., Mindess, S. and Lovell, J. (1983) Use of a Robinson Backscatter Detector and 'Wet Cell' for Examination of Wet Cement Paste and Mortar Specimens Under Load, **Cement and Concrete Research**, 13 (107-113).

15. Diamond, S. and Bentur, A. (1985) On the Cracking in Concrete and Fibre-Reinforced Cements, in S.P. Shah, (ed.), **Applications of Fracture Mechanics to Cementitious Composites**, Martinus Nijhoff Publishers, Dordrecht, pp. 87-140.

16. Bascoul, A., Ollivier, J.P. and Poushanchi, M. (1989) Stable Microcracking of Concrete Subjected to Tensile Strain Gradient, **Cement and Concrete Research**, Vol. 19, No. 1, pp. 81-88.

17. Bascoul, A., Detriche, C.H., Ollivier, J.P. and Turatsinze, A. (1989) Microscopical Observation of the Cracking Propagation in Fracture Mechanics for Concrete, in S.P. Shah, S.E. Swartz and B. Barr (eds.), **Fracture of Concrete and Rock: Recent Developments**, Elsevier Applied Science, London, pp. 327-336.

18. Eden, N.B. and Bailey, J.E. (1986) Crack Tip Processes and Fracture Mechanism in Hardened Hydraulic Cements, in Proceedings, **8th International Congress on the Chemistry of Cement**, Rio de Janeiro, Vol. III, pp. 382-388.

19. Stroeven, P. (1988) Characterization of Microcracking in Concrete, preprint, **RILEM Conference on Cracking and Durability of Concrete**, Saint Rémy le Chevreuses.

20. Stroeven, P. (1988) Some Observations on Microcracking in Concrete Subjected to Various Loading Regimes, Proc., **International Conference on Fracture and Damage of Concrete and Rock**, Vienna, (in press).

21. Chhuy, S., Cannard, G., Robert, J.L. and Acker, P. (1986) Experimental Investigations into the Damage of Cement Concrete with Natural Aggregates, in A.M. Brandt and I.H. Marshall (eds.), **Brittle Matrix Composites 1**, Elsevier Applied Science, London, pp. 341-354.

22. Reinhardt, H.W. and Hordijk, D.A. (1988) Various Techniques for the Assessment of the Damage Zone Between Two Saw Cuts, presented at the France-U.S. Workshop, **Strain Localization and Size Effect Due to Cracking and Damage**, Cachan, France, September, 12 pp.

23. van Mier, J.G.M. and Nooru-Mohamed, M.B. (1988) Geometrical and Structural Aspects of Concrete Fracture, Proceedings, International Conference on Fracture and Damage of Concrete and Rock, Vienna, (in press).
24. van Mier, J.G.M. (1988) Fracture Study of Concrete Specimens Subjected to Combined Tensile and Shear Loading, Proceedings, International Conference on Measurements and Testing in Civil Engineering, Lyon-Villeurbanne, 1988 (in press).
25. Stys, D. (1989) Numerical Analysis of the Stress Field Parameters in the Fracture Process Zone in Concrete, in A.M. Brandt and I.H. Marshall (eds.), Brittle Matrix Composites 2, Elsevier Applied Science, London, pp. 134-143.
26. Iori, I., Lu, H., Marozzi, C.A. and Pizzinato, E. (1982) Metodo per la Determinazione dei Campi dei Spostamento nei Materiali Eterogenei (Conglomerati Naturali ed Artificiali) a Bassa Resistenza Specifica a Trazione, L'Industria Italiano Del Cemento, No. 4, 275-280.
27. Cedolin, L., Dei Poli, S. and Iori, I. (1987) Tensile Behaviour of Concrete, Journal of Engineering Mechanics, 113 (3) 431-449.
28. Regnault, Ph. and Brühwiler, E. (1988) Holographic Interferometry for the Determination of Fracture Process Zone in Concrete, in Proceedings, Int. Conf. on Fracture and Damage of Concrete and Rock, Vienna, (to be published).
29. Dei Poli, S. and Iori, I. (1986) Osservazioni e Rilievi sul Comportamento a Trazione dei Calcestruzzi: Analisi di Risultanze Sperimentali, Studi e Ricerche, Politecnico di Milano, 8 35-62.
30. Ferrara, G. and Morabito, P. (1989) A Contribution of the Holographic Interferometry to Studies on Concrete Fracture, in S.P. Shah, S.E. Swartz and B. Barr (eds.), Fracture of Concrete and Rock: Recent Developments, Elsevier Applied Science, London, pp. 337-346.
31. Du, J.J., Kobayashi, A.S. and Hawkins, N.M. (1987) Fracture Process Zone of a Concrete Fracture Specimen, in Proceedings, International Conference on Fracture Concrete and Rock, Houston, pp. 280-286.
32. Du, J., Hawkins, N.M. and Kobayashi, A.S. (1989) A Hybrid Analyssi of Fracture Process Zone in Concrete, in Shah, S.P., Swartz, S.E. and Barr, B. (eds.), Fracture of Concrete and Rock: Recent Developments, Elsevier Applied Science, London, pp. 297-306.
33. Raiss, M.E., Dougill, J.W. and Newman, J.B. (1989) Observation of the Development of Fracture Process Zones in Concrete, in S.P. Shah, S.E. Swartz and B. Barr, Fracture of Concrete and Rock: Recent Developments, Elsevier Applied Science, London, pp. 243-253.
34. Schneider, U. and Diederichs, U. (1983) Detection of Cracks by Mercury Penetration Measurements, in F.H. Wittmann, (ed.), Fracture Mechanics of Concrete, Elsevier Science Publishers B.V., Amsterdam, pp. 207-222.
35. Swartz, S.E. and Go, C-G. (1984) Validity of Compliance Calibration to Cracked Concrete Beams in Bending, Experimental Mechanics, 24 (2) 129-134.

36. Swartz, S.E. and Refai, T.M.E. (1987) Influence of Size Effects on Opening Mode Fracture Parameters for Precracked Concrete Beams in Bending, in Proceedings, **International Conference on Fracture of Concrete and Rock**, Houston, pp. 403-417.

37. Bascoul, A., Kharchi, F. and Maso, J.C. (1987) Concerning the Measurement of the Fracture Energy of a Micro-Concrete According to the Crack Growth in a Three Point Bending Test on Notched Beams, in S.P. Shah and S.E. Swartz (eds.), **SEM/RILEM International Conference on Fracture of Concrete** and Rock, Houston, pp. 631-643.

38. Luong, M.P. (1986) Infrared Vibrothermography of Plain Concrete, in F.H. Wittmann, (ed.), **Fracture Toughness and Fracture Energy of Concrete**, Elsevier Applied Science Publishers B.V., Amsterdam, pp. 249-257.

39. Alexander, M.G. (1985) Fracture of Plain Concrete - A Comparative Study of Notched Beams of Varying Depth, **Ph.D. Thesis**, University of the Witwatersrand, Johannesburg, South Africa.

40. Alexander, M.G. (1988) Use of Ultrasonic Pulse Velocity for Fracture Testing of Cemented Materials, **Journal of Cement, Concrete and Aggregates**, ASTM, Vol. 10, No. 1, Summer, pp. 9-14.

41. Alexander, M.G., Tait, R.B. and Gill, L.M. (1989) Characterization of Microcracking and Crack Growth in Notched Conrete and Mortar Beams Using the J-Integral Approach, in S.P. Shah, S.E. Swartz and B. Barr (eds.), **Fracture of Concrete and Rock: Recent Developments**, Elsevier Applied Science, London, pp. 317-326.

42. Alexander, M.G. and Blight, G.E. (1986) The Use of Small and Large Beams for Evaluating Concrete Fracture Characteristics, in F.H. Wittmann, (ed.), **Fracture Toughness and Fracture Energy of Concrete**, Elsevier Science Publishers B.V., Amsterdam, pp. 323-332.

43. Alexander, M.G. and Blight, G.E. (1987) Characterization of Fracture in Cemented Materials, **private communication**.

44. Berthaud, Y. (1987) An Ultrasonic Testing Method: An Aid for Material Characterization, in Proceedings, **International Conference on the Fracture of Rock and Concrete**, Houston, pp. 644-654.

45. Berthaud, Y. (1988) Mesure de L'Endommagement du Beton Par Une Methode Ultrasonore, **Ph.D. Thesis**, Univerisité de Paris 6, France.

46. Mindess, S. (1980) Acoustic Emission in V.M. Malhotra and N. Carino (eds.), **Handbook of Nondestructive Testing of Concrete**, CRC Press, Inc., Boca Raton, Florida (in press).

47. Maji, A.K. and Shah, S.P. Process Zone and Acoutic Emission Measurements in Concrete, to be published in Experimental Mechanics (SEM Paper No. 3609).

48. Maji, A.K. and Shah, S.P. (1988) Application of Acoustic Emission and Laser Holography to Study Microfracture in Concrete, **ACI-SP-112 Nondestructive Testing**, American Institute, pp. 83-109.

49. Maji, A.K. and Shah, S.P. (1986) A Study of Fracture Process of Concrete Using Acoustic Emission, Proc., **Soc. Expt. Mech. Spring Conference**, New Orleans, June 8-13.

50. Maji, A. and Shah, S.P. (1988) Initiation and Propagation of Bond Cracks as Detected by Laser Holography and Acoustic Emission, in S. Mindess and S.P. Shah (eds.) **Bonding in Cementitious Materials, Materials Research Society Symposium Proceedings**, Vol. 114, Materials Research Society, Pittsburgh, pp. 55-64.

51. Maji, A.K. (1989) Study of Concrete Fracture Using Acoustic Emission and Laser Holography, **Ph.D. Thesis**, Northwestern University, Evanston, IL, June.

52. Maji, A.K., Ouyang, C. and Shah, S.P. (1990) Fracture Mechanisms of Quasi-Brittle Materials Based on Acoustic Emission, **Journal of Materials Research**, 5 (1) 206-217.

53. Izumi, M. Mihashi, H. and Nomura, N. (1986) Acoustic Emission Technique to Evaluate Fracture Mechanics Parameters of Concrete, in F.H. Wittmann, (ed.), **Fracture Toughness and Fracture Energy of Concrete**, Elsevier Science Publishers B.V., Amsterdam, pp. 259-268.

54. Chhuy, Sok, Baron, J. and Francois, D. (1979) Mecanique de la Rupture Applique au Beton Hydraulique, **Cement and Concrete Research**, 9 (5) 641-648.

55. Berthelot, J.-M. and Robert, J.-L. (1985) Damage Process Characterization in Concrete by Acoustic Emission, presented at the **Second International Conference on Acoustic Emission**, Lake Tahoe, Nevada, October.

56. Berthelot, J.-M. and Robert, J.-L. (1985) Application de l'Emission Acoustique aux Mecanismes d'Enjdommagement du Beton, <u>Bulletin de Liaison des Ponts et Chaussees</u>, 140, Nov.-Dec., 101-111.

57. Berthelot, J.-M. and Robert, J.-L. (1990) Damage Evaluation of Concrete Test Specimens Related to Failure Analysis, <u>Journal of Engineering Mechanics</u>, <u>116</u> (3) pp. 587-604.

58. Bensouda, M. (1989) Contribution a L'Analyse par Emission Acoustique de L'Endommagement dans le Beton, **Ph.D. Thesis**, Université du Maine, France.

59. Rossi, P. (1986) Fissuration due Beton: due Materiaux a la Structure, Appliation de la Mecanique Lineaire de Rupture, **Ph.D. Thesis**, Ecole Nationale des Ponts et Chaussees, France.

60. Rossi, P., Robert, J.L., Gervais, J.P. and Bruhat, D. (1989) Acoustic Emission Applied To Study Crack Propagation in Concrete, **Materiaux et Constructions** (Paris), Vol. 22, No. 131, pp. 374-383.

61. Karihaloo, B.L. and Nallathambi, P. (1989) An Improved Effective Crack Model for the Determination of Fracture Toughness of Concrete, **Cement and Concrete Research**, <u>19</u> (4) 603-610.

62. Karihaloo, B.L. and Nallathambi, P. (1990) Size Effect Prediction from Effective Crack Model for Plain Concrete, in **preparation**.

63. Kobayashi, A.S., Hawkins, N.M., Barker, D.B. and Liaw, B.M. (1985) Fracture Process Zone of Concrete, in S.P. Shah (ed.) **Applications of Fracture Mechanics to Cementitious Composites**, Martinus Nijhoff Publishers, Dordrecht, pp. 25-50.

286

64. Foote, R., Mai, Y. and Cotterell, B. (1987) Process Zone Size and Crack Growth Measurements in Fiber Cements, in S.P. Shah and G.B. Batson (eds.) **Fiber Reinforced Concrete Properties and Applications**, SP-105, American Concrete Institute, Detroit, pp. 55-70.
65. Hu, X.Z. and Wittmann, F.H. (1989) Fracture Process Zone and K_r-Curve of Hardened Cement Paste and Mortar, in S.P. Shah, S.E. Swartz and B. Barr (eds.), **Fracture of Concrete and Rock: Recent Developments**, Elsevier Applied Science, London, pp. 307-316.
66. Hu, X.Z. and Wittmann, F.H. (1990) An Experimental Method to Determine the Extension of the Fracture Process Zone, submitted to the **ASCE Journal of Materials in Civil Engineering**.
67. Hu, X.A. and Wittmann, F.H. (1989) An Analytical Method to Determine the Bridging Stress Transferred within the Fracture Process Zone, **private communication**.
68. Schorn, H. and Rode, U. (1987) 3-D Modelling of Process Zone in Concrete by Numerical Simulation," in Proceedings, **International Conference on Fracture of Concrete and Rock**, Houston, pp. 308-316.
69. Alvaredo, A.M., Hu, X.Z. and Wittmann, F.H. (1989) A Numerical Study of the Fracture Process Zone, in S.P. Shah, S.E. Swartz and B. Barr (eds.), **Fracture of Concrete and Rock: Recent Developments**, Elsevier Applied Science, London, pp. 51-60.
70. Gopalaratnam, V.S. and Ye, B.S. (1989) Numerical Studies of the Fracture Process Zone in Concrete, in S.P. Shah, S.E. Swartz and B. Barr (eds.), **Fracture of Concrete and Rock: Recent Developments**, Elsevier Applied Science, London, pp. 81-90.

CHARACTERIZATION OF THE FRACTURE BEHAVIOR OF CERAMICS THROUGH ANALYSIS OF CRACK PROPAGATION STUDIES

R.W. STEINBRECH, R.M. DICKERSON, G. KLEIST
Research Center Jülich
Institute for Reactor Materials
P.O. Box 19 13
5170 Jülich
FRG

ABSTRACT. The brittle fracture of structural ceramics, as analyzed from crack propagation studies under conditions of slow, quasi-static crack growth, is described. The experimental results allow one to partition ceramics in to those which have a constant toughness and those which exhibit increasing toughness upon crack extension. The latter effect, known as R-curve behavior is addressed in greater detail for long and short cracks in alumina and MgO-partially-stabilized zirconia. Correlations between microstructure, R-curve behavior and strength are outlined.

1. Introduction

In recent years, important progress has been achieved in the fields of material development and mechanical failure characterization of structural ceramics /1/. The knowledge gathered and accumulated in-service experience enabled a continous widening of the field of application. On the other hand, reliable use of ceramics still demands, for any individual case, knowledge about the interdependencies between application, loading conditions and damage mechanisms.

S. P. Shah (ed.), Toughening Mechanisms in Quasi-Brittle Materials, 287–311.

From the materials and mechanical point of view, characterization of the potential failure mechanisms plays the key role. Specifically, reliable data are needed to avoid catastrophic brittle fracture. Furthermore, information about subcritical crack growth, thermal shock, fatigue, corrosion, creep and high temperature chemical stability are important. Often, the failure of ceramic components is complex because more than one of the forementioned damage mechanisms may act simultaneously.

The present paper expands on previous review articles on structural ceramics /1-4/ and focusses on aspects of brittle fracture revealed through crack propagation studies. When using a Griffith-type crack instability criterion, ceramics can be divided into two groups; those which have a constant toughness and those which exhibit increasing toughness upon crack extension. The latter effect, known as R-curve behavior, is addressed here in greater detail for alumina and MgO partially-stabilized zirconia. Correlations between microstructure, R-curve behavior and strength are outlined.

2.   Crack Instability and Toughness Behavior

Brittle fracture of ceramics usually is assumed to be determined by pre-existing flaws which become unstable upon loading, thus causing catastrophic failure of the material. Using an energy concept /5/, the point of crack instability is given by the equation

$$\sigma_f = \frac{1}{Y} \sqrt{\frac{E \cdot R}{a_c}} \qquad (1a)$$

which relates fracture stress (σ_f) and material resistance (crack resistance force (R)) to the critical flaw size (a_c). The Young's modulus (E) and the parameter Y, which considers the finite size of components (specimens) with respect to flaw size, are included in Eq. 1a as well.

In the case of linear elastic fracture, the well known equation

$$\sigma_f = \frac{1}{Y} \frac{K_{IC}}{\sqrt{a_c}} \qquad (1b)$$

holds, where K_{IC} is the toughness parameter (critical stress intensity factor), which yields a desciption equivalent to Eq. 1a. In this paper, K and R formulations are used interchangeably, assuming that $K_{IC} = \sqrt{E \cdot R}$.

Using Eq. 1b, the strength (fracture stress in bending) is plotted in Fig. 1 as a function of toughness for various commercial ceramics /6/. In a crude approximation, increases of strength and toughness are positively correlated and the data indicate that a proportionality holds. Using the assumption of surface flaws (Y ≈ 1,3 /7/) a critical flaw size of about 60 μm could be deduced from the strength-toughness relationships.

However, such an interpretation of Fig. 1 should not be stretched too far, as will be shown by the crack propagation studies described in the following sections. Seperating ceramics into those which have a constant toughness upon crack extension and those which exhibit a rising crack length-dependent crack resistance (R-curve), some general aspects of fracture are outlined next.

2.1 CONSTANT TOUGHNESS

Upon loading, ceramics typically deform linear elastically until catastrophic failure occurs at crack instability (Fig. 2a) as described by Eq. 1 with the assumption of a constant toughness. In fact, quasi-static crack propagation studies reveal that, for many ceramics, toughness does not change over a wide range of crack extension. Some examples, shown in Fig. 3, are the fine grained materials: Si_3N_4 and SiC.

According to Eq. 1, the strength of ceramics can be increased by reducing the critical flaw size. Furthermore, narrowing the flaw size distribution helps to increase reliability (decrease in scatter of strength). The correlation between flaw size and microstructural dimensions urges the processing of fine grained ceramics. This concept of materials development has been sucessfully applied in recent years /8/. However, a more careful control of the individual processing steps is necessary to avoid material inhomogeneities which can act as stress concentrators and cause premature failure. Fig. 4 shows a large graphite inclusion in HIP-SiC and a pore in a Si_3N_4 body. Both were critical flaws for these fine grained non-oxide ceramics in bending strength tests.

The defect sensitivity of fine grained ceramics and their inherent failure potential through local stress concentrations or short overloads, favours an alternative concept of ceramic development which is based on R-curve behavior.

2.2 INCREASING TOUGHNESS (R-CURVE)

In the case of increasing toughness, the fracture stress can increase despite crack growth, if toughening over-compensates the strength decrease from crack growth (Fig. 2b). Ceramics which exhibit this crack propagation behavior are, to a certain extent, "flaw tolerant".

To date, R-curve behavior has been predominantly investigated in ZrO_2-containing ceramics /9, 10/ and in coarse grained oxide ceramics /11, 12/. Fig. 5 shows R-curves from measurements using thin, compact tension (CT-) like specimens, which highlight the strong transformation toughening effect of zirconia. The R-curve behavior of both monolithic ceramics is discussed in greater detail in sections 3 and 4.

Although it is out of the scope of this paper it should be mentioned that, in recent years, increasing efforts are under way in materials development to implant R-curve behavior in a large variety of ceramics with initially constant toughness by utilizing second phase toughening. Materials of specific interest are ceramic-ceramic composites /13/ with whisker-, fiber-, platelet reinforcement and ceramic-metal composites /14/.

3. Long Crack R-Curve Behavior

3.1 MEASUREMENT AND EVALUATION OF CRACK RESISTANCE

In principle, crack propagation studies of ceramics can be
performed with almost all of the testing geometries used
in linear elastic fracture mechanics. However, for stable,
quasi-static crack extension, specific conditions relating
crack growth-associated changes in crack driving and resistance forces must hold /15/. The change in crack driving force (dG/da), which depends on the energy stored in
the specimen and the testing device, has to be less than
dR/da (< 0 for constant R) to avoid crack instability.
When dG/da < dR/da holds, crack propagation is possible
only if an additional amount of external work is supplied.
In accordance with these considerations, constant
toughness materials need a long starter crack (deep notch)
which may reduce the available range of crack propagation
quite significantly. On the other hand, controlled crack
growth can be achieved for materials with steeply rising
R-curves over a much wider range. The experimental R-
curves of long cracks presented in this section were obtained from single edged notched bend (SENB), double
cantilever beam (DCB) and double torsion (DT) specimens
(Fig. 6). In section 4, R-curves for short "natural" surface cracks on the tensile surface of bend bars are
discussed.

In the SENB-geometry, the depth of the narrow ($\rho \approx$
70 μm), straight-through notches varied between $a_n/W = 0.2$
and $a_n/W = 0.8$, where W is the width in the direction of
crack growth. It should be noted that sharp starter cracks
generated by the bridging method /16/ are not appropriate
in the case of R-curve behavior because controlled crack
propagation starts at an already high resistance level.

The technique used for the propagation studies of long
cracks is illustrated in Fig. 7. Load (P), load-point displacement (d) and crack length (a) (monitored with a travelling microscope) were measured simultaneously to generate P-d-a curves, as shown in the lower part of Fig. 7.
In addition to continuous experiments under displacement
or cross head-position control, tests with several inter-

mediate unloading loading sequences can be run to gain insight into the underlying fracture mechanisms and/or to
provide further information for the evaluation of resistance data. Such P-d curves are shown for a coarse
grained Al_2O_3 (Fig. 8) and a high toughness Mg-PSZ (Fig.
9). Unloaded, both ceramics reveal a residual displacement
(d_r), but they significantly differ with respect to the
hysteresis upon reloading. Due to frictional wake effects
of rough serrated crack surfaces, hysteresis is very pronounced in Al_2O_3 /17/. Also, the residual displacement can
be increasingly reduced by stepwise renotching from notch
to crack tip. When the renotch approaches the crack tip,
no further d_r is left in Al_2O_3 /17/. In Mg-PSZ, the decrease in d_r depends on the width of the renotch cut /18/,
as well. The residual displacement only diminishes if the
complete transformation zone is machined away.

Which R-curve determination methodology to utilize when
analysing mechanical tests, particulary whether residual
displacements must be incorporated, is currently an unre-
solved topic of discussion.

Most often, the evaluation procedure follows a linear
elastic compliance approach, neglecting residual displacement effects. In the case of quasi-static crack growth,
G = R, and

$$R = \frac{P^2}{2B} \frac{dC}{da} \qquad (2)$$

where B is the width and C the compliance of the specimen.

The three possible methods used in applying Eq. 2 are:
A)   from experimental P-d-a data, the experimental com-
     pliance function C_{exp} (a_{exp}) is taken
B)   a theoretical compliance function C_{th} (a_{th}) is used
     after calculating a_{th} by equating C_{exp} and C_{th}
C)   the experimental crack length is combined with a
     theoretical compliance function ($C = C_{th}$ (a_{exp})).
     (Note that this method is equivalent to standard pro-
     cedures of stress intensity factor determination.)

These three methods can generate different crack resistance curves. For example, due to the wake interaction (discussed in section 3.2.1), alumina is less compliant than theoretically predicted. Thus, the crack length (a_{th}) calculated with method B, is shorter than the physical crack length (a_{exp}) measured with the travelling microscope.

If residual displacement is relevant for the specimen behavior during crack propagation then Eq. 2 has to be modified and extended /15/, which gives

$$R = \frac{P^2}{2B} \frac{dC}{da} + \frac{1}{B} P \frac{d(d_r)}{da} \tag{3}$$

where C is a compliance determined from unloading-loading sequences and d_r is the residual displacement on unloading.

Interestingly, in alumina, all 4 analysis methods yield almost the same R-curve over a wide range of crack extension (SENB: 2-3 mm) before significant deviations occur. The details of this comparison will be reported elsewhere /17/. For Mg-PSZ, the initial part of the R-curve is shifted towards a higher resistance if the contribution of residual displacement is considered. The curves from all evaluation methods then approach the same plateau value after about 2-3 mm of crack extension.

Note that the long crack R-curves shown in this paper were derived either from compliance method A (Mg-PSZ) or B (Al_2O_3, predominantly in SENB geometry).

3.2 R-CURVE BEHAVIOR

Controlled crack propagation studies using the standard specimens of LEFM allow one to trace the crack-microstructure interaction over a long range of crack extension. Thus, the full potential of a ceramic to resist crack growth can be explored with a long crack R-curve. Some features of the fracture- (resistance-) behavior of Al_2O_3 and Mg-PSZ resulting from such crack propagation studies are discribed in the following two subsections, respectively.

3.2.1. *Alumina*. In pure alumina, R-curve behavior is more pronounced in coarse-grained than in fine-grained materials (Fig. 10). In summarizing the various R-curve results from different testing geometries and grain sizes, a general shape can be deduced, which is given by an initially rising curve that plateaus out after a certain crack extension. Note, that, in Fig. 10 the specimen size (w = 7 mm) was not large enough to exhibit a plateau-regime. A detailed analysis of the grain size dependence reveals an interesting behavior of the R-curves. Fine grained materials start from a notch tip with a higher initial resistance (R_o) than coarse grained Al_2O_3 but experience almost no increase, unlike the coarse grained material. This crossing of R-curves, which is confirmed by recent work of Lawn and coworkers /18/, gave rise to some confusion in the past when K_{IC} values determined from different testing geometries (e.g. SENB and DT) were compared. DT measurements determine the plateau regimes of R-curves /11/ and, therefore, the K_{IC} values so calculated are relatively high. Also, the grain size dependence is inverted compared to K_{IC}-experiments with crack instability in the initial part of an R-curve, e.g. short cracks in SENB.

Crack propagation studies also probe the severe differences in R-curve behavior between the two modes of microfracture (Fig. 11). Nearly pure, coarse-grained alumina fractures predominantly intergranularly and exhibits a steeply rising R-curve. However, the R-curve of a material comparable in grain size but with more glassy phase and transgranular microfracture, is comparatively shallow. Although both curves start at about the same R_o-level, they differ significantly in their slope.

The information gained from crack propagation studies in alumina strongly favour wake controlled crack tip shielding as the basic R-curve mechanism /19, 20/. Our current understanding may be summarized with the aid of the schematic model in Fig. 12. Alumina, due to anisotropy in the thermal expension coefficients, contains residual stresses between neighbouring grains. They depend on grain

size and are large for pure coarse grained material. However, when a glassy grain boundary phase is present, the
stresses are assumed to be lower. The residual stresses
are relieved by the formation of a damage zone of
microcracks which is triggered by the stress field of a
macrocrack. The microcracks force the macrocrack to generate a rough, tortuous crack surface. In addition, the
microcracks become unconstrained when the macrocrack
advances, thus generating a layer of dilated material
along the crack surfaces. The crack surface roughness and
dilation give rise to frictional crack bridging effects
which shield the crack tip from the external stresses. The
maximum length of such a crack shielding wake zone is
reached when the local crack opening displacement is
larger than the interaction length of the bridging elements. A constant, steady state size of the wake zone,
which obtains thereafter, corresponds with the plateau of
the R-curve. In summary, it is important to note that both
microstructural properties and crack/specimen geometries
determine the actual shape of an R-curve. Quantitative
details for the model of the R-curve behavior of Al_2O_3 are
reported elsewhere, /17/.

3.2.2 *MgO partially stabilized Zirconia.* The effectiveness
of transformation toughening, the major toughening mechanism in zirconia-containing ceramics, is manifest in the
strong R-curve behavior of Mg-PSZ; with plateau toughnesses measured for long cracks of up to 18 MPa $\sqrt{m}$ /21/.

In Mg-PSZ, tetragonal precipitates transform in the
crack tip stress field to monoclinic symmetry. The frontal
transformation zone thus generated has no effect on the
crack tip field /24/. However, upon crack extension, the
constraint on the process zone unloads in the wake, giving
rise to residual dilatation-associated stresses and crack
shielding. A steady state level, which corresponds with
the plateau regime of the R-curve, is only attained after
substantial crack extension (2-3 mm in high toughness Mg-
PSZ /21/).

In the simplified case of supercritical transformation,
i.e. all particles within the transformation zone fully
transform, the achieveable plateau toughness depends on

the size (h) of the zone /23/. A Raman microprobe technique has been developed recently to measure h very accurately /24/. Qualitatively the zone can be detected easily on polished surfaces (Fig. 13) due to the surface uplifting triggered by the dilatation accompaning the t-m transformation. Nevertheless an exact correlation between the microstructure, zone size, and the R-curve shape cannot be specified to date /1/. Toughening can be optimized by special heat treatments of Mg-PSZ. Again, crack propagation studies with R-curve measurement probe the effectiveness of a given heat treatment.

Results for a 9.7 mole % Mg-PSZ are shown in Figures 14 and 15. After solution annealing at 1700°C and rapid cooling, samples were annealed at 1400°C for various times up to 10 hours. Additionally, some samples were annealed at 1100°C for up to 2 hours. Figure 14 shows the change in the precipitate microstructure due to the 1400°C heat treatment. The subeutectoid, 1100°C annealing did not change the precipitate size.

The R-curves determined after annealing reveal substantial differences, not necessarily at the beginning (R_0-value), but, with increasing crack extension (Fig. 15).

The R-curve results indicate that, for the given Mg-PSZ, toughening is optimized by annealing at 1400°C followed by 1100°C. The toughness in such samples changes from $K_R = 4$ MPa $\sqrt{m}$ in the solution annealed material to $K_{plateau} \approx 12$ MPa $\sqrt{m}$ after the treatment.

4.   R-Curve Behavior and Strength

The crack propagation studies and R-curve results presented in the previous section are very useful for material development in terms of better understanding the basic mechanisms. However, the high toughness values often cited for modern structural ceramics, which correspond to the plateau of long-crack R-curves, are not really relevant for most structural applications. After failure of ceramic components, no fracture causing cracks of millimeters in length are observed usually. Typically, critical crack sizes are at least one order of magnitude smaller. In order to determine the strength-toughness relationship of ceramics, crack propagation studies should focus on failure relevant short cracks rather than long cracks.

To observe and analyse the propagation of short cracks, stable growth must be achieved. This is possible when R-curve behavior obtains, as was shown by using the energy balance in section 3.1.

To date, only a few R-curve measurements for Al_2O_3 and Mg-PSZ with short, failure relevant surface cracks, are published. The short cracks either originated from indentation /12/ or initiated "naturally" on tensile surfaces of bend bars during increasing loading /25, 26/.

Figures 16 and 17 show surface cracks generated by stepwise loading of bend bars in coarse grained Al_2O_3 and high toughness ZrO_2, respectively. In both ceramics, "natural" surface cracks can grow from some tens of microns to several hundred microns. In order to calculate the toughness from such cracks, the crack profile must be determined. Dye penetrant techniques indicate that the surface cracks are almost semi-circular in Al_2O_3, whereas, serial sectioning in Mg-PSZ reveals that the cracks advance much further at the surface compared to depth (ratio $\approx 5/1$). With appropriate Y-functions, determined using the crack profile, K_R is determined in a straightforward manner from Eq. 1b.

The resulting short crack R-curves are plotted for comparison with those of long cracks in Figures 18 and 19.

The short crack R-curves start considerably lower in both ceramics and the plateau toughness level is not reached before instability. Unlike LEFM tests, where the long cracks are forced to initiate from a notch tip, the "natural" surface flaws originate from local heterogeneities, e.g. pores or larger grains. Often, more than one short crack forms on the tensile surface. In this case, the one with the shallowest R-curve slope becomes unstable first. The fracture stress at instability corresponds to the bending strength as determined from fracture tests with low loading rates. Thus, under technically relevant loading conditions, specimens or components cannot utilize the full potential of R-curve behavior. Clearly, toughness or resistance data referring to the plateau of long crack measurements overestimate the values pertainent for application of a ceramic material.

5. Conclusions

The quasi-static, controlled crack propagation studies reported here enable one to characterize the fracture behavior of structural ceramics in more detail than one parameter descriptions.

Furthermore, deeper insight into the correlation between micro-structure and mechanical behavior is possible.

Long-crack fracture tests using the standard specimens of LEFM reveal if toughness (equivalent: crack resistance) is a material property constant or increases with crack extension (R-curve behavior). The full potential of a microstructure for toughnening is enabled.

However, for failure relevant fracture characterization and a better understanding of strength-toughness relationship, crack propagation studies have to focus on short cracks. In the case of pronounced R-curve behavior, stable crack growth of natural surface flaws can be achieved experimentally by stepwise loading. As the resulting R-curves are significantly lower than those from long cracks, it must be emphasized that long crack plateau toughness values, often quoted in literature, may overestimate the toughness applicable in components considerably.

6. Acknowledgements

The authors would like to thank M.J. Readey, A. Reichl, G. Mundry, K.K.O. Bär and G. Dransmann who contributed with their experimental work to the present paper. The R-curve research was funded by the German Research Association (DFG).

References

/1/ A.G. Evans, "Perspective on the Development of High-Toughness Ceramics", J. Am. Ceram. Soc., 73 [2] 187-206 (1990)

/2/ S.W. Freiman, "Brittle Fracture of Ceramics", Ceramic Bulletin, Vol. 67, No. 2, 392-402 (1988)

/3/ T. Fujii and T. Nose, "Evaluation of Fracture Toughness for Ceramic Materials", IJIS International, Vol. 29, No. 9, 717-725 (1989)

/4/ R.O. Ritchie, "Mechanismus of Fatigue Crack Propagation in Metals, Ceramics and Composites: Role of Crack tip Shielding", Mat. Sci. Engng., A 103, 15-28 (1988)

/5/ D. Broek, "Elementary Engineering Fracture Mechanics", Noordhoff International Publishing, Leyden (1974)

/6/ Jahrbuch "Technische Keramik", Vulkan Verlag, Essen, (1988)

/7/ C. Mattheck, P. Morawietz, D. Munz, "Stress Intensity Factor at the Surface and at the Deepest Point of a Semi-elliptical Surface Crack in Plates and Stress Gradients", Int. J. Fracture, 23, 201-212 (1983)

/8/ DIA Research Institute, "Advanced Engineering Ceramics - Technology in Japan", Tokyo, July (1987)

/9/ D.J. Green, R.H.J. Hannik, M.V. Swain, "Transformation Toughening of Ceramics" CRC Press, Boca Raton (1989)

/10/ A.H. Heuer, "Transformation Toughening of ZrO_2-Containing Ceramics", J. Am. Ceram. Soc., 70 [10] 689-698 (1987)

/11/ F. Deuerler, R. Knehans and R.W. Steinbrech, "Testing-Methods and Crack Resistance Behavior of Al_2O_3", J. de Physique C1 [2] 617-620 (1986)

300

/12/ R.F. Cook, C.J. Fairbanks, B.R. Lawn and Y.W. Mai, "Crack Resistance by Interfacial Bridging: Its Role in Determining Strength Characteristics" J. Mater. Res. 2 [3] 345-356 (1987)

/13/ John J. Mecholsky, Jr., "Engineering Research Needs at Advanced Ceramics and Ceramic-Matrix Composites", Ceram. Bull. 68 [2] 367-75 (1989)

/14/ Minoru Taya and Richard J. Arsenault "Metal Matrix Composites - Thermomechanical Behavior", Pergamon Press, Oxford, England (1989)

/15/ A.G. Atkins and Y.W. Mai, "Elastic and Plastic Fracture", Ellis Horwood Limited, Chichester, England (1985)

/16/ R. Warren and B. Johannesson "Creation of Stable Cracks in Hardmetals Using Bridge Indentation", Powder Metallurgy 27 [1] 25-29 (1984)

/17/ H. Nickel and R.W. Steinbrech, Editors, "Mikrobruchvorgänge in Al_2O_3-Keramik", Abschlußkolloquium DFG-Forschungsvorhaben, KFA-Jülich (1990), Proceedings in print.

/18/ R.W. Steinbrech, E. Inghels, A.H. Heuer, "Residual Displacement Effects During Crack Propagation in High Toughness Mg-PSZ", J. Am. Ceram. Soc. 73 [7], 2016-2022 (1990)

/19/ R. Knehans and R.W. Steinbrech, "Memory Effect of Crack Resistance during Slow Crack Growth in Notched Al_2O_3 Bend Specimens", J. Mat. Sci. Letter, 1, 327-329 (1982)

/20/ Y.W. Mai and B. R. Lawn, "Crack-Interface Grain Bridging as a Fracture Resistance Mechanism in Ceramics: II, Theoretical Fracture Mechanics Model", J. Am. Ceram. Soc., 70 [4] 289-294 (1987)

/21/ A.H. Heuer, M.J. Readey and R.W. Steinbrech, "Resistance Curve Behavior of Supertough MgO-Partially-stabilized ZrO_2", Mat. Sci. Engng. A 105/106, 83-89 (1988)

/22/ R.M. McMecking and A.G. Evans, "Mechanics of Transformation Toughening in Brittle Materials", J. Am. Ceram. Soc. 65 [5] 242-47 (1982)

/23/ A.G. Evans and R.M. Cannon, "Toughening of Brittle Solids by Martensitic Transformations", Acta Metall. 34 [5] 761-800 (1986)

/24/ R. Dauskardt, D.K. Veirs and R.O. Ritchie, "Spatially Resolved Raman Spectroscopy Study of Transformed Zones in Magnesia-Partially-Stabilized Zirconia", J. Am. Ceram. Soc., 72, [7] 1124-30 (1989)

/25/ D.B. Marshall and M.V. Swain, "Crack-Resistance Curves in Magnesia Partially Stabilized Zirconia", J. Am. Ceram. Soc., 71 [6] 399-407 (1988)

/26/ O. Schmenkel and R.W. Steinbrech, "Crack-Resistance Curves of Surface Cracks in Alumina", J. Am. Ceram. Soc. 71 [5] C-271-273 (1988)

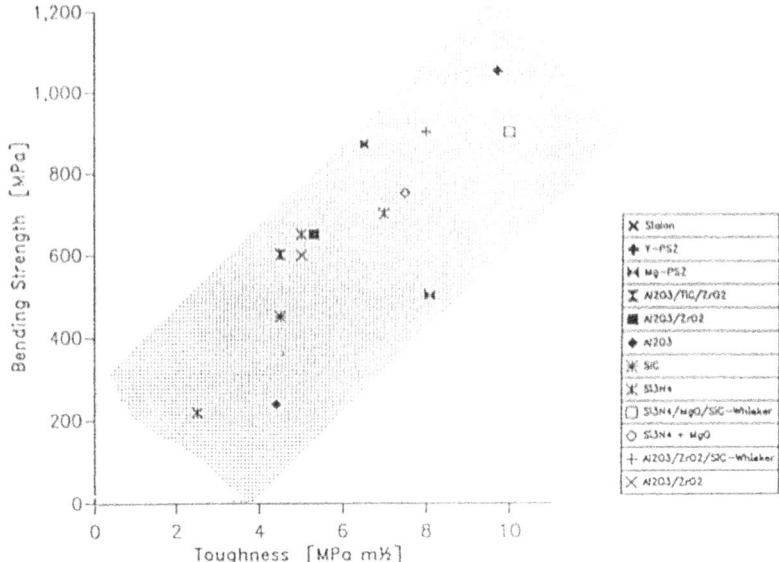

Fig. 1: Strength and toughness of commercial ceramics. Data from /6/

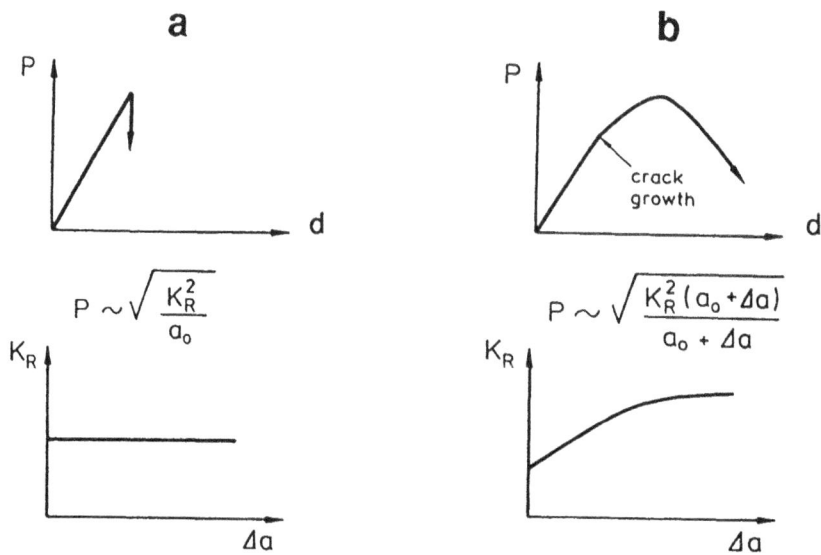

Fig. 2: Loading response for different crack resistance behavior (schematic) a) constant toughness b) increasing toughness (R-curve)

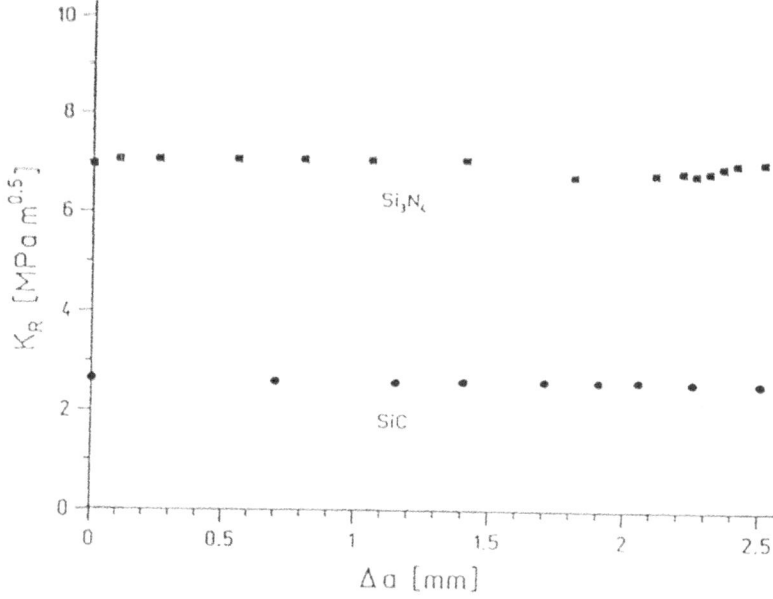

Fig. 3: Resistance behavior of SiC and Si₃N₄

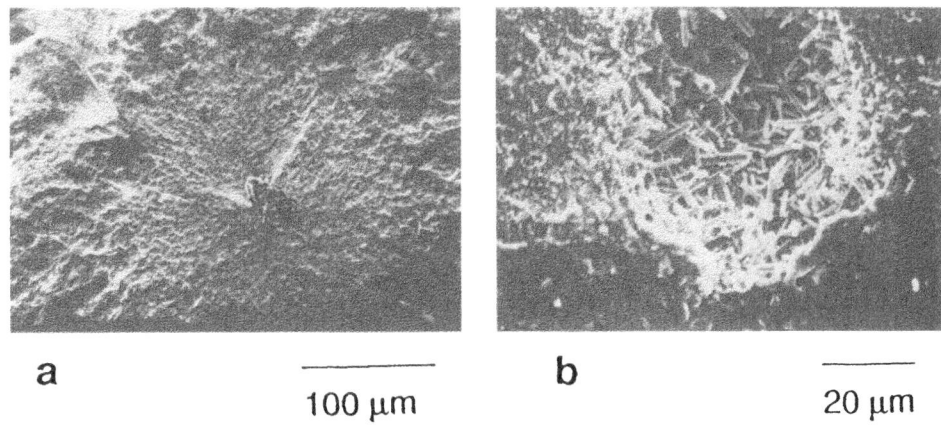

a   100 μm   b   20 μm

Fig. 4: Critical defects for fine grained non-oxide ceramics a) graphite inclusion in HIP-SiC b) pore in Si₃N₄

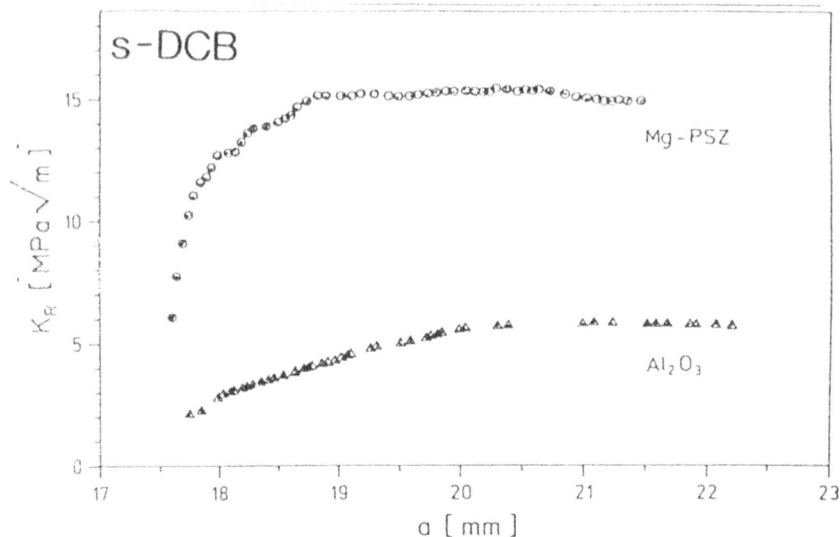

Fig. 5: R-curves of MgO-partially stabilized ZrO$_2$ and Al$_2$O$_3$

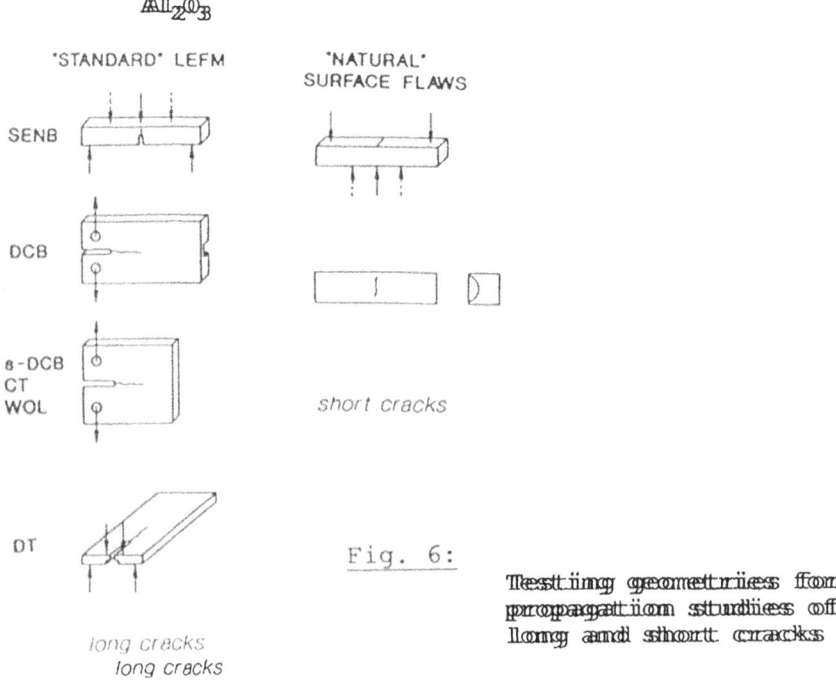

Fig. 6:

Testing geometries for propagation studies of long and short cracks

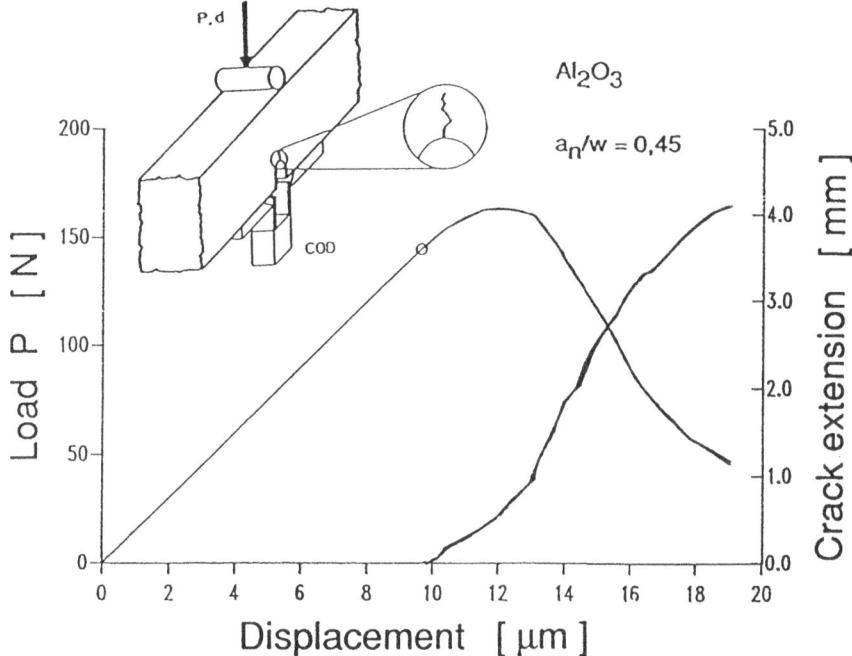

Fig. 7:   Experimental technique used for crack propa-
           gation studies. Simultaneous measurement of
           load, load-point displacement and crack length
           (with travelling microscope)

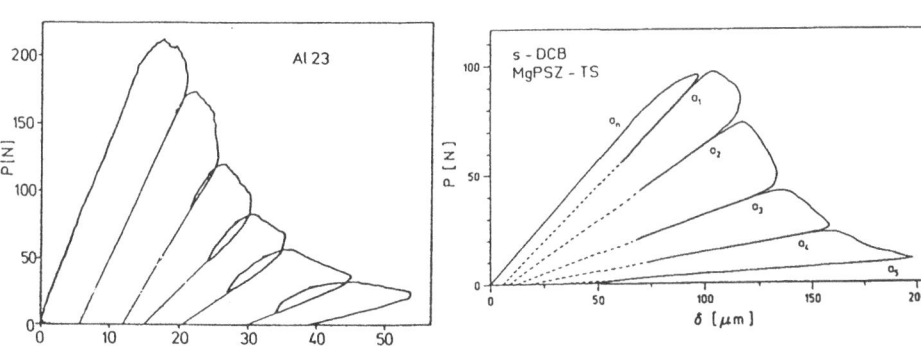

Fig. 8:
Intermediate unloading-
loading sequences in Al_2O_3

Fig. 9:
Intermediate unloading-
loading sequences in Mg-PSZ

306

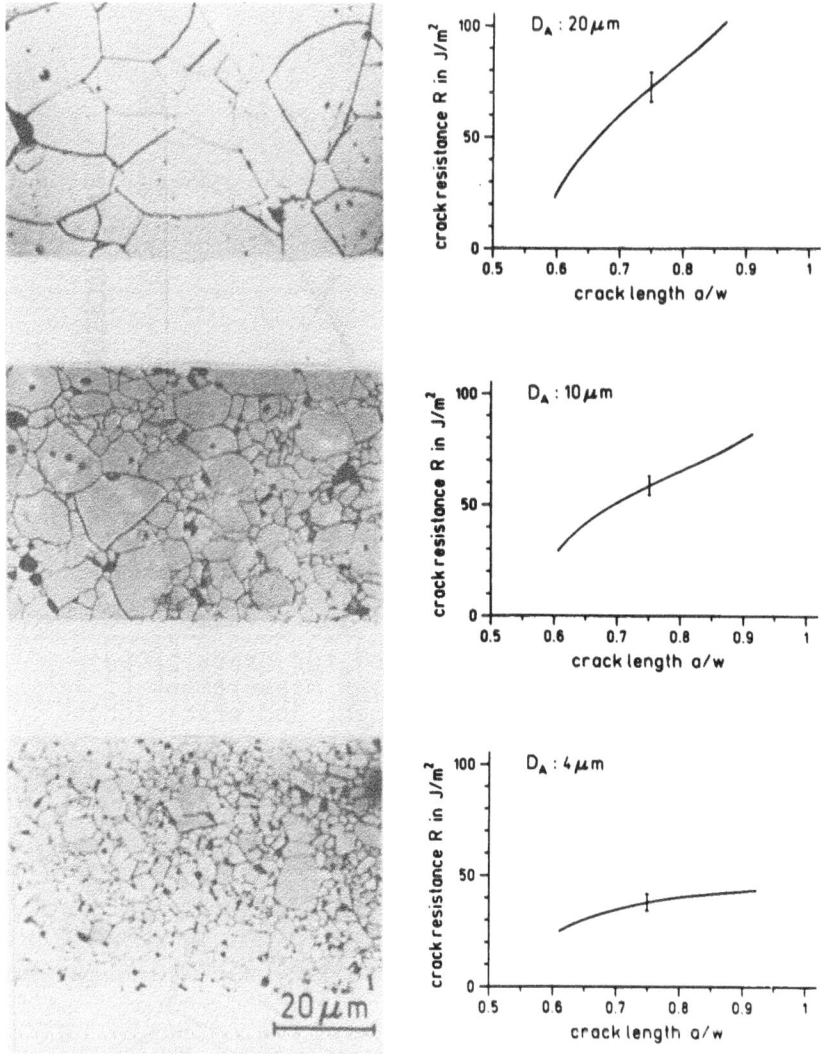

Fig. 10: R-curve behavior of long cracks in Al₂O₃.
SENB geometry. Intergranular microfracture.

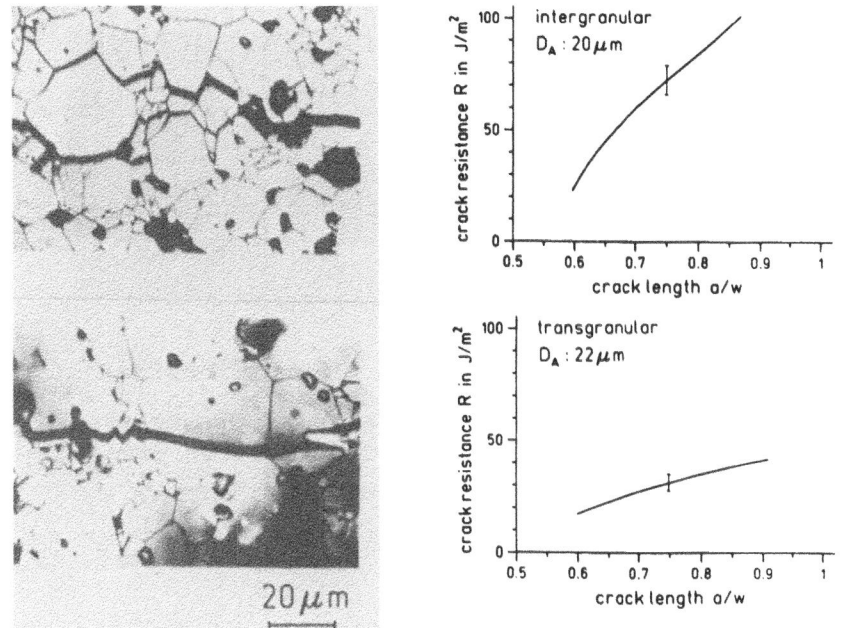

Fig. 11: Influence in mode (transgranular vs. intergranular) of microfracture on R-curve behavior

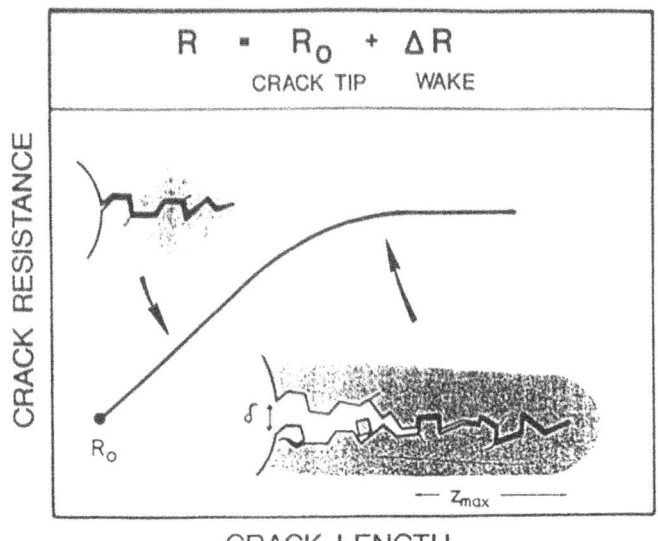

Fig. 12: Qualitative model of R-curve behavior in Al_2O_3

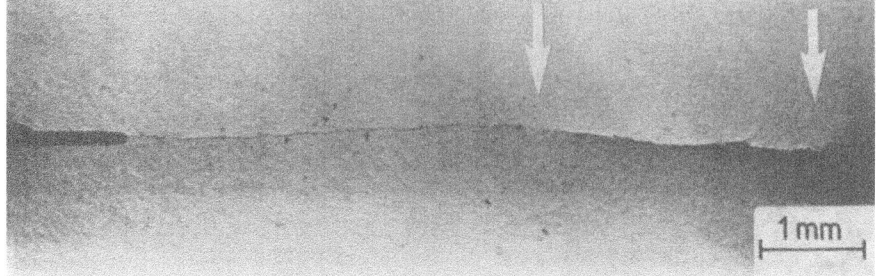

Fig. 13: Surface uplifting (between arrows) showing
the transformation zone around crack in
Mg-PSZ. The zone between the notch and first
arrow was annealed out at 1000°C after pre-
cracking

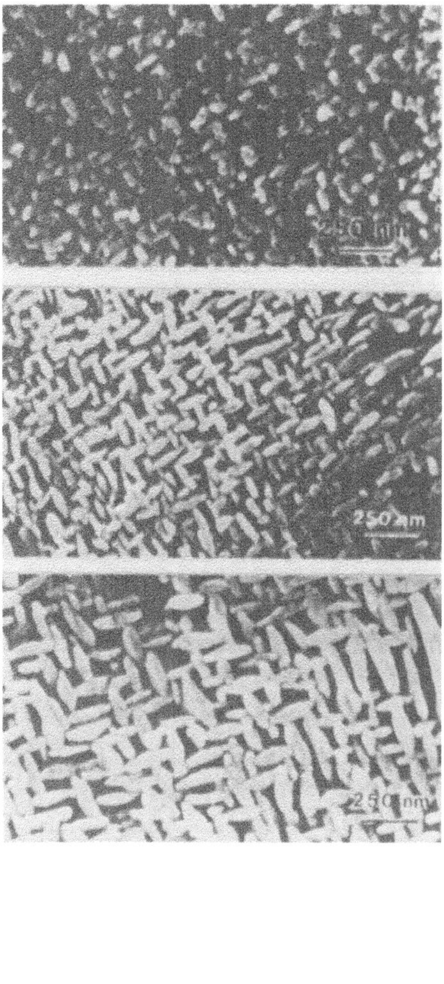

a)

b)

Fig. 14:
Aging of Mg-PSZ preci-
pitates at 1400°C
a) solution annealed
b) 2 h
c) 10 h
SEM-micrograph from
etched fracture
surfaces

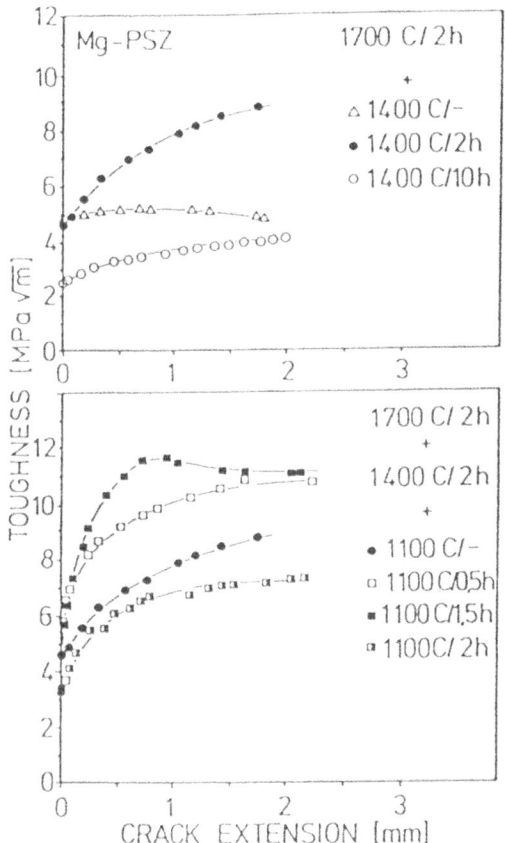

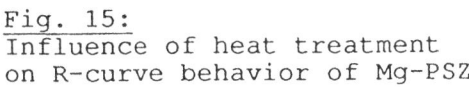

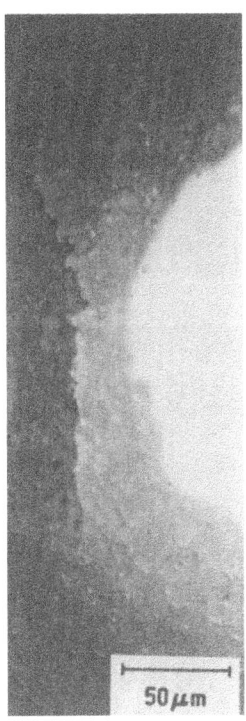

Fig. 15:
Influence of heat treatment
on R-curve behavior of Mg-PSZ

Fig. 16:
Surface crack in Al_2O_3
Micrograph contrast due
to oblique illumination

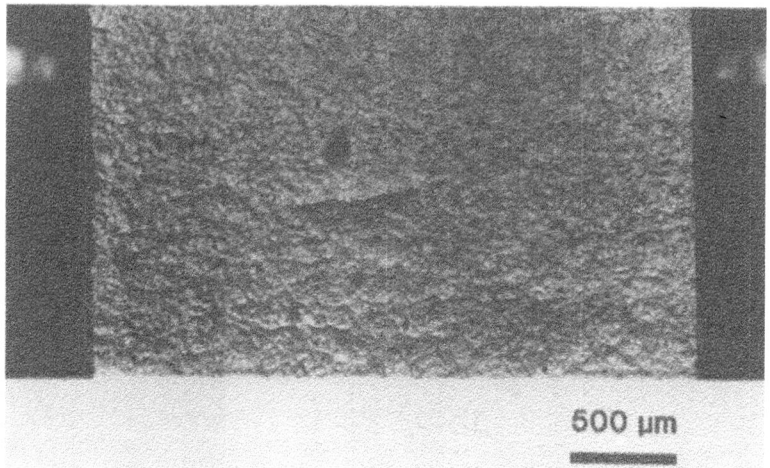

Fig. 17:   Surface crack in Mg-PSZ

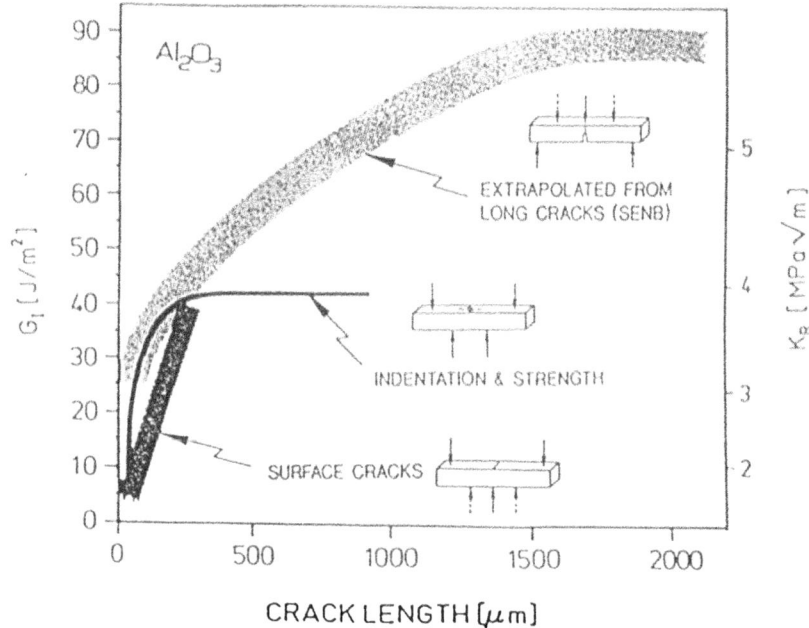

Fig. 18:   Comparison of short and long crack R-curve
behavior in Al$_2$O$_3$. Data for indentation and
strength method from /12/

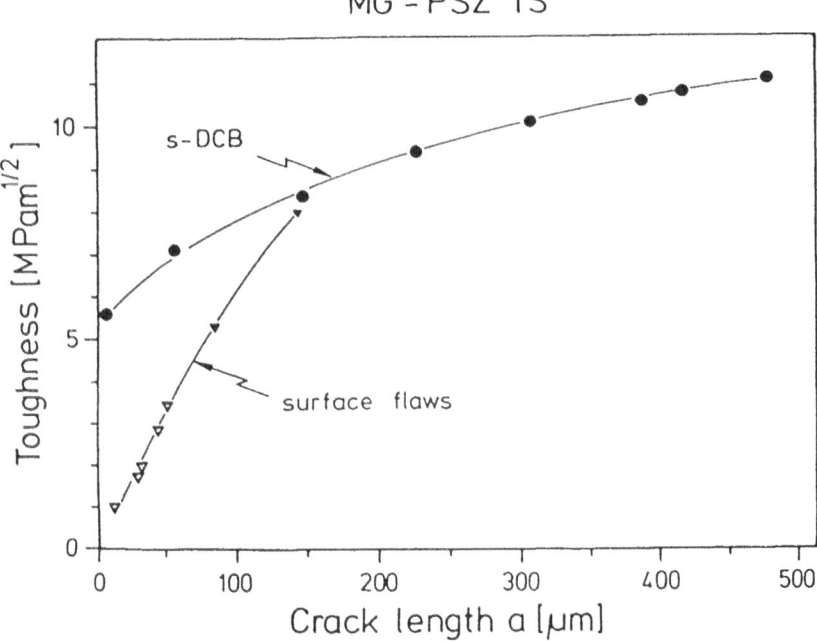

Fig. 19: Comparison of short and long crack R-curve behavior in Mg-PSZ

A REVIEW OF EXPERIMENTAL METHODS TO ASSESS DAMAGE
DURING FRACTURE OF ROCK, CONCRETE AND REINFORCED COMPOSITES

Hideaki Takahashi
Research Institute for Fracture Technology
Faculty of Engineering, Tohoku University
Aramaki Aoba, Aobaku
Sendai/980, Japan

ABSTRACT.   For   quasi-brittle materials like rock, concrete and their
composites it is usually adopted that a term of fracture process zone is
used as a measure of fracture damage during loading.   To assess the
fracture damage quantitatively, a significance of acoustic emission and
ultrasonic testing have been reviewed.

1. Introduction

    For quasi-brittle materials like rock, concrete and advanced
cementitious composites it is usually accepted that a term of fracture
process zone is commonly used as a measure of fracture damage during
loading. However, no physically reasonable explanation of fracture
process zone is not made uptil now, because there is no experimental
technique to characterize formation and extension of the fracture
process zone quantitatively. To investigate the fracture process at the
microscopic level, three kinds of NDE methods have been currently
developed at Tohoku University. These three techniques are photoelastic
coating method, acoustic emission rating method and ultrasonic time
difference method. The detail of these techniques are described in Refs
[1-5, 8].

2. Intense Microcracking and its Observation by Photo-elastic Coating
    Technique [1]

2-1. Materials and Photoelastic Coating Technique

    The materials examined in this study are rock (a granite with an
average grain diameter of 1.3 mm), graphite (a nuclear-grade graphite
designated PGX) and mortar. The compositions and mechanical properties
of these materials are shown in Table 1.
    Rectangular bend specimens (100 x 30 x 28 mm) were cut using a
diamond wheel saw. A notch with a root radius of 50 μm was machined
into each test specimen.
    A transparent ferroelectric ceramic, $(Pb_{0.91}, La_{0.09})$ $(Zr_{0.64},$
$Ti_{0.34})O_3$ was used as a photoelastic-coating material (designated PLZT).

313

S. P. Shah (ed.), Toughening Mechanisms in Quasi-Brittle Materials, 313–328.

TABLE 1—COMPOSITION AND MECHANICAL PROPERTIES OF THE MATERIALS USED

Material	Composition	Young's Modulus E (GPa)	Tensile Strength σ_{ult} (MPa)	Fracture Toughness J_{Ic} (J/m^2)	$K_{Ic} = \sqrt{E\,J_{Ic}}$ (MPa m$^{1/2}$)
Granite	Quartz (38 percent), Feldspar (56 percent), Mica (6 percent) average grain size: 1.3 mm	19.4	3.6	68.0	1.15
Graphite	Coke (70 percent), Binder (Pitch) (30 percent) max. grain size of coke: 1.0 mm	6.0	7.2	72.0	0.66
Mortar	Cement (42 percent), Sand (42 percent), Water (16 percent) max. grain size of sand: 2.5 mm	40.4	3.2	11.5	0.68

The material was obtained in the form of a circular disk, 50-cm diameter and 10-mm thick. Thin sections of the PLZT were cut from the disk. The foil thickness was in the range 110-150μm. For light-reflection purposes, an optical mirror was formed on one plane of the ceramic plate by evaporation of chromium. The prepared thin foil was glued with cyanoacrylate to the surface of the three-point-bend specimens. A load frame was designed to permit the specimen to be loaded by high compliance springs on the stage of a polarizing microscope.

Fracture-toughness tests were conducted on the three materials according to the ASTM standard test method using an Instron testing machine. Single-edge-notch specimens were tested in three-point bending.

The averaged load versus J-integral relationship obtained from fracture-toughness tests was used as a calibration to determine the load levels of specimens tested for strain-field observation in terms of J.

The calibration-test for the PLZT was carried out in terms of the relationship between the birefringence Δn and principal-strain difference Δ_ε.

The photoelastic sensitivity of the PLZT, defined as $\Delta n / \Delta_\varepsilon$ is 1.75, which is two or three orders of magnitude higher than common photoelastic-coating materials such as epoxy and Araldite.

2-2. Results and Discussions

Typical test records of load versus load-line displacement (P-V$_L$) are given in Fig. 1. The three materials exhibit different amounts of nonlinear deformation. The ratio of the secant specimen compliance at peak load to the initial compliance, Csec/Cini, can be taken as a measure of the degree of the nonlinearity (see Fig. 1). The ratio values are 2.22, 1.90, and 2.88 for mortar, graphite, and granite, respectively, indicating that the granite has the highest nonlinearity of deformation.

An example of the interference fringe pattern formed around the notch tip is illustrated for the granite in Fig. 2. The principal-strain difference is determined from the fringe observation using the calibration diagram. Figure 3 shows the distribution of the principal-strain difference along the notch plane at various load levels, where r is the distance from the notch tip, and the J integral is chosen to indicate the load conditions. It is noted that the fringe shape around the notch tip and the strain distribution shown in Figs. 2 and 3 are

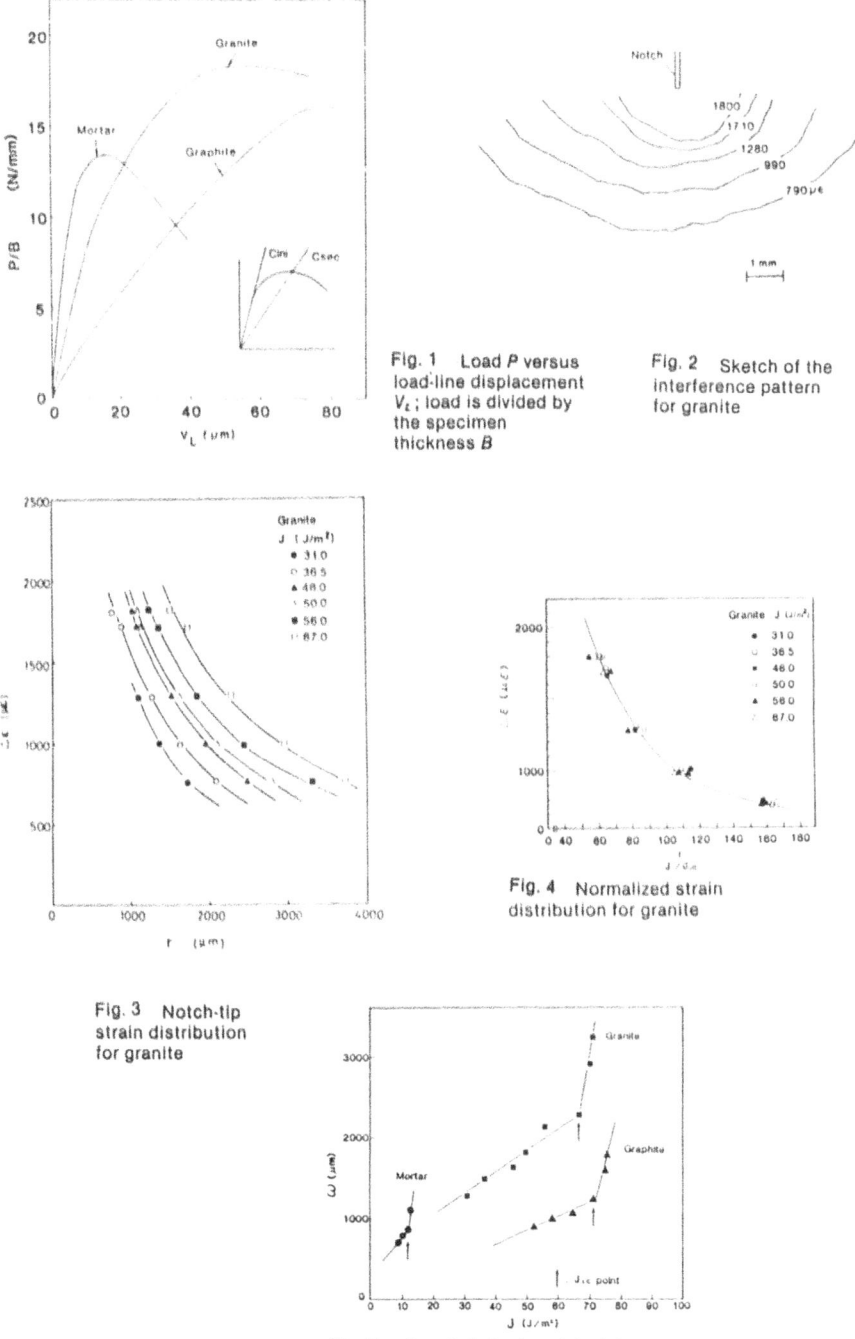

Fig. 1 Load P versus load-line displacement V_L; load is divided by the specimen thickness B

Fig. 2 Sketch of the interference pattern for granite

Fig. 3 Notch-tip strain distribution for granite

Fig. 4 Normalized strain distribution for granite

Fig. 5 Growth behavior of the intense microcrack zone

quite different from that observed for linear-elastic materials. The
deviation from linear-elastic-deformation behavior is considered to be
due to the formation of an extensive microcrack zone ahead of the notch
tip.

In order to characterize the strain field within the intense
microcrack region, the principal strain difference under several
different load conditions is plotted against the nondimensional distance
in Fig. 4, where the distance from the notch tip is normalized by the
length parameter J/σ_{ult}. It is noted tha the strain distribution around
the notch tip is well characterized by the J integral, irrespective of
load levels, and the logarithmic plot leads to the following empirical
relation.

$$\Delta\epsilon = \Delta\epsilon_0[(J/\sigma_{ult})/r]^m$$

where $\Delta_{\epsilon 0}$ and m are material constants. Looking at the normalized
strain distribution, we see that the value of m is close to 1.0 for the
three materials, and thus the principal-strain difference ahead of the
notch tip has the singularity of r^{-1}, equivalent to that of perfectly
rigid plastic materials.

We now examine the extension behavior of the intense microcrack
region. Here the length of intense microcrack region ω is taken as the
distance from the notch tip where the principal-strain difference
exceeds 760 $\mu\epsilon$. Figure 5 shows the development of the intense
microcrack region ω as a function of J. The intense microcrack region
increases linearly with increasing J integral at low load levels. Note
that the intense microcrack region extends rapidly with increasing load
at the point indicated by an arrow. The abrupt increase in ω is
interpreted to be due to the effect of macroscopic crack growth. In
this study, the J-integral value at the knee point is defined as the
critical J-integral value at the onset of macroscopi crack growth, and
it is denoted by J_{IC}.

As described above, the crack-tip strain fields in the brittle-
microcracking materials show the singularity of r^{-1}, unlike the strain
field expressed by the stress-intensity factor K. From this observation
the J integral was selected to characterize the strain fields in this
study. The fact that the crack-tip strain field and the extension
behavior of the intense microcrack region can be characterized by the J-
integral supports the use of the J_{IC} criterion for a valid measure of
the fracture toughness for these materials.

3. AE Evaluation of Fracture Damage in Rock and Concrete

3-1. Acoustic Emission Characteristics and Determination of Fracture
    Toughness Evaluation Point in Rock [2, 3]

The load versus displacement record (P-V_L) is shown in Fig. 6 for a

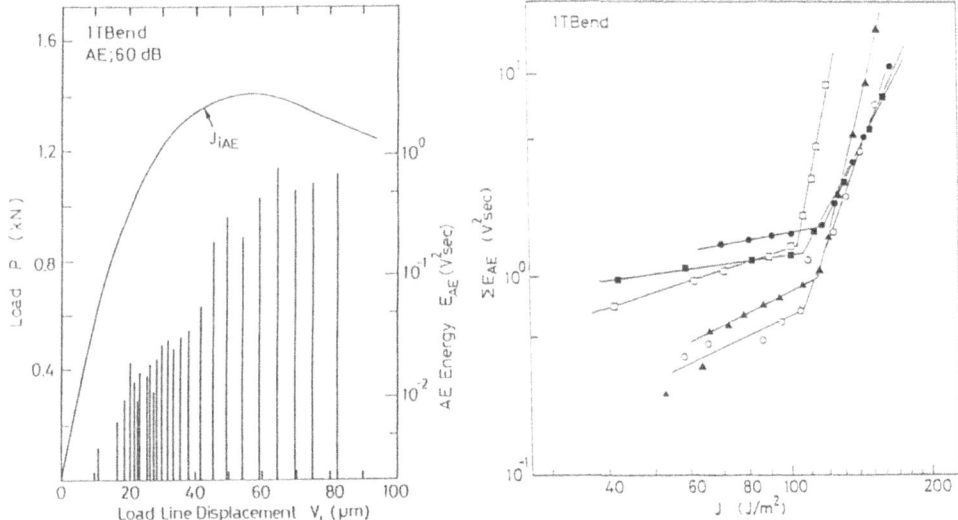

Fig. 6 P-V_L curve and AE behavior   Fig. 7 Accumulated AE energy VS J

small three point bend specimen (1TBend). The granite exhibits significant nonlinear deformation behavior well below the maximum load. The energy of the AE signal, E_{AE} for a 10 s interval at various stage of the bending test is plotted against the load-line displacement. AE signals are detected at early load level. The onset of AE occurs approximately at the load level at which the load-displacement curve becomes nonlinear. The AE activity increases with increasing load and a number of AE signals are emitted prior to the maximum load point. It can be considered from the AE behavior that the development of numberous microcracks around the initial notch tip precedes the macrocrack propagation in the rock. In order to correlate the AE behavior with the fracture process, the E_{AE} was summed with respect to load level, and the accumulated AE energy ΣE_{AE} is shown as a function of J-integral value in Fig. 7. The ΣE_{AE}-J curve can be divided into two regions. The first region, of lower slope, denotes the microcracking with little acoustic activity. The second region of the curve has a high value of slope, and consists of AE events of much higher amplitude than those detected earlier in the test.

To study the correspondence of the AE characteristics shown in Fig. 7 with the microfracture process at the crack tip spectral analyses of AE signals and microscopic observations at the notch tip were made. It is known that the measurement of the frequency spectrum of AE signals enables the identification and separation of individual sources of AE events. The spectral analyses were made on recorded AE signals and the variation in frequency content of the AE signals with respect to load level was examined. The results showed that AE signals observed during the tests can be classified into two groups, type I and II. This

318

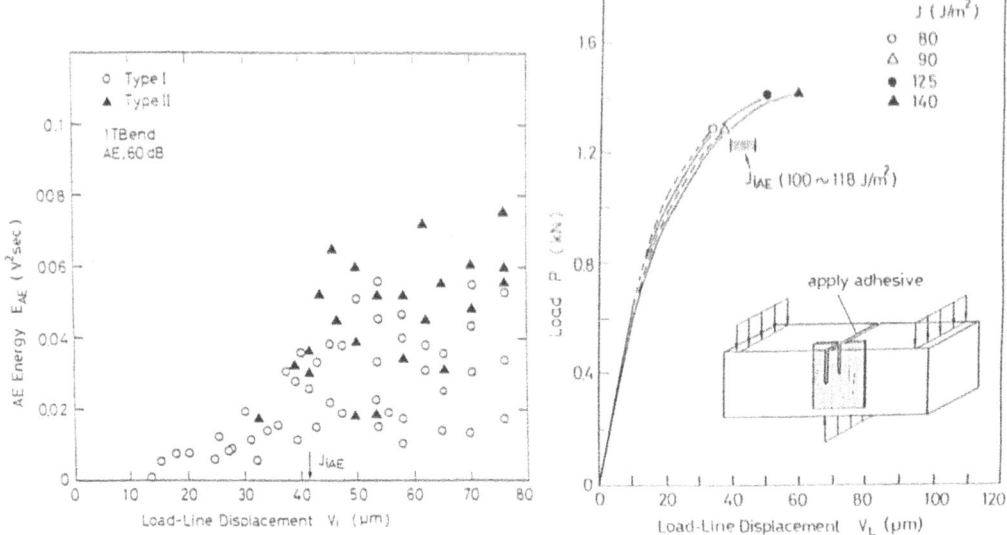

Fig. 8 Emission pattern of the classi-
         fied AE events

Fig. 9 P-V_L curves for the speci-
        men used for microscopic
        observation

classification is made on the basis of the difference in low frequency
content of AE signal. AE signals of the type II has a predominantly low
frequency content in the audible range compared to the type I signal.
Peak amplitude of the type II signal is observed to be generally larger
than that of the type I. In Fig. 8 is shown the load level at which
the classified individual AE event was detected during three-point bend
test of the specimen. AE signals of type I with small peak amplitude
are emitted at the early loading stage, and then the type II signals
start to appear as the load is further increased.   The occurrence
rates of type I and II signal are seen to be approximately the same
after the maximum load level. Similar trend in the emission behaviors
has been observed for other tests. It is noted that the abrupt increase
in the activity of type II signal corresponds to the knee point as shown
in Fig. 8.

     Microscopic observations of the notch tip region was made using
thin sections prepared as follows. Four three-point bend specimens with
identical dimensions(1TBend) were tested for this observation. As shown
in Fig. 9 two specimens were loaded to just beyond the load level which
corresponds to the J_{iAE} point, and the loads just below the J_{iAE} level
were applied to the remaining specimens. The predetermined load was
applied to the inverted specimen in such a manner that the precrack
mouth was located at the top, and maintained while adhesive
(cyanoacrylate) was injected to fix the local opening of fractured zone.
After the adhesive was cured the specimen were unloaded.   The loaded
specimen were sliced off in sections normal to the notch plane, polished
and thinned to a thickness of about 20 μm. The thin section was then
examined under a polarizing microscope. No crack initiation was
identified up to the loading stage of J = 90 J/m^2 , although some

microcracks were observed around the notch tips. However, when the load was further increased to just above the J_{iAE} point crack growth was found to occur from the notch tip throughout the specimen thickness. Based on the observations of the present study, the microfracture processes in the rock can be summarized as follows. On loading the prenotch, a few isolated microcracks are formed due to the heterogeneous nature of rock. On further loading the intensity of microcracking increases and deformation behavior around the crack tip region becomes nonlinear. Finally, macrocrack extension occurs because of the coalescence of microcracks in the nonlinear zone. The acoustic emission characteristics observed in the present experiments support the general picture of the microcrack-controlled fracture processes described above.

Let us discuss the source mechanisms of the classified AE events. The sequence in occurrence of type I and II signals suggests that the emission of type I signal is associated with the formation of microcrack. The coalescence of microcracks is expected to generate elastic stress waves having larger energy since the remaining ligament between microcracks possesses stronger linkage than the neighboring microcrack sites. Hence the intensive microcrack coalescence can be considered to produce the AE signal of type II. The critical event for macroscopic crack growth is the intensive coalescence of microcracks and linkage with the prenotch tip throughout the specimen thickness. We can say that the abrupt increase in the type II activity can pinpoint the critical stage of the notch tip region, and thereby the acoustic emission characteristics as shown in Fig. 7 is used to determine the fracture toughness evaluation point. The J-integral value corresponding to the knee point is hereater called J_{iAE}.

3-2. Effect of Concrete Strength on AE Behavior [4]

Concretes having different compressive strength have been tested in this study. The mix proportions are given in Table 2. Normal Portland Cement was used for all specimens. The maximum size of the coarse aggregate was approximately 20 mm. Water/cement ratio and

Table 2 **Mix proportions of Concretes**

sample	W/C (%)	s/a* (%)	unit weight (kg/m³)					
			water	cement	sand	gravel	silica -fume	w.r.a.** (cc)
A	60	42	188	313	721	1006	0	0
B	47	42	160	340	721	1006	60	12.1
C	25	42	111	444	721	1006	111	41.2

* s : sand, a : sand + aggregate
** w.r.a. : water reducing agent

Table 3 **Specimens dimensions of SR, CT and 3PB specimens**

specimen	loading configuration	a₀ (mm)	W (mm)	B (mm)	Bn (mm)	Ws (mm)	L (mm)	S* (mm)
SR	wedge splitting	90	170	100	60	2	-	-
SR	wedge splitting	105	170	100	60	2	-	-
CT	tension	40	100	100	60	5	-	-
CT	tension	50	100	100	60	5	-	-
CT	tension	60	100	100	60	5	-	-
3PB	bending	50	100	100	-	-	840	780

* S : span

contents of cement and silica-fume were controlled to obtain different
strengths. In order to fix the slump, furthermore, water-reducing agent
was added into high-strength concrete (B) and very high-strength
concrete (C). Each mixture was cast in the rectangular mold for 3 PB
specimen and in cylinder mold for SR specimen. All specimens were cured
together with the same condition (wet curing). The tests were carried
out in wet state on the twenty-eight days of curing time. The average
compressive strengths at this curing time were approximately 30, 60 and
95 MPa, respectively.

The geometry of the SR specimen tested is used, together with
those of two rectangular specimens; CT and 3PB. The length-to-diameter
ratio is set to be the same as that of the conventional compressive test
specimen. The specimen dimensions are listed in Table 3. An
artificial notch with width of 2 mm was machined into each specimen
using a diamond wheel saw. CT specimens were made from remaining halves
of tested 3PB specimen.

The wedge splitting method was employed for fracture tests of SR
specimens. The load was transmitted to the specimen by means of a
wedge and roller bearings. The two roller bearings are fixed to a load
box, which was placed on the top of the sample. A clip-gage was located
across the load application points, and used to measure and control
load-line displacement during tests.

AE signal were classified into eight threshold levels according to
the amplitude. AE event distributions plotted on load versus load-line
displacement curve of CT test are shown in Fig. 10. In all specimens,
the AE occurrence started below the maximum load and continued beyond
the maximum load. AE activity during the softening process shows the
clear tendency to increase with increasing strength. It can be
considered that AE signal was caused by nucleation and propagation of
microcrack.

The evidence suggests that a following AE parameter, high energy AE
ratio, Rn, can be used as a measure of microfracture resistance.

$$Rn = \frac{\Sigma N_{HI-AE}}{\Sigma N_{AE}}$$

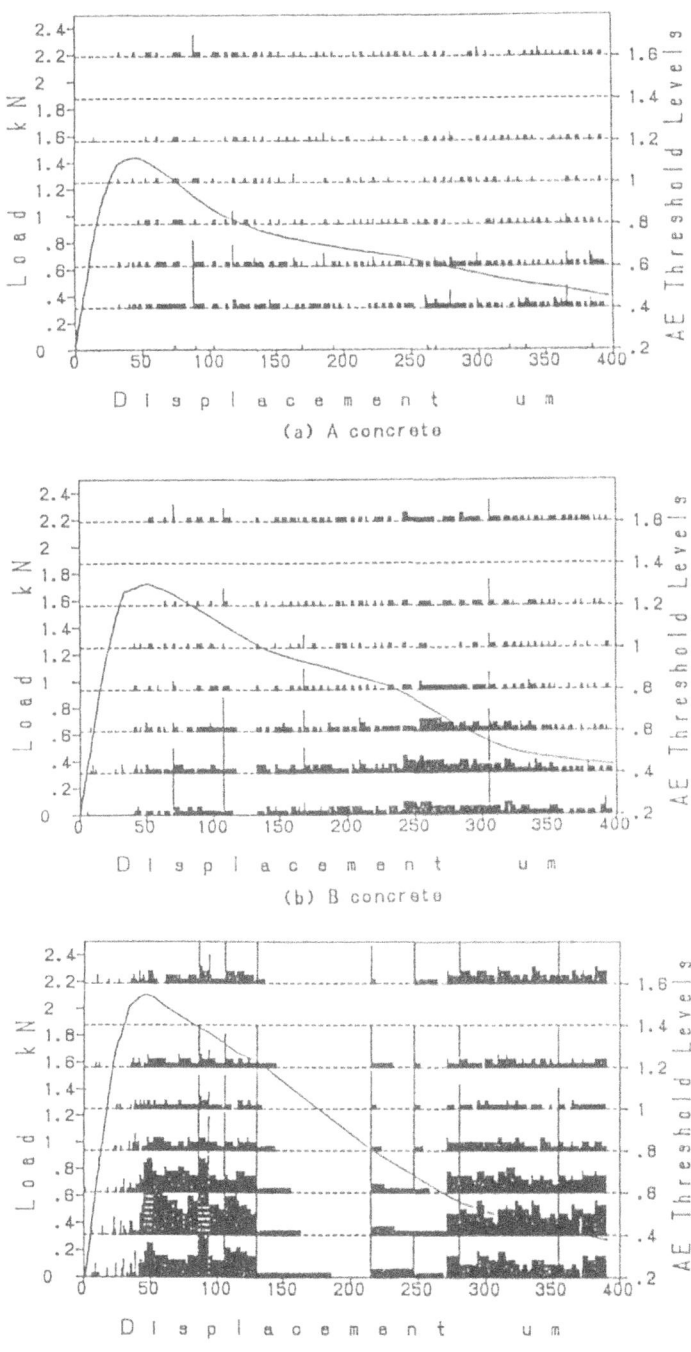

(a) A concrete

(b) B concrete

(c) C concrete

Fig. 10 AE event distributions plotted on load-displacement curve

where ΣN_{AE} is AE cumulative count and ΣN_{HI-AE} is cumulative count of AE signal with high energy. AE having energy above 1.6 V was taken as a high energy AE signal. It is considered that the frictional sliding due to crack bridging may produce high energy AE signals.

4. Application of AE to Fracture Evaluation in Advanced Concrete Composites [5]

4-1. Materials and Paper-laminate Composites

Table 4

(a) Starting materials for calcium silicate matrix

Silica	Lime	Cement	Gypsum (wt%)	Pulp fiber P/M(wt%)	Water W/M(wt%)
63	17	17	3	1 or 2	70

M=Silica+Lime+Cement+Gypsum

(b) Laminated spacing, molding process and specimen size

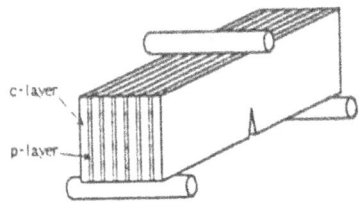

Divider

Fig. 11 **Section for fracture toughness test**

Laminated spacing (mm)	Fiber content in matrix (wt%)	Molding	Specimen size (mm)
—	0		
— 2.3 1.7 1.4	1.0	Cast	40x40x160
— 1.1 1.0 0.9	2.0	Press	20x80x160

Paper sheet laminated cementitious composite was made as shown Table 4 (a)(b).

Notched three point bending test with AE method was carried out on the divider section as shown in Fig. 11. On the press molded samples, also the crack growth resistance were measured by un-loading compliance method. The notch cut by carbon blade on the center of the beam had the tip of 0.15 mm curvature radius and one-third depth of the sample width. An AE sensor was glued near the tip of notch. A clip gage for the measurement of the displacement was attached on the shoulder of the notch.

4-2. Results.

A typical compliance curve is shown in Fig. 12. The deformation behavior at un-loading and reloading of calcium silicate/woodpulp laminates was irreversible similar as that of rocks and concretes. Therefore, the determination of compliance by 10 % unloading measurement, which established to examine the compliance of these laminates. In the case of these laminates, as the deformation at reloading had linear behavior, a gradient at unloading as shown dotted lines in Figure 12 was adopted.

The normalization load-deformation curves of all samples are shown in Fig. 13. The displacement on the shoulder of the notch (Vg) was measured. On the normalization for different sample size, the load was provided by the bending strength, and the deformation by Vg/Vgcal. Vgcal. was calculated by formulas obtained by Tada [6];

J integral values at maximum load, J_{pmax} was more than 30 times larger than that of non-laminate. It is difficult to examine the tendency of J_{pmax} values with laminated spacing because the maximum load point could not be determined as shown in P-Vg curves. Figure 14 shows the relationship between the J integral value and the crack growth length obtained by unloading compliance method on the press-molding laminates (J-R curves).

Khan et al. [7] proposed an AE rating parameter Tac by using AE events energy after the onset of the crack growth and fracture energy, and reported that Tac was an effective parameter evaluating the fracture toughness in various alloys. Tac is defined as follows;

$$T_{ac} = \frac{\Sigma E_{AE} / B}{\Delta J}$$

where ΣE_{AE} is AE cumulative energy and ΔJ is the difference of J integral value after the on-set of the crack growth. The relationship between Tac and the crack growth resistance $\Delta J/\Delta a$ obtained by un-loading

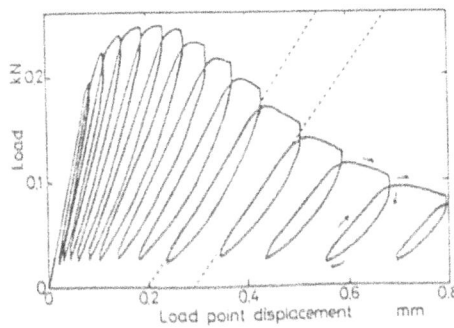

Fig. 12 Typical compliance monitoring curve

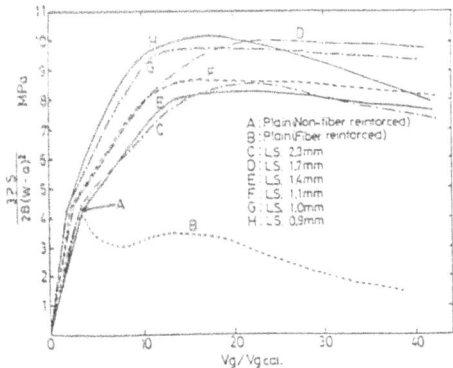

Fig. 13 Load-deformation curves

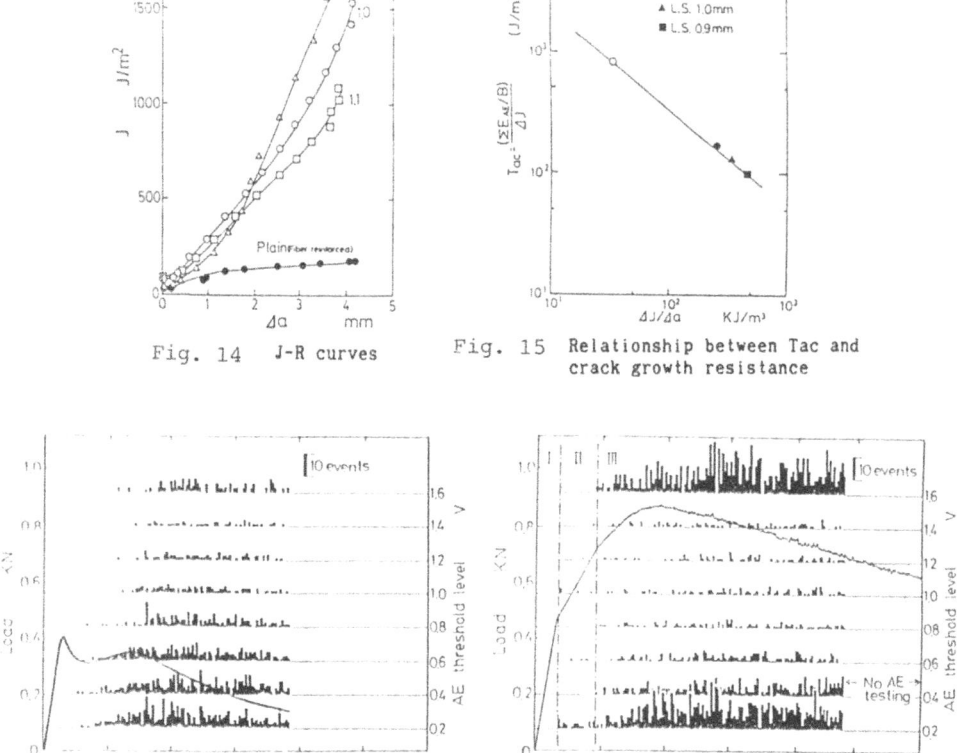

Fig. 14　J-R curves

Fig. 15　Relationship between Tac and crack growth resistance

(a)　non-laminate

(b)　laminate of 2.3mm laminated spacing

Fig. 16　AE event counts distributions on the load-deformation curves according to threshold level

compliance method (Figure 14) is shown in Figure 15. This figure shows the similar relationship to alloys. Also in calcium silicate/woodfiber comosites, therefore, the crack growth resistance can be predicted by Tac.

The AE event counts distributions plotted on the load-deformation curves according to threshold level were illustrated in Figure 16. These representative illustrations are the nonlaminate and the laminates of 2.3 mm spacing. In the case of laminates, most AE events had low energy in the Area II, while many AE events had high energy in the nonlinear behavior area (Area III). The curves had the load tremble which was caused by the pull-out and/or the failure of the fibers in the non-linear behavior. It was observed that high energy AE events occurred at every short load-drops. Therefore, it is considered that most high energy AE events were caused by the full-out and/or the failure of the reinforcing fibers.

AE parameters, high AE relative frequency R_E and R_N were defined respectively as follows;

$$R_E = \frac{\Sigma E_{1.6V-AE}}{\Sigma E_{AE}}$$

and

$$R_N = \frac{\Sigma N_{1.6V-AE}}{\Sigma N_{AE}}$$

where $\Sigma E_{1.6V-AE}$ is AE cumulative energy over 1.6 V threshold level and $\Sigma N_{1.6V-AE}$ and ΣN_{AE} are AE cumulative count over 1.6 V threshold level and AE cumulative count, respectively. The relationships between R_E and crack growth resistance and between R_N and crack growth resistance is shown in Figure 17. The relationship between R_E and crack growth resistance had large dispersion with the difference of AE event rates in low ∿ middle threshold levels, because the lower energy AE events were treated lightly in energetic evaluation. On the other hand, R_N increased with increasing of the crack growth resistance. It is suggested that R_N is an effective parameter for reinforcement effect, whether laminating or non-laminating. Figure 18 shows the relationship between R_N and the laminated spacing. R_N clearly increased with decreasing of the laminated spacing. In other words, it is suggested that the crack growth resistance increased with increasing of lamination number. From the other aspect of this figure, R_N values of the laminates asymptotically approached to R_N value of the non-laminate with increasing of laminated spacing. R_N obtained by AE measurement can be used as a measure of effectiveness of lamination reinforcement.

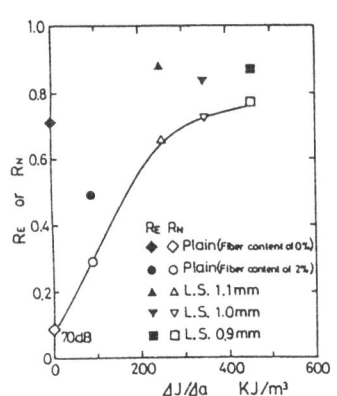

Fig. 17 Relationship between high AE relative frequency and crack growth resistance

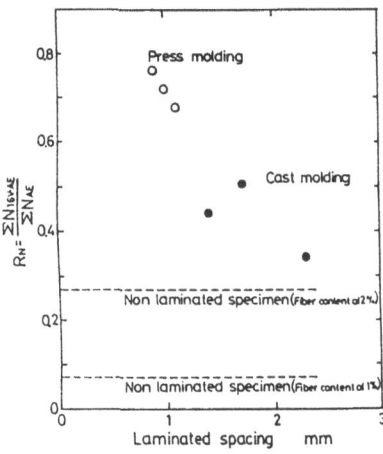

Fig. 18 Relationship between R_N and laminated spacing

326

5. Application of Ultrasomic Timing Method to Fracture Process zone On-line Monitoring in Rock [8]

Labuz et al.[9] and Swanson [10] investigated an extension behavior of the fracture process zone by use of ultrasonic attenuation method, where there exists no quantitative relationship between UT data and process zone size.

In this section the usefulness of ultrasonic timing method for monitoring the formation and extension of the fracture process zone in granite is described. The specimen geomery used is a CT type of 50 mm thickness and 125 mm width.

During the fracture toughness test, the travel time of longitudinal wave prapagation through the specimen is monitored using 500 kHZ transduser as shown in Fig. 19. Based upon an experimental calibration curve of travel time and prenotch length, the extension of fracture process zone can be estimated. Figure 20 shows a typical load-displacement curve and variation of travel time Δt, where the increase of Δt corresponds directly to the length of process zone extension (Δlp). In addition, compliance macrocrack monitoring was also made.

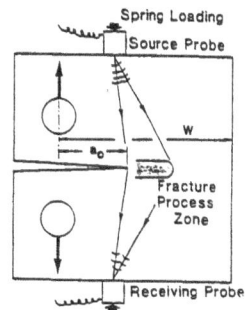

Fig. 19 Illustration of
ultrasonic tim-
ing method

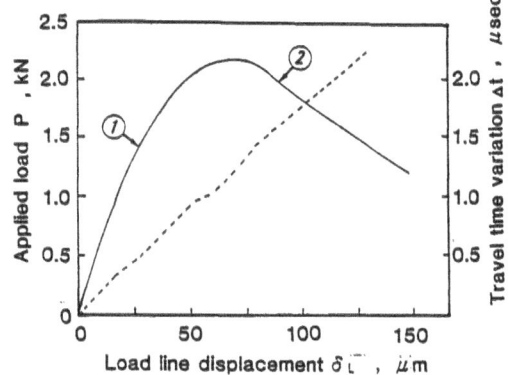

Fig. 20 Load-displacement curve
and time-difference (Δt)

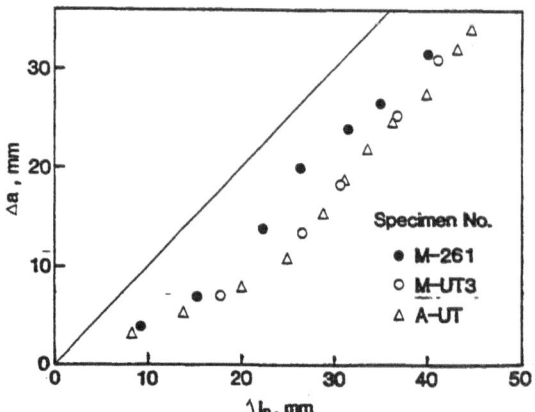

Fig. 21 Macrocrack ex-
tension (Δa) vs
process (Δlp)

Figure 21 is a relationship between fracture process zone length (Δlp) and the amount of crack growth (Δa). It is clearly shown that although there is a non-linear relation of Δlp - Δa at an initial loading stage, a linear Δlp - Δa relation exists during steady propagation stage.

Atkinson proposed an idea of "cloud" of microcracks and macrocrack extension to discuss an extension behavior of the fracture process zone quantitatively, as shown in Fig. 22, where the actual extension of physical macrocrack (Δa) and the length of fracture process zone (Δlp) are indicated.

This illustration is supported by experimental finding that there is nonlinear Δlp - Δa at the beginning of loading, whereas there exists the unique Δlp - Δa relation during the macrocrack extension, as shown already in Fig. 21.

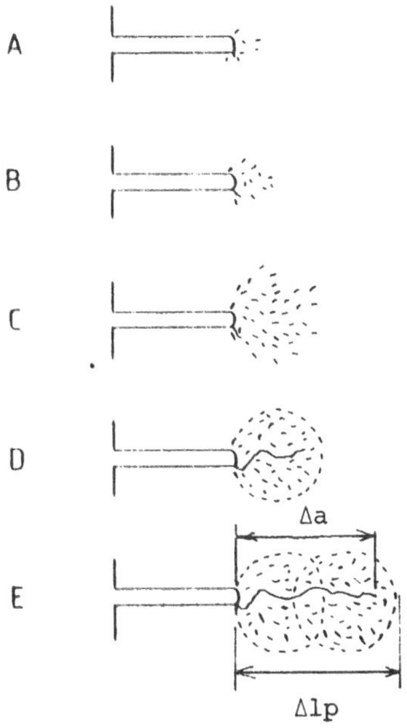

6. Concluding Remarks.

Fig. 22 Sketch of macrocrack growth and process zone extension (modification of the picture by Atkinson [11])

Although there exists several experimental techniques like AE or UT reviewed in this paper to evaluate fracture damage in brittle-microcracking materials, it is still lacking to understand a quantitative relationship between a macroscopic crack extension and microscopic cracking behavior. Extensive efforts for development of reasonable and quantitative NDE procedure to evaluate microcracking behavior in brittle materials are highly encouraged.

Acknowledgements: The author wishes to express his gratitute to Dr. T. Hashida, Tohoku University and Dr. S. Teramura, Onoda ALC Co., for their continuous and fruitful discussions.

References

[1] Hashida, T. (1989) 'Characterization of Crack-tip Strain Singularity in Brittle-microcracking Materials by Means of the Photoelastic-coating Method', Experimental Mechanics, vol.29, 307-311.

[2] Hashida, T. et al. (1982), 'Determination of Granitic Rock by Means of AE Technique' in M. Onoe, K. Yamaguchi and T. Kishi (eds.), Progress in Acoustic Emission (I), Proc. 6th International

328

AE symposium, JSNDI, Tokyo, 78-89.

[3] Hashida, T. and Takahashi, H. (1990), 'Significance of AE Crack Monitoring in Fracture Toughness Evaluation and Nonlinear Rock Fracture Mechanics', submitted to Int. J. Rock Mech. Min. Sci. & Geomech. Abstr.

[4] Teramura, S. et al. (1990) 'Development of a Core-based Testing Method for Determining Fracture Energy and Tension-softening Relation of Concrete', Proc. SEM Int. Conf. on Micromechanics of Failure of Quasi-brittle Materials, Albuquerque, (in press).

[5] Teramura, S. and Takahashi, H. (1988) 'Evaluation of Fracture Toughness on Autoclaved Calcium Silicate/Woodfiber Laminate', in K. Yamaguchi, I. Kimpara and Y. Higo (eds.), Progress in Acoustic Emission (IV), Proc. 9th International AE Symposium, JSNDI, Tokyo, 748-756.

[6] Tada, H., Paris, P. and Irwin, G. (1973) 'Stress Analysis of Cracks Handbook', Del Research Co., 2-7.

[7] Khan, M. A., et al., (1982), 'Acoustic Emission Rating Parameter for Prediction of Tearing Instability in Structural Materials', Engineering Fracture Mechanics, vol. 16, 645-658.

[8] Hashida, et al., (1990), 'Ultrasonic Monitoring of the Fracture Process Zone in Fracture Toughness Tests of Granite' (in Japanese), Trans. JSME, Series A, vol. 56, 1177-1182.

[9] Labuz, J. F., Shah, S. P. and Dowding, C. H. (1987), 'The Fracture Process Zone in Granite: Evidence and Effect', Int. J. Rock Mech. Min. Sci. & Geomech. Abstr., vol. 24, 235-246.

[10] Swanson, P. L. (1987), 'Tensile Fracture Resistance Mechanism in Brittle Polycrystals, an Ultrasonic and In-situ Microsocopy Investigation', J. Geophys. Res., vol. 92, 8015-8036.

[11] Atkinson, B. K., (1987), 'Introduction to Fracture Mechanics and its geophysical Applications', in B. K. Atkinson (ed.), Fracture Mechanics of Rock, Academic Press, London, 1-26.

SIMILARITIES BETWEEN FRACTURE PROCESSES IN CONCRETE, ROCK
AND CERAMICS: RECORDERS REPORT TO SESSION 5 'EXPERIMENTAL
METHODS TO ASSESS DAMAGE'

J.G.M. VAN MIER
Delft University of Technology
Department of Civil Engineering
Stevin Laboratory
P.O. Box 5048
2600 GA Delft
The Netherlands

ABSTRACT. In the present paper, which is based upon the discussion that took place
during session 5 "Experimental methods to assess damage" of the NATO Advanced
Research Workshop on Toughening Mechanisms in Quasi-Brittle Materials, the similarities and differences in fracture response between the various classes of quasi-brittle
materials (concrete, rock, ceramics) are discussed. Additional recent proof of crack
face bridging is presented, and it is concluded that much of the toughening can be traced
back to a particular flexural type of crack bridging. The bending of the 'intact' ligaments
between overlapping interacting cracks introduces a length scale in the material, and
justifies the introduction of so called gradient models for localization.

1. Introduction

In session 5 of the workshop three papers were presented. In the first contribution
Mindess (1990), discussed the existence and nature of the fracture process zone in
concrete. Following the early work by Hillerborg and co-workers (1976) the process
zone was defined as "*a region of discontinuous microcracking ahead of the continuous
(visible) crack*". However, following common practice in ceramics, the process zone may
either be regarded as a zone of microcracking in front of the macrocrack tip or as a
bridging zone behind the crack tip. Irrespective of which view is taken, the definition
of the *crack tip* seems of major concern in both approaches. In current mechanics
practice, as argued by Bažant, the sufficient knowledge of the fracture process zone is
the correct energy dissipation and correct stress vs. displacement relation in an *elastically equivalent* crack system. As a consequence of all these different definitions an
enormous confusion exists about the fracture process zone in concrete.

Mindess reviewed a large number of experimental techniques which can broadly be
classified in three groups: (a) surface measurement techniques, (b) techniques for

329

S. P. Shah (ed.), Toughening Mechanisms in Quasi-Brittle Materials, 329–335.
© 1991 *Kluwer Academic Publishers.*

measurement in the specimens interior and (c) indirect methods. In using either of these techniques it should be realized that the resolution of a specific method determines the amount of detail which can be detected and thus, indirectly, the conclusions that are drawn. The reported sizes of fracture process zone varried considerably, which may be traced back to the resolution of the various measuring techniques. Furthermore, quite often phenomena are observed which we *believe* to exist. It is emphasized here that some of the differences which are reported do not necessarily contradict each other. Instead the point of view should be taken that a combination of measurement techiques may lead to an improved understanding of the fracture mechanism (Van Mier 1990). Not a single 'best' technique can be identified. In the discussion the opinion was expressed that the exact size of the process zone is not important, rather the fracture mechanism either in the wake or in a zone in front of the visible macrocracks seems of major concern.

The second contribution was presented by Steinbrech on fracture of ceramics, Steinbrech et al. (1990). The relation between crack propagation, material structure (viz. grain size and phase compositions), crack and specimen geometries, pre-existing natural flaws and R-curve behaviour was discussed. It was concluded that the full potential for toughening of the microstructure is completely employed when long cracks can develop. Therefore, material development can be based on R-curve and crack propagation studies. On the other hand, the existence of small natural cracks will lead to a rapid decrease of toughening. In the discussion it was mentioned that when the initial slope of the R-curve is steep, there might still be some increased toughness for small cracks as well. Next to this some attention was given to toughening mechanisms in transforming ceramics. Some of the discussion of the paper focussed on the surface uplift in the transforming materials.

In the third paper presented by Takahashi (1990), the results of a study of toughening mechanisms in concrete, rock and paper laminate cementitious composites using different measurement techniques were presented. The fracture process in the various materials was followed using a surface technique (photoelastic coating method) and techniques suitable for exploring structural changes in the specimens interior (ultrasonic time difference method and accoustic emission rating method). In addition to this, thin sections were studied with a polarizing microscope, and it was tried to match the accoustic results with direct observations. The accoustic emission technique was used in the paper laminate composite, and the AE activity was determined as a function of the laminate spacing. During the discussion of the paper most attention was given to the paper laminate cementitious composite, especially on the application of this material. Some questions were raised regarding the physical interpretation of the ultrasonic method.

Following these three presentations, short contributions were given by Swartz on the use of dye to study the internal fracturing in concrete beams (see Swartz and Refai (1989)), by Maji on the application of accoustic emission techniques for studying internal cracking in concrete (see Ouyang et al. (1990)), by Horii who presented some theoretical results concerning the contribution of crack bridging and pre-tip microcracking (Horii & Nirmalendran (1990)) and finally by Mazars who gave some examples of the application of damage mechanics.

2. Crack Interface Grain Bridging

In ceramics the toughening is associated with crack interface grain bridging in the wake of a crack. Noteworthy are Figs. 10 and 11 in Steinbrech et al. (1990). Cracking in fine grained material starts from a notch tip with a higher initial resistance than in coarse grained Al_2O_3 but experiences almost no increase, unlike the coarse grained material. Similar observations were recently reported for cement-based composites by Van Mier (1990a,b). In Fig. 1 the relation between the size of the crack face bridges and bridging stress as determined in uniaxial displacement controlled tensile experiments using single edge notched concrete and mortar specimens is shown. It seems as if the bridging stress decreases almost linearly with the bridge size. Some care should be taken however, as the number of bridges in the fine and coarse grained material may not be the same. Furthermore, the results revealed the existence of small grain bridges in the coarse grained mix as well, and most likely bridges along a variety of size scales may exist in a cement-based composite, depending on the actual material structure, especially the grading of the aggregates.

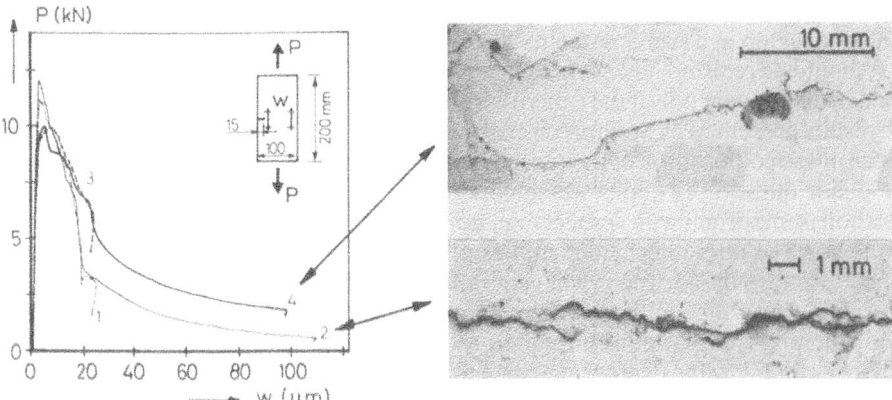

Figure 1. Crack interface grain bridging in cement-based composites. Curves no. 1 and 2 are results of experiments on 2 mm mortar, nos. 3 and 4 the results of tests on 16 mm concrete. The increased bridging stress is directly related to the size of the material bridges. After Van Mier (1990a,b).

The fracture mechanism is shown schematically in Fig. 2. Some bending of the ligament may occur, but final failure seems to proceed through extension of one of the crack tips (Fig. 2b). The type of flexural bridging justifies the introduction of higher order strain terms in continuum theories for localization. For example the use of Cosserat theory, as introduced recently by Mühlhaus & Vardoulakis (1987) and others seems to correspond with these results. However, the failure mode depicted in Fig. 2b, indicates that a further extension of such theories is needed.

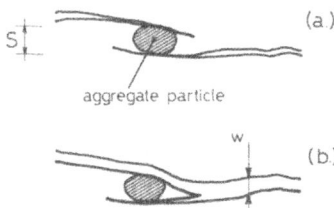

Figure 2. Interacting overlapping cracks near large aggregates (a) and hypothesized failure mechanism of the ligament (b), after Van Mier (1990b).

The tensile tests on mortar and concrete revealed also the apearance of overlapping cracks at the specimen level (Van Mier 1990b). In addition to these and the ceramics results presented by Steinbrech et al. (1990, Fig. 11), flexural bridges between interacting overlapping cracks are observed along a variety of size scales in different quasi-brittle materials such as glass (bridge size S = 20 mm), in granite (S = 25 cm), in Mancos shale (S = 250 m) and also at the ocean floor where overlapping spreading centers develop at a size S = 2.5 to 10 km (see Sempere & Macdonald 1986).

It should be realized however that the type of crack interactions descibed above (Fig. 1 and 2) may appear only if the (micro)structure of the material allows this. For example the results presented by Steinbrech et al. (1990, Fig. 11) clearly indicate that depending on the phase composition of the ceramics, toughening may decrease considerably. Thus the major question is under which conditions overlapping cracks can develop. The question is important with respect to materials engineering, which is considered as one of the goals of micromechanics studies of fracture. When for example high strength concrete is tested, specimens will generally fracture through the aggregates. The type of bridging described in the previous paragraph will then not develop, and a decrease of toughening is observed. Consequently, the increase of strength in these concretes may be nullified by the decreasing toughening. This may limit the practical applications of the high strength concrete.

The crack interface grain bridging as described above does not necessarily omit pre-tip microcracking as additional fracture mechanism. In fact the results of Van Mier (1990b) clearly showed that some internal microcracking must take place before the continuous surface cracks start to develop. In ceramics, the existence of a microcrack zone in front of the macrocrack tip was shown by Steinbrech et al. (1990) who used an X-ray scattering technique. The theoretical results obtained by Horii & Nirmalendran (1990) indicate that the contribution to crack toughening is primarily due to crack interface grain bridging and to a smaller extent to pre-tip microcracking.

It should be mentioned that crack bridging can be demonstrated also using different measurement techniques such as the double cutting technique applied by Hu & Wittmann (1989), or indirectly by using a combination of surface techniques and accoustic techniques. Yet, in all cases a combination with direct observational techniques prevails because this is the only technique which may reveal the true nature of the crack face bridging mechanism.

3. Effect of Residual Stress

One argument which was raised for explaining the fact that microcracks in front of a continuous crack tip can more easily be visualised in ceramics than in concrete was the existence of residual stresses in ceramics which would prevent microcracks from closing. However, it should be noted that as a direct result of the hydration of cement residual stresses may be present in concrete as well. Additional shrinkage of the cement paste may lead to microcracking around stiff aggregate particles as shown by various researchers. Similar effects may be expected in rocks (with thermal mismatch in their constituent minerals) which have been subjected to thermal load cycles. Nevertheless the character and level of residual stresses may be quite different in the various materials. Noteworthy is the surface uplift in transforming ceramics which seems to be a rather extreme case.

In concrete non-uniform drying may cause tensile eigenstresses at the specimens surfaces. As a result, crack extension due to mechanical loading may proceed more easily at the specimens surface. Impregnation techniques have indeed shown that crack fronts in concrete and mortar are curved, with a larger crack extension near the surfaces (see Bascoul et al. (1989), Swartz & Refai (1989) and Van Mier (1990a)). Similar effects may be caused by differential temperatures through a specimens cross-section. Clearly the extent of surface cracks is not a measure of the internal cracking of a specimen. In response to this the argument can be raised that if the crack front's shape is known, surface crack measurements can be used for establishing damage in specimens.

4. Size Effects

During the workshop several presentations dealt with the size effect in concrete (e.g. Bažant (1990)). If crack interface grain bridging is accepted in ceramics and concrete, and likely also in rock, non-statistical size effects will probably also occur in the other materials (ceramics and rock). Surprisingly not much attention was given during the workshop to the size effect in either rock of ceramics. On the other hand if the micromechanics of failure are better understood, the size effect will be dealt with implicitly. The impression is that micromechanical studies, directed towards a fundamental knowledge of fracture behaviour is more common in the fields of rocks and ceramics.

5. Concluding Remarks

Crack interface grain bridging in the wake of a continuous crack seems the major mechanism of load transfer in cracks in quasi-brittle materials such as concrete, ceramics and rock. The overlapping cracks are found at a variety of size levels, and acceptance of the mechanism implies that non-statistical size effects cannot be refused in the various materials. In assessing the micromechanisms of failure different experimental methods should be combined. No single technique seems capable of revealing all details of the fracture process. In all circumstances it should be realized that the resolution of the experimental technique should correspond with the size of the phenomenon investigated. This is often not recognized and may lead to the search of a 'holy grail'.

References

Bascoul, A., Kharchi, F. and Maso, J.C. (1989) 'Concerning the measurement of the fracture energy of a micro-concrete according to the crack growth in a three point bending test on notched beams', in S.P. Shah and S.E. Swartz (eds.), *Fracture of Concrete and Rock*, Springer Verlag, New York, 396-408.

Bažant, Z.P. (1991) 'Rate effect, size effect and non-local concepts for fracture of concrete and other quasi-brittle materials', this volume.

Hillerborg, A., Petersson, P.E. and Modeer, M. (1976) 'Analysis of crack formation and crack growth in concrete by means of fracture mechanics and finite elements', *Cement and Concrete Research*, 6(6), 773-782.

Horii, H. and Nirmalendran, S. (1990) 'Roles of microcracking and bridging in fracture of quasi-brittle materials', in S.P. Shah, S.E. Swartz and M.L. Wang (eds.), *Micromechanics of Failure of Quasi-Brittle Materials*, Elsevier Applied Science Publishers, London/New York, 569-578.

Hu, X.Z. and Wittmann, F.H. (1989) 'Fracture process zone and K_r-curve of hardened cement paste and mortar', in S.P. Shah, S.E. Swartz and B.Barr (eds.), *Fracture of Concrete and Rock - Recent Developments*, Elsevier Applied Science Publishers, London/New York, 307-316.

van Mier, J.G.M. (1990a) 'Internal crack detection in single edge notched concrete plates subjected to uniform boundary displacement',in S.P. Shah, S.E. Swartz and M.L. Wang (eds.), *Micromechanics of Failure of Quasi-Brittle Materials*, Elsevier Applied Science Publishers, London/New York, 33-42.

van Mier, J.G.M. (1990b) 'Mode I fracture of concrete: discontinuous crack growth and crack interface grain bridging', *Cement and Concrete Research* (in print).

Mindess, S. (1991) 'The fracture process zone in concrete', this volume.

Mühlhaus, H.-B. and Vardoulakis, I. (1987) 'The thickness of shear bands in granular materials', *Géotechnique*, 37, 271-283.

Ouyang, C., Maji, A. and Shah, S.P. (1990) 'Damage evaluation of mortars based on accoustic emission', in S.P. Shah, S.E. Swartz and M.L. Wang (eds.), *Micromechanics of Failure of Quasi-Brittle Materials*, Elsevier Applied Science Publishers, London/New York, 600-609.

Sempere, J.-C. and Macdonald, K.C. (1986) 'Overlapping spreading centers: implications from crack growth simulation by the displacement discontinuity method', *Tectonics*, 5, 151-163.

Steinbrech, R.W., Dickerson, R.M. and Kleist, G. (1991), 'Characterization of the fracture behavior of ceramics through analysis of crack propagation studies', this volume.

Swartz, S.E. and Refai, T. (1989) 'Cracked surface revealed by dye and its utility in determining fracture parameters', in H. Mihashi, H. Takahashi and F.H. Wittmann (eds.), *Fracture Toughness and Fracture Energy*, Balkema, Rotterdam, 509-520.

Takahashi, H. (1991) 'A review of experimental methods to assess damage during fracture of rock, concrete and reinforced composites', this volume.

Theoretical Micromechanics Based Models

A review of some theories of Toughening mechanisms in Quasi-Brittle materials

C. Atkinson
Department of Mathematics,
Imperial College,
London

ABSTRACT. Various theories of toughening mechanisms in quasi-brittle materials are reviewed and a common theme identified. Essentially each of them uses a partially compressively loaded flat crack as a means of accounting for the toughening effect, the partial loading being modelled by linear or nonlinear springs. Micromechanical models are invoked to determine the spring characteristics. Models which combine energy dissipation via phase transformation in addition to internally induced stress fields are also briefly discussed.

1   Introduction

To properly account for fracture processes in quasi-brittle materials such as ceramics, rocks or fibre reinforced composites consideration must be given in the first place to the propagation of discrete cracks. However these materials have many microcracks either inherent in the microstructure or created by the inhomogeneous nature of the induced stresses of applied or residual loads produced by fabrication, joining or wear. Thus subsequent steps in the analysis must account for microcrack arrays and statistical methods have been applied to this and other probabilistic aspects of the fracture problem. Although it is important to transfer from analysis of discrete cracks to arrays of cracks, including important interaction effects, we will limit attention here to a single macroscopic crack and discuss various toughening mechanisms in terms of it.

Various toughening possibilities have been suggested which include

(a) Martensitic toughening processes (b) Controlled micro-fracture in a crack tip process zone (c) Ceramic-metallic systems (e.g. cermets such as tungsten carbide cobalt) and (d) Fibre toughening. Although the intrinsic toughness of most ceramics is of the order of 2 to 3 MPa$\sqrt{m}$, the maximum toughness provided by the

339

S. P. Shah (ed.), Toughening Mechanisms in Quasi-Brittle Materials, 339–353.

above mechanisms can be as much as four or five times the intrinsic toughness.

Cases (b), (c) and (d) above have usually been discussed in terms of a mechanism which reduces the effective load seen by the crack. Thus for case (c) the metallic inclusions are assumed to toughen the ceramic by a crack bridging mechanism in which the faces of the advancing crack are pinned together by intact particles for some distance behind the crack tip. These particles provide an additional resistance to the crack opening at the tip hence reducing the effective stress intensity factor of the crack. In case (d) as a crack advances through a fibre composite fibres bridge the crack tip again causing a resistance to the crack opening. For short cracks the fibres may span the whole crack. Each of these situations can be modelled as a crack which in addition to the external applied stress has a compressive internal stress acting on the crack faces which is some function of the crack opening. This stress acts to oppose the external stress and hence reduces the effective crack tip stress intensity factor or an equivalent energy release rate. Such models which have received a fair bit of attention in the last few years have a sentimental interest for the writer because of a review in Applied mechanics reviews (1971) of the paper Atkinson (1970) . The paper considered a penny shaped crack in which the faces of the crack were acted upon by a normal traction which was an arbitrary specified function of the crack opening displacement and subsequently derived a way of solving it by an iterative numerical method. The motivation came from the fibre composites work going on at N.P.L. at the time (1968 when the work was done) but the method wasn't restricted to any particular model and I said so. The review said " Author claims no compelling physical motivation which led him to consider this problem and indeed reviewer fails to see any problem wherein this analysis could be meaningfully applied ". It is a pity that we no longer have published named reviews of papers, Applied mechanics reviews now only publishes abstracts I believe.

Case (a) above, martensitic toughening , is somewhat different to the other cases since here the increased fracture toughness has two fairly distinct ingredients. First there is the work supplied to effect the martensitic transformation which reduces the energy available to produce fracture hence causing an increase in toughness and secondly there is the effect of the transformed inclusions as sources of internal stress. These internal stresses may increase or decrease the effective loads seen by a crack depending on the particular distribution of internal stresses produced. Recently some models of the interaction of these two effects have been discussed .

It is of course a well known result, for a plane crack with the usual definition of

stress intensity factor ($\sigma \sim K/\sqrt{(2r\pi)}$) , that for a non-uniform loading σ^{app} the stress intensity factor can be represented by

$$K = (a\pi)^{-.5} \int_{-a}^{a} (a + x_1)^{.5}(a - x_1)^{-.5}\sigma^{app}(x_1)dx_1 \qquad (1)$$

With this expression and a suitable model leading to the non-uniform loading seen by the crack a number of situations can be considered.Thus the well known Dugdale model has the non-uniform loading as a small "process" zone at the crack tip in which there is a fixed compressive stress resulting from plastic flow in addition to the tensile influence of the applied stress. (We are assuming here that the remote stresses have been subtracted out of the problem so the above formula applies , one can then think of the applied tensile stresses as being equivalent to an appropriate internal pressure acting on the crack faces.) The model then requires that the net stress intensity factor produced by these competing stresses is zero and hence a relation between the extent of a plastic zone to the crack length is obtained. Similar models have been invoked to describe craze formation. For the more sophisticated models considered here the effective stress seen by the crack is not independent of the shape of the crack,thus in general one has to solve a (possibly) non-linear integral equation to determine the crack shape. Models of this kind will be discussed in section 2 where we pay most attention to the problem of a plane crack in plane strain or stress although the case of a penny shaped crack will be considered briefly.

2   Crack bridging by fibres, particulate reinforced ceramics, line spring models.

The cases considered here (b), (c) and (d) of the introduction have in common the fact that the crack can be modelled as reinforced internally by some mechanism which results in a resistance to the crack opening which is some known function of the opening itself. Thus the crack can be imagined as being supported internally by springs which resist its opening in either a linear or non-linear manner and over an extent of the crack which is localised near the tip or extends over the whole crack. It is , of course, essential to determine the particular relationship between the crack opening and the internal stress acting on the crack from a detailed model of the microstructural processes which are active for particular composites. This will be considered in the next section, here attention is given to the stress analysis of the internally loaded crack.

2.1   Mathematical statement of the discrete crack problem.

(A) THE PENNY SHAPED CRACK.

If the crack is assumed to have a penny shaped planform and the fibres or particles bridge the crack normal to it then using cylindrical polar coordinates (ρ, θ, z) with the crack lying on $z = 0, 0 < \rho < a$ one can write an integral over the crack faces

342

as (Collins (1962))

$$u(\rho) = \frac{2(1 - \nu)}{\pi \mu} \int_\rho^a \frac{dt}{(t^2 - \rho^2)^{.5}} \int_0^t \frac{w\sigma(w)dw}{(t^2 - w^2)^{.5}} \tag{2}$$

this being valid for $0 < \rho < a$. Changing the order of integration we can write this in the form

$$u(\rho) = \frac{2(1 - \nu)}{\pi \mu} (\int_0^\rho w\sigma(w)I_1(w, \rho)dw + \int_\rho^a w\sigma(w)I_2(w, \rho)dw) \tag{3}$$

where

$$I_1(w, \rho) = \int_\rho^a dt / ((t^2 - \rho^2)^{.5}(t^2 - w^2)^{.5}) \tag{4}$$

and

$$I_2(w, \rho) = \int_w^a dt / ((t^2 - \rho^2)^{.5}(t^2 - w^2)^{.5}) \tag{5}$$

In the above integrals u is the half opening of the crack of radius a, and σ is the normal stress acting internally on the crack faces. Thus σ positive means the crack opening up under a given pressure σ negative denotes the resistance to the crack opening due to the micromechanics. The above forms are suitable for the determination of u or equivalently σ once the relationship between u and σ has been determined from the micromechanics. Once σ on the crack has been determined the stress intensity factor follows as some weighted integral over the crack faces as discussed in the introduction (equation (1)).The corresponding formula for the penny shaped crack can be written as

$$K = 2(a\pi)^{-.5} \int_0^a \frac{\sigma(\rho)\rho d\rho}{(a^2 - \rho^2)^{.5}} \tag{6}$$

As stated above this formulation was originally given by Atkinson (1970), a similar formulation has been presented more recently by Marshall, Cox and Evans (1985) to describe the mechanics of matrix cracking in brittle-matrix fiber composites. The expression they derive for the expression of $\sigma(u)$ has a resistance to the crack opening which is proportional to $\sqrt{u}$. We will return in the next section to the details of their model we merely note that the above equations together with the expression for $\sigma(u)$ enable the calculation of the stress intensity factor of the crack in the composite to be determined and comparing this with the corresponding stress intensity factor of a pure matrix crack or a crack in an assumed composite medium enables the toughening effect to be determined.

(B) THE PLANE CRACK.

As in the penny shaped crack case considered above the crack opening at a given position is determined by the entire distribution of surface tractions and can be written as

$$u(x) = \frac{2(1-\nu)}{\pi\mu}(\int_x^a \frac{t\,dt}{(t^2-x^2)^{.5}} \int_0^t \frac{\sigma(w)dw}{(t^2-w^2)^{.5}})$$ (7)

Changing the order of integration enables a single integral expression to be obtained with a weak logarithmic singularity i. e.

$$u(x) = \frac{2(1-\nu)}{\pi\mu} \int_0^a \sigma(w)dw \log(\frac{(a^2-x^2)^{.5}+(a^2-w^2)^{.5}}{\sqrt{(\,|\,(x^2-w^2)\,|\,)}})$$ (8)

Here it has been assumed that the stresses on the crack are symmetric about its centre. An integral equation similar to this is used by Budiansky , Amazigo and Evans (1988) to analyse a model of small scale crack bridging in particulate reinforced ceramics.Essentially the model considers various relationships between u and σ and for each relationship there results an integral equation to be solved. A number of authors have considered some aspects of this problem e. g. Walton and Weitsman (1984) gave an analytical and a numerical treatment of the problem when the resistance σ(u) was assumed to be linear in u as a model of a craze tip zone also called a "mackeral region". Also Rose (1987) has given a fairly extensive treatment of the linear spring model and given useful asymptotic relations. As part of a "soft" contact theory in colloid science Hughes and White (1979) give an account of some non-linear models. In addition to these fairly extensive treatments it is possible to get some simple results using simple physical arguments and path independent integrals .

2.2   Simple qualitative results

Perhaps the simplest system from which to derive qualitative results is that where a plane crack has a bridged region localised at the crack tip and this region is small compared to the crack length (figure. 1). Since in this case the bridging zone length ℓ is such that $\ell/a \ll 1$ the stress field viewed on a length scale such that a is of order one sees to a first approximation the crack without the bridging zone. The stresses near the crack tip would then have the distribution

$$\sigma_{\alpha\beta} \approx K f_{\alpha\beta}(\theta)/(2r\pi)^{.5}$$ (9)

344

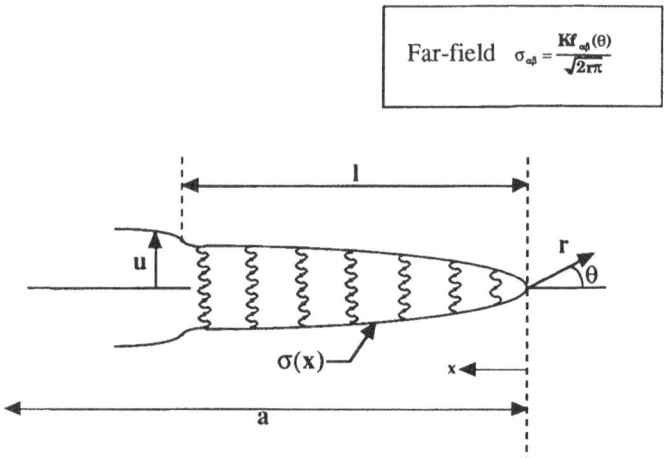

Figure 1: Spring model

where K is the stress intensity factor due to the applied loading alone on an unbridged crack and the functions $f_{\alpha\beta}$ define the well known θ dependence of the crack tip stress field. A standard procedure of fracture mechanicsis to consider the problem with the bridging zone in a new coordinatesystem scaled on the zone length ℓ . In this coordinate systemnin the limit $\ell/a \to 0$ the crack plus bridged zone appears to be semi-infinite with the stress field (9) as the boundary condition at infinity. This near crack stress distribution can be expressed as

$$\sigma_{\alpha\beta} \approx K_m f_{\alpha\beta}(\theta)/(2r\pi)^{.5} \tag{10}$$

where K_m is now the stress intensity factor of the complete bridged crack problem in the presence of the applied stresses. A relation between K_m , K the "applied" stress intensity factor (i.e. with no bridged zone) and the spring characteristics can be obtained via the J integral (Rice (1968) , Eshelby (1951)) . This can be written as

$$J = \oint (W n_1 - \sigma_{\alpha\beta} u_{\alpha,1} n_\beta) ds \tag{11}$$

taken around the path shown in figure 2 . (Here W is the strain -energy density and u_α components of the displacement vector). The following result is obtained

$$(1 - \nu^2)K^2/E = (1 - \nu^2)K_m^2/E + \int_0^\ell \sigma u_{,1} \, dx \tag{12}$$

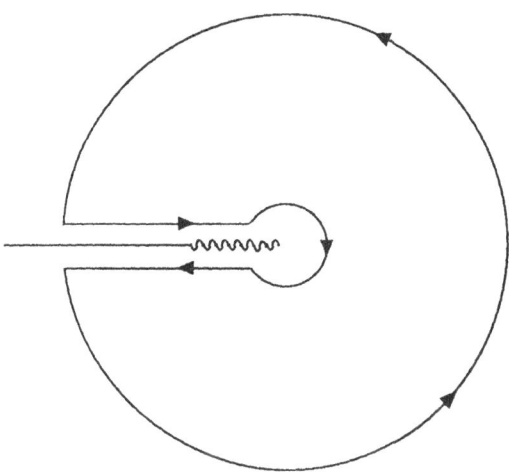

Figure 2: Path for integral

The result on the left of the equation comes from the integral around the large circle which picks up the stress field of (8) , the outer solution, and hence gives the energy release rate (per unit crack advance) of the unbridged crack. The first term on the right of the equation is from the integral around a small contour at the tip giving the bridged crack energy release rate . Further since the crack surfaces are unloaded except in the bridged region where the only non zero stress for mode one deformation is σ_{22} (σ in our notation) one gets the second expression on the right of (12) .

(1) LINEAR SPRINGS

In the special case when the bridged region is modelled by linear springs and the spring stress σ is written as

$$\sigma = kEu/(1 - \nu^2) \tag{13}$$

in terms of u (the normal crack tip displacement for mode 1 deformation), Youngs modulus E , Poissons ratio ν and a spring stiffness k, equation (12) reduces to

$$(1 - \nu^2)K^2/E = (1 - \nu^2)K_m^2/E + (1 - \nu^2)\sigma^2(\ell)/(kE) \tag{14}$$

where $\sigma(\ell)$ is the spring stress at the edge (furthest from the crack tip) of the bridged zone (Rose (1987) , Budiansky et al (1988)). An implicit assumption has been made here that $K > K_m > 0$ and that the bridged region has opened up with $u > 0$. With this proviso equation (14) provides a relation for the " toughening ratio " $\lambda = K/K_m$ (Budiansky et al (1988)) with the following interpretation . It is supposed that K_m represents the critical stress intensity factor for crack growth in the matrix and that new springs connecting the crack faces emanate from the crack tip whenever the crack propagates i.e. the crack runs into a self similar bridging zone . With the peak spring stress $\sigma(\ell)$ set equal to the spring breaking strength σ_Y crack propagation with simultaneous fracture of the last spring occurs for

$$\lambda = [1 + (\sigma_Y^2/kK_m^{-2})]^{1/2} \tag{15}$$

Of course equation (15) is nothing more than a reformulation of (14) since we do not know K_m without solving the integral equation (8) even though k and σ_Y would be specified from the micromechanics. Various expressions are given by Budiansky et al (1988) and Rose (1987) including full numerical solution of the integral equation. A useful alternative is to rewrite equation (14) in the form

$$\lambda = [1 - (\sigma_Y^2/kK^2)]^{-1/2} \tag{16}$$

since K, the applied stress intensity factor, is easily deduced from the energy input provided from the applied stress field. It is also readily available for a given crack length and loading geometry from handbooks of stress intensity factors. The expression (16) does not require solution of an integral equation it is merely a consequence of the energy balance argument and the assumption that $\ell/a << 1$. This is analogous to the situation with the Dugdale model where the condition of no net stress intensity factor enables the length of the plastic zone to be removed from a crack opening displacement growth condition in terms of the ratio of applied load to yield stress. However for comparison with experimental observations of bridging zone lengths relations have been derived which involve this length. Thus Budiansky et al (1988) and Rose (1987) give relations between λ and α defined by

$$\alpha = 4k\ell/\pi \tag{17}$$

which are derived from numerical solution of the integral equation (8). Simple asymptotic solutions which are useful are

$$\lambda \approx (1 + \alpha) \tag{18}$$

for α tending to zero, and

$$\lambda \approx \pi(\alpha/2)^{.5} \tag{19}$$

for α tending to infinity. The above results equations (13) to (19) are all for the linear spring model whereas equation (12) applied to any stress displacement law acting in the bridged zone. Thus for application to particulate toughening when plastic yielding of the particles is important other relationships are neccessary, these are considered by the above mentioned authors and outlined below.

(2) ELASTIC PLASTIC SPRINGS

If for increasing u, the springs obey the elastic ideally plastic constitutive law

$$\sigma = kEu/(1 - \nu^2),$$

for

$$u < u_Y$$

where

$$u_Y \equiv \sigma_Y(1 - \nu^2)/kE$$

and

$$\sigma = \sigma_Y,$$

for

$$u \geq u_Y$$

Then if $u(\ell)$ at the end of the bridged zone exceeds u_Y the result (12) gives

$$(1 - \nu^2)K^2/E = (1 - \nu^2)K_m^2/E + (1 - \nu^2)\sigma_Y^2/(kE) + 2\sigma_Y[u(\ell) - u_Y] \qquad (20)$$

If failure of the last spring is now assumed to occur when $[u(\ell) - u_Y]$ attains a critical plastic value u_P, then the toughening ratio becomes

$$\lambda = [1 + \sigma_Y^2(1 + 2u_P/u_Y)/(kK_m^2)]^{1/2} \qquad (21)$$

Alternatively writing this in terms of K gives

$$\lambda = [1 - \sigma_Y^2(1 + 2u_P/u_Y)/(kK^2)]^{-1/2} \qquad (22)$$

Results from the numerical solution of integral equations which give the lengths of the zones where plastic yielding occurs are given in Budiansky et al (1988) .

(3) RIGID PLASTIC SPRINGS

This case follows from the above result when u_P is considered very large the elastic contribution to the spring energy is then ignored and (20) reduces to

$$(1 - \nu^2)K^2/E = (1 - \nu^2)K_m^2/E + 2\sigma_Y u_P \qquad (23)$$

The toughening ratio is now

$$\lambda = [1 + 2E\sigma_Y u_P/(K_m^2(1 - \nu^2))]^{1/2} \qquad (24)$$

or equivalently

$$\lambda = [1 - 2E\sigma_Y u_P/(K_m^2(1 - \nu^2))]^{-1/2} \qquad (25)$$

For this case a simple relation can be presented which gives λ in terms of the bridge length ℓ as

$$\lambda = 1 + 2(2/\pi)^{1/2}\sigma_Y L^{1/2}/K_m \qquad (26)$$

All of the above results are for the case when the zone length is small compared to the crack length. For situations where the zone extends over the whole crack length such as occurs for short cracks in composites the integral equation formulation is still applicable.

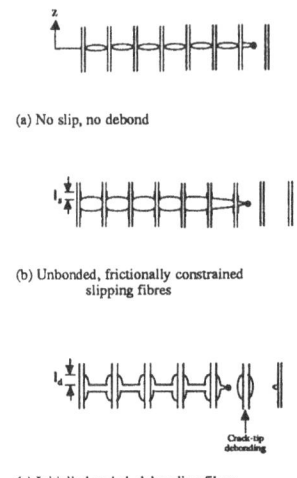

(a) No slip, no debond

(b) Unbonded, frictionally constrained
slipping fibres

Crack-tip
debonding

(c) Initially bonded, debonding fibres

Figure 3: Steady state matrix cracking

3 Micromechanical models leading to "spring" properties.

We give a rather brief survey of some of the models which have been used to produce stress displacement relations for the region modelled by springs. As we have indicated earlier such considerations are ,of course, an essential part of the models. The basis of these models is a more detailed accounting of the deformation processes on the scale of individual fibres or particles and the transfer of this information to concentrations of such fibres or particles. For example in the case of fibre composites models of fibre pull out are used to give a relationship between σ and u for use in the spring model. Similarly for crack bridging by particles a model relating the average crack opening produced by the particle pinning to the average energy released on a microcrack level can be used to relate average σ values to u values for use together with the spring models for the main macroscopic crack analysis. We give a cursory account of some recent models below.

3.1 Brittle matrix fiber composites, fibre pullout

The nature of the debonding process between an individual fibre and the matrix can be quite complicated. For example experiments on pulling out a glass rod from a polyeurethane matrix and observing the interface fracture process (e.g. Atkinson et al (1981)) showed a debond to occur initially either at the stress singularity where the rod met the free surface or at a preexisting microscopic flaw at the interface. This latter situation usually occured at the rod base and as the rod was pulled the defect propagated around the interface finally stopping as it met the compressive stresses seen by the rod sides. In this paper the details of the observed failure process was explained in terms of debond fracture mechanics and the induced stress fields. In principle the more detailed the knowledge of this process the more meaningful the application to the spring models . For what it is worth we mention here two theoretical papers (Atkinson and Avila (1982,1983)) that may be of interest to those involved with the detailed stress analysis of the fibre problem . In these detailed calculations we take into account the influence of the singular stress fields of the fibre ends on subsequent crack initiation and propagation. Figure 3 illustrates what might happen to a matrix crack as it proceeds across a fibre composite. If there is enough frictional resisrance no slip will occur at the fibre matrix interface (cf. figure 3 (a)). When slip does occur the situation shown in figure 3 (b) may occur, this is the situation considered below where an equivalent spring model is derived. We consider a simple model of this here see e.g. (Marshall et al (1985)) . The main assertion is that the application of tractions T to the end of a fibre cause sliding between the matrix and fibre over a distance ℓ_1 say and allows the fibre to pull out of the matrix a distance u. Such behaviour does of course depend on the nature of the matrix bond. For a purely frictional bond the sliding distance is determined by the length over which the interface shear stresses exceed the friction stress τ. Marshall et al (1985) deduce the following approximate relation between T and u.

$$T = 2[uE_f\tau(1+\eta)/R]^{1/2} \tag{27}$$

where $\eta = E_fV_f/E_mV_m$ with E_f and E_m the modulus of fibres and matrix respectively and V_f ,V_m the volume fractions. πR^2 is the fibre cross sectional area. It remains to relate T to σ acting in thespring model. This is done by writing

$$\sigma = TV_f \tag{28}$$

Thus from this model σ is proportional to the square root of u.

The application of the spring model with the above value of σ leads, via solution of the integral equation , to a value for K^L the stress intensity factor of the loaded spring crack model i.e. including both the effect of applied load and the springs. This intensity factor thus characterises the composite stress and strain fields in the region immediately ahead of the matrix crack. Because in this region no relative displacements between fibres and matrix are permitted the matrix and fibre strains must be compatible. Hence

$$\sigma^M/E_m = \sigma/E_c \tag{29}$$

where σ^M is the matrix stress, σ is the composite stress and E_c is the composite modulus

$$E_c = E_mV_m + E_fV_f \tag{30}$$

The matrix and composite stress intensities scale with the stresses such that

$$K^L = K^M E_c/E_m \tag{31}$$

where K^M is the stress intensity in the matrix. The condition for equilibrium crack growth (in the absence of environmental effects) is given by setting K^M equal to the critical stress intensity factor, K_c^M for the matrix. The criterion for crack growth can then be expressed in terms of K^L as

$$K^L = K_c^L = K_c^M E_c/E_m \tag{32}$$

This model has been further developed recently by Budiansky et al (1986) and Budiansky and Amazigo (1989) and some simpler energy balance arguments such as discussed in the last section apply to this situation also. These authors thus attempt to generalise the earlier model of Aveston et al (1971).

3.2 Particulate toughening

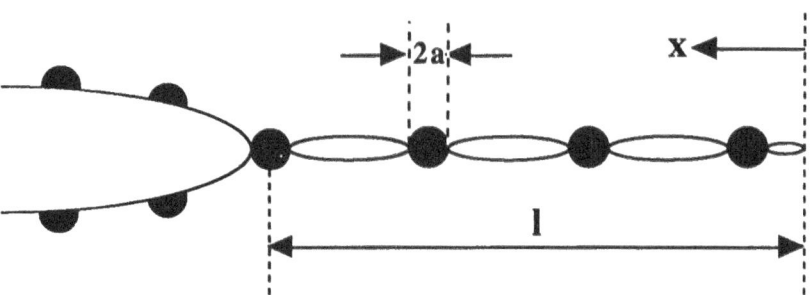

Figure 4: Particulate reinforcement

Budiansky et al (1988) assume spherical particles and suppose that the faces of an advancing plane crack are pinned by particles at their equators. Such an assumption interpreted strictly would require that the matrix area on the bridged zone were reduced by a fraction that exceeds the volume concentration c of particles. As an approximation they assume that the matrix area concentration on each crack face has the value (1-c) which corresponds to an arbitrary plane cross section of the composite material. To apply the spring models of section 2 the spring stress σ at a given location on the crack plane is identified with smeared out particle stresses $c\sigma_p$ where σ_p is the average stress at that location i.e. the average over all particles in the thickness direction perpindicular to the plane of figure 4 for the plane crack model. The following estimate for the spring constant k is given by Budiansky et al (1988), they write

$$k = (\frac{2c}{a\pi\beta})(\frac{E_m}{1 - \nu_m^2})(\frac{1 - \nu^2}{E}) \qquad (33)$$

where E and ν are the effective elastic constants of the composite material consisting of the ceramic matrix containing a randomly distributed concentration c of particles. a is the particle radius and $\beta(c)$ is given approximately by the formula

$$\beta \approx (1 - c)(1 - c^{.5}) \qquad (34)$$

The equation (14) denoting the energy balance is modified to

$$(1 - \nu^2)K^2/E = (1 - c)(1 - \nu_m^2)K_m^2/E_m + (1 - \nu^2)\sigma^2(\ell)/(kE) \qquad (35)$$

Here they argue that since the first term on the right denotes energy release into the real crack tip and this tip advances only into matrix material then E and ν should be replaced by E_m and ν_m and further this term should be multiplied by (1-c) to take into the fact that the bridging particles reduce the the length of the advancing crack front by a factor c. The reader is referred to the original paper for details of comparisons with experiment. The observation is also made by these authors that to be aneffective toughening mechanism a matrix crack should tend· to be attracted to the particles and a neccessary condition for this might be that the elastic stiffness of the particles should be less than that of the matrix. Such an effect occurs when one considers the interaction of a plane crack and a circular inclusion of different elastic constants see e.g. Atkinson (1971).

4   Concluding remarks

In this rather brief account of some current theories of toughening we have not treated case (a) of the introduction in any detail .We merely note that recently Budiansky and Amazigo (1988) have discussed the role of both ductile crack bridging particles and phase transforming particles witha view to the possible enhancement of toughness by these two effects. Rose (1985) has also discussed the transformation toughening problem. It is clear of course that a transforming particle induces an internal stress field in the matrix which will influence crack growth. Precisely how it does this requires the interaction of the spring model and the inclusion problem and a definitive treatment of this perhaps still remains to be done. Although Budiansky and Amazigo (1988) claim to show a synergistic effect in certain parameter ranges. Another interesting paper we have not mentioned hitherto is that of Foote et al (1986) who derive crack growth resistance curves for a fibre cement composite in a double cantilever beam geometry. They look at the effect of various softening indexes (n) of a power law relationship between σ and u .Thus again a σ u relationship holds across the faces of the crack so perhaps Atkinson (1970) was not so useless afterall.

5 References

Amazigo , J. C. and Budiansky, B., (1988), Interaction of particulate and transformation toughening,J. Mech. Phys. Solids, 36, 581-595

Atkinson, C. , (1970) , An iterative scheme for solving problems relating to cracks opening under a displacement -dependent internal stress , Int. Jnl. of Frac. Mechs. 6, 193 -197.

Atkinson, C. , (1971) , A simple approximation for calculating the effect of inclusions on fracture , Scripta Met , 5, 643-650.

Atkinson, C. and Avila, J. , (1982) , Special analytical solutions for use in debond stress analysis , IMA Jnl of Appl. Maths ,29, 143-171

Atkinson, C. and Avila, J. , (1983) , A comparison of exact and model solutions for the initiation of debond fracture , Int. J. Engng Sci. , 9, 1053-1059

Atkinson, C. , Avila, J., Betz , E. and Smelser, R. E. ,(1982), The rod pull out problem ,theory and eaperiment., J. Mech. Phys. Solids, 30, 97-120

Aveston, J., Cooper, G. A. and Kelly, A. ,(1971) , The properties of fibre composites,15-26, Conference Proceedings, NPL , Guildford., IPC Science and technology press ltd.

Budiansky, B. , Amazigo, J. C. and Evans, A. G. , (1988), Small scale crack bridging and the fracture toughness of particulate-reinforced ceramics, J. Mech. Phys. Solids, 36, 167-187

Budiansky, B. , Hutchinson, J. W. and Evans, A. G. , (1986), Matrix fracture in fiber reinforced ceramics , J. Mech. Phys. Solids, 34, 167-189

Budiansky, B. and Amazigo, J. C. , (1989), Toughening by aligned frictionally constrained fibres,J. Mech. Phys. Solids, 37, 93-109

Collins, W. D. , (1962), The penny shaped crack, Proc. Roy. Soc. A, 266 , 359-386

Eshelby, J. D. , (1951), The force on an elastic singularity, Phil. Trans. Roy. Soc. Lond. ,A 244, 87 -111

Eshelby, J. D. , (1961), Elastic inclusions and inhomogeneities, Prog. Solid. Mech. 2, 89-140

Foote, R. M. L. , Mai, Y. and Cotterell, B. , (1986) , Crack growth resistance curves in strain softening materials,J. Mech. Phys. Solids, 34 , 593-607

Hughes, B. D. and White, L. R. , (1979) , "Soft" contact problems in linear elasticity , Q. J. Mech. Appl. Math. 32, 445-471

Marshall, D. B. , Cox, B. N. and Evans A. G. , (1985) , The mechanics of matrix cracking in brittle-matrix fibre composites. , Acta Met. 33, 2013-2021

Rice, J. R. , (1968) , A path independent integral and the approximate analysis of strain concentration by notches and cracks , J. Appl. Mechs. 35, 379-386

Rose, L. R. F., (1985), Theoretical aspects of reinforcement and toughening, Fundamentals of deformation and fracture (Ed B.A.Bilby et al) C. U. P. Press

Rose , L. R. F. , (1986) , The size of the transformed zone during steady state cracking in transformation toughened materials,J. Mech. Phys. Solids, 34, 609-616

Rose , L. R. F. , (1987) , Crack reinforcement by distributed springs , J. Mech. Phys. Solids, 35 ,383-405

Shah, S. P. (Ed), (1985), Application of fracture mechanics to cementitious composites, Martinus Nijhoff, Dordrecht.

Walton, J. R. and Weitsman, Y. , (1985) , Deformations and stress intensities due to a craze in an extended elastic material, J. Appl. Mechs. 51, 84-92

ON THE FORM OF MICROMECHANICAL MODELS
OF THE BRITTLE DEFORMATION OF SOLIDS

D. KRAJCINOVIC and M. BASISTA[1]
Mechanical and Aerospace Engineering
Arizona State University, Tempe, AZ 85287-6106
U.S.A.

ABSTRACT. The present paper offers a cursory review of the state-of-art in analytical modelling of brittle response. The discussion is formally divided to considerations of discrete (statistical) and continuum (micromechanical and phenomenological) models. An effort was made to point out the relations between these three classes of models and ascertain the manner of using these relations in improving currently existing modeling techniques.

1. Introduction

The objective of this paper is to review and scrutinize the methodologies for analytical modelling of the response of elastic material weakened by a diffuse ensemble of crack-like micro-defects. The ultimate goal is to examine the methods leading to the establishment of a constitutive relation between the macro-stresses and macro-strains reflecting the physics of irreversible changes of the microstructure of the solid. Since the utility of an analytical model, and its acceptance in engineering practice, are proportional to its simplicity a due consideration must be directed towards the introduction of a set of physically justified simplifying assumptions. Blurring the inconsequential and experimentally not reproducible details of the mechanical response these assumptions emphasize the salient aspects and essential nature of the underlying phenomena. On the other hand, the introduced simplifications by their very nature place limits on the utility and applicability of the model which must be recognized in analyses.

The discussion in this paper is limited to the response of an elastic specimen containing a large number of microcracks distributed over a significant fraction of its volume. It will be assumed that most, if not all, of the energy is dissipated on the formation and propagation of microcracks. Focusing on the predominantly brittle response the macro-strains and attendant deformations may be regarded as infinitesimal rendering the linear theory of elasticity applicable.

For the sake of simplicity assume that all microcracks are planar and elliptical in shape. The stress and strain fields in the specimen containing N of these cracks can be determined, at least in principle, solving N coupled integral equations (Kunin 1983). Naturally, since N is, by definition, a very large number such a strategy will in most cases

---

[1] Permanent address: Institute of Fundamental Technological Research, 00-049 Warsaw, Poland

S. P. Shah (ed.), Toughening Mechanisms in Quasi-Brittle Materials, 355–372.
© 1991 *Kluwer Academic Publishers.*

not be a feasible one. Moreover, in the case of a typical engineering material (such as concrete) the shape of these cracks is irregular and their size and position random. Therefore, a rigorous determination of stress and strain fields within the specimen will not be possible. Additionally, the growth pattern of these microcracks depends not only on the fluctuations of the local stress field but also on the randomness of the material toughness on the microscale. In other words, the evolution of damage (understood herein as a gradual change in the number of cracks and their size) is a problem with an infinite number of degrees of freedom (Bolotin 1989).

Since a rigorous analytical solution of the problem is not possible it becomes necessary to rely upon a phenomenological theory, construct an approximate micromechanical model or resort to computational methods in conjunction with statistical considerations. First of these strategies, most frequently used in engineering analyses, will be given only a cursory attention within this paper. Even then, phenomenological modelling will be addressed only in relation to the micromechanics. The approximate micromechanical models are commonly based on idealized crack geometry (considering them, for example, penny shaped in form) and involve an averaging (homogenization) process within the framework of the effective continua models (Mura 1982, Kunin 1983, Krajcinovic and Sumarac 1987, Sumarac and Krajcinovic 1988, Ju 1989, etc.). Except for few cases (M. Kachanov 1987), this approach, by its very nature, disregards the irregularities in the microcrack geometry and their interaction leading to tractable, local analytical models involving acceptable levels of computational effort in applications. The last of these three approaches, popular among statistical physicists, fully acknowledges the randomness in the defect morphology and the distribution of fracture energy on the microscale. However, this class of algorithms is purely numerical in form involving numerous and repetitious large scale computations. In fact, the computations are so extensive that all of the existing analyses were restricted to two-dimensional lattices.

The emphasis of this paper will be placed on the analytical micromechanically based models. The statistical models will be discussed solely in the context of the limitations and possible modifications of the analytical models.

2. Discrete (Statistical) Models

For a better understanding of the problem it is, perhaps, advisable to start with a very simple model such as a loose bundle of parallel rods carrying an external tensile load F (Krajcinovic and Silva 1982, Hult and Travnicek 1983, Krajcinovic 1989). The analysis of the parallel bar system, shown in Fig.1, is based on the following simplifying assumptions:

- all rods are elastic, perfectly brittle and arranged in such a manner that they equally share in carrying the external load F,
- all N rods have identical stiffness K/N and elongation,
- the strengths of the rods are different.

The system is supplied by a rigid beam (bus) at both ends serving as a device preventing unequal elongation of individual bars.

Despite the initial (quenched) disorder, introduced by unequal bar strengths, the analysis is thereafter deterministic. It is trivial to prove that the equilibrium is satisfied when

$$F = Kx(1 - D) \tag{1}$$

where x is the elongation of the system and

$$D = (n/N) \tag{2}$$

the damage variable selected as the ratio between the number of the already ruptured bars n and the total number of bars N.

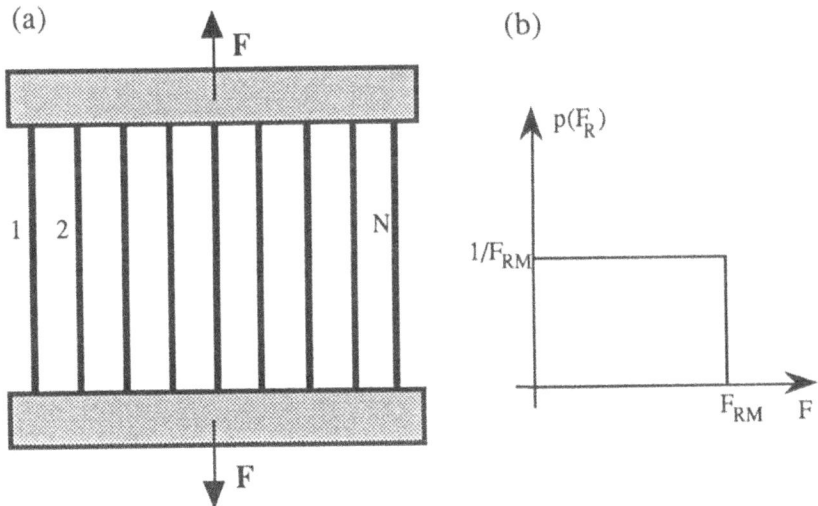

Fig.1. (a) System of parallel loose bars. (b) Probability density function of bar strengths

Assuming that the strength probability density function of individual bars is $p(F_R)$, and that the number of bars tends to infinity the equilibrium equation (1) can be recast in form of an integral

$$F = Kx \int_{Kx}^{\infty} p(F_R)\, dF_R \tag{3}$$

The damage variable can now be written as the cumulative probability function $P(Kx)$ of the given strength probability density function $p(F_R)$

$$D = \int_{0}^{Kx} p(F_R)\, dF_R = P(Kx) = pr(F_R < Kx) \tag{4}$$

In the case when the probability density function of bar strengths is uniform $p(F_R) = (1 / F_{RM})$, where F_{RM} is the strength of the strongest bar, it can readily be demonstrated that

$$D = Kx / F_{RM} \quad \text{and} \quad F = Kx [1 - (Kx / F_{RM})] \tag{5}$$

The force-displacement curve is a quadratic parabola with apex at

$$F = F_M = (1/2) Kx_M = N F_{RM} / 4 \tag{6}$$

At the apex of the force-displacement curve (Fig.2) the damage is

$$D_M = 0.5 \quad \text{at} \quad x = x_M \tag{7}$$

For different strength distributions the damage at maximum stress and the maximum stress itself will be different (see, Krajcinovic 1989).

Using conventional expressions $F = A_0\sigma$, $\varepsilon = x/L$ and $K = EA_0/L$, where E, A_0 and L are the elastic modulus, initial (undamaged) cross-sectional area and length of the macro-bar, the expressions (5) can be rewritten in the conventional form

$$\sigma = E(1 - D)\varepsilon \quad \text{and} \quad D = A_v / A_0 \tag{8}$$

where A_v is the area initially occupied by the ruptured bars which is not available for the transmission of forces.

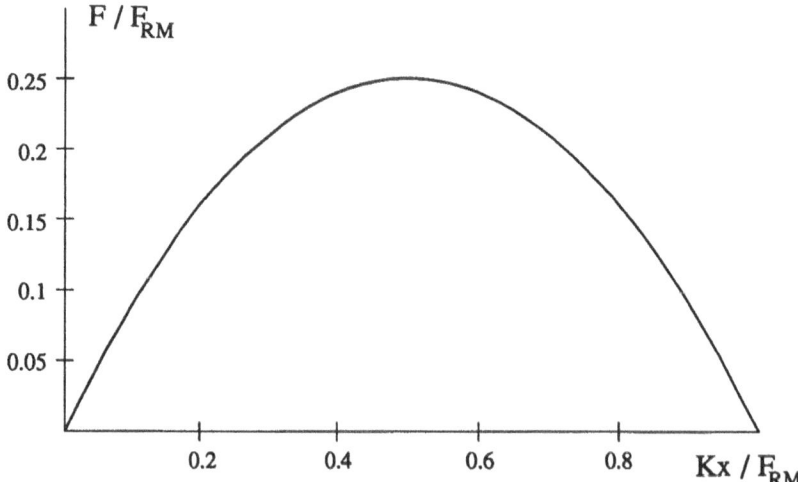

Fig.2. Force-elongation curve for parallel bar model

A computationally much more intensive analysis of a triangular lattice (Fig.3) was lately performed by Herrmann, et al. (1989) and Hansen, et al. (1989). The central idea was to examine the influence of the initial disorder and the size of the lattice on the mechanical response. The first task was accomplished assigning different strengths to different bars of the lattice and repeating computations for each physical realization (initial disorder). In each case the computations of forces in the lattice members for a given displacement of the rigid end members are performed in the routine manner. Naturally, the strength probability density function (taken as uniform as in Fig.1) was identical for all cases allowing for a meaningful comparison of results. Computed quantities were then averaged over different lattices keeping the number n of ruptured bars constant ("history" averaging). The influence of the lattice size was then assessed considering lattices for which L (Fig.3) was taken as 4, 8, 16 and 24.

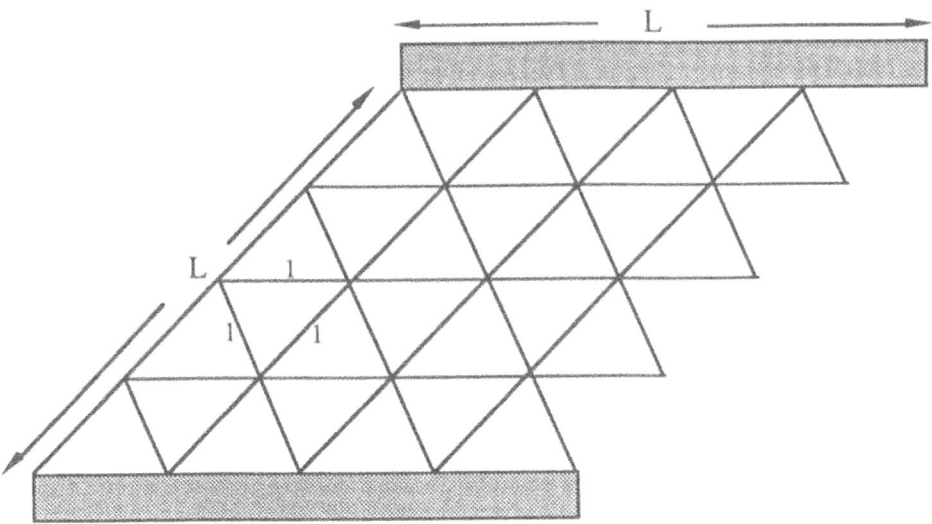

Fig.3. Central-force triangular lattice (L = 4).

According to the reported results of these computations the relationship between macro-stresses and macro-strains (translated into more traditional form) was found to be

$$\sigma = E\varepsilon \, (1 - \alpha\varepsilon \, \sqrt{3}) \tag{9}$$

where the parameter α, proportional to $L^{1/4}$, is determined in such a manner that it fits the results for all lattice sizes equally well. The damage variable can be deduced to be of the form

$$\omega = (3/4) \, \sqrt{3} \, L^{1/4} \, \varepsilon \tag{10}$$

It is notable that the corresponding expressions for the loose bundle parallel system and the lattice are (except for the size effect embodied in L) identical in form for the pre-critical regime. In other words, the response is during the hardening phase of the deformation process not very sensitive to the level of sophistication (such as type of lattice, distribution of the disorder, etc.) in modelling.

It is interesting that these computations demonstrate that in the pre-critical regime:

(a) the average stress-strain curves for all four lattice sizes and all distributions of the initial disorder can be collapsed on a single (parabolic) master-curve, and

(b) the relation between damage and strain (10) is linear.

(c) the damage at the apex of the stress-strain curve was again equal to 0.5 as in the previously discussed case of the parallel bar model

In concert with the loose bundle parallel bar system this obviously means that in the pre-critical regime the crack interaction has little influence on the overall response which is well described by simple volume averages of the involved fields. In other words, the pre-critical regime depends only on the volume averages of the initial disorder distribution (initial defects and toughnesses). The crack interaction and even the redistribution of stresses is not crucial for the determination of the overall response. The fractality, i.e. dependence of the overall response on the size of the lattice (10) is, perhaps, associated with a rather small number of triangular elements and should disappear for larger lattices. In a sharp contrast, the distribution of forces in the post-peak regime, just before the rupture, is strongly multi-fractal. The inability to identify a single length scaling parameter should be related to the fact that the post-peak regime strongly depends on the distances between the adjacent cracks. It seems reasonable to expect that more than one length parameter must be introduced to model that behavior. Additionally, the post-peak response is found to depend strongly on the details (highest statistical moments) of the initial disorder distribution. Consequently, since the initial disorder distribution in engineering materials is random the post-peak behavior cannot be deterministic.

It is very important to notice that neither of two models required an additional kinetic law describing the rate of the damage accumulation with either stresses or strains. This law was, indeed, derived from the initial disorder, i.e. assumed distribution of initial defects and fracture strengths. The damage evolution law and consequently the macro-response were, however, found to be different for different strength distributions (Krajcinovic and Silva 1982).

3. Continuum Models

3.1 MICROMECHANICAL MODELS

One of the first tasks related to the introduction of an effective continuum is to define the representative volume element (RVE). In a very general sense, a RVE is the smallest volume of material which with respect to a given random variable has the same properties as the macro-specimen. More specifically, the first (n) statistical moments of the distribution of a particular random variable taken over the volume V of the RVE should to a desired level of accuracy equal the corresponding statistical moments of the distribution of the same variable taken over all volumes larger than V. These requirements were further reduced by Nemat-Nasser and Hori (1990) who, roughly speaking, considered a RVE to be statistically representative of the macro-response if the macro-stress is a volume average of micro-stresses corresponding to the same elastic macro-strain. An analogous definition has been advanced by the same two authors for the case of imposed macro-strains.

Assuming that these conditions for the RVE are satisfied it becomes possible to establish a relation between a field on the micro-scale and the same field on the macro-scale. All pertinent informations regarding the structure of the materials and the defects within the RVE will form the configurational space attached to a material point of the effective continuum. Therefore, the RVE allows for the mapping of the micro-structure of the material on the material point of the continuum. Naturally, since the RVE must contain a statistically valid sample of microcracks a configurational space describing all defects must contain a large number of data. Assuming, for simplicity, all N cracks to be planar and elliptical in shape an all inclusive configurational space should comprise of 8N scalars (2N half-axis lengths, 3N Euler angles and 3N scalars defining the position of the microcrack center). In the case of inhomogeneous state of macro-stress and strain the bookkeeping associated with updating the configurational space in every material point of the specimen for each increment of the externally applied tractions or displacements and each time step can obviously present a significant and often unsurmountable problem.

The conventional method of establishing the governing equations is typically prefaced by introducing the Helmholtz $\Phi(\varepsilon)$ and Gibbs $\Psi(\sigma)$ energies of the specimen (or RVE) with volume V such that

$$\Phi(\varepsilon) + \Psi(\sigma) = V(\sigma : \varepsilon) \tag{11}$$

where σ and ε are the volume averages or, in sense of Nemat-Nasser and Hori (1990) definition, the macro-fields. The macro-stresses and macro-strains are defined as

$$V \sigma = \partial \Phi / \partial \varepsilon \qquad \text{and} \qquad V \varepsilon = \partial \Psi / \partial \sigma \tag{12}$$

When the elastic strains in the matrix are small, and the concentration of inhomogeneities dilute, the macro-stresses are mapped on the macro-strains by a linearized relation of a general form

$$\varepsilon = \varepsilon^p(H) + S(H) : \sigma \tag{13}$$

where H is used to denote a set of scalar and tensor variables defining the irreversible rearrangements of the material microstructure (recorded history). Also, $S(H)$ is a fourth rank tensor known as compliance which, in the present case, depends on the properties of the matrix and the given distribution of microdefects. From equations listed above the compliance is

$$S = (1/V) (\partial^2 \Psi / \partial \sigma \, \partial \sigma) \tag{14}$$

Additionally,

$$\varepsilon^p = \varepsilon(\sigma = 0) \tag{15}$$

is the plastic (or residual) strain in the specimen.

The incremental form of the relation between the macro-stresses and macro-strains (13) necessary for inelastic analyses is

$$d\varepsilon = d\varepsilon^p + S : d\sigma + dS : \sigma \tag{16}$$

The last term on the right hand side of the equation (16) represents the inelastic strain associated with the change in the compliance of the specimen. Naturally, the compliance can change only if the recorded history is changed, i.e. if the energy is dissipated.

Substituting (13) into (11) it also follows that

$$\Psi(\sigma, H) = -\Phi^o(H) + V(\sigma : \varepsilon^p) + (1/2) V(\sigma : S : \sigma) \tag{17}$$

where Φ^o is the locked in energy at vanishing state of stress.

Assuming that the irreversible rearrangement of the microstructure can be described by a finite set of thermodynamic fluxes $d\xi_i$ ($i = 1,2,...,n$), the inelastic increment of the Gibbs energy can always be written as a scalar product of fluxes and their conjugate thermodynamic forces (Rice 1975),

$$d^i\Psi = \Psi(\sigma, H + dH) - \Psi(\sigma, H) = \Sigma f_j d\xi_j \qquad (j = 1,2,...,n) \tag{18}$$

Combining (17) and (18) it further follows that

$$\Sigma f(\sigma,H) d\xi = - d\Phi^o(H) + V(\sigma : d\varepsilon^p) + (1/2) V(\sigma : dS : \sigma) \tag{19}$$

Concentrating on the defects in the form of microcracks the history recording set of parameters H consists of data related to the size, shape, orientation and position of all microcracks within the volume V of the RVE. The inelastic change in the Gibbs energy is then (Rice, 1975)

$$d^i\Psi = \sum f_j d\xi_j = \sum \int_L \left[(G - 2\gamma) \, d\ell \right] dL > 0 \tag{20}$$

where G is the elastic energy release rate, γ the surface (fracture) energy, and $d\ell$ the local advance of the crack front L in the direction of its normal. The integral in (20) is taken along the perimeter of each active microcrack. The sign of the inequality in (20) is the consequence of the second law of thermodynamics (entropy production). It is important to realize that no energy is dissipated unless at least on of the cracks changes its size (such that $d\ell > 0$), i.e. unless additional atomic bonds in the material are ruptured.

The energy release rate of a crack embedded in an elastic matrix can be written in form (Rice 1975)

$$G = (\pi/4) k_i C_{ij} k_j \tag{21}$$

where **k** is the vector having the stress intensity factors as components (i = 1,2,3) and C(H) a second rank tensor depending on the elastic properties of the matrix and the recorded history H. From the expressions (19) to (21) the inelastic change in the compliance tensor is now

$$dS_{ijmn} = \frac{1}{V} \sum \int_L \left[\frac{\pi}{2} C_{qr} \frac{\partial k_q}{\partial \sigma_{ij}} \frac{\partial k_r}{\partial \sigma_{mn}} d\ell \right] dL$$

(22)

with integration extended over the perimeters of all active cracks within the RVE.

It is important to notice that:

(a) the deformation process is elastic (non-dissipative and reversible dH = 0) in the case when the crack surface area does not change (dℓ = 0), regardless of possible changes in the stress and strain fields or even crack opening displacements and/or stress intensity factors,

(b) the overall compliance changes as the history is recorded. Correspondingly, the slopes of the unloading segments at different values of H will not be parallel among themselves or with the initial, elastic part of the loading curve (at H = 0).

The plastic macro-strain ε^p can readily be derived from the known eigenstrains within the inhomogeneous inclusions (see Mura 1982)

$$\varepsilon^p = \langle \varepsilon^p \rangle = f \varepsilon^{**} = f (\varepsilon^i + \varepsilon^*)$$

(23)

where ε^i is the eigenstrain within the inhomogeneity (due to phase transition, expansion of reaction product, etc.), while ε^* is the equivalent eigenstrain simulating the mismatch in material parameters between the matrix and inhomogeneous inclusion. The angular brackets here denote averaging over the entire volume of the RVE. Also

$$f = \Sigma \Omega_\alpha / V$$

(24)

is the fraction of the volume occupied by inhomogeneous inclusions. The sum in (24) is extended over all inhomogeneous inclusions α.

The expressions for the rate of the inelastic change in the compliance tensor can also be derived using the inclusion method in conjunction with averaging (Horii and Nemat-Nasser 1983, Krajcinovic and Fanella 1986, etc.). An elliptical crack is modeled as a limiting case of an ellipsoidal inclusion of vanishing thickness 2a₃ in the direction of the crack normal. The eigenstrain can then be written as (Mura 1982, Horii and Nemat-Nasser 1983, Krajcinovic and Sumarac 1987, etc.)

$$\lim_{a_3 \to 0} (a_3 \varepsilon) = \frac{1}{2} \int_A \left\{ \mathbf{n} \otimes [\mathbf{u}] + [\mathbf{u}] \otimes \mathbf{n} \right\} dA$$

(25)

In (25) **n** is the normal to the crack surface and [**u**] the displacement discontinuity across the surface of the crack. The integration is extended over the entire crack surface A. The symbol $\otimes$ in (25) denotes the dyadic (tensor) product of two vectors.

Introducing the expressions relating the stress intensity factors and crack opening displacements (Hoenig 1978) it can be shown that the inelastic change of the compliance can be written (Krajcinovic 1989)

$$dS = \frac{3}{V} \sum \int_L (a B \delta a) \, dL \tag{26}$$

where a is a characteristic length (for instance, the length of one of the axes of the ellipse) and **B**(H) a fourth rank tensor containing information about the crack sizes and orientations (Euler angles ϕ and θ). Since the components of the tensor **B** are computed from the crack opening displacements they, in principle, depend on the presence of other cracks within the RVE. Analytical expressions for the components of the tensor **B** are, for certain cases, available in the literature (see, for example, Krajcinovic and Sumarac, 1987, etc.). Since the perimeter of the crack can be normalized by the length a from (26) it finally follows that

$$dS = (N/V) < (3a^2 \, \delta a) \, B(\phi, \theta) > \tag{27}$$

where N is the total number of cracks within the volume V of the RVE. If the crack sizes and orientations are not correlated the expression (27) can be rewritten in a simpler form as

$$dS = \{ (N/V) <3a^2 \, \delta a> \} <B(\phi, \theta)> \tag{28}$$

where the term in the first brackets on the right hand side of (28) is the increment of the Budiansky and O'Connell (1976) damage variable. Consequently, expressions (27) and (28) confirm that the change in compliance is nothing but an orientation weighted volume average (i.e. macro-structural analogue) of the Budiansky-O'Connell damage variable. In all of the above expressions the change of the compliance is inelastic. Since the elastic change of the compliance does not exist it is not considered necessary to use the superscript "i" in equations (26) to (28).

According to the above discussion the constitutive law (21) relating macro-stresses and macro-strains can be written in an explicit form whenever the expressions for the rate of change of the volume averages of the inelastic strains de^i and compliance tensor dS are available. In view of the expressions (21), (22) and (25) this is the case when the geometry, loading, crack shapes and material properties are simple enough to admit analytical solutions for the stress intensity factors, components of the second rank tensor **C** and the crack opening displacements [**u**] (needed to compute the components of the tensor **B**). In general this restricts the analyses to the planar and penny-shaped cracks embedded in a homogeneous isotropic or at most transversely isotropic and infinitely extended elastic matrix. In all other cases the determination of the stress intensity factors and/or crack opening displacements requires use of approximate expressions (Nemat-Nasser and Horii 1982, Fanella and Krajcinovic 1988, Nemat-Nasser and Obata 1988) or nontrivial quadratures of complicated integrals (Mura 1982) unsuitable for the considered analyses. The above mentioned restrictions severely limit the applicability of these micromechanical models since the anisotropy will, in a general case, be introduced by microcracks into an otherwise isotropic matrix.

The expressions such as (22) and (28) imply summation of contributions of each of the N microcracks needed to compute the change of the compliance of the RVE mapping on the material point. This process is further complicated by the fact that both the stress intensity factor and the crack opening displacement of the observed crack will be influenced by the presence of the other (N-1) cracks within the RVE. Even though these stress intensity factors and crack opening displacements can be, as already mentioned, determined solving a system of N coupled integral equations, such a strategy is by no means feasible since the computations would have to be repeated for every increment of the stress field (during which the history changes) at each material point.

A much more promising strategy is associated with the application of the effective continuum models discussed at length in Kunin (1983), Nemat-Nasser and Hori (1990) and many other sources. The basic idea of this approach consists in reducing the problem of determining the stresses and strains in an elastic solid weakened by many cracks by a series of much simpler problems of computing the stresses and strains for the case of a single crack embedded in an effective continuum. The simplest, and most frequently applied, model belonging to this family is based on the so-called self-consistency property. More specifically, the self-consistent model implies that:

(a) the external field of an observed defect weakly depends on the exact location of the other defects within the RVE, and that

(b) the external field of each crack is equal to the average (macro, or far-field) stress in the RVE.

The assumption (b) eliminates the need to solve the system of integral equation while the assumption (a) reduces the configuration space by eliminating from considerations the location of cracks within the RVE. More importantly, it becomes possible to place each crack into the center of the RVE occupied by an effective solid which in the sense of the overall energy reflects the presence of all other cracks. Since a microcrack is typically small compared to the size of the RVE, the stress intensity factors and the crack opening displacements of a given microcrack (assuming it to be penny-shaped or elliptic in form) can readily be determined if the effective solid is either isotropic or transversely isotropic. The simplest solution, to be referred to as the Taylor's model, is to neglect the other cracks altogether and assume that each crack is embedded in the virgin (undamaged) solid. This approximation is, obviously, justified only in the case of a dilute crack concentration. In all of these cases, regardless of the approximation level the overall solution still implies a superposition of a large number of simple problems. The superposition is typically replaced by quadratures introducing the probability density functions for the crack sizes and orientations and integrating over the corresponding space (M. Kachanov 1982, Horii and Nemat-Nasser 1983, Krajcinovic and Fanella 1986, Krajcinovic and Sumarac 1989, etc.).

It is notable that in its conventional form the self-consistent model (Budiansky and O'Connell 1976) predicts that the elastic moduli will vanish at some microcrack density. For example, for the case of isotropic damage (perfectly random orientation of cracks) the elastic modulus vanishes when

$$\omega = (Na^3/V) = 9/16 \qquad (29)$$

It is very interesting, and perhaps not entirely fortuitous, that the damage at the maximum stress is again close to 0.5 as found in the case of the parallel bar model (no interaction) and lattice systems.

This result provided motivation for several modifications and improvements of the original method such as double embedding suggested by Christensen and Lo (1979), a more recent method promoted by Mori and Wakashima (1990). Additionally, the application of the differential scheme (Hashin 1988) attracted substantial interest. All of these models significantly improve the estimates of the elastic moduli for moderate to high microcrack densities defined as the range $0.2 < \omega < 0.5$ of the Budiansky-O'Connell variable (see, Nemat-Nasser and Hori 1990). However, a question remains whether the direct interaction of microcracks at these densities already becomes significant. A more rigorous level of modelling (Kanaun 1974) appears to be too complex to be appealing. An approximation along the lines suggested by M. Kachanov (1987) seems to present a feasible alternative even though the configurational space attached to each material point of the continuum becomes substantially larger since it becomes necessary to keep track of the exact position of each crack within the RVE. Numerical simulations within the framework of the percolation theory indicate that the expression (29) underestimates the critical crack density at which the elastic moduli vanish by a factor of three. This is especially true if the microcracks are allowed to intersect.

The most difficult problem in the process of the formulation of a continuum theory consist in a rational definition of the kinetic equations, i.e. in prescribing the law describing the advance of the crack front $\delta\ell$ as a function of the stresses. It was already noted that this was not necessary in the case of the disordered discrete models discussed in the Section 2 above. The kinetic law was just a consequence of the initial disorder, i.e. of the fact that the resistance to the crack growth was not uniform. An essentially similar idea was advanced in Krajcinovic and Fanella (1986), Krajcinovic and Sumarac (1987), Krajcinovic (1989), etc. who actually derived the damage evolution laws from the consideration of statistical distribution of strengths associated with the hierarchy of fracture energies on the microstructural level, and the morphology of the microstructure itself. The performed computations did not prove to be very sensitive within the considered range of initial disorders. This was to be expected since all performed computations were based on the Taylor's and/or self-consistent models and were, therefore, valid only in the pre-critical regime. The post-critical regime will, as indicated in Section 2 above, strongly depend on the details of the morphology of the microstructure (grain size, initial disorder and initial defects).

A rational formulation of kinetic equations, analogous to the procedure commonly adopted in slip theories, was suggested by Krajcinovic and Fanella (1986), Krajcinovic and Sumarac (1987), Krajcinovic (1989), etc.. This formulation is based on the Griffith's criterion, derivable from (20), according to which an observed microcrack will commence its growth ($\delta\ell > 0$) when

$$G - 2\gamma > 0 \qquad (30)$$

The elastic energy release rate G of a given crack depends on the crack size a, orientation (Euler angles ϕ), recorded history (accumulated damage) H and the state of stress. Thus, the expressions (28) written for all N cracks within the RVE, represent a family of surfaces in the (a, ϕ, σ, H) space. The inner envelope of this family of surfaces (which contains the unstressed state $\sigma = 0$) represents the locus of points (thermodynamic states) at which the history can change ($\delta\ell > 0$).

The fracture (surface) energy γ must be considered as being a random variable. The distribution of this random variable depends on the morphology of the micro-scale (coarse aggregate grading in concrete, grain size distribution in ceramics, etc.) as well as the hierarchy of strengths of the constituent phases.

The analyses using the described models (see a summary in Krajcinovic 1989, Nemat-Nasser and Obata 1988, or Ju 1989) proved to be reasonably straightforward in the case of simple, homogeneous states of stress. Most, if not all, material parameters were directly identifiable reducing the ambiguity to a minimum. Despite the simplifying assumptions introduced into the model the results proved to be sufficiently accurate for all practical purposes. In all considered cases of uniaxial tension and compression of brittle and semi-brittle materials, time dependent and time independent deformation of polycrystalline ceramics these models were restricted to the pre-critical regime. Nevertheless, by relating the macro-response to the micro-structure of the specimen the micromechanical theories will, eventually, provide a rational basis for optimization of materials for a specific set of circumstances. This aspect of micromechanical modeling was further emphasized in the

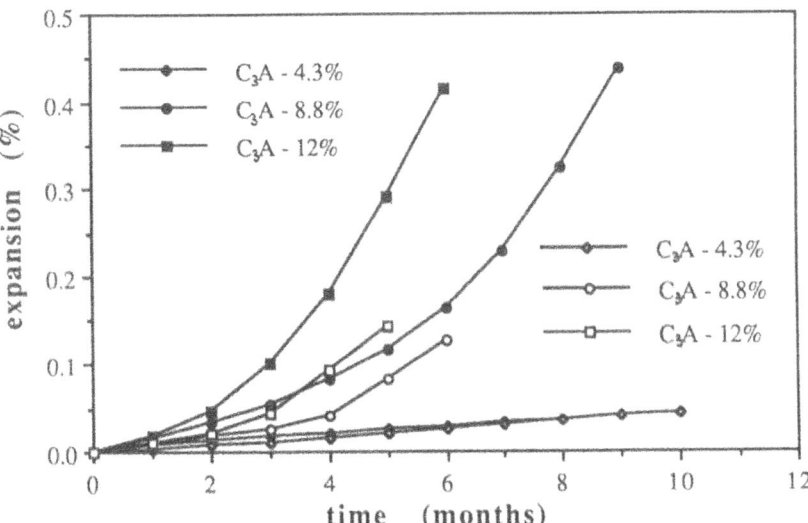

Fig.4. Expansion of mortar bars caused by external sulphate attack (solid symbols - experiments by Ouyang, et al. 1988)

formulation of an analytical model for the prediction of the gradual degradation of concrete exposed to the sulphate attack, Fig.4 (Krajcinovic, et al., t. a. 1).

In this particular case it was possible to go a step further, examine the equations governing the kinetics of chemical reactions, and establish a rational connection between the available volumes of the reactants and the eigenstrains related to the expansion of the reactive products directly from stoichiometric considerations. As a result, the overall strain, and the failure mode, was found to depend on the chemical composition of the concrete, its microstructure (diffusivity and porosity), time of exposure and the concentration of the diffusing solute. The results shown in Fig.4 clearly indicate the power of the adopted modelling technique in establishing the causal relationship between the microstructure and chemical composition of a solid and its macro-response and ultimately its functional or structural failure.

3.2. PHENOMENOLOGICAL MODELS

Despite their great advantage in modelling physical reality with a minimum of ambiguity and arbitrariness the micromechanical models are, in general, computationally inefficient for practical applications. Conversely, the ambiguities inherent in macro-modelling of micro-processes often result in emergence of many different and often contradictory phenomenological models. This was certainly true in the case of the damage mechanics (see Krajcinovic, 1984) which was from the very beginning plagued by a plethora of models claiming to be rightful generalizations of the original Kachanov's (1958) model.

It seems, therefore, useful to settle some of the outstanding points using the already known results following from the micromechanics of the analyzed phenomenon. Concentrating on the perfectly brittle processes on the basis of the expression (18) can be rewritten in form of a scalar product,

$$d^i \Psi = Q_i \, dq_i \tag{31}$$

between the vector the thermodynamic macro-flux defined as

$$dq = \{dS_{1111}, dS_{2222}, ...\} \tag{32}$$

and the corresponding (or conjugate) vector of affinities

$$Q = (1/2)\{\sigma_{11}\sigma_{11}, \sigma_{22}\sigma_{22}, ...\} \tag{33}$$

The micro-fluxes and affinities can be derived from (20) as

$$f_j = (L/2A) <(G - 2\gamma)_j> \quad \text{and} \quad d\xi_j = (1/V)(2A \, dA/L)_j \tag{34}$$

and related to the corresponding macro fields (32) and (33) in a manner derived in Rice (1971). Thus, as argued in Krajcinovic, et al. (t.a.) the thermodynamic force is the excess energy release rate integrated along the perimeter of the crack and averaged over its surface. Also, the thermodynamic flux is recognized as the increment of the Budiansky-O'Connell damage variable (29).

Following Rice (1971) it can be further shown that if the increase in the surface area of a crack depends on the external stress only via its own affinity $d\xi_n = F(f_n, H)$ the macro-potential (Krajcinovic, et al., t.a.2)

$$\Omega(Q, \xi) = \frac{1}{V} \int_0^{f_n (Q,\xi)} \xi_m (f_m, H) \, df_m$$

(35)

can be obtained superimposing all micro-potentials $F(f_n, H)$. The macro-fluxes are then obtained as in the theory of plasticity from the normality property of the macro-potential.

The most important aspect of the outlined micro-to-macro transition, discussed in considerable detail in Krajcinovic, et al. (t.a.2) are that the potential exists as long as the propagation of a crack depends only on its own energy release rate. All other cracks influence the observed crack only through the properties of the effective medium. Consequently, the potential exists whenever the self-consistent approximation is justifiable. It is also noticed that while the change of the compliance can be selected as the macro-flux it is not always possible to identify an instantaneous magnitude of the compliance as the internal variable. This situation is analogous to the theory of plasticity which also does not provide means necessary to determine the initial the already accumulated plastic strains present in a material specimen at the onset of a deformation process. In other words the outlined phenomenological damage model is related to its micromechanical counterpart in the manner completely analogous to one existing between the theory of plasticity and the slip theory.

4. Conclusions

The objective of the present paper was to review the state-of-art in modelling of brittle response of engineering solids. Without dwelling on details of the discussed models it seems reasonable to conclude that despite the significant progress achieved during the last decade the task of formulating a realistic, reliable and comprehensive analytical model has not as yet been fully accomplished. The mere fact that the three discussed classes of models have not as yet been properly related and reconciled is just one of the reasons for the above statement.

The statistical (discrete) modelling was limited to expensive and time consuming computations which will become even more extensive and protracted with transition to the three-dimensional problems, more complex loading programs and more realistic lattices. Nevertheless, the physical insight into the phenomenon of damage evolution and its dependence on the random properties of the microstructure, brought to light with this class of models, should not be underestimated.

The micromechanical modelling seems to work exceedingly well for the homogeneous states of stress and strain and the pre-critical regime. The versatility of this class of models and the ease with which it can be adopted to different types of problems has been established beyond any doubt. The most important accomplishment of these models is associated with their ability to relate the macro-response to the micro-structure and even chemical composition of the solid. The extension of this type of models to the post-critical response (see the papers in Mazars and Bazant, 1989) is still a matter of considerable disagreements. In view of the results of the statistical modelling it appears doubtful whether a deterministic model of the post-peak response presents a realistic objective at all.

It was finally argued that a reasonably simple phenomenological model can be derived from the micromechanical theory in a rigorous manner via a micro-to-macro transition (similar to those establishing connection between the slip and plasticity theories). This approach seems to be very helpful in settling the dispute centering on a "proper" selection of the internal (damage) variable which plagued the adolescent period of the development of the damage mechanics. However, insufficient data and relative inexperience related to the application of this model to a wide spectrum of problems prevents any definitive conclusions and assessments of its utility at the present time.

The most far reaching conclusions is, perhaps, that the deterministic nature of the pre-critical response lends itself well to the micromechanical and phenomenological modelling alike. The volume averages of the involved fields are sufficient indicators of the mechanical response and further refinements of these models will lead to an accurate assessment of many related phenomena. The modelling of the response in the post-critical (or softening) regime is an altogether different problem. The response becomes by its nature strongly non-local and non-deterministic. Direct crack interaction, involving considerations of the details of the defect distribution, assumes a dominant role. In the language of the statistical physics the involved fields are multifractal. Consequently the representation of this fields will involve more than one scale (length) parameter. In view of the random nature of the response, large deviations from the expected values and large skewness of the results, the volume averages have much less significance than in the pre-critical regime. Hence, formulation of a rational analytical model for this regime of the overall response presents a serious challenge which must be addressed in all of its complexity.

Acknowledgement

It is a pleasant duty to acknowledge the financial support in the form of research grants from the U.S. Department of Energy, Office of Basic Energy Sciences, Engineering Research Program and the U.S. Air Force Office of Scientific Research, Directorate of Aerospace Sciences, Civil Engineering Program which made this study possible.

References

Bolotin, V. V. (1990) 'Prediction of Service Life for Machines and Structures', ASME Press, New York, NY.

Budiansky, B. and O'Connell, R. J. (1976) 'Elastic moduli of a cracked solid', Int. J. Solids Structures, **12**, pp. 81-97.

Christensen, R. M. and Lo, K. H. (1979) 'Solutions of effective shear properties in three phase sphere and cylindrical models', J. Mech. Phys. Solids, **27**, pp. 315-330.

Fanella, D. and Krajcinovic, D. (1988) 'A micromechanical model for concrete in compression', Eng. Fract. Mech., **29**, pp. 49-66.

Hansen, A., Roux, S. and Herrmann, H.J. (1989) 'Rupture of central-force lattices' J. Phys. France, **50**, pp. 733-744.

Hashin, Z. (1988) 'The differential scheme and its application to cracked materials', J. Mech. Phys. Solids, **36**, pp. 719-734.

Herrmann, H. J. (1989) 'Fracture of disordered, elastic lattices in two dimensions', Phys. Rev. B, **39**, pp. 637-648.

Hoenig, A. (1978)' The behavior of a flat elliptical crack in an anisotropic body', Int. J. Solids Struct., **14**, pp. 925-934.

Horii, H. and Nemat-Nasser, S. (1983) 'Overall moduli of solids with microcracks:load induced anisotropy', J. Mech. Phys. Solids, **31**, pp. 155-171.

Hult, J. and Travnicek, L. (1983) 'Carrying capacity of fibre bundles with varying strength and stiffness,' J. Mec. Theor. Appl., **2**, pp. 643-657.

Ju, J. W. (1989) 'On energy based coupled elastio-plastic damage theories: constitutive modeling and computational aspects', Int. J. Solids Struct., **25**, pp. 803-833.

Kachanov, M. (1982) 'A microcrack model for rock inelasticity', Mech. Mater., **1**, pp. 19-27.

Kachanov, M. (1987) 'Elastic solids with many cracks: a simple method of analysis', Int. J. Solids Struct., **23**, pp. 23-43.

Kanaun, S. K. (1974) 'A random crack field in an elastic continuum', Isled. po Uprugosti i Plastichnosti, **10**, pp. 66-83.

Krajcinovic, D.(1984) 'Continuous Damage Mechanics', Appl. Mech. Rev., **37**, pp.1-6.

Krajcinovic, D. (1989), 'Damage Mechanics', Mech. Mater., **8**, pp. 117-197.

Krajcinovic, D., Basista, M. and Sumarac, D. (to appear 1), 'Micromechanics of brittle composites exposed to chemically aggressive ambients', in Proc. Symp. Microcracking Induced Damage in Composites, ASME Winter Annual Meeting 1990, Dallas, TX.

Krajcinovic D., Basista, M. and Sumarac, D. (to appear 2) 'Micromechanically inspired phenomenological damage model', J. Appl. Mech.

Krajcinovic, D. and Fanella, D. (1986) 'A micromechanical damage model for concrete', Eng. Fract. Mech., **25**, pp. 585-596.

Krajcinovic, D. and Silva, M.A.G. (1982) 'Statistical aspects of the continuous damage mechanics', Int. J. Solids Structures, **18**, pp. 551-562.

Krajcinovic, D. and Sumarac, D. (1987) 'Micromechanics of damage processes', in D. Krajcinovic and J. Lemaitre, (eds.), Continuum Damage Mechanics: Theory and Application, Springer Verlag, Wien, pp. 135-194.

Kunin, I. A. (1983) Elastic Media with Microstructure II, Three-Dimensional Models, Springer-Verlag, Berlin.

Mazars, J. and Bazant, Z. P. (1989) Cracking and Damage, Elsevier , London, UK.

Mori, J. and Wakashima, K. (1990) 'Successive interaction method in the evaluation of average fields in elastically inhomogeneous materials', in G. J. Weng, M. Taya and H. Abe, (eds.), Micromechanics and Inhomogeneity - The T. Mura Anniversary Volume, Springer-Verlag, New York, pp. 269-282.

Mura, T. (1982) Micromechanics of Defects in Solids, M. Nijhoff , The Hague.

Nemat-Nasser, S. and Hori, H. (1990) 'Elastic solids with microdefects', in G. J. Weng, M. Taya and H. Abe, (eds.), Micromechanics and Inhomogeneity - The T. Mura Anniversary Volume, Springer-Verlag, New York, NY, pp. 297-320.

Nemat-Nasser, S. and Horii, H. (1982) 'Compression induced non-planar crack extension with application to splitting, exfoliation and rockburst', J. Geophys. Res., **87**, pp. 6805-6821.

Nemat-Nasser, S. and Obata, M. (1988) 'A microcrack model of dilatancy in brittle materials', J. Appl. Mech., **55**, pp. 24-35.

Ouyang, C., Nanni, A. and Chang, W. F. (1988) 'Internal and external sources of sulphate ions in portland cement mortar: two types of chemical attack', Cement. and Concrete Res., **18**, pp. 699-709.

Rice, J. R. (1971) 'Inelastic constitutive relations for solids: an internal variable theory and its applications to metal plasticity', J. Mech. Phys. Solids, **19**, pp. 433-455.

Rice, J. R. (1975) 'Continuum mechanics and thermodynamics of plasticity in relation to microscale deformation mechanisms', in Constitutive Equations in Plasticity, A. S. Argon, (ed.), The MIT Press, Cambridge, MA, pp. 23-79.

Sumarac, D. and Krajcinovic, D., (1987) 'A self-consistent model for microcrack weakened solids', Mech. Mater., **6**, pp. 39-52.

ON THE RELATIONSHIP BETWEEN FRACTURING OF A MICROCRACKING SOLID AND ITS EFFECTIVE ELASTIC CONSTANTS

Mark KACHANOV
Department of Mechanical Engineering
Tufts University
Medford, MA 02155
USA

ABSTRACT. It is argued that, contrary to the spirit of many damage models, there is no simple correlation between fracturing of a brittle microcracking solid and the change of its effective elastic moduli. Physically, the absence of such a correlation is explained by the fact that the fracture-related properties (like stress intensity factors) are determined by <u>local fluctuations</u> of the microcrack field geometry whereas the effective elastic constants are the <u>volume average</u> quantities, relatively insensitive to such fluctuations.

1. INTRODUCTION

We discuss whether a correlation exists between the effective elastic moduli of a solid with multiple cracks and the fracture-related properties (like stress intensity factors, SIFs). This discussion is relevant for a number of damage models (see reviews [1,2]) which are aimed at description of fracturing of a brittle microcracking solid but in fact deal with its effective elastic constants. This substitution is done either explicitly, by assuming that the tensor of effective elastic compliance S_{ijkl} can be used as a damage parameter, or implicitly, by constructing an elastic potential f which is quadratic in stresses (or strains) and contains, in addition, a damage parameter D (scalar, vectorial or tensorial). Such a construction, aside from the statement that the derivative $\partial f/\partial D$ can be interpreted as an energy release rate associated with damage, reduces to a certain model for the effective elastic constants.

The underlying idea - that progression towards failure is uniquely correlated with the change of effective elastic constants - may seem intuitively reasonable; it appears particularly attractive because the effective constants can be easily measured.

An objection can be raised that, as is well known, a small crack in a brittle material has a very small impact on the effective elastic constants but drastically reduces the tensile strength. One may argue, however, that, after a certain initial

S. P. Shah (ed.), Toughening Mechanisms in Quasi-Brittle Materials, 373–378.

set of cracks has been nucleated and started to grow, the progression towards failure is uniquely correlated with the change of the effective elastic moduli. This idea is examined below from several points of view. The analysis is based on a direct solution of the problem of many interacting cracks for a number of sample crack arrays (using the method of [3]).

2.   "PARADOXICAL" EXAMPLE

To demonstrate that this idea is far from obvious, we start with a simple example when the relation between the SIFs and the effective elastic properties is "paradoxical". Consider a plate containing stacks of parallel cracks (solid lines) as shown in fig. 1. Suppose that additional cracks - are introduced in-between the neighboring cracks in stacks (dashed lines).

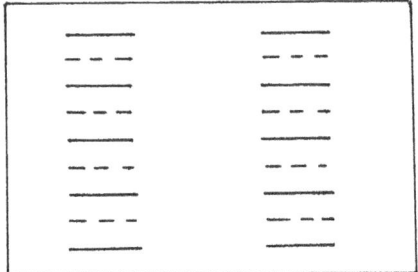

Fig. 1. Crack array with paradoxical relation between SIFs and effective stiffness

Introduction of these new cracks, obviously, reduces elastic stiffness in the direction normal to cracks (since it produces additional displacements at remote boundaries when a load is applied) - material elastically "softens". However, the SIFs decrease (due to an increased shielding) so that the critical load at which the cracks start to propagate increases - material is strengthened by new cracks.

3.   RANDOM CRACK ARRAYS

It may be argued that the example above does not represent any realistic crack statistics (although such parallel crack patterns do occur in rocks and certain composites) and that for the more "random"crack arrays the progression towards failure is uniquely correlated with the reduction of elastic stiffness.

To demonstrate that such a correlation is not obvious, we did the following computer experiments. A number of two-dimensional crack arrays containing randomly oriented cracks of the same length were generated; statistics of crack centers was also random (subject to the restriction that cracks were not allowed to intersect; this was achieved by generating cracks successively and discarding a newly generated crack if it intersects the already existing ones). For each sample array, using the method of [3], we calculated

(1) the effective Young's modulus E_{eff};
(2) maximal, among all the crack tips, value of $K_I^2 + K_{II}^2$ induced by a
      uniaxial loading.

Strictly speaking, the value of $K_I^2 + K_{II}^2$ determines the initiation of crack propagation only in the case of a rectilinear crack extension; when both K_I and K_{II} are present, a certain other combination of K_I, K_{II} is a relevant parameter. Typically, however, this combination does not differ much (numerically) from $K_I^2 + K_{II}^2$ so that the latter quantity can be used for an approximate estimate of proximity of a crack to the onset of propagation.

We found that the value of E_{eff} was quite stable, differing by not more than a few percent from one statistical sample to another, whereas the value of Max ($K_I^2 + K_{II}^2$) fluctuated significantly (reaching high values in the samples containing two closely spaced crack tips). Moreover, Max ($K_I^2 + K_{II}^2$) tends to increase with the size of the sample, at the same (or even decreasing!) crack density reflecting a higher probability of occurrence of closely spaced cracks, i.e. at the same (or even increasing!) effective stiffness E_{eff}.

Thus, the fracture-related parameter Max ($K_I^2 + K_{II}^2$) is statistically unstable and depends on the specimen's size; it does not appear to be correlated with the statistically stable effective stiffness E_{eff}.

4.   STRONG AND WEAK INTERACTIONS. CLUSTERING OF CRACKS

As discussed in [4], crack interactions that are strong in terms of their impact on SIFs may be weak in terms of their impact on the effective elastic constants.

Indeed, as is well known, contribution of a given crack (having a unit normal n and length 2ℓ) into the overall strain of a solid with cracks is $\epsilon = (<b>n + n<b>)(2\ell/A)$ where $<b>$ is the average displacement discontinuity across the crack line and A is a representative area. Thus, the contribution of a given crack into the reduction of the effective stiffness is proportional to $<b>$. In many configurations, crack interactions produce a significant increase in SIFs whereas the $<b>$-values and, therefore, the effective moduli, are affected by interactions insignificantly.

A closely related issue is that the SIFs in a crack array are highly sensitive to clustering of cracks. Fig. 2 illustrates this statement: Max ($K_I^2 + K_{II}^2$) is, typically, substantially higher in the configurations of the type 2b as compared with configurations of the type 2a.

(a)   (b)

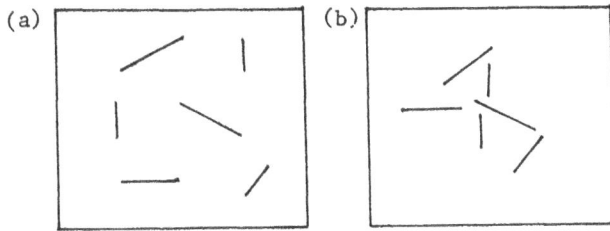

Fig. 2. Clustering of cracks.

At the same time, the effective elastic moduli's sensitivity to clustering is quite low, i.e. the difference between the values of S_{ijkl} for the configurations 2a and 2b is generally insignificant (provided the overall crack density is the same).

Thus, monitoring the change in effective elastic constants may not necessarily detect the onset of strong crack interactions and clustering of defects - events that are crucially important from the point of view of progression towards failure.

5.   EFFECT OF CRACK SHAPE

Consider a 3-D elastic solid containing elliptical cracks. As is well known, the profiles of SIFs along the crack edges and their maximal values depend on the aspect ratios of the ellipses. At the same time, the crack shapes (aspect ratios) have a very small impact on the overall effective elastic stiffness [8] (provided the overall crack density is the same).

6.   CRACK-MICROCRACK   INTERACTION PROBLEM

Fracture   propagation   in brittle   inhomogeneous   materials   is   often accompanied by microcracking (see, for example, experimental observations on high temperature fracture of ceramics [9]). Elastic interactions with microcracks may either reduce or increase the SIFs at the main crack (stress shielding or stress amplification) thus producing either toughening or enhancing effect. The absence of correlation between the effective elastic constants and the fracture-related properties of a cracked solid is particularly apparent in the problem of crack-microcrack interactions. As discussed in [5-7], the microcracks located in the short range interaction zone (several microcracks closest to the main crack tip) produce a dominant impact on the SIFs at the main crack tip, as compared with the total impact of all the microcracks located farther away. Due to a high sensitivity of the interaction effect to the exact positions of microcracks in the short range zone, the impact of microcracks on the SIFs fluctuates significantly and even qualitatively (from shielding to amplification) from one sample of the microcrack statistics to another.

This means that there is no statistically stable effect of stress shielding, at least, if the locations of microcrack centers are random. On the contrary, being conservative, one may conclude that the overall effect of interactions with microcracks is the one of stress amplification, particularly in the 3-D configurations where the local peaks of SIFs along the crack front are almost always present (in spite of the fact that the average along the front SIF may experience a mild shielding); these local peaks enhance local crack advances [5,7]. (This does not exclude, of course, the possibility of toughening by microcracking due to mechanisms other than elastic interactions, like expenditure of energy on nucleation of microcracks; neither does it exclude the possibility of stress shielding due to interactions with some special microcrack arrangements, like an extremely dense array of microcracks parallel to the main crack [10] or for the statistics of the microcrack centers that is not random but "biased" towards shielding configurations, as seems to be the case in the experimental observations of [9]). At the same time, replacing the microcracked region by an effective elastic material of reduced stiffness would produce a different result - that the effect of interactions is the one of stress shielding.

Another important effect of interactions is the appearance of "secondary" modes on the main crack, due to stochastic asymmetries in the microcrack field, i.e. K_{II} under mode I remote loading (or K_I under shear remote loading). Since the appearance of mode II SIF promotes crack kinking, this may be partially responsible for an irregular shape of crack paths in brittle microcracking materials. This effect is, obviously, missed if the microcracked zone is replaced by an effective elastic material.

Thus, modelling of the microcracked region (at least, of its short range zone that produces the dominant effect) by an effective elastic material may not be adequate, in the sense that there is no apparent correlation between the reduction of stiffness of the damage zone and its impact on the SIFs at the main crack tip.

7. CASES WHEN PROGRESSION TOWARDS FRACTURE CAN BE MONITORED BY THE CHANGE OF EFFECTIVE ELASTIC STIFFNESS

The basic underlying reason for the absence of one-to-one correspondence between progression towards fracture and change of the effective elastic constants is that the fracture-related quantities (like SIFs) are determined by the local fluctuations of the microdefect field geometry whereas the effective elastic properties are the volume average quantities, relatively insensitive to such fluctuations.

Therefore, in the cases when evolution of the defects' population follows a deterministic reprocucible pattern, the progression towards fracture may indeed be monitored by the reduction of the effective elastic stiffness. Such situations occur in certain laminated composites where, under certain loading conditions, formation of microcracks follows well established patterns (called "characteristic damage states" in [11]). It is interesting to note, however, that, as the data of [11, fig. 7] shows, the correlation between the fraction T of the remaining lifetime of a specimen and its effective stiffness E holds only for a given arrangement of layers: if the ply orientations are changed,(and, therefore, the crack patterns are changed), the E - T curve becomes entirely different. The drastic change of this curve with the change of the ply orientations demonstrates that the E - T curve is not universal, even if the material components of a composite remain the same.

8. CONCLUSION

A short discussion above points to the absence of a direct correlation between the effective elastic constants and the progression towards fracture of a solid with multiple cracks (except for the cases mentioned above). Physically, the absence of such correlation has a clear explanation: the fracture-related quantities (like SIFs) are highly sensitive to local fluctuations of the crack array geometry.

378

The effective elastic constants on the other hand, are the <u>volume average</u> properties that are relatively insensitive to such fluctuations.

ACKNOWLEDGEMENT

This work was supported by the U.S. Department of Energy (Grant DE-FG02-86ERi3607) and by the U.S. Army Research Office (Grant DAAL02-88-K-0027). The research was conducted using the Cornell National Supercomputer Facility.

REFERENCES

1. M. Kachanov (1987), "On Modelling of anisotropic damage in elastic-brittle materials - a brief review", in A. Wang and G. Haritos (eds.), Damage Mechanics in Composites, ASME, New York, pp. 99-105.
2. D. Krajcinovic (1989), "Damage mechanics", Mechanics of Materials, 8, pp. 117-197.
3. M. Kachanov (1987), "Elastic solids with many cracks: a simple method of analysis", International Journal of Solids and Structures, 23, pp. 23-43.
4. M. Kachanov and J.P. Laures (1989), "Three-dimensional problems of strongly interacting arbitrarily located penny-shaped cracks", International Journal of Fracture, 41, pp. 289-313.
5. J.P. Laures and M. Kachanov (1990), "Three-dimensional interactions of a crack front with arrays of penny-shaped microcracks", International Journal of Fracture, in press.
6. E. Montagut and M. Kachanov (1988), "On modelling a microcracked zone by weakened elastic material and on statistical aspects of crack-microcrack interactions", International Journal of Fracture, 37, pp. R55-R62.
7. M. Kachanov, E. Montagut and J.P. Laures (1990), "Mechanics of crack-microcrack interactions", Mechanics of Materials, in press.
8. B. Budiansky and R.J. O'Connell (1976), "Elastic moduli of a cracked solid", International Journal of Solids and Structures, 12, pp. 81-97.
9. L.X. Han and S. Suresh (1989), "High temperature failure of an alumina-silicon carbide composite under cyclic loads", Journal of American Ceramic Society, 72, pp. 1233-1238.
10. A. Chudnovsky and S. Wu (1990), "Effect of crack-microcrack interaction on energy release rates", International Journal of Fracture, 44, pp. 43-56.
11. K. Reifsnider and W. Stinchcomb (1983), "Stiffness change as a fatigue damage parameter for composite laminates", in U. Yuceoglu et al (eds), Advances in Aerospace Structures, Materials and Dynamics, ASME, New York, pp. 1-6.

Report of Session 6: THEORETICAL MICROMECHANICS BASED MODELS

H. Horii
Department of Civil Engineering
University of Tokyo
Bunkyo-ku, Tokyo, Japan

First, a review of some theories of toughening mechanisms in quasi-brittle materials was presented by Professor C. Atkinson. The spring model that includes the bridging mechanism was explained. Micromechanical models for various bridging mechanisms are summarized. As an example of studies on micromechanics of bridging, analyses of crack growth along the interface between short fiber and matrix are presented. In the discussion, mechanisms other than bridging such as crack deflection, crack branching, and microcracking are discussed.

Following the presentation by Professor C. Atkinson, a short presentation was given by Professor S.J. Bennison. It was shown that a relationship between bridging stress σ and crack opening displacement w is obtained from the crack profile measured in situ and a Green function. Then, by integrating it, a resistance curve, a relationship between J-value and COD, is obtained.

One of interesting issues on the bridging model is the generalization of the spring model to nonplaner crack growth under general loading. The bridging property that is represented by $\sigma - w$ curve may be different under pure mode I loading and mixed-mode loading. In fracture of concrete it has been clarified that the bridging is the dominant mechanism and fracture of concrete has been modeled by Dugdale-Barenblatt type model with a tension-softening curve. Crack growth phenomena under pure mode I loading are reproduced by the model with reasonable accuracy. It is an interesting question whether the analytical model developed for pure mode I loading is applicable to crack growth under mixed-model loading. Even under mixed-mode loading a crack (or bridging zone) propagates normal to the maximum tensile direction. The difference is that crack does not propagate straight but kinks at the tip of an initial crack and curves as it propagates. The following is an example of a study of crack growth under mixed-mode loading where deviation from pure mode I is considered to be relatively small.

Pullout tests of a two dimensional anchor bolt are carried out on concrete specimen. Cracks are initiated at the upper corner of the anchor and grow along curving path toward the position of support. Surface displacements are measured with laser speckle method, and distribution of normal and tangential displacement gaps and the location of the tip of macrocrack (bridging zone) are identified at different loading levels. Measured crack lengths at different pullout load are plotted in Fig. 1.

A BEM code of Dugdale-Barenblatt type model with a linear tension-softening relation is employed for the analysis of crack growth in the pullout test. It reproduces crack growth phenomena under mode I loading with reasonable accuracy. The same value of critical opening displacement $w_{cr} = 25\,\mu m$ is used. The results are plotted in Fig. 1 and compared with experimental observation. It is surprising that the observation and the prediction show very different tendencies. This discrepancy seems to indicate a significant difference in fracture mechanisms between crack growth under pure mode I loading and mix-mode loading even though the ratio of mode II component is relatively small.

S. P. Shah (ed.), Toughening Mechanisms in Quasi-Brittle Materials, 379–382.
© 1991 *Kluwer Academic Publishers.*

380

Careful investigation of results by laser speckle method revealed a noticeable fact. The tangential component of displacement gap is comparable to the normal component at the very early stage of crack formation although the crack propagates in the direction normal to maximum tensile stress. After fracture proceeds and displacement gap becomes large, the tangential component is negligible and direction of displacement gap is almost normal to the crack path.

If we consider an application of small shear deformation to a specimen under uniaxial tension after the peak stress and formation of fracture plane, it is expected that the bridging stress is suddenly cut off and further application of shear may lead to compressive stress due to the clenching of rough surfaces. Therefore along the bridging zone propagating under mixed-mode loading, the full bridging according to the tension-softening curve for pure mode I may not be materialized. The bridging stress may be cut off at much smaller opening displacement.

To check this idea we carried out the BEM analysis with a very small value of

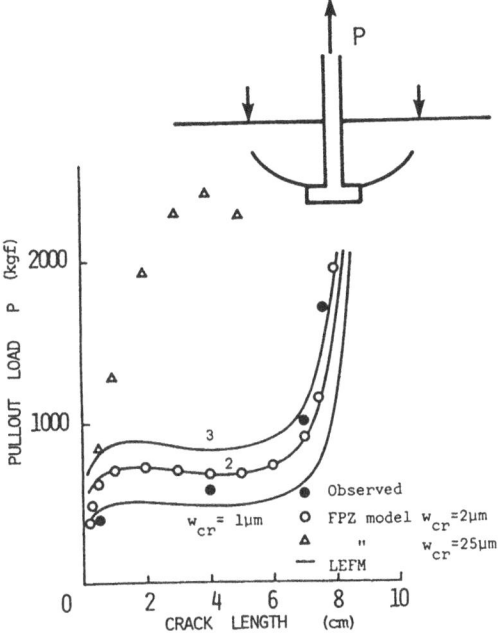

Fig. 1. Relation between pullout load and length of crack emanating from anchor bolt.

the critical opening displacement; $w_{cr} = 2\mu m$. Since the critical opening displacement is very small, the bridging zone is fully developed in an early stage and traction free crack extends keeping the fully developed bridging zone at its tip. The results are plotted in Fig. 1. The agreement with experimental observation is noteworthy. This agreement seems to support the idea mentioned above while it requires further examination from different aspects to lead to general consensus. For such a small value of critical opening displacement, the length of fully developed bridging is small and linear elastic fracture mechanics (LEFM) is considered to be valid. The results by LEFM are also plotted in Fig. 1 with the values of fracture toughness, K_{IC}, corresponding to fracture energy calculated for $w_{cr} = 1, 2, 3\mu m$. The result by LEFM for $w_{cr} = 2\mu m$ agrees with that of the Dugdale-Barenblatt type model.

Professor D. Krajcinovic presented works on micromechanical model of brittle deformation. Statistical models are presented, which simulate behaviors in uniaxial tensile test. Contrasting nature of behaviors before and after the peak load was demonstrated. While the pre-peak behavior can be modeled by a continuum model, the post-peak behavior is non-deterministic and strongly dependent on initial disorder. Effective continuum model of cracked material and other topics are also presented.

Related to the statistical model, short presentations are made by Professor Z.P. Bažant and Professor P. Rossi on simulation of uniaxial tension tests. Professor Bažant showed that the size effect law is obtained from the simulation.

A paper on the relation between effective elastic and fracture properties of microcracked solids and mechanics of crack-microcrack interactions was presented by Professor M. Kachanov. It was pointed out that effective compliance and stress intensity factor do not have correlation because stress intensity factor is local quantity. Corresponding to this point the relationship between the effective compliance and energy release rate (and accordingly the stress intensity factor) was pointed out in the discussion.

A method of solution for multicrack problems that is powerful for three dimensional problem was also presented. Results of numerical simulation of crack-microcrack interaction with randomly generated distribution of microcracks were shown. It was pointed out that no statistically stable effect of shielding or amplification was observed. There was in the discussion one comment that the effect of residual strain is much larger than that of reduction in stiffness in toughening due to residual strain induced microcracking. Change in stress intensity factor of a stationary crack due to microcracks cancels out if microcracks are induced only by stress concentration without residual strain.

One of interesting issues coming up in this session is on the method of approach for description of mechanical phenomena: computer simulation versus micromechanics (or micromechanics based continuum approach). Both approaches are based on behavior of microstructures. However, ways of bridging microscopic behavior and macroscopic behavior are different in two approaches.

In computer simulation relationship between micro and macro is treated through numerical computation. Hence it is sometimes called numerical experiments. The arrangement of microstructures in each computation is important and sometimes the method is incorporated with statistical technique.

On the other hand, micromechanics (or micromechanics based continuum approach) aims at, at least as the final goal, establishing continuum model through homogenization. One of main tasks of micromechanics approach is the analysis of behavior of microstructure. Many studies with micromechanics approach are finished at this stage. However, our final goal in many cases is the use of those results at microscopic scale for the prediction of macroscopic phenomena.

In general, it is not an easy task to bridge microscopic phenomena and macroscopic behavior of materials by homogenization or averaging. One typical example of such process is the derivation of overall moduli of cracked solids. Another example is found in the analysis of fracture of fiber-reinforced materials shown by the first speaker of this session; see Fig. 2. Pullout behavior of a single fiber with debonding and cracking is a microscopic behavior. By modelling the microscopic mechanism one may reproduce microscopic phenomena. Then through averaging or homogenization process, the relationship between bridging stress and crack opening displacement is obtained. It is a macroscopic behavior that is used as the constitutive equation in the bridging zone for the analysis of crack growth in fiber-reinforced material.

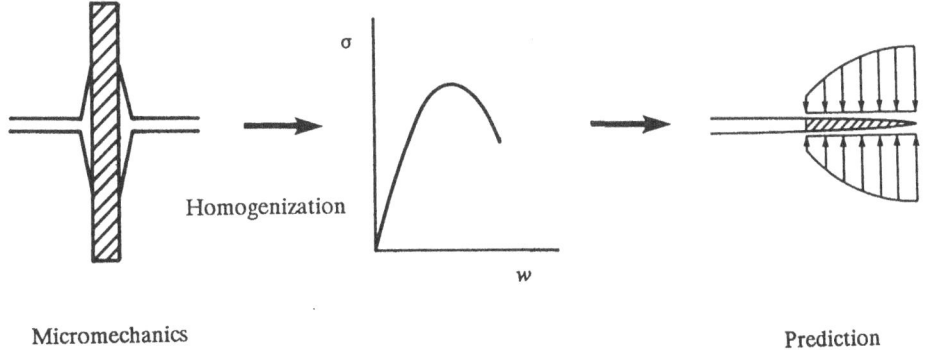

Fig. 2. Example of micromechanics based continuum approach.

382

As a matter of course it is important to clarify the objective of study. What is the objective when studies on material behaviors are carried out paying attention to microscopic phenomena? One answer may be the clarification of the governing mechanism. It would be phrased as to develop an improved understanding of process of deformation or fracture. Another answer may be the establishment of analysis tools for prediction or reproduction of phenomena in actual problems. The appropriate way to approach the problem should be different for different objectives.

Computer simulation is suitable for clarification of the governing mechanism. If the governing mechanism is modelled properly, the phenomena are simulated under general conditions. The agreement of prediction with observation is a support of the modeled mechanism. However, one should be careful since the simulation may not work if the most dominant mechanism is not included or it is lost in the process of simulation. It is too optimistic to expect that real phenomena are reproduced by introducing imperfections. The buckling of a column cannot be predicted by giving initial imperfections unless geometrical nonlinearity is included. A typical example in solid mechanics is localization phenomena. In localization of any type, the interaction between microstructures is important. Unless it is treated properly, localization phenomena may not be simulated.

Computer simulation is not suitable for analysis of actual problems. In actual problems each microstructure cannot be treated separately. We have to use continuum quantities and governing equations for them. Micromechanics approach (or micromechanics based continuum approach) is suitable not only for identification of the governing mechanism but also for development of analysis tools. However, careful consideration is also necessary for micromechanics approach because important information may be lost in the process of homogenization.

For example, growth of two parallel cracks shown in Fig. 3 depends on the loading condition. In the case (a) only one crack grows when stress reaches its critical value; localization. In case (b) two cracks grow equally; delocalization. When we introduce homogenization and use a continuum model as shown in Fig. 3c, the information that makes the distinction between (a) and (b) must be kept in the homogenization process. Otherwise the prediction by the continuum model could be very different from reality. It would be the future task to find the governing mechanism of localization phenomena and to establish proper procedure of homogenization.

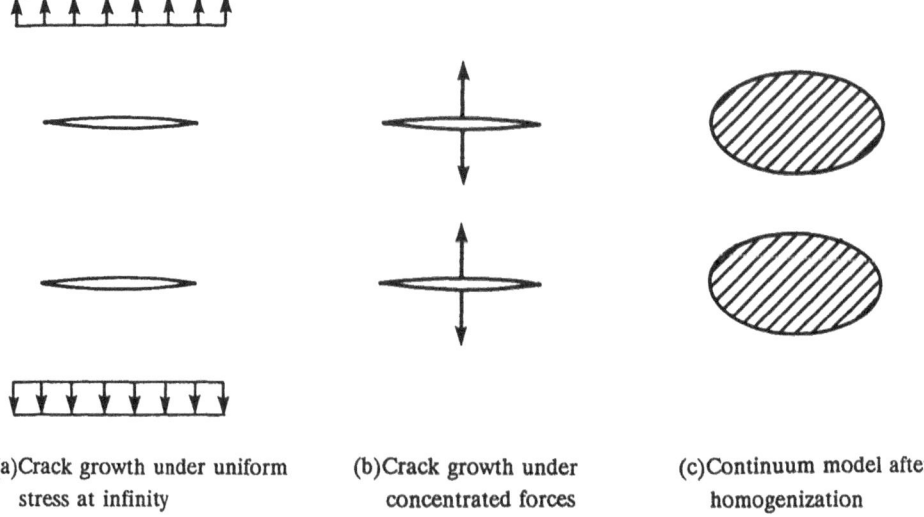

(a)Crack growth under uniform stress at infinity   (b)Crack growth under concentrated forces   (c)Continuum model after homogenization

Fig. 3. Growth of two equal parallel cracks.

Fracture Process in Fiber Reinforced Ceramics

DETERMINATION OF FIBER-MATRIX INTERFACIAL PROPERTIES OF IMPORTANCE TO CERAMIC COMPOSITE TOUGHENING

E. R. FULLER, Jr., E. P. BUTLER, and W. C. CARTER

Ceramics Division
National Institute of Standards & Technology
Gaithersburg, MD 20899

ABSTRACT

The fiber-matrix interface is investigated for its role during fiber debonding and pull-out and its effect on the toughening of ceramic composites. Possible interface debonding criteria (critical interface shear strength; interface toughness, or critical energy-release rate, and asperity interlock) are considered. The interfacial and mechanical properties of a model SiC-fiber, borosilicate-glass-matrix composite system are measured by a single fiber pull-out test. Initial stresses are determined by photoelastic methods.

1. Introduction

The fracture toughness of brittle ceramics has been enhanced greatly in the last decade by the development of composite microstructures which lead to the phenomenon of crack bridging [1]. In such material systems, or composites, the incorporation of ceramic reinforcements in a brittle ceramic matrix results in the deflection of macro-cracks along and around the reinforcements, thereby leaving them as bridging elements in the wake of the macro-cracks. There they apply local closure forces to retard crack advance. Advantages of reinforcement toughening over other toughening mechanisms are that: (1) a variety of reinforcements (fibers, whiskers, platelets, etc.) can be incorporated into a myriad of host materials; (2) directional properties can be obtained by texturing the reinforcements; and (3) toughening can be retained at elevated temperatures. Reinforcement toughening depends critically upon the properties of the fiber-matrix interface. Initially, the interface must be "weak" enough to deflect the crack along or around the reinforcements; otherwise no toughening can occur. The resulting toughening is governed by the amount of energy that is dissipated at the "debonding" interface. The toughening increment from fiber bridging is simply the product of the volume fraction of fibers, V_f, and the integral of the force-displacement curve for fiber debonding and frictional pull-out divided by the cross-sectional area of the fiber. This is the rate of energy dissipation by the bridging elements per unit area of crack surface. In terms of the J-integral [2], this toughening increment is:

$$J_T = V_f \int [T(\delta)/\pi R^2] d\delta , \tag{1}$$

where $T(\delta)$ is the force-displacement relation for a single-fiber pull-out and R is the fiber radius. The important interfacial parameters for fiber toughening lie in this force-displacement

385

S. P. Shah (ed.), Toughening Mechanisms in Quasi-Brittle Materials, 385–403.
© 1991 *Kluwer Academic Publishers.*

relation. These parameters are amenable to direct experimental measurement via a single-fiber pull-out test. This test was initially developed for testing polymeric composites [3], but is now widely applied to all composite materials. Furthermore, this test provides a systematic way for investigating the influence of fiber-matrix properties and processing conditions on the toughness of the final manufactured composite. In this paper we discuss both the theoretical aspects of fiber pull-out with a perspective more towards a physical description of the important phenomena rather than a rigorous treatment, and experimental techniques for determining the relevant interfacial properties. In Section 2, a simplistic analysis is developed to describe the physical processes that govern fiber pull-out, namely, frictional stress transfer and interfacial debonding. Fiber pull-out is treated as a micro-fracture problem in which a Mode II interface crack with frictional wake tractions propagates down the fiber-matrix interface. Various debond criteria are discussed for the propagation of this interface crack. Finally, extensions to more complete analyses are discussed, particularly in relation to Poisson's contraction, a phenomena that must be utilized in the experimental studies to deconvolve the frictional shear resistance into its component parameters. In Section 3, experimental studies are described for determining the relevant interfacial properties. Particular emphasis is given to the initial "clamping" stress on the fiber for which three independent techniques are available.

2. Theoretical Aspects of Fiber Pull-out

2.1 GENERAL CONCEPTS

Fiber pull-out is treated here as a micro-fracture problem. A cylindrical crack of radius R (also the fiber radius) is propagated down the fiber-matrix interface as a pure Mode II or shear crack as the fiber is pulled from the matrix. See Fig. 1. Frictional tractions along the crack wake serve both to transfer the axial stress from the fiber to the matrix and to retard the propagation of the "debond crack". Analyses of this problem, thus far (see, for example, [4-7,16,17]), have been axially symmetric elasticity analyses in which the details of the stress singularities at the debond (crack) tip have not been considered. In general, we take the same approach. Details of the crack-tip mechanisms, with one exception, are taken as a global debond criterion (for example, an interfacial shear strength or an interfacial critical strain-energy-release rate). The one exception is our considerations of the shear stress necessary to "break" asperity interlocks for a micro-mechanical frictional model.

2.2 ZERO POISSON'S RATIO (ZPR) MODEL

Analysis of fiber debonding and subsequent frictional pull-out is treated here as an interfacial fracture problem [8,9] with frictional wake tractions [4-7]. A simplistic analysis [6], i.e., one with no Poisson's contractions, is illustrated schematically in Fig. 2. A cylindrical crack of radius R (also the fiber radius) is propagated down the fiber-matrix interface as a pure Mode II crack. The axial stress, σ, is transferred from the fiber to the matrix along the debonded-crack flanks by interfacial friction. The differential stress transfer with axial distance z is:

$$d\sigma/dz = -2\tau/R, \tag{2}$$

where τ is the frictional shear stress on the interface. Friction, in general, is treated as Coulomb friction, where the frictional shear stress on the interface, τ, is the product of a coefficient of friction, μ, and the total radial, matrix (clamping) stress, $-\sigma_{rr}$:

$$\tau = -\mu\sigma_{rr} \tag{3}$$

for $\sigma_{rr} < 0$. In the plane strain approximation with no Poisson's contractions, the radial matrix (clamping) stress, σ_{rr}, is only the initial clamping pressure, q_o, which results from the

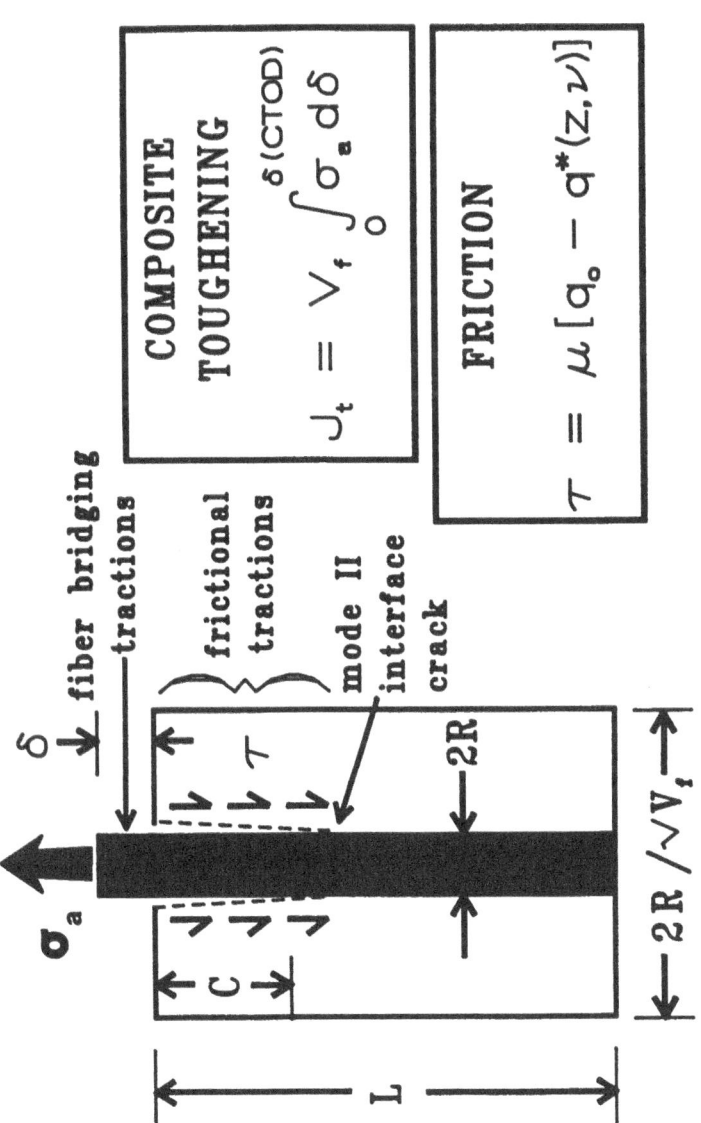

387

COMPOSITE TOUGHENING

$$J_t = V_f \int_0^{\delta(CTOD)} \sigma_a \, d\delta$$

FRICTION

$$\tau = \mu[q_o - q^*(z, \nu)]$$

□ Figure 1

Schematic representation of the fiber pull-out process both as a bridging traction for the macro crack and as a Mode II interface crack with frictional wake tractions.

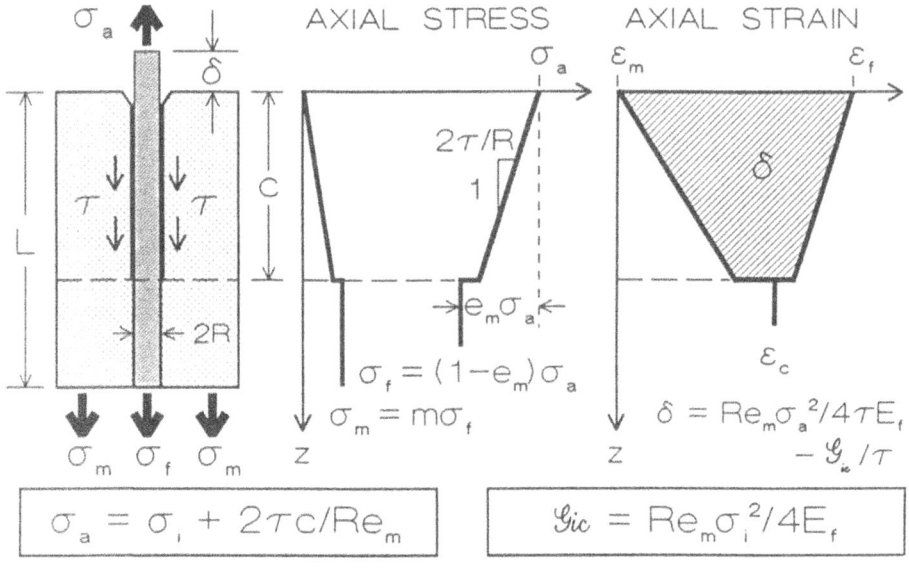

□ **Figure 2**
**Schematic drawing of the ZPR model of fiber debonding and subsequent frictional fiber
pull-out: an interfacial fracture problem with frictional wake tractions.**

thermal expansion mismatch strains that occur during fabrication. The frictional shear stress, $\tau = \tau_o$, is accordingly independent of z, so that the axial fiber stress, as determined from Eqn. (2), decreases linearly in z, as depicted in Fig. 2, as this stress is transferred to the matrix. The interface is presumed to be "cohesively" bonded at the crack tip. Accordingly, the axial fiber stress does not fall entirely to its average composite value, σ_f, at the crack tip, but rather has a stress discontinuity. A more rigorous analysis would include both a non-frictional shear-stress transfer ("shear-lag" analysis) to σ_f beyond the crack tip (see Section 2.3.1.) and a crack-tip stress singularity. Another physical criterion is required to determine the magnitude that the stress discontinuity can reach before equilibrium crack propagation or fiber debonding ensues. Several possibilities exist, and these are discussed in the next section.

2.3 INTERFACE DEBOND CRITERION

2.3.1 *Interface Shear Strength*. The most commonly used debond criterion is that the interfacial shear stress in the bonded region exceeds some critical value, τ_S, the interfacial shear strength. The ZPR model of Section 2.2., however, has a zero interfacial shear stress in the bonded region. Inclusion of a simple shear-lag analysis in the bonded region "smooths" the axial stress drop to an exponential decay with decay constant, $\beta = R^{-1}[2m/\ln(V_f^{-1})]^{1/2}$, where $m = E_m/E_f$, [the ratio of the Young's modulus of the matrix to that of the fiber]. See Fig. 3. This yields an interfacial shear stress along the bonded interface, which is a maximum at the "crack" tip and decreases exponentially ahead of the tip. The resulting criterion is

$$\tau_s = \beta R \sigma_s /[2\tanh(\beta L)] \tag{4}$$

where σ_s is an axial debond stress. The equilibrium debond relation is

$$\sigma_a = \sigma_s + 2\tau c /R \tag{5}$$

where c is the debond length.

2.3.2 *Interface Toughness*. A alternative debond criterion is that the strain-energy-release-rate for the crack system exceeds some critical value, G_{ic}, corresponding to an interfacial (Mode II) fracture toughness. To determine the magnitude that this stress discontinuity can reach before equilibrium crack propagation ensues, one could use a strain-energy-release-rate analysis similar to that of Marshall and Oliver [4] and Gao et al. [5]. An alternate, and more elegant approach is to use a three-dimensional contour, or J-integral [10]. The result is the same; the strain-energy-release-rate, G_{II}, of the debond crack tip is

$$G_{II} = J_a (\sigma_i /\sigma_a)^2 \tag{6}$$

where the applied strain-energy-release-rate, J_a, is given by

$$J_a = \frac{Re_m \sigma_a^2}{4E_f} \tag{7}$$

and $e_m \sigma_i$ is the size of the crack-tip stress discontinuity, $e_m = m/(m+f)$, and $f = V_f/(1 - V_f)$ [the fiber-to-matrix volume fraction ratio]. Equating this G_{II} to a critical interface toughness, G_{ic}, we obtain the following relation between external stress, σ_a, and debond crack length, c, for equilibrium crack propagation:

$$\sigma_a = \sigma_i + 2\tau_o c /Re_m, \tag{8}$$

where the "frictionless debond" stress, σ_i, is related the interface toughness, G_{ic}, through the relation: $\sigma_i = [4E_f G_{ic}/Re_m]^{1/2}$.

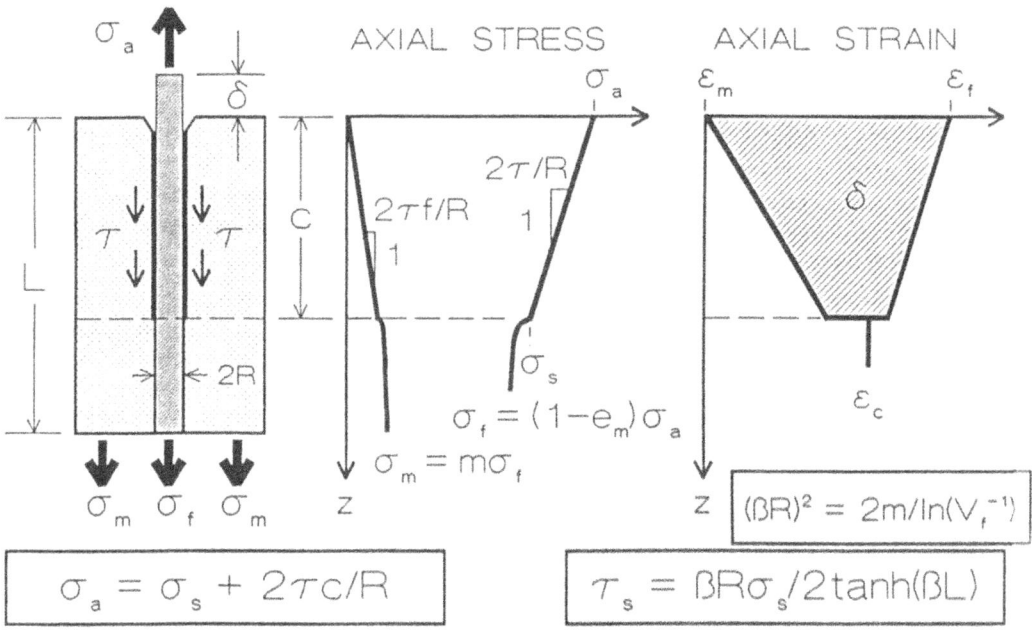

□ **Figure 3**
Schematic drawing of the ZPR model of fiber debonding and subsequent frictional fiber
pull-out with a shear-lag addition in the "bonded" region.

We note that the equilibrium debond relation is similar for both the shear strength and the toughness criterion, cf., Eqns (5) and (8). The important difference is the dependence of σ_s and σ_i on other parameters, particularly R. Since τ_s is a physical constant in that approach, σ_s is only weakly dependent on R (through V_f). In contrast, σ_i is proportional to $R^{-1/2}$.

2.3.3 *Asperity Interlock Strength.* The debond criterion for asperity inter-lock is in some sense a micro-mechanical model for the interfacial shear strength of Section 2.3.1. The importance of asperity interlock was first noted by Jero [11] and Jero and Kerans [12] when they reported the magnitude of the "re-seating" load-drop as a fiber was pushed back into its original "inter-lock" configuration. These observations have recently been confirmed for fiber pull-out as well [13]. Using a pull-out specimen with a fiber protruding from both ends, the fiber was first pulled in one direction, and then pulled back though the matrix from the opposite side. A load-displacement curve for one cycle of loading is shown in Fig. 4.

The present analysis of asperity interlock strength is a synopsis of a more complete analysis [14]. We propose the following model to account for asperity interlock and the associated "seating drops". Spherical asperities are presumed to be randomly located on both the fiber surface and the opposing matrix surface. Tractions between the fiber and matrix are transmitted via asperity contact so that the model reduces to that of Hertzian contact[†]. Initially, the asperities of one surface are wedged into the "valleys" of the other. The wedging force on each asperity, or the force which acts normal to the global plane of contact, is such that the average net force per unit area matches the clamping pressure derived from the differential thermal expansion contractions. This initial configuration is illustrated in Fig. 5. Resistance to fiber pull-out has two components: elastic deformation of the asperities; and local frictional resistance given by the product of the friction coefficient, μ, and the local normal force, F_h. The forces due to elastic deformation alone can either hinder pull-out when an asperity first contacts another asperity; or, assist pull-out when an asperity slips past the "summit" of another asperity. Friction, however, always opposes pull-out. In the initial, "locked," configuration, all elastic deformation forces oppose pull-out, thereby requiring the largest tractions to be applied initially. After the fiber has slid out of the initial configuration, it is presumed that asperities randomly sample the opposing interface; in this case, the elastic part of the forces average to zero leaving only frictional forces. To calculate this effect, we consider the system illustrated in Fig. 5 with the additional simplifying assumption that the Poisson's contraction is zero. The upper body (fiber) in Fig. 5 slides past the lower body (matrix) in a completely rigid fashion, except for the deformation of the spherical asperities. If the axial coordinate of the center of the sphere in the fiber is z, then the distance between the centers is:

$$c^* = (z^2 + (2R^* - h)^2)^{1/2} \tag{9}$$

where R^* is the radius of the asperity, and h is the "overlap" distance which is calculated from the initial clamping pressure, q_o as follows: The force between two asperities is given by,

$$F_h = E_h(R^*/3)^{1/2}(R - c^*/2)^{3/2} \tag{10}$$

where E_h is an effective modulus, given by,

[†] A particularly relevant account of this Hertzian problem is given by Mindlin [15].

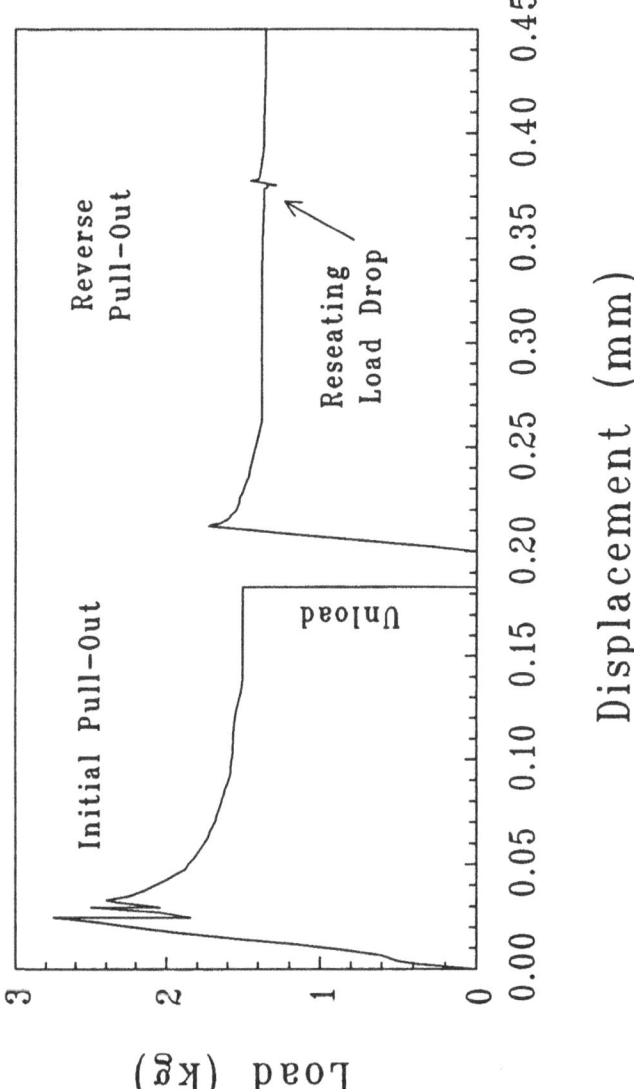

□ **Figure 4**

Force-displacement curve showing load drop on asperity interlock re- seating.

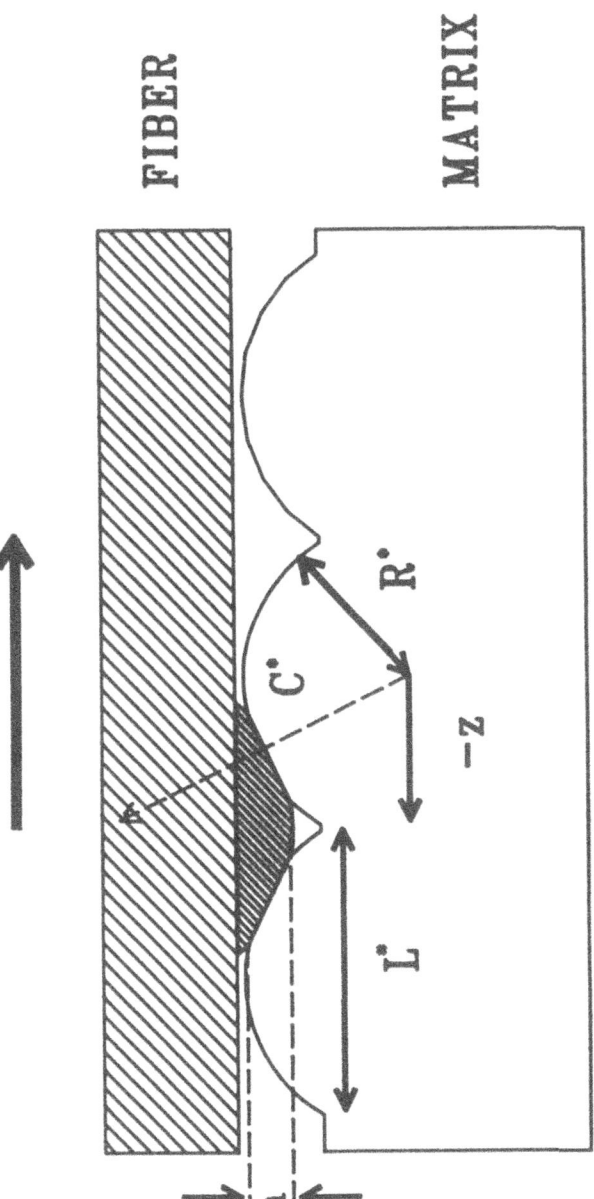

□ **Figure 5**

Schematic representation of the asperity friction model: interlocking, spherical asperities with an interlock distance of h.

$$\frac{1}{E_h} = \frac{1 - \nu_1^2}{E_1} + \frac{1 - \nu_2^2}{E_2} \tag{11}$$

One asperity must bear half the load in the radial direction, or,

$$\frac{q_o A_b}{2} = F_h \frac{2R^* - h}{c^*} \tag{12}$$

where A_b is the area per bump. Eqn. (12) must be solved for h. The axial force required to quasi-statically move one asperity past another as a function of the axial coordinate, z, of one sphere with respect to another is given by:

$$F_{ax} = \frac{F_h}{c^*}[\mu(2R^* - h) - z] \tag{13}$$

The overall effect of one cycle (moving from the initial to an adjacent valley) is obtained by summing over all three asperities in Fig. 5. The shearing stress, or sliding stress is:

$$\tau = \Sigma F_{ax}/A_b \tag{14}$$

with Σ indicating the relevant sum over all asperities. One particular case is plotted in Fig. 6 for $\mu = 0.05$, $E_h = E_{glass}/[2(1 - \nu_{glass}^2) = 30$ GPa, $q_o = 65$ MPa, $R^* = 1$ μm, $A_b = 6.7$ μm^2. The stress oscillates between a peak sliding stress of about 60 MPa and -45 MPa (negative stresses indicate the fiber is being elastically pushed out). The system is unstable to dead load when the stress is decreasing with slide-out distance. The average stress, about 8 MPa, is illustrated at the end of an entire cycle to represent an average sampling of asperities.

2.4 POISSON'S CONTRACTION

In most of the preceding analysis Poisson's ratios of both the matrix and the fiber have been asssumed to be zero. Accordingly, no Poisson's contractions, or expansions, were considered. As discussed, this is a simplifying assumption that permits the development of a completely analytical analysis, the ZPR model. For considerations of composite toughening this assumption is probably not an important limitation. However, as we discuss in Section 3, inclusion of the Poisson's effect in the analysis and the attainment of experimental conditions to see its influence is vital for deducing a full complement of interfacial parameters. When Poisson's effects are included, the resulting contractions reduce the radial clamping pressure and accordingly the effective distance over which the frictional shear stresses are transferred for a given debond crack length. (We note that the applied stress must also be reduced, since the larger stress cannot now be transferred over a shorter distance with reduced friction.) This phenomenon results in an asymptotic, or steady-state value for the fracture resistance, and debond stress. An analysis that includes fiber-debonding, frictional pull-out, and Poisson's contractions is that of Gao et al. [5]. Although their treatment is a plane-strain approximation to the exact problem, it is fully analytical and expressed in terms of materials' interface and processing parameters. An analysis by Hsueh has the same general features [16,17]. Using here the analysis of Gao et al., the generalization of Eqn. (8), is:

$$\sigma_a = \sigma_i e^{-\lambda c} + \tilde{\sigma}(1 - e^{-\lambda c}), \tag{15}$$

where λ is a reciprocal length giving the effective frictional shear-stress transfer distance and $\tilde{\sigma}$ is the asymptotic pull-out stress for long cracks. The experimentally determined parameters σ_i, λ and $\tilde{\sigma}$ are related, respectively, to the interface fracture toughness, G_{ic}, the fiber-matrix friction coefficient, μ, and the initial fiber-clamping pressure, q_o, by:

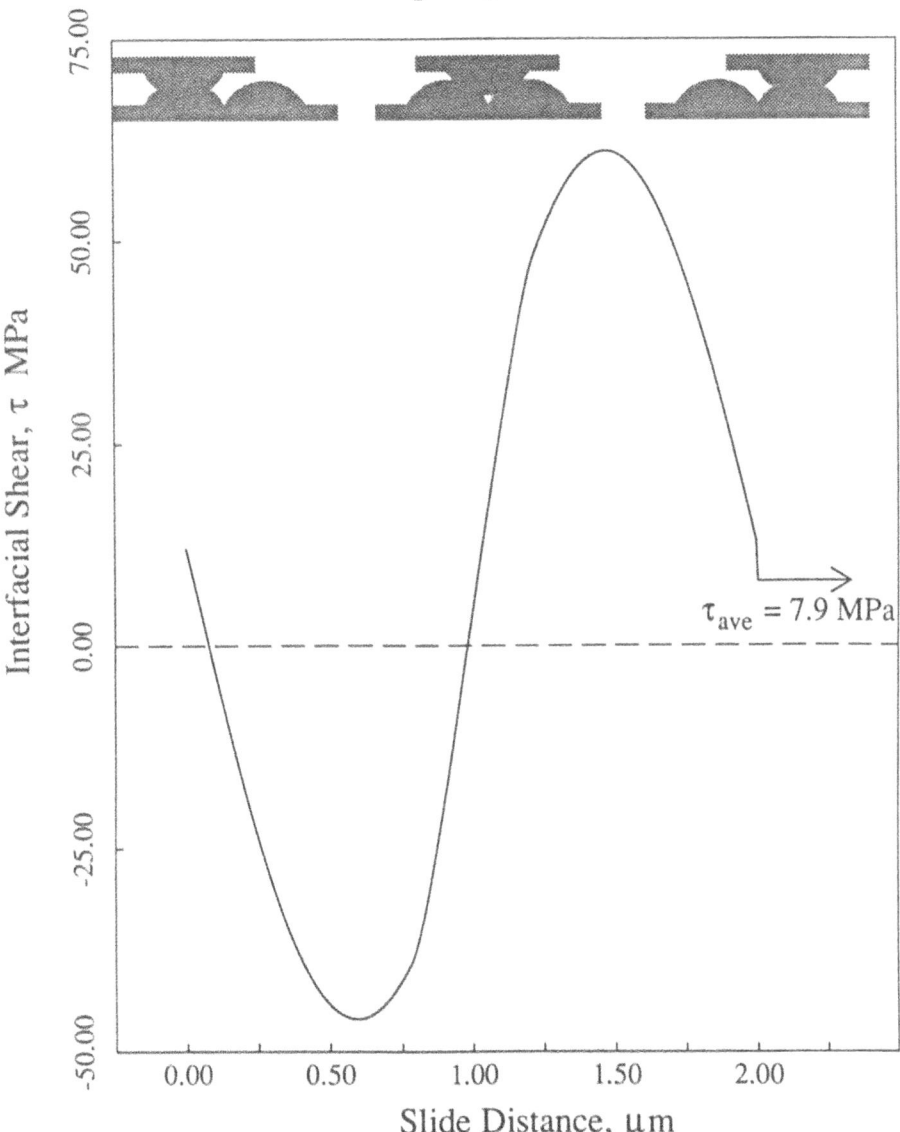

□ **Figure 6**
Calculated interfacial shear resistance from asperity deformation, interlock and friction
from the Hertzian model for $\mu = 0.05$, $q_o = 65$ MPa, $R^* = 1$ µm. The process leading a
seating drop is illustrated above the curve. Elastic forces push the asperity along initially and hinder asperity motion as it climbs out of the "valley." The elastic stresses average
to zero over one cycle; friction is always positive, so the only contribution to τ_{ave} is from
friction.

$$\sigma_i = \left\{ (4E_f G_{ic}/R) \frac{1 + (f/m)(1-2k\nu_m)/(1-2k\nu_f)}{1-2k\nu_f} \right\}^{\frac{1}{2}} \qquad (16)$$

$$\lambda = 2\mu k/R \qquad (17)$$

$$\tilde{\sigma} = \frac{q_o}{k}[1 + (f/m)(\nu_m/\nu_f)] \qquad (18)$$

where $k = (m\nu_f + f\nu_m)/[m(1-\nu_f) + 1 + \nu_m + 2f]$, and ν_m and ν_f are the matrix and fiber Poisson's ratio, respectively. These results reduce to the simplistic analysis of Fig. 2, when $\nu_m = \nu_f = 0$.

3. Determination of Interfacial Properties

3.1 MODEL COMPOSITE SYSTEM

The SiC fibers (AVCO SCS-6) are 142 μm diameter monofilaments of chemically vapor-deposited SiC on a carbon core and have two carbon-rich surface layers. The matrix material is Corning borosilicate glass #7740.

3.2 SINGLE-FIBER PULL-OUT TEST

3.2.1 *Specimen Fabrication*. Specimens for the single-fiber pull-out test were fabricated from SiC fibers sandwiched between two plates of borosilicate glass. The fibers were first degreased by washing them in 1,1,1-tri-chloro-ethelyne, followed by rinsing them in acetone and then ethanol. The sandwich assembly was diffusionally bonded by hot pressing in vacuo at a temperature of 728° C and under a stress of 3.4 MPa. The bonded assembly with fiber protruding from both ends was then cut to give two specimens of different embedded fiber lengths. To enhance fiber-matrix bonding, some fibers were washed with a 50 vol% nitric acid solution following the degreasing procedure to remove the carbon-rich surface layers [18].

3.2.2 *Experimental Procedure*. Fibers were pulled from the matrix using an universal testing machine with a 5 kg load cell. The sample-grip arrangement is shown in Fig. 7. The cross-head displacement speed was 10 mm/min, a rate fast enough to prevent environmentally assisted fracture, but slow enough for the response time of the strip-chart recorder and load cell. Effects of system compliance on experimental results were evaluated and found to be negligible. A principal experimental difficulty in many fracture mechanics measurements is the determination of the crack length. This difficulty is particularly pronounced for the fiber-matrix debond crack due to its three-dimensional nature. A simple stratagem was used in the current experiments to resolve this difficulty. The debond crack was propagated to the end of the specimen, at which point the crack length is known explicitly, namely the embedded fiber length, L. See Fig. 2 or 7. Since the applied stress, σ_a, increases with crack length, c, according to Eqn. (15) [i.e., the "rising fracture resistance" behavior of the micro-mechanical pull-out problem], this "break-through" fracture condition corresponds to a maximum fiber-debond stress, σ_d:

$$\sigma_d = \sigma_i e^{-\lambda L} + \tilde{\sigma}(1 - e^{-\lambda L}) \qquad (19)$$

After "break-through", the interface toughness no longer contributes to Eqn. (19), i.e., $\sigma_i = 0$, and the applied stress drops to that of the maximum frictional stress, σ_f:

$$\sigma_f = \tilde{\sigma}(1 - e^{-\lambda L}) \qquad (20)$$

Accordingly, by using specimens of varying embedded fiber lengths, empirical relations can be developed between σ_d and σ_f versus $c=L$. The stress drop upon "break-through" is

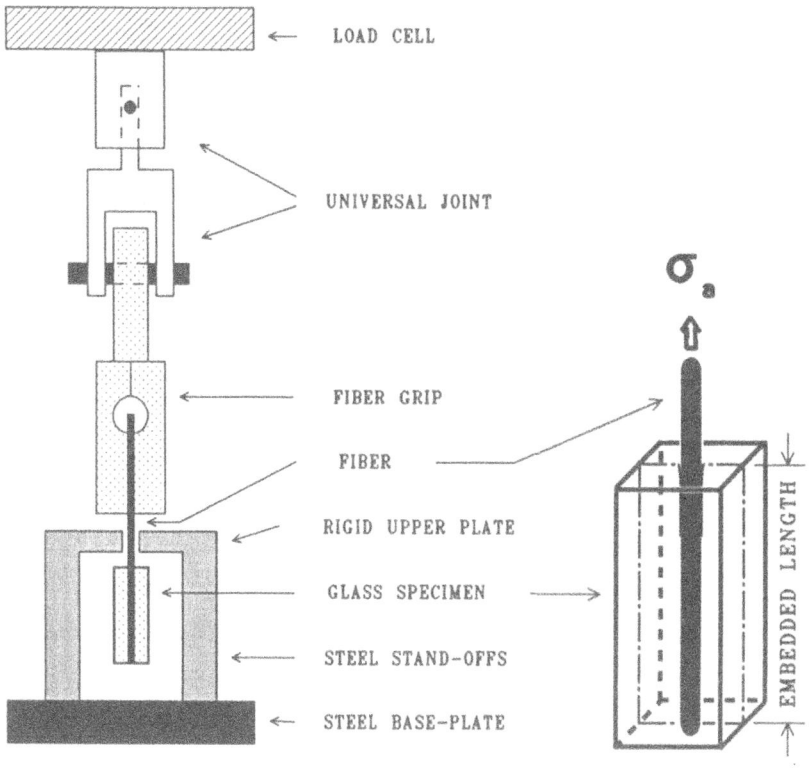

LOAD CELL

UNIVERSAL JOINT

σ_a

FIBER GRIP

FIBER

RIGID UPPER PLATE

GLASS SPECIMEN

STEEL STAND-OFFS

STEEL BASE-PLATE

EMBEDDED LENGTH

□ **Figure 7**
Schematic diagram of the single-fiber pull-out test geometry.

an accurate way to determine, σ_i, or G_{ic}:

$$\Delta\sigma = (\sigma_d - \sigma_f) = \sigma_i e^{-\lambda L} \tag{21}$$

Upon continued loading after the stress drop, the applied stress monotonically decreases to zero as the fiber pulls from the matrix. Theoretically this decrease should be represented by Eqn. (20) with L replaced by the remaining embedded length. Experimentally, a reverse curvature to that expected (i.e., positive rather than negative) is occasionally observed.

3.2.3 *Experimental Results.* The maximum debond stress and the maximum frictional stress were both measured as a function of embedded fiber length. The results are presented in Fig. 8 for as-received fibers and nitric acid-washed fibers in a borosilicate glass matrix. For each fiber treatment, both the debond and frictional stress data show an initial linear behavior and then turn over into a common asymptotic stress, $\eth$, at long embedded fiber lengths. The debond stress is always greater than the frictional stress at a given L. In the linear region of the debond curve, the initial gradient is equal to $\lambda(\eth - \sigma_i)$ and the stress-axis intercept is the frictionless debond stress, σ_i.

Table I.

Single-fiber pull-out parameters and intrinsic interface
properties of a SiC fiber-reinforced, borosilicate glass
matrix composite system with varied fiber surface treatments.

SINGLE-FIBER PULL-OUT PARAMETERS AND INTRINSIC COMPOSITE INTERFACE PROPERTIES	Surface Treatment of SiC Fiber			
	As-Received		Acid-Washed	
Frictionless Debond Stress, σ_i (MPa)	149	± 36	285	± 52
Asymptotic Pull-Out Stress, $\eth$ (GPa)	2.9	± 0.3	3.3	± 0.6
Inverse Stress Transfer Length, λ(m^{-1})	31	± 5	48	± 12
Interface Toughness, G_{ic} (J/m^2)	1.0	± 0.5	3.6	± 0.1
Frictional Shear Resistance, τ_o (MPa)	3.3	± 0.6	5.5	± 0.2
• Friction Coefficient, μ	0.05	± 0.01	0.08	± 0.02
• Initial Clamping Pressure, q_o (MPa)	65	± 6	72	± 12

The σ_d and σ_f data for each fiber treatment were fitted simultaneously to Eqns. (19) and (20), respectively, using a non-linear least squares analysis to determine the three independent parameters: σ_i, λ, and $\eth$. Since the value for σ_i was typically very small, this simultaneous fit was not an accurate way to determine σ_i (i.e., this procedure occasionally gave a negative value for σ_i, which is not physical, since experimentally the stress differences in Eqn. (21) were all greater than or equal to zero). Accordingly, both data sets were initially fit to Eqn. (20) to give a common value of λ and $\eth$. The difference between the debond stress and the frictional stress at each value of L was then fit to Eqn. (21) with the known value of λ to determine σ_i. The values for the parameters thus determined are given in Table I for both the as-received and nitric acid-washed fibers. The error given for each parameter represents only that error resulting from the least squares fit. The intrinsic interface properties, G_{ic}, μ, and q_o, were calculated from these parameters using Eqns. (16), (17), and (18), respectively.

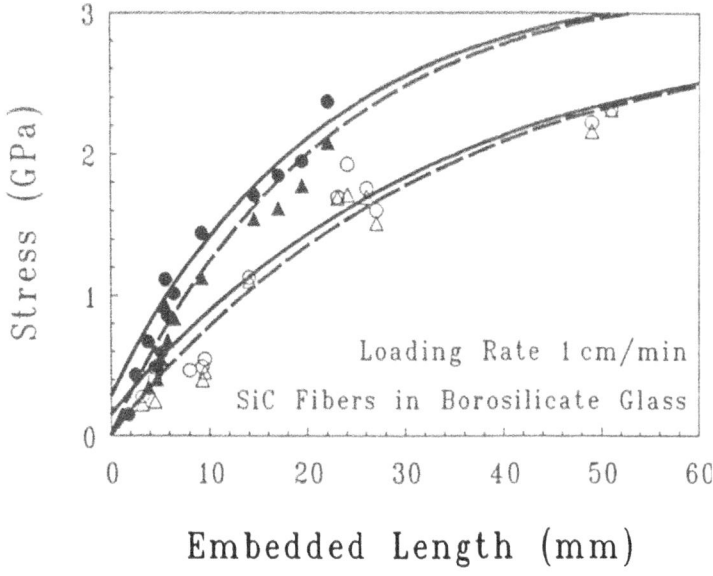

Loading Rate 1 cm/min

SiC Fibers in Borosilicate Glass

Embedded Length (mm)

□ **Figure 8**
Debond stress (circles) and maximum frictional stress (triangles) as a function of embedded fiber length for as-received fibers (open symbols) and nitric acid-washed fibers (filled symbols) in a borosilicate glass matrix. The solid and dashed lines are graphs of Eqns. (19) and (20), respectively, with the parameters σ_i, λ and $\tilde{\sigma}$ determined from the least squares fit.

These results are also given in Table I. Pull-out measurements at long embedded lengths are required to deconvolve the interfacial shear resistance, τ_o, into its components, μ and q_o. However, τ_o by itself can be determined from the initial slope, $\lambda\eth$, of frictional pull-out curve:

$$\tau_o = \mu q_o = (\lambda\eth R/2)/[1 + (f/m)(v_m/v_f)]. \tag{22}$$

Comparing the results of the as-received and nitric acid-washed fiber pull-out tests, we observe that the fiber-matrix bond, as characterized by the interface toughness and friction coefficient, were both enhanced by the nitric acid treatment. Little change was observed in the initial clamping pressure within estimated errors. The acid wash facilitated the removal or etching of the outermost carbon layer(s) of the SCS-6 fibers [18]. These layers are important in two ways: as an interlayer to prevent strong fiber-glass bonding and as a frictional lubricant at the interface. Thus, removal of these layers would cause an increase in both interfacial bonding and interfacial friction. The initial clamping stress is only dependent on the processing temperature and difference in linear expansion coefficients between the fiber and matrix. As these parameters are not affected by the nitric acid wash, no change is expected in the clamping stress.

3.3. INITIAL "CLAMPING" STRESS

3.3.1 *Thermal Expansion Mismatch* The initial clamping stress on the fiber, as well as the stress in the matrix away from the interface, can be independently calculated from the known processing conditions. When the thermal expansion coefficient of the matrix is greater than that of the fibers, the matrix contracts onto the fibers upon cooling from the fabrication temperature. The residual (biaxial) clamping pressure on the fiber, q_o, that results from a cooling differential of ΔT is [19]:

$$q_o = E_{eff}(\alpha_m - \alpha_f)\Delta T \tag{23}$$

where α_m is the thermal expansion coefficient of the matrix and α_f is the (radial) thermal expansion coefficient of the fiber. The effective elastic modulus, E_{eff}, is given by:

$$E_{eff} = E_m k\frac{1 + (m + f)/(mv_f + f v_m)}{(m + f) + 2kf(v_f - v_m)} , \tag{24}$$

where the parameters m, f, and k are defined above. Using the coefficient of thermal expansion for borosilicate glass, $3.2 \times 10^{-6}/^\circ C$, and the value for the fiber, $2.0 \times 10^{-6}/^\circ C$ from the manufacturer's data sheet [20], q_o was determined from Eqn. (23) to be 47 MPa, in good general agreement with the value determined from the pull-out experiments. However, using the radial fiber coefficient of $2.63 \times 10^{-6}/^\circ C$ measured by Goettler and Faber [21], this value drops to 22 MPa.

3.3.2. *Stress-Induced Birefringence* Accordingly, the clamping stress can also be measured from the radial stress distribution in the borosilicate-glass matrix. This stress-induced retardation of light travelling through glass is directly proportional to the difference in principal stresses, $(\sigma_{rr} - \sigma_{\theta\theta})$, through the stress optical law

$$\Delta = Ct |(\sigma_{rr} - \sigma_{\theta\theta})| \tag{25}$$

where Δ is the relative phase shift of light waves polarized along the radial and tangential directions, C is the stress optic coefficient for the material, and t is the axial specimen thickness (and the optic path length). The principal stress difference in the matrix, $(\sigma_{rr} - \sigma_{\theta\theta})$, is related to the initial clamping pressure on the fiber, q_o, by the elasticity relation [5]

$$\sigma_{rr} - \sigma_{\theta\theta} = -2q_o(1 + f)(R^2/r^2) \tag{26}$$

where R is the fiber radius, r is the polar coordinate measured from the fiber axis, and f is the volume fraction ratio, $V_f/(1 - V_f)$. Thus, the initial clamping pressure is related to the relative phase shift, Δ, by

$$q_o = \Delta\frac{(r/R)^2}{2Ct(1 + f)} . \tag{27}$$

The phase shift Δ was measured by the Tardy method [22] using a circular polariscope with both polarizer and analyzer and two quarter-wave plates in the crossed arrangement. The quarter-wave-plate wavelength was 546 nm. A fractional fringe order of 0.609 ± 0.003 was observed at a radial distance of 142 ± 7 µm. Literature values for the stress optical coefficient, C, of borosilicate glass are between 2.8 TPa^{-1} [23] and 3.0 TPa^{-1} [24]. Using these values, we calculate the principal stress difference in the matrix to be 41.4 MPa $\pm$ 0.2 MPa, or 38.6 ± 0.2 MPa, respectively. Corrected to the fiber-matrix interface, the value of q_o determined for the as-received fiber composite is 83 ± 8 MPa, or 77 ± 8 MPa, respectively. These values are in reasonable agreement, although slightly higher, with those for the pull-out results.

4. Summary

Reinforcement toughening depends significantly upon fiber-matrix interfacial properties. The important relation for fiber toughening is the force-displacement behavior for fiber pull-out, which is controlled by the fiber-matrix interfacial properties. Fiber-matrix interfacial properties are amenable to direct experimental measurement by a single-fiber pull-out test. Long debond-crack lengths (embedded fiber lengths), however, are required to deconvolute the interfacial shear resistance into a friction coefficient and an initial fiber-clamping stress. Interfacial properties were measured for a model SiC fiber-borosilicate glass system with reasonable results. The initial clamping stress on the fiber is amenable to determination by three different techniques which yield similar results. The overall experimental procedure provides an effective way to determine micro-mechanical properties of composite interfaces, to infer composite toughening increments resulting from fiber reinforcement, and to monitor influences of processing variations on composite properties.

5. Acknowledgments

The authors gratefully acknowledge the support of the U. S. Dept. of Energy, Advanced Research and Technology Development (AR&TD) Fossil Energy Materials Program managed by Dr. R. R. Judkins of Oak Ridge National Laboratory, under the NIST-DOE interagency agreement: DE-AI05-800R20679.

6. References

1.   D. B. Marshall and J. E. Ritter (1987), "Reliability of Advanced Structural Ceramics and Ceramic Matrix Composites," Am. Ceram. Soc. Bull., 66 [2], 309- 317.

2.   J. R. Rice (1968), "A Path Independent Integral and the Approximate Analysis of Strain Concentration by Notches and Cracks," J. Appl. Mech., 35, 379-386.

3.   L. J. Broutman (1969), "Measurement of the Fiber-Polymer Matrix Interfacial Strength,"

402

4. in Interfaces in Composites, edited by M. J. Salkind, STP 452, Am. Soc. for Testing and Materials, Philadelphia, pp. 27-41.

5. D. B. Marshall and W. C. Oliver (1987), "Measurement of Interfacial Mechanical Properties in Fiber Reinforced Ceramic Composites," J. Am. Ceram. Soc., 70, [8], 542-48.

6. Y.-C. Gao, Y.-W. Mai and B. Cotterell (1988), "Fracture of Fiber-Reinforced Materials," J. Appl. Math. and Phys. (ZAMP), 39, 550-572.

7. E. P. Butler, E. R. Fuller, Jr., and H. M. Chan (1990), "Interface Properties for Ceramic Composites From a Single-Fiber Pull-Out Test," in Interfaces in Composites, edited by C. G. Pantano and E. J. H. Chen, MRS Symp. Proc., Vol. 170, (MRS, Pittsburgh, PA) pp. 17-24.

8. J. W. Hutchinson and H. M. Jensen (1990), "Models of Fiber Debonding and Pullout in Brittle Composites with Friction," Mech. of Matls., in press; also Harvard University Publication, MECH-157.

9. C. Gurney and J. Hunt (1967), Proc. Roy. Soc. (London), A299, 508.

10. H. Stang and S. P. Shah (1986), J. Mater. Sci., 21, 953.

11. G.P. Cherepanov, Mechanics of Brittle Fracture, translated by A.L. Peabody, edited by R. deWit and W.C. Cooley, McGraw Hill, New York (1979), chapts. 5,9

12. P. D. Jero (1990), "Interfacial Properties of SiC/Borosilicate Glass Systems by Indentation Pushout," American Ceramic Society, 92th Annual Meeting, Dallas, TX, April 24, 1990; Am. Ceram. Soc. Bull. 69 [3], 484 (1990)

13. P.D. Jero and R.J. Kerans, "Effect of Interface Roughness on the Frictional Shear Stress Measured Using a Push-Out Test,", J. Amer. Ceram. Soc., submitted.

14. E. P. Butler, private communication.

15. W. C. Carter, E. P. Butler and E. R. Fuller, Jr., to be published.

16. Mindlin (1949), "Compliance of Elastic Bodies in Contact," J. App. Mech. 16, 259.

17. C.H. Hsueh, "Elastic Load Transfer from Partially Embedded Axially Loaded Fiber to Matrix," J. Mater. Sci. Lett., 7 [5] 497-500 (1988).

18. C.H. Hsueh, "Interfacial Debonding and Fiber Pullout Stresses of Fiber-Reinforced Composites," Mater. Sci. and Eng., A123 [1], 1-11 (1990).

    J. H. Cranmer, G. C. Tesoro, and D. R. Uhlmann (1982), "Chemical Modification of Carbon Fiber Surfaces with Organic Polymer Coatings," Ind. Eng. Chem. Prod. Res. Dev., 21, 185-190.

19. M. Vedula, R. N. Pangborn and R. A. Queeney (1988), "Fiber Anisotropic Thermal Expansion and Residual Thermal Stress in a Graphite/Aluminum Composite," COMPOSITES, 19, [1], 55-60.

20. Technical Data Sheet for Textron Silicon Carbide Fibers, TEXTRON Specialty Materials, Lowell, MA, March 1988.

21. R. W. Goettler and K. T. Faber (1988), "Interfacial Shear Stresses in SiC and AL_2O_3 Fiber-Reinforced Glasses," Cer. Eng. Sci. Proc., 9, [7-8], 861-870.

22. A. Kuske and G. Robertson (1974), Photoelastic Stress Analysis, John Wiley & Sons, New York, pp. 111-114.

23. W. Balmforth and A. J. Holland (1945), "The Stress Optical Coefficient of Glasses," J. Soc. Glass Tech. 29, 111-123.

24. Roy M. Waxler and A. Napolitano (1957), "Relative Stress-Optical Coefficients of Some NBS Optical Glasses," J. Res. NBS, 59 [2], 121-125.

QUASI-DUCTILE BEHAVIOUR OF CARBON-REINFORCED CARBON

A. WANNER*, G. RIZZO**, K. KROMP*
*Max-Planck-Institut für Metallforschung,
Institut fuer Werkstoffwissenschaft,
Seestrasse 92, D-7000 Stuttgart, FRG
**Dipartimento di Ingegneria Chimica dei
Processi e dei Materiali, Università di Palermo
Viale delle Science, I-90128 Palermo, Italy

ABSTRACT. Firstly a brief review of the problem of a "ductilization" of ceramic materials by endless-fiber reinforcement is given. The problems, especially the influence of the frictional stress in the fiber-matrix interface, are pointed out for the unidirectionally reinforced ceramic matrix composite (CMC) system LAS/SiC, which is well documented in the literature.

Secondly displacement controlled three-point bending experiments with unloading-reloading cycles in edgewise loading (parallel to the plies) with notched specimens are presented as a means to characterize the "quasi-ductile" behaviour of CMC laminates. The effectiveness of the method is demonstrated for carbon-reinforced carbon (C/C) laminates in two different states of graphitization and for a silicon infiltrated C/C, both in a bidirectional layup.

1. Introduction

Great efforts have been made in the last years to develop tough ceramics which are not susceptible to fracture during impact or under stress in the presence of flaws or notches. In particular in the aerospace industry specific strengths, temperature resistance and failure characteristics are demanded, which cannot be fulfilled except by ceramic composites containing high performance fibrous reinforcements.

The general problem of ceramic materials is their extreme brittleness. The aim is to "ductilize" these materials, to increase their toughness. Several principles exist for increasing fracture toughness either through incorporation of particles of second phase, short whiskers or platelets or transformation toughening particles of zirconia. Such efforts increase the thoughness by typically a factor of 2-3. By far the most successful method of improving toughness is to reinforce a matrix with strong carbon or ceramic filaments. The fracture toughness can be increased by nearly an order of magnitude and the fracture energy by two orders of magnitude with these materials.

S. P. Shah (ed.), Toughening Mechanisms in Quasi-Brittle Materials, 405–423.

A lot of experience has been acquired in the past with carbon fiber reinforced polymer matrix and with metal matrix composites. Efforts have been made to extend this experience to materials which can withstand high temperatures. The carbon-reinforced carbon composites (C/C) were a step in this development, although they possess the disadvantage of not being applicable in arduous surroundings without coating. Fresh impulses came from the development of new fibers, such as the organometallic derived silicon carbide yarn.

2. The Problem of a "Ductilization" of Ceramic Materials

As is already known from metal and resin matrix composites, high elastic modulus fibers have to be incorporated into lower elastic modulus matrices to achieve reinforcement and ductile behaviour.

Beside the main condition for a ductilized ceramic matrix composite with high toughness, that is, the failure strain of the fibers must be significantly greater than that of the matrix, the fiber should not be too strongly bonded to the matrix, it should only be frictionally bonded. In recent years, since the Nicalon-SiC fiber has widely been available [1], the general features of this "ductilization" of brittle matrix composites were studied predominantly in the LAS/SiC system, that is lithium aluminosilicate glass-ceramic matrix, reinforced with silicon carbide fibers (Nicalon-fibers). The general problems of "ductilization" will be discussed with this material, considering an unidirectional fiber reinforcement (1-D material).

3. Unidirectionally, Single-fiber Reinforced Ceramic Matrix Composites

The "ductile" load response for an unidirectionally single-fiber re-inforced CMC, obtained from the LAS/SiC system [2], is schematically shown in Fig. 1. In axial loading a linear elastic region ends at a stress σ_o, defined by the onset of matrix cracking. This point is similar to a "yield" point in alloys, here originated by periodic cracking of the matrix. A nonlinear region follows, which depends on the frictional behaviour of the fiber in the matrix. This region is accompanied by multiple matrix cracking, with fibers remaining intact and then by fiber pull-out. At the stress σ_u, an "ultimate tensile strength", fiber bundle failure occurs. Then the load drops slowly - not a brittle, but a delayed fracture is observed.

From an energy balance analysis, given earlier [3] and a fracture mechanics approach [4,5,6] a relation between the matrix cracking stress σ_o and microstructural parameters was given:

$$\sigma_o = [6(1-\nu^2)K_{mc}{}^2 \tau E_f \ f^2 (1-f)(1+\eta)^2/E_m R]^{1/3}, \qquad (1)$$

where K_{mc} is the matrix toughness, ν Poisson's ratio, $\eta = E_f \cdot f/E_m(1-f)$, R the fiber radius, τ the sliding frictional stress at the interface and f the volume fraction of fibers.

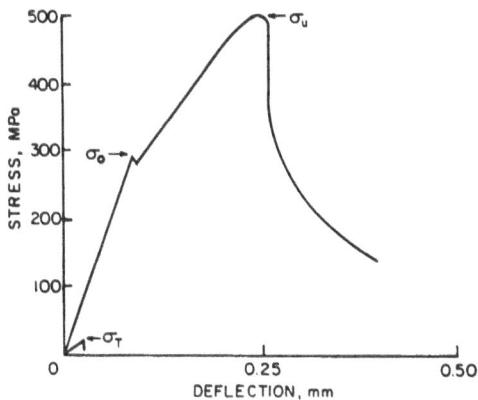

Figure 1. Schematic graph of an uniaxially endless fiber-reinforced composite, indicating longitudinal and transverse stress-deflection behaviour [2].

This situation is not to be confused with that of unreinforced materials, in which the stress at the tip of the crack increases with increasing crack length (equation (1) does not directly depend on crack length). The important result is that in these composite systems the stress at the tip of the crack becomes independent of crack length once the crack is longer than a length c_o necessary in order to reach the equilibrium situation (Fig. 2). This holds as long as the fibers in the

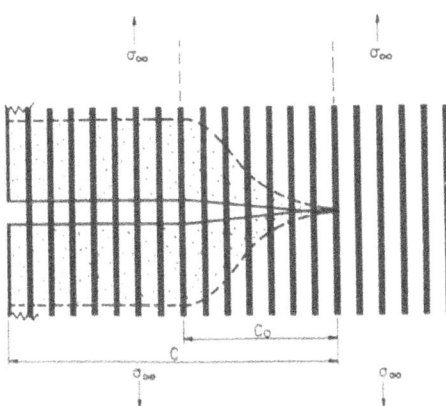

Figure 2. Crack perpendicular to the fibers. In the dotted area, slipping between matrix and fibers has occurred (schematically).

408

wake of the crack front remain intact. Thus matrix cracking is impossible below the "threshold" stress σ_o of equation (1), no matter how
large a crack or a preexisting defect! The equilibrium-stress/crack-size
function for straight cracks is plotted in Fig. 3 (for straight cracks
$\sigma_o/\sigma_m = 1.02$ and $c_0/c_m = 1.88$). From Fig. 3 it can be seen that the
stress required to propagate a matrix crack is independent of crack
length for cracks larger than about $c_m/3$. The crack response in this
region differs from the behaviour of cracks in unreinforced brittle
materials, in which the strength decreases with $c^{-1/2}$, Fig. 3.

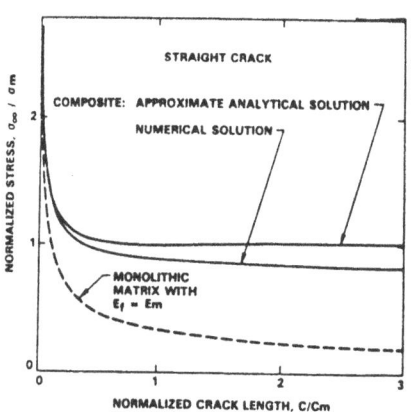

Figure 3. Equilibrium-stress/crack size functions for matrix cracks in
composites and in a monolithic ceramic [4].

$c_m/3$ was evaluated to be several fiber spacings long and this implies
that the stress for matrix cracking is not substantially reduced by introduction of larger flaws during fabrication or in service.
It is obvious from the analysis that the attainment of steady state
cracking at high stress levels is restricted to a narrow range of microstructures. The limiting value is the level of σ_o that can be achieved
without causing fiber failure; optimum values of all three parameters
K_{mc}, τ and f increase the level of σ_o in equation (1). Additionally, the
influence on c_m has to be considered. A significant microstructural
restriction concerns the volume fraction of fibers f because optimum
properties can be achieved only in unidirectional composites whereas in
multidirectional systems the volume fraction of fibers is in any case
less.
In bending tests the load-deflection behaviour is similar to that in
tension [2]. High failure strains and notch insensitivity are characteristic for such composites. These qualities are explicable in terms of
the slip processes occurring at the fiber/matrix interface or the frictional stress τ [2,4].
Any change in the interface layer between fiber and matrix has a
decisive influence on the "quasi-ductile" response of the CMC. This was

proved for the system LAS/SiC, discussed as an example here. Even short exposures to high temperature can reembrittle the material: Bending experiments in Ar and in air at high temperatures resulted in the same load level, but the morphology of fracture changed completely from pull-out in Ar to rather brittle fracture in air [7,8]. For an unitape 2-D laminate of the same system LAS/SiC this behaviour was investigated systematically [9,10]. It was found that the reembrittlement after an exposure of 2-4 hours at high temperatures in air resulted from a change in the C-interface layer to SiO_2. This change provoked an increase in the frictional stress between fiber-matrix and thus a decrease in the pull-out lengths.

4. Measurement of Interfacial Mechanical Properties

The question is, whether only frictional forces act in the interface, as assumed for the steady state cracking in the matrix, or fiber displacement involves also debonding followed by frictional sliding. The general case will be a weak bonding and debonding in addition to frictional forces.

The frictional stresses were measured by pushing a pyramidal indenter on the end of an individual fiber and measuring the resulting displacement of the surface of the fiber below the matrix surface [2,11]. This method was refined recently by using an ultralow-load indentation instrument (Nano Instruments, Inc. Braintree, MA) to measure force and displacement continuously during loading, unloading and load cycling. Frictional sliding and combined debonding and frictional sliding at the interface were analyzed for a LAS/SiC material [12,13]. In [13] a method was proposed to achieve relative frictional stress values in the special cases before and after heat treatment at 1000°C in air on the system LAS/SiC. For this purpose, specimens were cut into blocks and by erosion of a portion of the matrix, fibers were disclosed, resulting in a distribution of protruding fibers across a flat matrix surface. A load was applied by means of a sapphire hemisphere of 1 mm diameter on the protruding fibers. The load-displacement curves were measured with respect to the matrix.

5. Laminated Ceramic Matrix Composites

It was demonstrated in Fig. 1 that when unidirectionally reinforced composites are loaded along the fiber axis, matrix cracking occurs at a stress σ_o, followed by fiber failure at the ultimate stress σ_u. The transverse properties are substantially inferior, load response is linear elastic and fracture occurs at a stress $\sigma_T \ll \sigma_o$, Fig. 1.

This considerable anisotropy requires that the most structural composites be laminated.

The C/C and C/SiC laminates discussed in the next chapters consist of bundles of 3000 single carbon fibers, the bundles are woven to prepregs and these plies are stapled orthogonally. Inside such a fiber bundle considerations and measurements of the frictional stress, as discussed in the last chapters may be valid. But in the laminates there also exist interbundle interfaces between parallel bundles and orthogonal bundles, in-plie delamination and inter-plie delamination will occur.

The response in loading will depend on the loading mode to a high degree: e.g. in three-point bending with loading perpendicular to the plies, because of shear deformation, the result in general depends on the span to height ratio. For laminates with low frictional stresses such as C/C the result additionally depends on the plie thickness to height ratio [14].

6. How to Characterize the "Quasi-ductile" Behaviour of CMC Laminates?

As previously discussed the frictonal stress in the interfaces fiber-matrix has a decisive influence on the quasi-ductile behaviour of uni-directionally single fiber reinforced CMCs. For laminates not only interfaces single fiber-matrix inside a fiber bundle, but also inter-bundle interfaces parallel and interbundle interfaces orthogonal may in-fluence the quasiductile behaviour. The methods to measure the fric-tional stress in the interfaces single fiber-matrix mentioned in the last chapter seem not to be applicable to those of fiber bundles. There-fore, the authors tried to get information on the frictional behaviour of laminates by a special kind of experiments: Prismatic specimens of rectangular cross-section were loaded edgewise in displacement con-trolled three-point bending experiments.

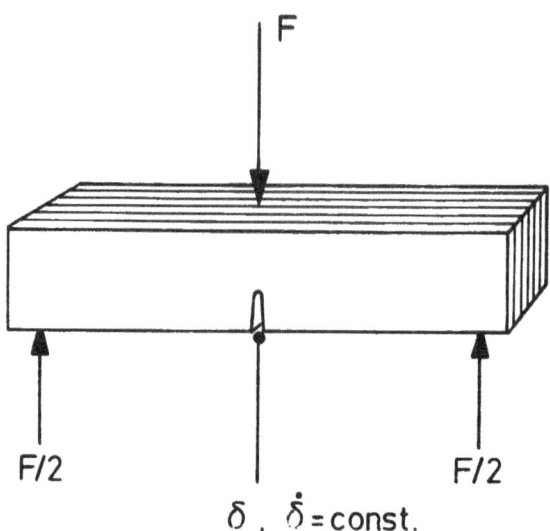

Figure 4.   Edgewise loading of notched prismatic bars in displacement controlled three-point bending.

The loading principle is demonstrated schematically in Fig. 4. The chosen span was S = 90 mm, the displacement rate in the closed loop, measured and controlled directly at the notch by a SiC-pushrod, was δ = 200 μm/min. The SiC loading rods had a radius of curvature of 5 mm. The experiments could be performed at high temperatures by induction heating indirectly by means of a carbon susceptor tube. The whole loading system, inside a vacuum vessel, could be evacuated to $2 \cdot 10^{-5}$ mbar [15].

Using this equipment load-displacement curves in monotonic loading and with unloading-reloading cycles were achieved. With this kind of experiment changes in the ductile response of a C/C (carbon-reinforded carbon) laminate by silicon infiltration and the influence of the temperature of graphitization on the ductile behaviour of another C/C laminate were investigated.

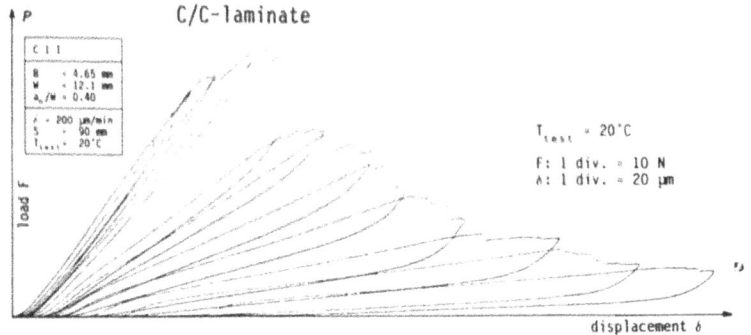

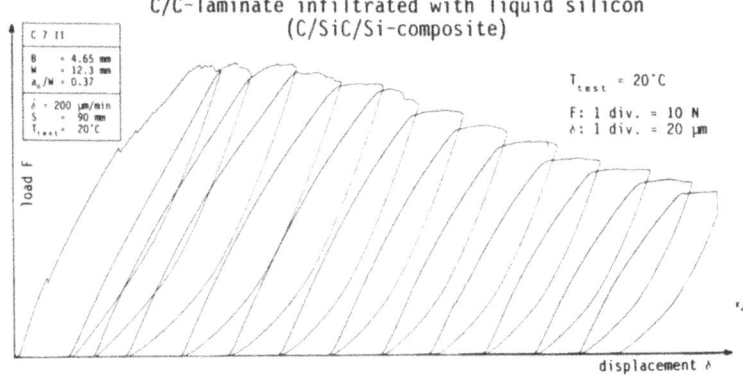

Figures 5a,b. Unloading-reloading cycles in displacement controlled three-point bending experiments
a) (C/C) laminate carbonized only
b) laminate carbonized + infiltrated by silicon (C/SiC/Si).

7. The Change in the "Ductile" Response of a C/C Laminate After Infiltration with Silicon

C/C laminates of 5H-satin prepregs, woven from tows of 3000 PAN-based fibers were laminated in a [0/90°] layup. All the material was kindly supplied and the processing performed by Sintec-Keramik, Buching, FRG. The material was only carbonized but not graphitized. After the carbonization the material was not reimpregnated and was thus highly porous. This material was infiltrated by silicon as a first step for the development of an oxidation protected type. In Figs. 5a,b the load-displacement curves after cyclic loading are shown [16]. The enveloping curves of the unloading-reloading cycles are nearly identical to the load-displacement curves in monotonic loading.

In Fig. 5a it can be seen for the C/C laminate carbonized only:
• for the enveloping curve:
- only small deviation from linearity at a relative high level of load;
- a sharp maximum in load;
- subsequently rapid decay in load;
• for the unloading in successive cycles:
- low amounts of irreversible displacement.

For the carbonized and silicon-infiltrated (C/SiC/Si) laminate in Fig. 5b:
• for the enveloping curve:
- the deviation from linearity starts already at a low level of load;
- no sharp maximum in load, but a smooth decay in sustained load (the load-displacement curve could be continued beyond the displacement range shown in Fig. 5b);
• for the unloading in successive cycles:
- the displacement is nearly completely irreversible.

After the experiments the specimens were broken completely open and the pull-out behaviour was studied in the SEM, see Figs. 6a,b [16].

In Figs. 6a,b it can be observed that:
- in the only carbonized material the fiber-bundles failed in a brittle manner without pull-out of single fibers inside the bundle, but complete bundles were pulled out;
- in the siliconized material single fibers inside the bundles failed and were pulled out, but no complete bundle was pulled out;
- the surface of the single fibers appears rough because of adherent siliconized carbon matrix (see Fig. 7).

This behaviour may be interpreted as follows:
• For the carbonized only material it is concluded from the shape of the load-displacement cycles and the pull-out behaviour that
- the friction in the interbundle interfaces is very low (the bonding may be not much more than van der Waals), thus resulting in good gliding behaviour of complete bundles and low irreversible displacement during cycling;
- because of the low bonding in the interfaces most of the load has to be sustained by the fiber bundles and these fail in a brittle manner and thus the load drops rapidly from a sharp maximum.
→Because of these facts, the fracture behaviour is estimated to be brittle rather than ductile.

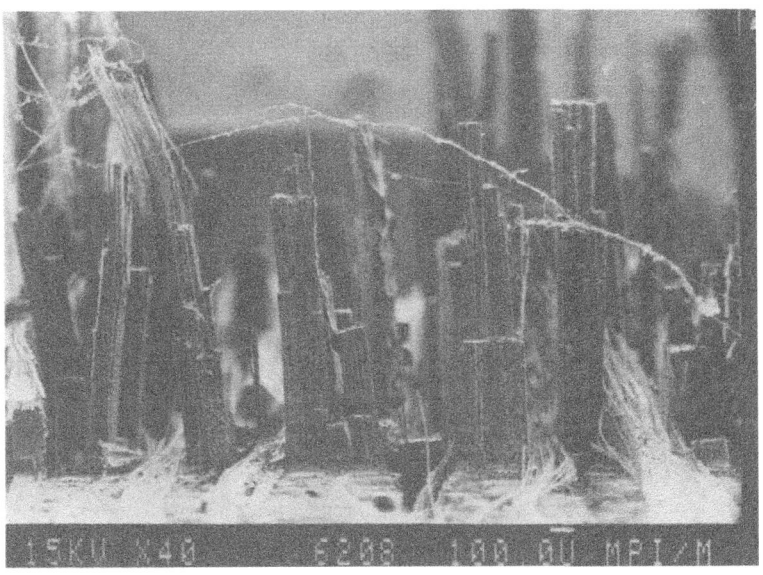

a) laminate carbonized only: pull-out of complete bundles

b) laminate carbonized + infiltrated by silicon:
   pull-out of single fibers in the bundles,
   no pull-out of complete bundles.

Figures 6a,b. Pull-out of the fracture surface of notched specimens in
edgewise loading

414

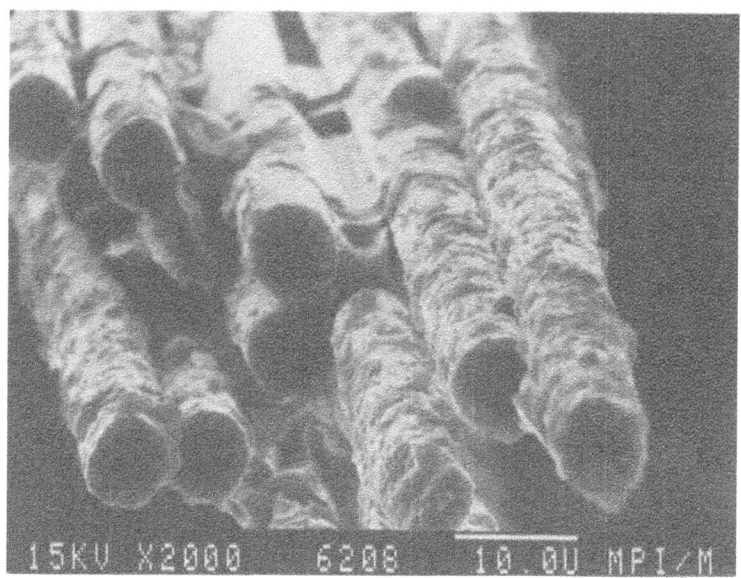

Figure 7. Laminate, carbonized + infiltrated by silicon:
single fibers with adherent siliconized carbon matrix.

• For the silicon-infiltrated material it is concluded that
- the friction in the interbundle interfaces and partly between the
  fibers inside the bundles is high, thus the displacement during
  cycling is nearly completely irreversible;
- because of the relatively high bonding in the interfaces the load is
  distributed to the volume, thus the load drops slowly.
→ Because of these facts the fracture behaviour is estimated to be
  rather "quasi-ductile".
An estimation of the areas below the load-displacement curves of
Figs. 5a,b as a measure for the work of fracture may confirm these
statements.

8. The Change in the "Quasi-ductile" Behaviour of a C/C Laminate by a Variation in the Temperature of Graphitization

C/C laminates of 8H-satin prepregs, woven from tows of 3000 PAN-based
fibers were laminated in a [0/90°] layup. The materials were kindly
supplied and processed by Schunk-Kohlenstofftechnik, Gießen, FRG. After
four cycles of reimpregnation and carbonization the materials were
graphitized at temperatures of 2100 ("laminate A") and 2400°C ("laminate
B").

As already described notched specimens of both the materials were
loaded edgewise in displacement controlled experiments at room temperature and at 1000°C. In Figs. 8a,b load-displacement curves with numerous

unloading-reloading cycles are shown for both the materials at room temperature. The dimension and the loading condition can be read from the diagrams. Again the enveloping curves of the unloading-reloading cycles are nearly identical to the load-displacement curves in monotonic loading. For both the materials the deviation from linearity occurs at a high load level, the maximum load appears sharp, followed by a rapid decay in sustained load. A difference between the materials is evident in the amount of irreversible displacement after unloading in successive

C/C-laminate "CF222", graphitized at 2100°C

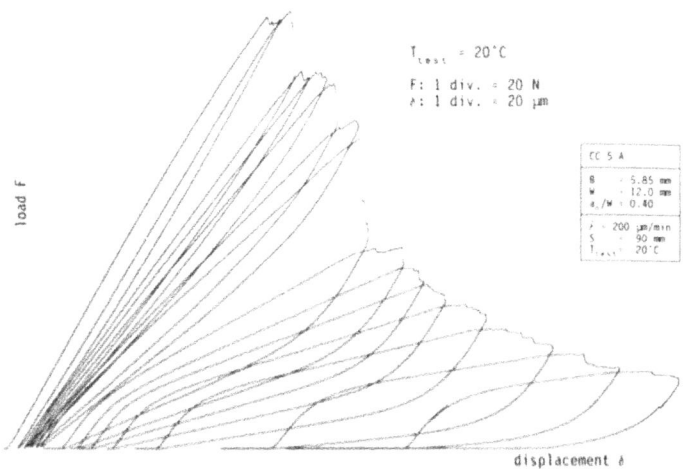

C/C-laminate "CF222", graphitized at 2400°C

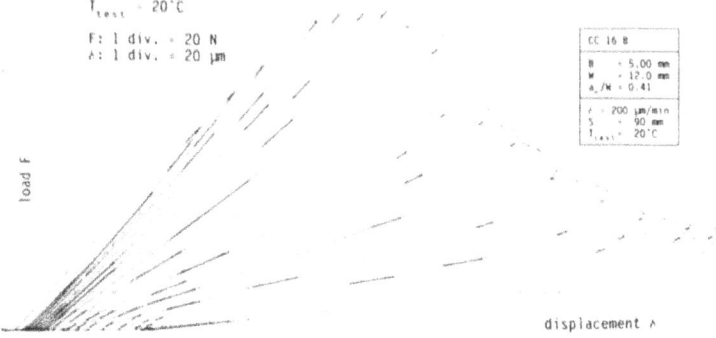

Figures 8a,b. Unloading-reloading cycles in displacement controlled three-point bending experiments at room temperature
a) laminate A, four times reimpregnated and carbonized, finally graphitized at 2100°C
b) laminate B, four times reimpregnated and carbonized, finally graphitized at 2400°C.

cycles: Material B exhibits only small amounts of irreversible displacement (Fig. 8b), while material A shows remarkable amounts of irreversible displacement (Fig. 8a).

Also the shape of the cycles differs: If one observes one of the cycles in Fig. 8a at a low load level (high irreversible displacement), the first slope (starting the reloading) and the slope at the end of the cycle (starting the unloading) are steep, nearly as steep as the initial loading. This behaviour is sketched in Fig. 9. These initial slopes also appear for material B in Fig. 8b, but only slightly. These slopes are interpreted as a measure of "bonding" friction at the moments were the loading stops and changes its direction. At a certain load level this slope drops suddenly, friction turns to "gliding" friction. The same occurs in unloading, where the stored elastic energy now "loads" the specimen in the opposite direction. When the energy is exhausted, the cycle stops at a certain irreversible displacement, see Fig. 9.

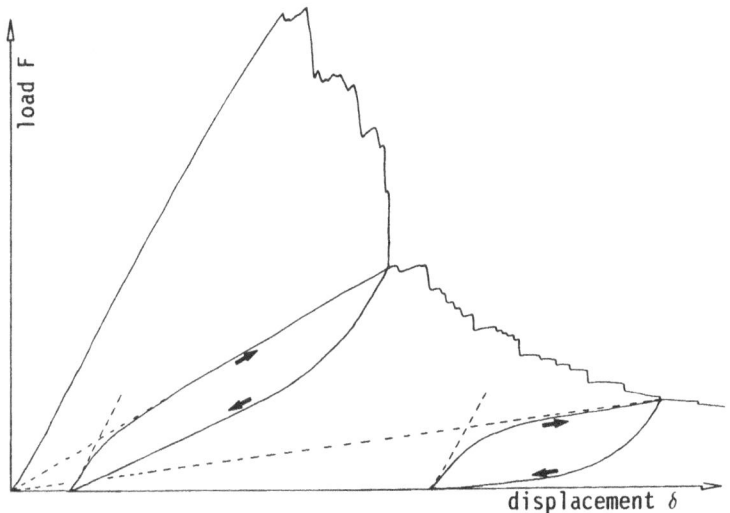

Figure 9. Change in slope of unloading-reloading cycles, schematically analogous to Fig. 8a for laminate A.

This difference in bonding and gliding friction nearly vanishes for material B (Fig. 8b), it is supposed that the frictional stress in the interfaces is lowered by the increase in temperature of graphitization from 2100 to 2400°C. A look at the structure of the materials confirms these statements (Figs. 10a,b): Laminate B with the higher temperature of graphitization exhibits a higher content of macropores in the inter-bundle-regions (Fig. 10b), compared to laminate A with the lower temperature of graphitization (Fig. 10a).

After breaking open the specimens, the pull-out behaviour was observed in the SEM. The appearance for both the laminates A and B was similar to that shown in Fig. 6a: Pull-out of complete fiber-bundles and brittle fracture of the bundles, see Figs. 11a,b. For the first three rows of

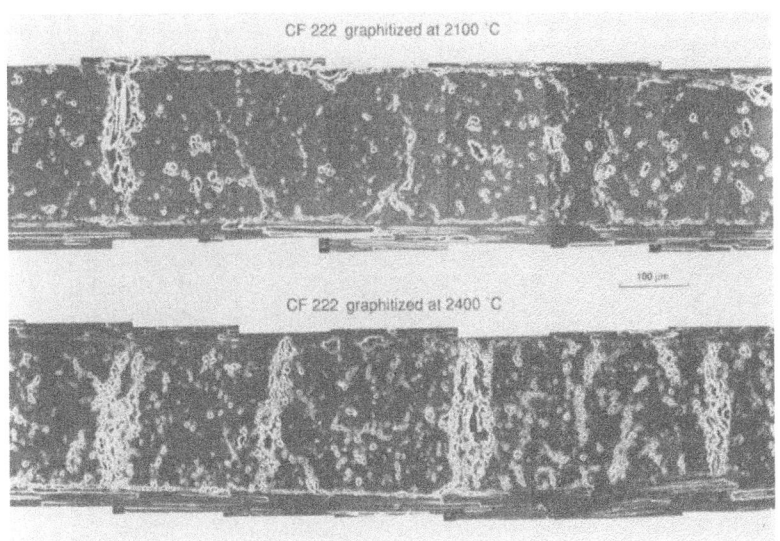

Figures 10a,b. Polished cut perpendicular to the plies
a) laminate A
b) laminate B.

pulled out bundles beyond the notch root (in Fig. 11b from the left) the pull-out lengths of the fiber-bundles were measured. The mean pull-out length increased by 64% with increasing the temperature of graphitization.

From the response to the cycling and the estimation of the pull-out lengths it is concluded that the "quasi-ductile" behaviour decreases with increasing the temperature of graphitization from 2100 to 2400°C.

The same cycling experiments were performed at a testing temperature of 1000°C in a vacuum of $2 \cdot 10^{-5}$ mbar for both the laminates A and B, see Figs. 12a,b: The shape of the enveloping curves and that of the cycles is similar to those gained at room temperature. Except the amount of irreversible displacement seems to be higher for both the laminates, compare Figs. 8a,b to Figs. 12a,b. The pull-out appearance was similar to that in Figs. 11a,b for the room temperature tests.

A measurement of the pull-out lengths again showed the mean pull-out length to increase by 69% with increasing the temperature of graphitization from 2100 to 2400°C. This means that the "quasi-ductile" behaviour decreases with the increase in temperature of graphitization, as for the tests performed at room temperature.

Compared to the tests at room temperature the mean pull-out lengths for the tests at 1000°C were shorter by $\sim$ 20%! The results of the measurements of the pull-out lengths are summarized in Table 1.

418

a) parallel to the fracture surface (notch root in front)

b) perpendicular to the fracture surface (notch root left).

Figures 11a,b. Pull-out of the fracture surface of notched specimens in edgewise loading, laminate A

Table 1:	Pull-out lengths for laminates A and B, loaded at room temperature and at 1000°C, [μm]	
	T_{graph} = 2100°C	T_{graph} = 2400°C
	+64%	
T_p = 20°C	840 ± 319 (±38%)	1381 ± 545 (±39%)
	- 22%	- 19%
T_p = 1000°C	657 ± 300 (±46%)	1115 ± 560 (±50%)
	+69%	

According to the results in Table 1 it is supposed that the frictional stress in the interbundle interfaces is increased by an increase in testing temperature for both the laminates.

To prove this the temperature dependence of the Young's modulus of laminate A parallel to the 0°-fiber direction was investigated with the resonant's beam method [17]. The modulus, at room temperature 83±2GPa, increased with increasing temperature to reach a plateau value at ~ 1000°C, see Fig. 13. In the same range of temperature the material was observed in a SEM with a heating equipment besides the macropores already shown in Fig. 10a, at room temperature cracks were observed inside the bundles, see Fig. 14, left. These cracks were observed to close completely in the temperature range, where the Young's modulus reaches its plateau value.

It is proposed that during the processing of the materials cooling down from the high temperatures to room temperature residual stresses in tension perpendicular to the bundles arise, which cause cracking as observed in Fig. 14. Approaching the equilibrium temperature these stresses relax, cracks inside the bundles close and the friction between the bundles increases [18].

The relaxation of the residual stresses may explain the higher irreversible displacement observed in Figs. 12a,b at the testing temperature of 1000°C compared to Figs. 8a,b at the testing at room temperature and thus the decrease in the mean pull-out lengths by ~ 20% (Table 1) from room temperature to 1000°C.

It can be stated here that an increase in testing temperature (temperature of application) improves the "quasi-ductile" behaviour of the investigated C/C laminates A and B.

420

C/C-laminate "CF222", graphitized at 2100°C

T_{test} = 1000°C

F: 1 div. = 20 N
δ: 1 div. = 20 µm

load F

displacement δ

CC 9 A

B   = 5.85 mm
W   = 11.5 mm
a_0/W = 0.41

ŝ = 200 µm/min
S   = 90 mm
T_{test} = 1000°C

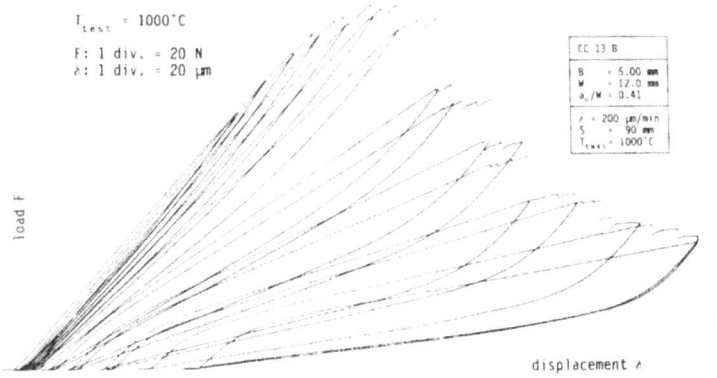

C/C-laminate "CF222", graphitized at 2400°C

T_{test} = 1000°C

F: 1 div. = 20 N
δ: 1 div. = 20 µm

load F

displacement δ

CC 13 B

B   = 5.00 mm
W   = 12.0 mm
a_0/W = 0.41

ŝ = 200 µm/min
S   = 90 mm
T_{test} = 1000°C

Figures 12a,b. Unloading-reloading cycles in displacement controlled
three-point bending experiments at 1000°C
a) laminate A
b) laminate B.

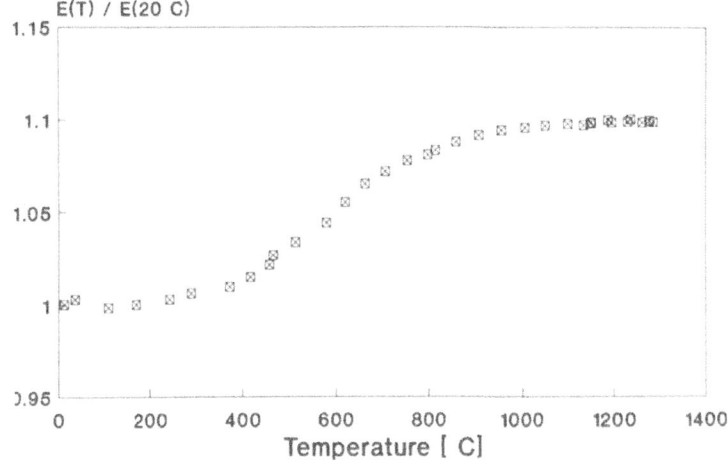

Figure 13. The dependence of Young's modulus on temperature for laminate A.

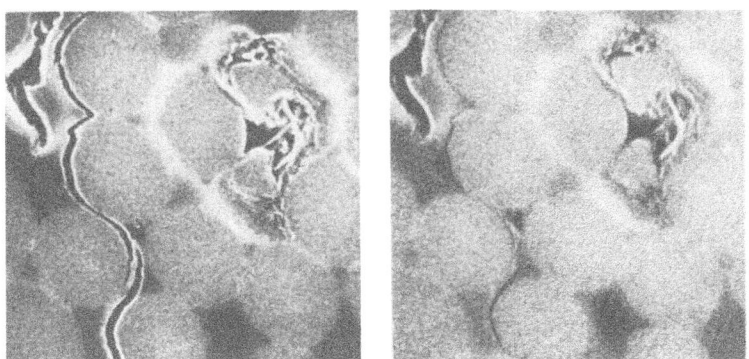

Figure 14. Cut perpendicular to the plies; pictures by SEM with heating
equipment: room temperature (left), 780°C (right).

Summary and Outlook

Firstly, a brief overview of the problem of the "ductilization" of ceramics by endless fiber-reinforcement and of the importance of the frictional stress in the interface fiber-matrix was given. The problems for the unidirectionally endless-fiber reinforced LAS/SiC were discussed, a material well documented in the literature.

A special kind of experimental procedure is proposed to characterize the "quasi-ductile" behaviour of C/C-(carbon fiber reinforced) and C/SiC-(carbon fiber reinforced silicon carbide)-laminates in [0/90°] layup: Notched prismatic specimens are loaded edgewise (parallel to the plies) in displacement-controlled three-point bending experiments with intermediate unloading-reloading cycles.

A purely carbonized, porous [0/90°]-laminate and the same material in a siliconized version were loaded to demonstrate the effectiveness of the method. Two batches of another [0/90°]-laminate four times re-impregnated and carbonized, were graphitized at two different temperatures (2100 and 2400°C) and then loaded in the same way.

Though the "quasi-ductile" behaviour of laminates depends to a high degree on the mode of loading, e.g. notched-unnotched or edgewise (parallel to the plies) in summary the following statements may be made:

From the shape of the enveloping curve, the amount of irreversible displacement of successive unloading cycles, and from the shape of the cycles themselves and the pull-out appearance, it is concluded that these materials exhibit "quasi-ductile" behaviour. Siliconizing improves the quasi-ductility of the purely carbonized material, an increase in the temperature of graphitization decreases that of the reimpregnated material.

The proposed method seems sensitive enough to characterize the quasi-ductile behaviour of C/C and C/SiC laminates. The next steps will be to give an analytical evaluation of the unloading-reloading cycles and for the bonding and gliding friction in the individual cycles. TEM investigations of the interfacial regions are now in progress.

References

[1] S. Yajima, K. Okamura, J. Hayashi, M. Omori: Synthesis of con-
     tinuous SiC fibers with high tensile strength; J. Am. Ceram. Soc.
     59/7-8 (1976) 324-327.
[2] D.B. Marshall, A.G. Evans: Failure mechanisms in ceramic-fiber/
     ceramic-matrix composites; J. Am. Ceram. Soc. **68**/5 (1985) 225-231.
[3] J. Aveston, G.A. Cooper, A. Kelly in: The properties of fiber com-
     posites, Conf. Proc. pp.15-26; National Physical Laboratory, IPC
     Science and Technology Press (1971).
[4] D.B. Marshall, B.N. Cox, A.G. Evans: The mechanics of matrix
     cracking in brittle-matrix fiber composites; Acta metall. 33/11
     (1985) 2013-2021.
[5] B. Budiansky, J.W. Hutchinson, A.G. Evans: Matrix fracture in fiber
     reinforced ceramics; J. Mech. Phys. Solids 34/2 (1986) 167-189.
[6] L.N.McCartney: Mechanics of matrix cracking in brittle-matrix fiber
     reinforced composites; Proc. R. Soc. Lond. A **409** (1987) 329-350.
[7] J.J. Brennan, K.M. Prewo: Silicon carbide fiber reinforced glass-
     ceramic matrix composites exhibiting high strength and toughness;
     J. Mater. Sci. **17** (1982) 2371-2383.
[8] K.M. Prewo: Tension and flexural strength of silicon carbide fiber-
     reinforced glass ceramics; J. Mater. Sci. **21** (1986) 3590-3600.
[9] E. Bischoff, M. Rühle, O. Sbaizero, A.G. Evans: Microstructural
     studies of the interfacial zone of a SiC-fiber-reinforced lithium
     aluminium silicate glass-ceramic; J. Am. Ceram. Soc.**72** (1989)
     741-45.
[10] M.D. Thouless, O. Sbaizero, L.S. Sigl, A.G. Evans: Effect of inter-
     face mechanical properties on pullout in a SiC-fiber-reinforced
     lithium aluminium silicate glass-ceramic; J. Am. Ceram. Soc. **72**
     (1989) 525 .
[11] D.B. Marshall: An indentation method for measuring matrix-fiber
     frictional stresses in ceramic composites; J. Am. Ceram. Soc. 67/12
     (1984) C259-260.
[12] D.B. Marshall, W.C. Oliver: Measurement of interfacial mechanical
     properties in fiber-reinforced ceramic composites; J. Am. Ceram.
     Soc. **70**/8 (1987) 542-548.
[13] E.Y. Luh, A.G. Evans: High-temperature mechanical properties of a
     ceramic matrix composite; J. Am. Ceram. Soc. **70**/7 (1987) 466-469.
[14] C. Rief, M. Lindner, K. Kromp: Experimental documentation and model
     calculation of damage mechanisms in loading a RCC-material in a
     bidirectional layup; ASTM STP (1989) 565-579.
[15] A. Bornhauser, K. Kromp, R.F. Pabst: R-curve evaluation with
     ceramic materials at elevated temperatures by an energy approach
     using direct observation and compliance calculation of the crack
     length; J. Mater. Sci. **20** (1985) 2586-96.
[16] I. Appel: Thesis, University of Stuttgart, FRG; 1990.
[17] A. Wanner, K. Kromp: Young's and shear moduli of laminated carbon/
     carbon composite by a resonant beam method; in: "Brittle Matrix
     Composites 2", ed. by A.M. Brandt and I.H. Marshall, Elsevier Appl.
     Sci. Publ., London-New York (1988) 280-89.
[18] J. Jortner: Macroporosity and interface cracking in multi-
     directional carbon/carbon; Carbon **24**/5 (1986) 603-13.

THE FRACTURE RESISTANCE OF BRITTLE MATRIX COMPOSITES

FRANK ZOK
Department of Materials
College of Engineering
University of California
Santa Barbara, CA 93106

ABSTRACT. The fracture resistance of brittle matrix composites has been examined through a combination of experimental and modelling studies. R-curve measurements on both fiber-reinforced and metal-reinforced composites have been correlated with models of crack bridging. An important feature of the measurements and calculations is the strong influence of specimen size on the apparent fracture resistance. The influence of other geometric effects is also discussed.

1. Introduction

It has now been well established that many brittle matrix composites exhibit R-curve behavior, i.e. their fracture resistance increases with crack extension. The R-curve is due mainly to bridging processes in the crack wake. Specifically, intact particles or fibers exert closing tractions on the crack faces, thus reducing the stress intensity factor at the crack tip. This behavior has been observed in a broad range of materials including fiber-reinforced ceramics [1-5] and cements [6,7], metal-reinforced ceramics [8-10] and intermetallics [11-12], and some monolithic ceramics [13-19].

The fracture resistance K_R can be expressed as the sum of two components,

$$K_R = K_o + K_b \tag{1}$$

where K_o is the initial fracture resistance and is typically of the same order as the fracture toughness of the matrix itself. The term K_b is the component due to the bridging ligaments and is described by [20]

$$K_b = \frac{2}{\sqrt{\pi a}} \int_0^{\Delta a} \sigma(x) \, F\left(\frac{x}{a}, \frac{a}{W}\right) dx \tag{2}$$

where a is the total crack length, x is the distance behind the crack tip, W is the specimen width, Δa is the crack extension, $\sigma(x)$ is the spatial variation in the bridging stress and F is the weight function applicable to the geometry under consideration. When the crack opening displacement becomes sufficient to break the ligaments furthest from the crack tip, the length of the bridging zone reaches a saturation level, L, and thus the upper limit on the integral in Eqn. (2) must be replaced by L.

In most cases the stress distribution $\sigma(x)$ cannot be measured directly and so is inferred from experimentally measured R-curves and equations (1) and (2). The bridging stress so obtained is then compared with the results of micromechanical models which take into account the details of the bridging process. To conduct such comparisons, an appropriate expression for F must be selected. It should be noted that the expression applicable to small scale bridging conditions

425

S. P. Shah (ed.), Toughening Mechanisms in Quasi-Brittle Materials, 425–439.

$(F = \sqrt{a/2x})$ is not valid for specimen geometries in which the bridging zone length is a significant fraction of other in-plane specimen dimensions. Neglecting the geometric effects leads to erroneous estimates of the bridging stresses, as discussed in [21].

The purpose of this paper is to demonstrate that R-curve behavior is indeed dominated by bridging processes in both fiber-reinforced and metal-reinforced composites. This is accomplished through a combination of experimental measurements, micromechanical modelling and microstructural observations. Emphasis is placed on understanding the effects of specimen geometry on the measured fracture resistance. Furthermore, the effect of reinforcement orientation in *metal-reinforced* composites is examined. These effects are important in composites containing randomly oriented metal fibers, such as those recently developed by Lange et al. [22].

2. Fracture Resistance Measurements

Standard testing geometries have been employed for the R-curve measurements, including compact tension specimens and pre-notched 3 point and 4 point bending beams, designed in accordance with ASTM standards [23]. The fracture resistance calculations were based on the usual linear elastic formulations, using the applied loads and the length of the matrix crack. Some tests were interrupted to observe microstructural changes.

A variety of composite materials have been studied, including a metal-reinforced ceramic, a model system consisting of a brittle polymer matrix and metal reinforcements, and a fiber-reinforced glass ceramic. Pertinent details of the microstructures are described in subsequent sections; additional details can be found in Refs. [5], [21] and [24].

3. Metal-Reinforced Composites

Two metal-reinforced composites were examined. The first was Al_2O_3 reinforced with randomly oriented, continuous fibers of an Al 4% Mg alloy, with a metal volume fraction of $\approx 23\%$. The processing route is described in detail in Ref. [22]. The second was a model system consisting of a polymethylmethacrylate (PMMA) and continuous Al wire reinforcements. It was fabricated by aligning the wires between the PMMA sheets and hot pressing at 200°C for several hours. The PMMA/Al composite has numerous advantages, including ease of fabrication and good control of the orientation and volume fraction of reinforcements. Furthermore, the wires can be extracted from the composite by dissolving the matrix and thus the *in-situ* flow properties of the reinforcements can be measured directly. Typical micrographs of both composites are shown in Figs. 1 and 2.

Figures 3 and 4 show the R-curves for the Al_2O_3/Al composite and two of the PMMA/Al composites. The reinforcements in the PMMA/Al composites were aligned perpendicular to the crack plane. The curves for both materials are characterized by an initial fracture resistance K_0 equivalent to the fracture toughness of the matrix (3 MPa $\sqrt{m}$ for the Al_2O_3 and 1 MPa $\sqrt{m}$ for the PMMA), an intermediate region where the fracture resistance increases gradually and finally a region in which the resistance apparently increases very rapidly with crack extension. As shown later, the rising portion is symptomatic of large scale bridging (analogous to large scale yielding in metals) and is therefore not representative of the true fracture resistance. Microscopic examinations

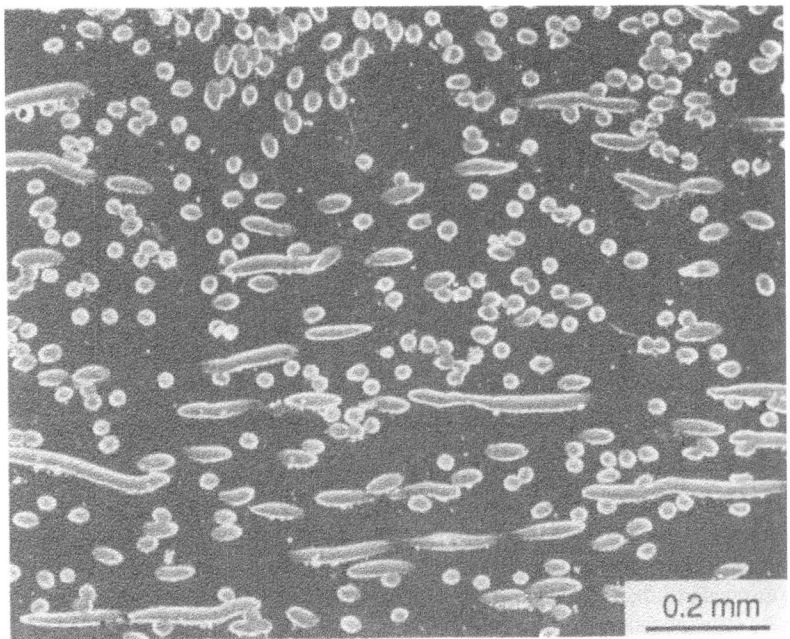

Fig. 1 Micrograph of the Al₂O₃/Al composite.

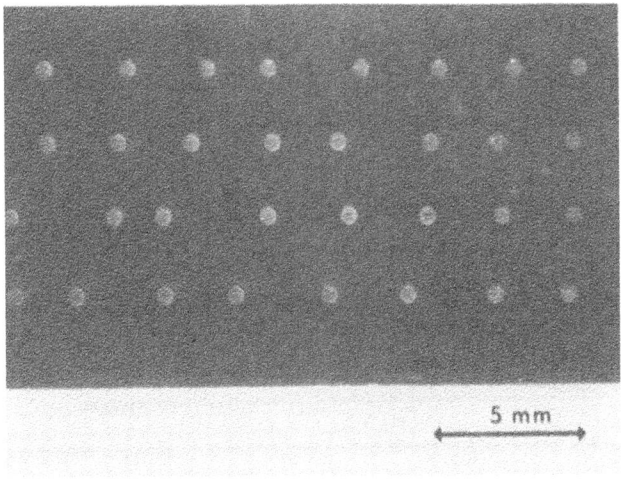

Fig. 2   Transverse section through the PMMA/Al composite ($\theta = 0^{\circ}$).

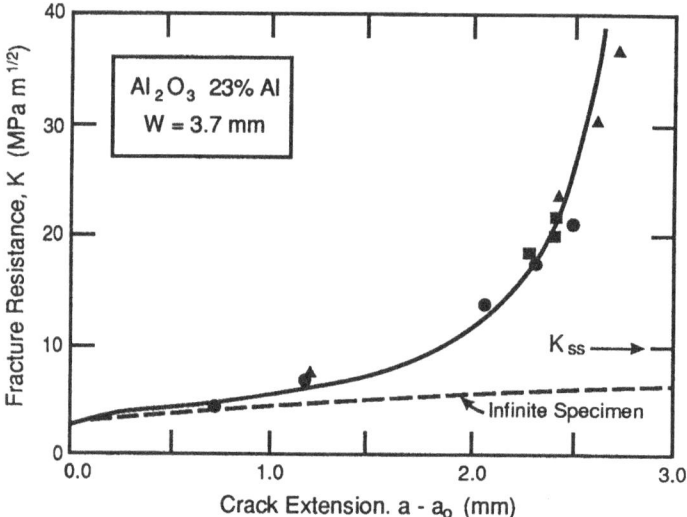

Fig. 3   R-curve for the Al₂O₃/Al composite. The solid line is the computed fracture
          resistance, taking into account the finite specimen geometry, whereas the dashed line
          corresponds to the fracture resistance under small-scale bridging conditions. The
          steady state toughness K_{ss} was determined from the work of rupture test.

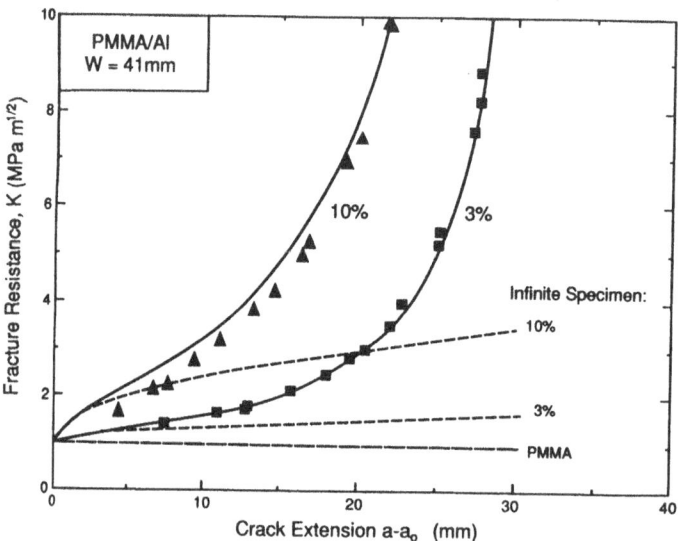

Fig. 4   Measured and calculated R-curves for the PMMA/Al composites containing aligned
          reinforcements ($\theta = 0^{\circ}$).

of the tested specimens reveal the presence of ductile ligaments across the crack faces, an example of which is shown in Fig. 5.

A simulation of the R-curves for the PMMA/Al composites was conducted assuming that the bridging tractions are uniform (as in the Dugdale zone model) and equal to the ultimate tensile strength of the Al wire reinforcements (80 MPa), as measured in a tensile test of the wire itself. The choice of the bridging stress can be justified on the basis that the flow stress of the Al wire increases rapidly from the yield point to the ultimate strength and subsequently remains constant over most of the plastic straining history (Fig. 6). Furthermore, the debond length is much larger than the wire radius (Fig. 5) and thus plastic constraint effects can be neglected [25,26]. The fracture resistance is calculated from equations (1) and (2) using an expression for F applicable to the present specimen geometry [20,21]. The agreement between theory and experiment is excellent for composites containing 3% and 10% Al (Fig. 4), thus indicating that the toughness enhancement is indeed attributable to a crack bridging mechanism.

For comparison, the computed R-curves for infinitely large specimens of the same materials are also shown on Fig. 4. These correspond to the behavior under small-scale bridging conditions i.e. when the bridging zone length is small relative to all other in-plane specimen dimensions. Evidently the R-curves for the finite specimen geometries are appreciably larger than those corresponding to the infinite specimen, even at relatively small crack extensions. The discrepancy is attributable to *large scale bridging* in the finite geometry and emphasizes the need to account for size effects in interpreting R-curve behavior in composite materials.

A similar simulation was conducted for the Al_2O_3/Al composite, again assuming uniform bridging tractions, and is shown on Fig. 3. In this case, selection of the bridging stress is complicated by the fact that the metal reinforcements are oriented *randomly* in space and thus do not contribute equally to the fracture resistance. Consequently, equations (1) and (2) were used along with the experimental data to infer an average ligament stress of 110 MPa: this corresponds closely to the average between the initial yield stress (70 MPa) and the ultimate tensile strength (170 MPa) of the Al-Mg alloy [21]. The shape of the computed R-curve is in good agreement with the experimental data, provided the finite geometry is taken into account. The R-curve corresponding to small-scale bridging conditions is also shown and again differs substantially from the experimental data obtained on the finite specimen geometry.

Though the R-curve data for the Al_2O_3/Al composite are seemingly described adequately by the Dugdale zone model, cognizance must be taken of the fiber orientation distribution and its role in the crack bridging process. To examine orientation effects in metal-reinforced composites, a series of PMMA/Al composites containing a symmetric arrangement of *inclined* wires was tested. The composites contained 4 layers of equally spaced wires with a metal volume fraction of $\approx 3\%$. The wires in the two outer layers were aligned parallel to one another but were inclined at an angle 2θ to the wires in the two inner layers. The compact tensions specimens were then machined such that the wires were inclined symmetrically at angles of $\pm \theta$ to the crack plane, as shown schematically in Fig. 7. Figure 8 shows the measured R-curves for values of θ ranging between $0°$ and $65°$. Though they exhibit characteristics of large scale bridging, the curves provide a relative measure of toughening due to bridging since all specimens are the same size. The data show that the toughening contribution K_b decreases substantially with θ: it drops by a factor of ~2 as θ goes from $0°$ to $45°$ and is negligible for $\theta \gtrsim 65°$.

The reduction in K_b with θ is attributable to two sources. First, the number of wires intercepted by the crack front per unit area of crack, N_A, *decreases* with θ as:

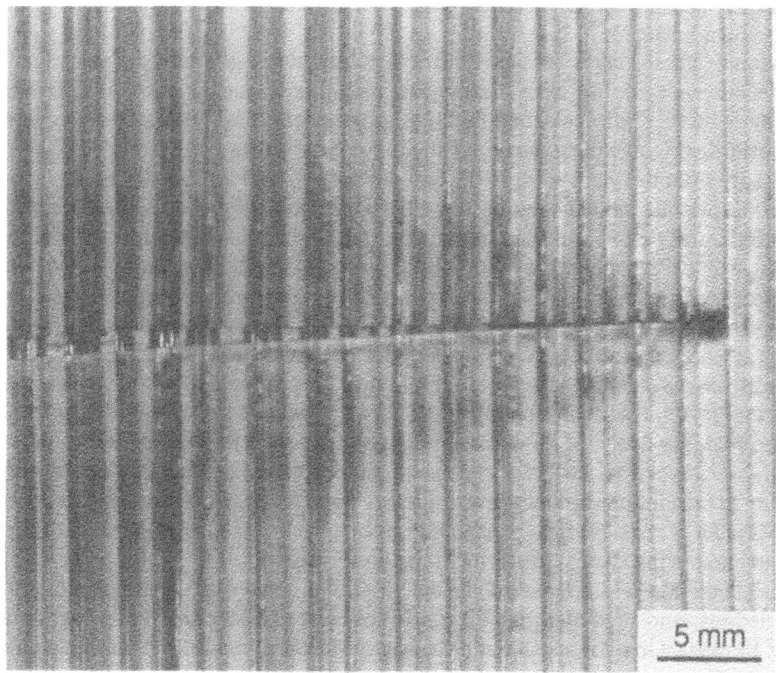

Fig. 5   Macrophotograph of a tested PMMA/Al specimen showing intact Al wires in the crack wake. To observe the degree of debonding at the fiber-matrix interface, a red dye penetrant was inserted into the crack after testing. The debonded region appears dark in the photograph.

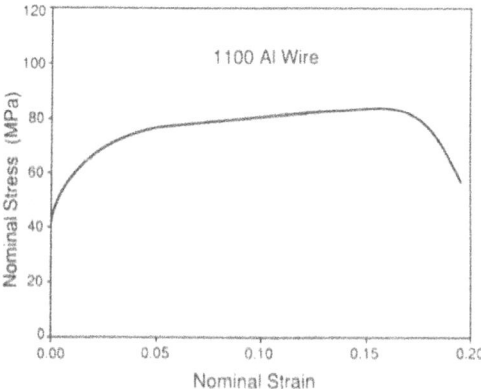

Fig. 6   Nominal tensile stress-strain curve for the 1100 Al wire reinforcement.

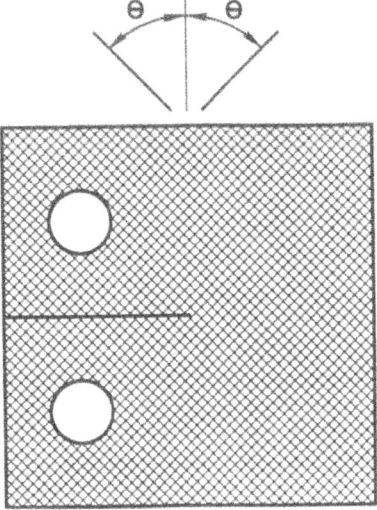

Fig. 7   Schematic diagram of the PMMA/Al specimens containing inclined wires.

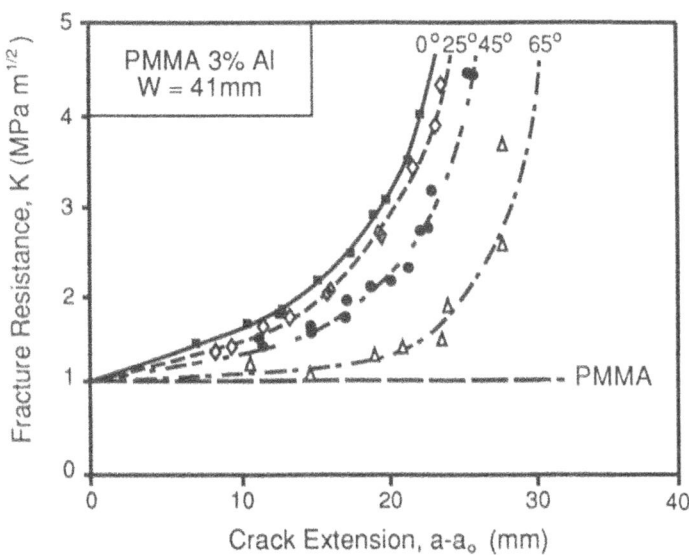

Fig. 8   A diagram showing the effect of orientation on the fracture resistance of the PMMA/Al composite ($f \approx 3\%$).

$$N_A = \frac{f}{\pi R^2} \cos \theta$$

(3)

where f is the metal volume fraction and R is the wire radius. Second, the inclined fibers require a smaller load to be plastically deformed than those which are aligned perpendicular to the crack plane because of the imposed bending moment. At the extreme, the wires inclined at a steep angle act essentially as end-loaded cantilever beams and thus contribute minimally to the crack closing tractions and the corresponding toughness K_b. A more detailed examination of orientation effects in both the PMMA/Al and Al_2O_3/Al composites is currently in progress and will be presented elsewhere [24]. For present purposes it is sufficient to note that such effects are important in the fracture resistance behavior of ductile-reinforced composites and must therefore be incorporated into micromechanical models of crack bridging.

An important feature of the present measurements is the absence of a steady-state fracture toughness, K_{SS}, expected to occur when the ligaments furthest from the crack tip begin to fail. This behavior arises because of the large scale bridging effects at large crack lengths. Similar trends have been observed in other materials exhibiting R-curve behavior [16,17] and suggest that the present techniques are not suitable for obtaining steady-state toughness values. An alternate technique, known as the "work of rupture test" [27], was thus employed in this study to evaluate G_{SS} (the steady-state strain energy release rate) for the Al_2O_3/Al composite. The measurements yield an average value of $G_{SS} \approx 400 \, Jm^{-2}$, which corresponds to $K_{SS} \approx 10 \, MPa \, m^{1/2}$ (assuming an elastic modulus of 250 GPa) [24]. This value is significantly lower than the apparent fracture resistance at large crack lengths measured in the previous tests but is likely more representative of the true asymptotic fracture resistance. The magnitude of the discrepancy again demonstrates the problems associated with large scale bridging in the R-curve measurements.

4. Fiber-Reinforced Ceramics

A similar series of R-curve measurements were made on a lithium aluminosilicate (LAS) glass ceramic reinforced with continuous SiC (Nicalon) fibers. The composite contained a symmetric arrangement of 0°/90° cross plies and a fiber volume fraction of 0.44. Beams suitable for bend testing were machined from the composite plates, with the notch front oriented perpendicular to the plane of the laminate (Fig. 9). The beams were annealed in air at 800° C for periods ranging from 1 h to 16 h [5].

Figure 10 shows a summary of the R-curves for the LAS/SiC composites. The curves are similar to those of the ductile reinforced composites in the sense that they start at a relatively low level of fracture resistance ($\sim$5 to 6 MPa $m^{1/2}$) and increase rapidly with crack extension. Once again, however, the data are influenced by large scale bridging effects, particularly at crack extensions $\gtrsim$ 1mm.

Insight into the origin of the rising R-curve is provided by microscopic observations of the side surfaces of the tested specimens (Fig. 11). Evidently some of the fibers fracture along the plane of the matrix crack whereas others fail at some distance away. Examples of the latter are indicated by arrows in Fig. 11. The embedded fibers are then pulled out of the matrix during crack growth and thus exert tractions on the crack faces.

Near the crack tip the bridging stress associated with fiber pull-out can be expressed as [5]

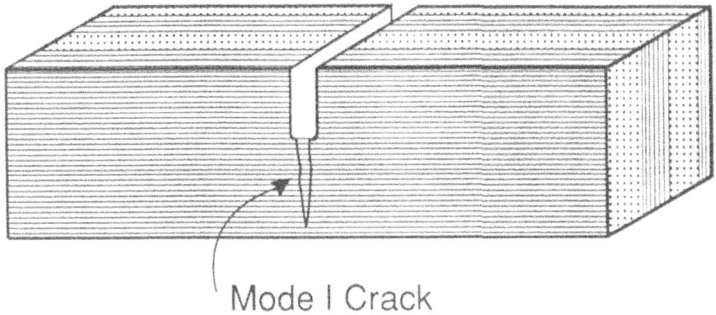

Fig. 9   Schematic diagram showing notch orientation in the LAS/SiC specimens.

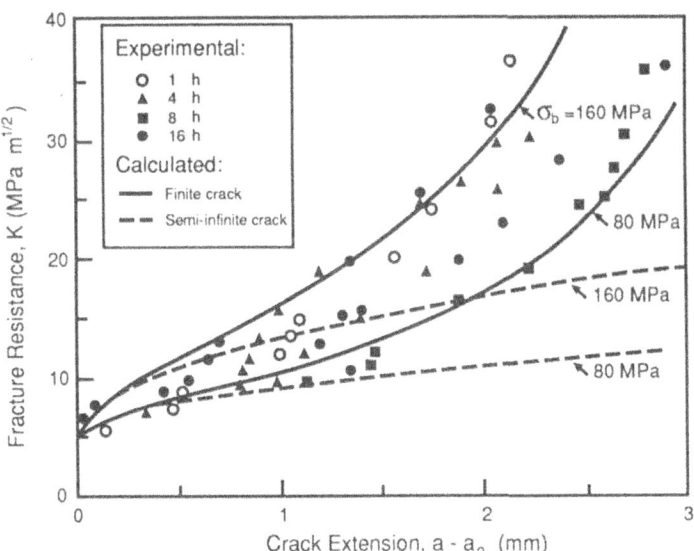

Fig. 10   Measured and computed R-curves for the LAS/SiC composites.

434

Fig. 11   Micrograph of a notch tip region in the LAS/SiC composite. The arrows show fibers which have failed away from the plane of the matrix crack and subsequently been pulled out of the matrix.

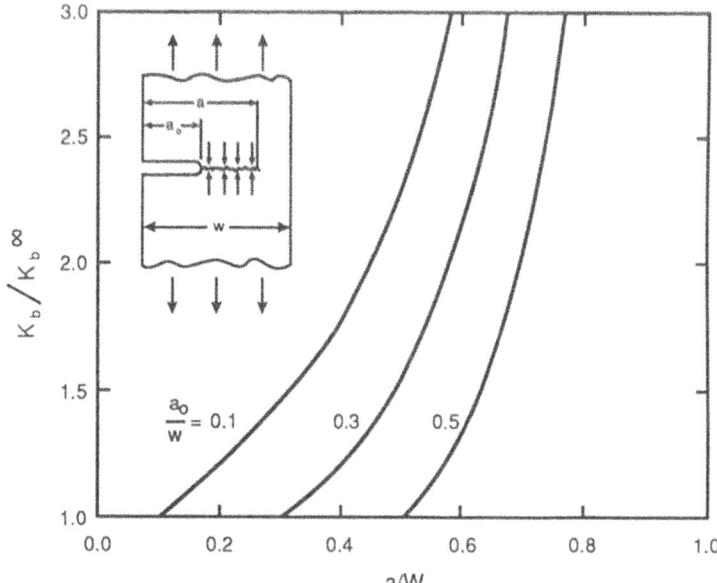

Fig. 12   A diagram showing the effects of large scale bridging, assuming uniform bridging tractions, for three different values of notch depth, a_0/w.

$$\sigma(x) = \frac{2\tau f h}{R} \left[1 - \frac{4\left(1-v^2\right)}{E h} \sqrt{\frac{x}{2\pi}} K_0 \right]$$

(4)

where τ is the sliding resistance of the fiber-matrix interface, h is the average pull-out length, K_0 is the critical value of the crack tip stress intensity factor and f is the volume fraction of fibers oriented perpendicular to the crack plane (f = 0.22). Provided the pull-out length is sufficiently large, the second term in the square brackets in equation (4) can be neglected and the bridging stress then taken as

$$\sigma_b \approx \frac{2\tau f h}{R}$$

(5)

This approximation allows for a preliminary comparison between theory and experiment, taking into account the large scale bridging effects. Fig. 10 indicates that the measurements are consistent with the computed R-curves for a bridging stress in the range of 80 MPa to 160 MPa. Combining this result with measurements of pull-out length [28], equation (5) gives a sliding stress of ~75 MPa to 250 MPa. These values are broadly consistent with those inferred from pull-out measurements [28], but somewhat higher than those measured by push-through tests [29].

5. Discussion

The correlation between the computed and measured R-curves in both the metal-reinforced and fiber-reinforced composites along with the microstructural observations indicate that the toughness enhancement is attributable to bridging processes in the crack wake. The important parameters governing the resistance behavior are the spatial variation in the bridging stress and the specimen geometry via the function F. In fiber-reinforced composites the bridging stress depends on the interfacial sliding stress and the average pull-out length. In metal-reinforced composites the bridging stress is governed by the flow characteristics of the metal, the degree of debonding, and the orientation of the reinforcements with respect to the crack plane. An important feature of the measurements is the marked influence of large scale bridging on fracture resistance. These effects have been incorporated into the simulations through the use of appropriate functions.

In general, the magnitude of the large scale bridging effects can be described by the ratio K_b/K_b^∞, where K_b is the increment of toughness due to bridging in the finite specimen and K_b^∞ is the corresponding increment in an infinite specimen. Equation [2] gives

$$\frac{K_b}{K_b^\infty} = \sqrt{\frac{a}{2L}} \int_0^{L/a} F\left(\frac{x}{a}, \frac{a}{W}\right) d\left(\frac{x}{a}\right)$$

(6)

The ratio $K_b/K_b^\infty \to 1$ when small-scale bridging conditions exist; conversely, $K_b/K_b^\infty \gg 1$ when the bridging zone length is on the same order as the crack length and specimen width. Figure 12 shows a plot of K_b/K_b^∞ against the normalized crack length, a/W, for various values of notch depth, a_0/W, assuming uniform bridging tractions. The curves show a strong dependence of apparent fracture resistance on the relative size of the bridging zone.

The results of the preceding calculations can be used as guidelines for the design of appropriate specimen geometries for R-curve measurements, analogous to those of the ASTM Standards for ductile materials. In particular, the specimen dimensions required to maintain small scale bridging conditions can be determined. First, an estimate of the steady state bridging zone length is required: this is usually available from information about the nature of the bridging process and the scale of the microstructure. From the curves in Fig. 12, appropriate combinations of a and W are then selected such that K_b/K_b^∞ remains below a prescribed level for all resistance measurements up to the saturation level. As a general rule, if the allowable error in K_b is 5 %, i.e. $K_b/K_b^\infty \leq 1.05$, and the initial notch depth a_0/W is between 0.1 and 0.5, then the specimen width should be at least 40 times the bridging zone length. This figure is essentially the same as the minimum ratio of specimen width to *plastic zone* size in ductile materials, as prescribed by ASTM Standard E399 [23]. Though the present calculations are based on a uniform stress distribution in the crack wake, similar trends are expected when the bridging stresses are more complex.

In addition to the problems of large scale bridging, there exist other effects associated with specimen geometry and mode of loading which may influence resistance measurements in composites. First, the crack opening profile in bending is generally greater than that corresponding to a uniform remote tension, as shown schematically in Fig. 13. Thus, in bending, the bridging stress reaches a maximum at smaller distances from the crack tip, the steady state bridging zone length is shorter, and the slope of the R-curve is higher. Second, the rotational effects in bending result in a crack opening profile which depends on the total crack length. Specifically, the increase in specimen compliance associated with crack growth causes the crack opening displacement to increase more rapidly with distance from the crack tip, as shown schematically in Fig. 14a. Consequently, the spatial variation in the bridging stress is altered, as is the length of the bridging zone at steady state (Fig. 14b). The resultant R-curve may then exhibit a maximum at the point at which the crack opening effects become larger than those due to large scale bridging (Fig. 14c). Such behavior has recently been observed in TiAl/TiNb composites and has been modelled accordingly [30].

6. Conclusions

The results of the present study demonstrate that R-curve behavior in composites is attributable to bridging processes in the crack wake. Good correlation between experiment and theory is obtained when the finite specimen geometry is incorporated into the model of crack bridging. However, issues regarding the effects of fiber orientation and specimen geometry on R-curve behavior still need to be resolved.

Acknowledgement

The author gratefully acknowledges research support from the College of Engineering, University of California, Santa Barbara.

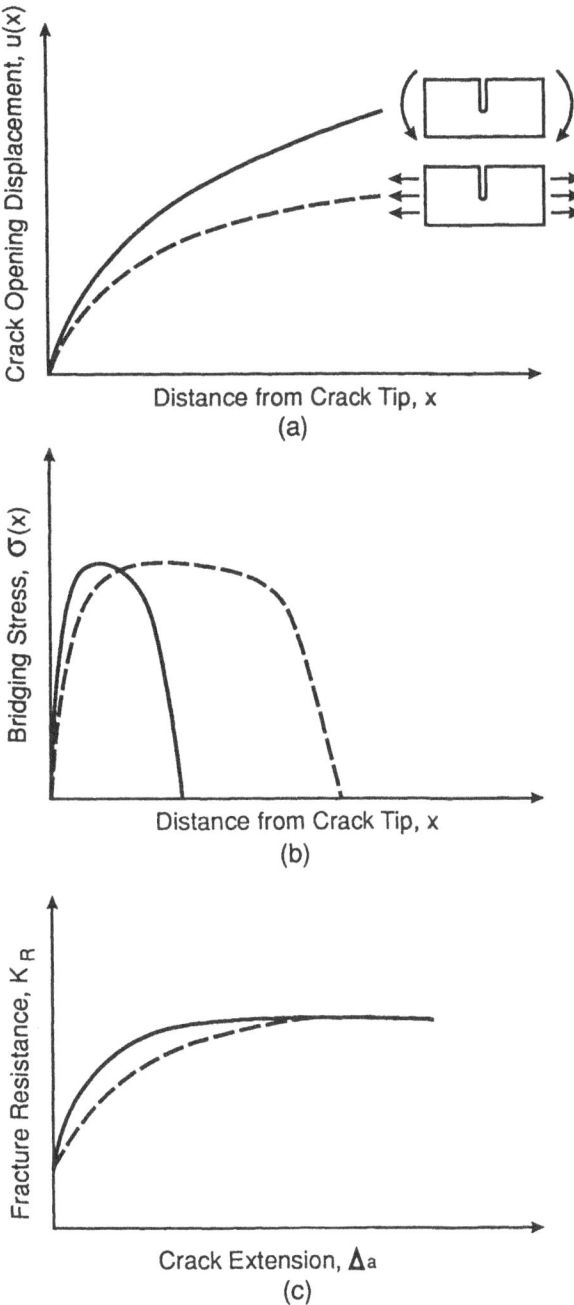

Fig. 13   A schematic diagram showing the effects of mode of loading (tension vs bending) on
          the crack opening profile, the bridging tractions and the R-curve.

438

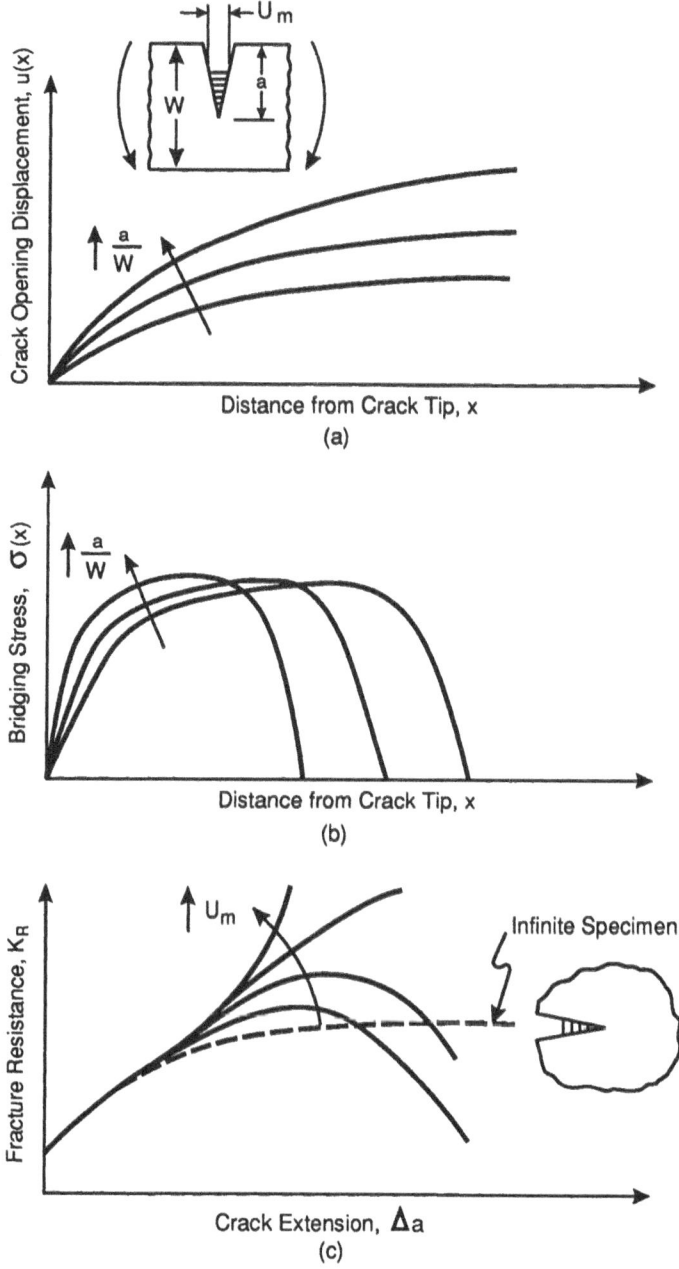

Fig. 14   A schematic diagram showing the influence of crack length on the crack opening
profile, the bridging tractions and the R-curve in a bend specimen. Here u_m is the
crack opening displacement necessary to fracture the bridging ligaments.

References

1. A. G. v and D. B. Marshall, *Acta Metall.*, **37**, 2567 (1989).
2. J. J. Brennan and K. M. Prewo, *J. Mater. Sci.*, **17**, 2371 (1982).
3. P. F. Becher and G. C. Wei, *Comm. Am. Ceram. Soc.*, C-267 (1984).
4. P. Becher, C. H. Hsueh, P. Angelini and T. N. Tiegs, *J. Am. Ceram. Soc.*, **71**, 1050 (1988).
5. F. Zok, O. Sbaizero, C. L. Hom and A. G. Evans, *submitted to J. Am. Ceram. Soc.* (1990).
6. R. M. L. Foote, Y. Mai and B. Cotterell, *J. Mech. Phys. Solids*, **34**, 593 (1986).
7. B. Cotterell and Y. Mai, *J. Mater. Sci.*, **22**, 2734 (1987).
8. L. S. Sigl and H. E. Exner, *Metall. Trans.*, **18A**, 1299 (1987).
9. L. S. Sigl and H. F. Fischmeister, *Acta Metall.*, **36**, 887 (1988).
10. B. O. Flinn, M. Rühle and A. G. Evans, *Acta Metall.*, **37**, 3001 (1989).
11. H. E. Deve, A. G. Evans, G. R. Odette, R. Mehrabian, M. L. Emiliani and R. J. Hecht, *Acta Metall.*, in press (1990).
12. C. K. Elliott, G. R. Odette, G. E. Lucas, and J. W. Sheckherd, *Mat. Res. Soc. Symp. Proc.*, **120**, 95 (1988).
13. M. Sakai, J. Yoshimura, Y. Goto and M. Inagaki, *J. Am. Ceram. Soc.*, **71**, 509 (1988).
14. M. V. Swain, *J. Mater. Sci. Letters*, **5**, 1313 (1986).
15. H. Wieninger, K. Kromp and R. F. Pabst, *J. Mater. Sci.*, **21**, 411 (1986).
16. R. Knehans and R. Steinbrech, *J. Mater. Sci. Letters*, **1**, 327 (1982).
17. R. Steinbrech and O. Schmenkel, *J. Am. Ceram. Soc.*, **71**, C271 (1988).
18. P. L. Swanson, C. J. Fairbanks, B. R. Lawn, Y. Mai and B. J. Hockey, *J. Am. Ceram. Soc.*, **70**, 279 (1987).
19. Y. Mai and B. R. Lawn, *J. Am. Ceram. Soc.*, **70**, 289 (1987).
20. H. Tada, P. C. Paris and G. R. Irwin, "The Stress Analysis of Cracks Handbook," Del Research Corp., Pennsylvania (1973).
21. F. Zok and C. L. Hom, *Acta Metall.*, in press (1990).
22. F. F. Lange, B. V. Velamakanni and A. G. Evans, *J. Am. Ceram. Soc.*, **73**, 388 (1990).
23. ASTM E399-83, "Plane Strain Fracture Toughness of Metallic Materials," Annual Book of ASTM Standards, ASTM, Philadelphia (1988).
24. F. Zok and B. D. Flinn, unpublished work.
25. M. F. Ashby, F. J. Blunt and M. Bannister, *Acta Metall.*, **7**, 1847 (1989).
26. P. A. Mataga, *Acta Metall.*, in press (1990).
27. H. G. Tattersall and G. Tappin, *J. Mater. Sci.*, **1**, 296 (1966).
28. M. D. Thouless, O. Sbaizero, L. S. Sigl and A. G. Evans, *J. Am. Ceram. Soc.*, **72**, 525 (1989).
29. T. P. Weihs, C. M. Dick and W. D. Nix, *MRS Proceedings*, **120**, 247 (1988).
30. G. R. Odette and G. E. Lucas, unpublished work.

SESSION 7 DISCUSSION

N. J. PAGANO
WRDC/MLBM
Wright-Patterson Air Force Base, Ohio 45433

In discussion following the various presentations, Kobayashi asked Fuller why pull-out tests, rather than the more popular push-out tests, were conducted. Fuller replied that there is an initial stiffening caused by the Poisson effect. Under these conditions, one cannot determine the coefficient of friction and radial stress separately, only their product. Fuller also commented, on response to a comment by Reinhardt regarding crack closing in Kromp's work, that bundle pull-out tests should be conducted to characterize Kromp's bundle fracture observations. Fuller thought he would try such a test. Shah claimed that the improved ductility caused by silicon infiltration of carbon-carbon is analogous to that which takes place in the curing of cement in the presence of a certain additive. He also remarked that a time-dependent bundle pull-out test has been developed at the Center. In response to another question by Shah, Fuller stated that optimum fiber coefficient of friction has yet to be established for ceramic composites. Bazant asked if Kromp had studied size effect in geometrically similar carbon-carbon specimens. Kromp replied that only the size effect on modulus had been examined.

Kobayashi asked if Zok had used orthotropic material properties in his model of fiber bridging. Zok admitted that he did not, but suggested that for low anisotropic ratios correction factors would be small. (Ed. comment. Because of the manner in which fiber bridging is modeled, it is not clear whether isotropic matrix properties or effective composite moduli would be more appropriate. Recall that the crack tip is assumed to lie in the matrix and also slip between fibers and matrix must occur near the crack surfaces.) Li wondered why Zok's R-curves did not saturate for large crack lengths. No good answer. Bazant was unimpressed by the results since K due to large scale bridging is not a material property. Zok discusses this in his report and provides infinite plate calculations to characterize small scale bridging. Someone else asked if there is a limit on K or ΔK due to fiber bridging. Answer - this has not been settled yet due to specimen size effects. In response to Karihaloo, Zok stated that normalization with respect to specimen size has been attempted without success.

At this point, a brief presentation was given by Dr Mobasher on work conducted by Dr Shah and himself. This research involved the fracture of fiber reinforced cement-based composites. Reinforcement was provided by a high volume

S. P. Shah (ed.), Toughening Mechanisms in Quasi-Brittle Materials, 441–443.

loading of polypropylene fibers. Holographic measurements of surface microcrack density were made. Significant strength improvement at equal Young's modulus was observed. This is an example of an R-curve determination for a composite having an R-curve matrix. Improved toughness was provided by the fiber-bridging mechanism. The work inspired a barrage of questions by Mai regarding the two-parameter fracture model employed, test method sensitivity, and test details. Apparently, such questions are premature at this point. Li queried regarding the comparison of the ACK model prediction and the experiment result for matrix cracking strain. Shah commented that the stress prediction was reasonable, but not the strain, and that the calculation is very sensitive to the frictional stress τ. The BOP (Bend-over point) could be correlated with ACK by curve fitting different values of τ for different conditions. Shah also stated that, as in many ceramic composites, the initial matrix failure did not correspond to the steady state cracking mode.

A general discussion followed and Becher questioned Zok regarding strain-hardening of his aluminum wires. Zok replied that the assumed value of bridging stress was equal to the (constant) yield point of the wires.

A small discussion took place regarding the modeling of fibers within a bundle. This was obviously inspired by Kromp's research. Karihaloo, Fuller, Krajcinovic, and Becher were involved, concluding that the bundle itself could be modeled as a composite, which is true. The difficulty in bundle models involves the knowledge or lack of knowledge of constituent material properties within the bundle.

Atkinson raised the issue of our knowledge regarding the mode of fracture in which the crack front grows around the fiber, like a person hugging a tree trunk. Fuller noted that tests demonstrate complex behavior but is not aware of any analytical model. Zok observed that only planar models (fibers and matrix represented as layers) are available, which are not satisfying, but he felt the influence is a second order effect. Li wondered how the details of this failure mechanism effect the R-curve behavior. Fuller also suggested the occurrence of similar phenomena in rock which have not been subjected to detailed analysis.

Becher questioned whether Fuller had inadvertently influenced the residual stress state in his procedures of modifying the interfacial zone. Fuller responded that his purpose was to remove the carbon layer, but this does not answer the question.

There was a brief discussion regarding fatigue behavior of ceramic matrix composites. Mai asked how fatigue is characterized in these materials, e.g., by use of crack growth measurements? Zok stated that the fatigue mechanism differs from that in static loading, i.e., the bridging cracks are on the order of single fiber spacing in fatigue. Karihaloo wondered if microcracks were counted. Zok answered this has not been done yet as the fatigue research was just beginning.

Shah raised the issue of the importance of such things as bond slip and frictional energy dissipation mechanisms in toughening ceramic composites. Zok responded that fiber bridging is the dominant mechanism by far and his attention to this phenomenon is consistent with the maturity of ceramic composites research.

Chan was concerned that Zok's aluminum and plexiglas constituents would not bond so the success of those composites would depend on aluminum having the lower thermal expansion coefficient. Both Zok and Daniel confirmed that this is the case.

In the reporter's (N. J. Pagano) summary of the session, he pointed out that there is a lot of confusion and misstatements in the literature regarding the initial damage mode in ceramic matrix composites. Extensive testing at the Air Force Materials Lab (confirmed by I. Daniel at Northwestern) has shown that damage begins much earlier than the proportional limit of the $0°$ stress-strain curve and this damage takes the form of diffuse distributed matrix cracks rather than one single dominant crack bridging all the fibers. Conventional strain gage techniques are also incapable of detecting damage initiation.

The reporter pointed out a major weakness of all shear lag models of the fiber pullout process, namely, that a reversal of stress can occur in a narrow boundary layer region near the free surface. Shear lag models do not capture this behavior, which may actually be responsible for initiation of debonding and/or matrix cracking.

Pagano also noted an important peculiarity of ceramic matrix composite laminate behavior, i.e., the so-called ply properties (elastic moduli) cannot be considered as unique quantities as these properties are strongly influenced by fiber-matrix interface debonding, which in turn is affected by the state of stress in the ply. It should also be noted that the latter state of stress depends both on applied loading and residual stresses.

Finally, the reporter observed that most concrete research reported in the workshop involved testing and modeling of the concrete itself, rather than the reinforced composite. This is quite different from the classical treatment of composites in other material systems, such as graphite-epoxy, ceramic composites, etc., where emphasis is placed on the composite itself (macromechanics). The reporter was informed that this is the new trend in concrete research and that the macromechanics was being accomplished by another community.

Fracture Toughness of Fiber-Reinforced Cement Composites

RESEARCH CHALLENGES IN TOUGHNESS DEVELOPMENT OF FIBER REINFORCED CEMENTITIOUS COMPOSITES

VICTOR C. LI
Advanced Civil Engineering Materials Research Laboratory
Department of Civil Engineering,
University of Michigan,
Ann Arbor, MI48109-2125

ABSTRACT. This paper presents a discussion of the research challenges in fiber reinforced cementitious composites within the context of the Performance-Processing-Property-Structure framework. Specific examples on the inter-relationships between two types of quasi-brittleness and the material structure are described, and their links to pseudo-ductility and processing conditions are briefly mentioned. The need to bring the disciplines of micromechanics and materials processing closer together underlines a fundamental challenge to the research community in advancing the engineering design of high performance fiber reinforced cementitious composites.

1. Introduction

Research in fiber reinforced cementitious composites (FRCC) has been ongoing for at least two decades. Yet the use of this material at the present time, while growing, is still relatively limited. Although FRCC has been repeatedly demonstrated to be an improved structural material over plain concrete, its present commercial 'push' is more on 'reducing shrinkage cracks'. In other words, after curing, the presence of fibers in concrete is not expected to contribute to the properties of this material. While this usage of fibers is indeed important, especially when structural durability is considered, the structural properties of FRCC appear to be underutilized at present.

The present design of FRCC is governed mainly by the need of a workable mix and by economics. Both of these considerations leads to very low fiber loading, typically less than one volume percent. In the long run, it seems prudent to design FRCC on the basis of long term performance on a unit cost basis. This price/performance approach would require knowledge in (1) the required performance for a given structure and its relation to the

447

S. P. Shah (ed.), Toughening Mechanisms in Quasi-Brittle Materials, 447–466.
© 1991 *Kluwer Academic Publishers.*

unique properties of FRCC, and (2) route of optimization of this performance per unit volume fraction of fibers.

In recent years, there appears to be a rising interest in FRCC, partly because of the commercial need of finding substitute materials for asbestos cement which has been found to be carcinogenic and is being phased out in most industrialized countries, and partly because of the increasing availability of a wide variety of fiber types and geometries for use as reinforcements. In addition, the increasing ease of making high strength concrete and the recognition of the brittleness and difficulty in quality control of this material is begging for a solution in the form of fiber reinforcement. These developments present an opportunity for realizing the structural utilization of fiber reinforced cementitious composites. Despite advances in processing and in property studies, there remains a large gap in our knowledge of the linkage between performance, property, processing and material structure of FRCC. These linkages are the key to successful optimization and hence the achievement of advanced composites suitable for a price sensitive construction industry and market. Although it is not the intention of this paper to review the complete picture of performance-property-process-structure relationships, it is still useful to have a brief overview of these relationships to place the more detailed discussions to follow in the context of this framework.

2. The Performance-Property-Process-Structure Relationship

Performance, property, process and material structure form the apexes of a tetrahedron schematically shown in Figure 1. Their relationship to one another was proposed as a general framework for the study of modern engineered materials by the National Research Council (1990). For our present purpose, we shall restrict the use of this framework to FRCC.

The **performance** of FRCC may include pseudo-ductility, durability, reliability and safety, manufacturability, seismic resistance (energy absorption capacity), flexural and shear capacity and other desirable features of the specific structure which utilize this material. These are elements which the end user and the structural engineer normally would like to see improvements in. In certain circumstances, trade-off in performances may necessitate a critical evaluation of required performance of a structure under given loading and environmental conditions.

The important **properties** of FRCC may include stiffness, strength, fracture toughness

449

and others. In the past, most construction codes are concerned with stiffness and (usually compressive) strength. However, the effectiveness of fiber reinforcement in improving these properties are usually insignificant. This may be one of the reasons for the lack of appreciation for the unique structural property which FRCC offers. These include for example, the tensile first and post-cracking strength and fracture energy which have a direct bearing on several of the performance parameters such as reliability, seismic resistance, and shear and flexural capacities.

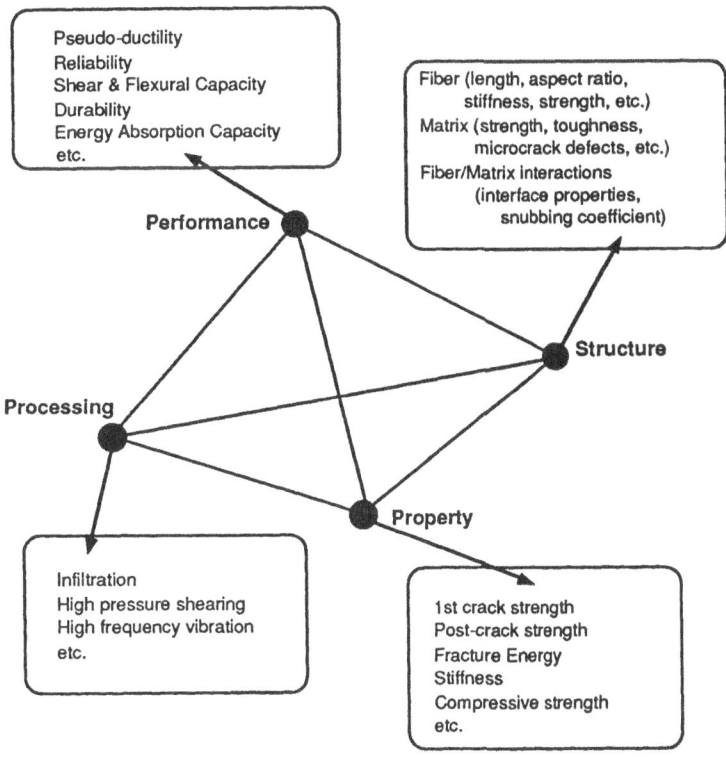

Fig. 1: The Performance-Processing-Property-Structure Tetrahedron for FRCC

Until recently, both **processing** and **material structure** received little attention from researchers. For FRCC, the use of pultrusion, intense shear rolling, infiltration, high frequency vibration processing techniques, and controlled curing processes have been attempted with varying successes. They are also responsible for some of the more high performance modern FRCC materials such as SIFCON and reinforced MDF and DSP cement. The material structure of FRCC includes the fibers and their orientation, the cement or concrete matrix and the pores in them, and the fiber/matrix interface.

Observational studies of interface microstructure (see e.g. Mindess and Shah, 1988) have received much attention in recent years, but quantitative links between interface observations and bond properties are rare.

The performance-processing-property-structure relationship suggests a useful framework offering a wide variety of challenges to the research community. The links between these four cornerstones affords significant opportunities for materials engineering of FRCC which may lead to advanced cement based structures overcoming many of the problems we currently experience in our infrastructures.

3. Two types of Quasi-Brittleness

Quasi-brittleness in FRCC is a desirable property imparted by the inclusions of fibers in a brittle cementitious matrix. Depending on the volume fraction and fiber aspect ratios (among other parameters), two very different types of quasi-brittleness can be achieved. For high fiber loading, and usually in the form of continuous fibers, a type of pseudo-strain-hardening can be accomplished. This pseudo strain-hardening can be in the form of a high critical strain or first cracking strength property of the FRCC, or it can be in the form of multiple cracking. In some cases, both of these forms are exhibited in the same composite. High critical strain or high first crack strength is a direct result of microcrack stabilization by fibers. That is, the microcracks in the matrix propagate against an increasing closing pressure exhibited by the stretched fibers, thus allowing a large number of microcracks to grow stably (volumetrically) resulting in large strain capacity, before any one microcrack grows unstably to form a macrocrack. If the material is unloaded prior to formation of this macrocrack, the microcracks can probably close with no visible signs of material damage. This type of pseudo-strain hardening effect is somewhat analogous with the stabilization of dislocation growth by pinning at grain-boundaries in metals. An example of such a FRCC is a cement composite reinforced with 12% volume fraction of continuous Polypropylene fibers (Figure 2, from Krenchel and Stang, 1988) produced by a pultrusion process. This composite has a first crack strength of about 15 MPa (about five times normal strength concrete) and a critical strain up to .001 (about ten times normal strength concrete). It also exhibits multiple cracking to as much as 2% strain. Another example of a high performance FRCC is a so called Compact Reinforced Composite (Figure 3, from Bache, 1987) produced with more than 6% discontinuous fine steel fibers and then heavily reinforced with continuous steel bars. For this composite within a composite, a critical strain of .003 (about 30 times normal strength concrete) and an increased flexural strength of 150 MPa (about 5 to 10 times that of normal steel reinforced concrete). This particular composite shows significant deformability but does not appear to fail with any sign of multiple cracking. Although multiple cracking allows even higher strain capacity, the resulting damage is permanent, with sub-parallel visible macrocracks.

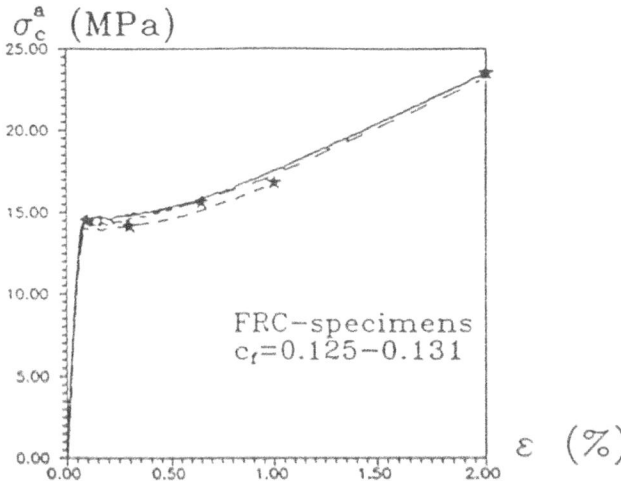

Fig. 2: A Tensile Stress-strain Relationship for a Continuous Polypropylene FRCC (Krenchel and Stang, 1988)

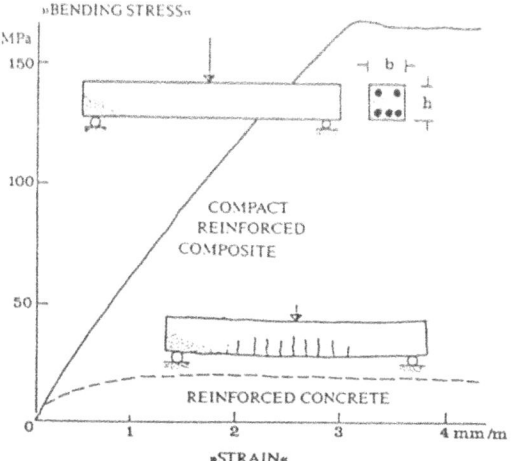

Fig. 3: Flexural Stress-strain Relationship for a CRC Beam (Bache, 1987)

452

A second type of quasi-brittleness is achieved with fiber pull-out as a macrocrack opens. The energy absorbed by frictional work on interface sliding can be significant, sometimes up to several orders of magnitude in comparison to the fracture energy of the unreinforced material. This type of FRCC exhibits a softening stress-displacement curve as shown in Figure 4 for a mortar matrix reinforced with various types of polymeric fibers (processed with a vacuum equipped bladeless Omni-mixer, Wang et al, 1989). The current commercial FRCC (using usually steel or polypropylene fibers) is of this type. The low volume percentage makes processing relatively easy. However, the large gain in fracture energy is sometimes deceptive since to access the full amount, the crack opening would have to be of the order of half the fiber length which can be unacceptable in engineering practice. Even so, there is no doubt that such material property can be utilized in energy absorption members of earthquake resistant structures or structures which need to withstand impact loads.

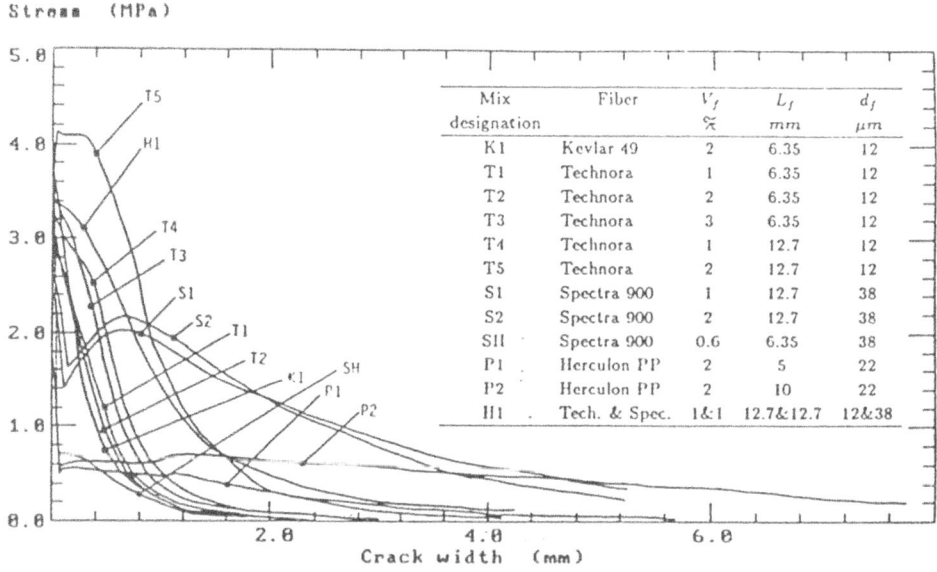

Mix designation	Fiber	V_f %	L_f mm	d_f μm
K1	Kevlar 49	2	6.35	12
T1	Technora	1	6.35	12
T2	Technora	2	6.35	12
T3	Technora	3	6.35	12
T4	Technora	1	12.7	12
T5	Technora	2	12.7	12
S1	Spectra 900	1	12.7	38
S2	Spectra 900	2	12.7	38
SII	Spectra 900	0.6	6.35	38
P1	Herculon PP	2	5	22
P2	Herculon PP	2	10	22
H1	Tech. & Spec.	1&1	12.7&12.7	12&38

Fig. 4: Stress-Displacement Relationship for Low V_f FRCC (Wang et al, 1989)

4. Concept of Bridging Stress-Crack Width Relation

The concept of bridging stress-crack width relation may be illustrated with the help of Figure 5. In Figure 5c, the matrix has fully cracked. For this configuration, if the specimen is loaded from zero load, the complete bridging stress-crack width relation (σ_B-δ) of Figure 6a can be measured. In reality, figure 5c follows the two previous stages of loading in Figure 5a which shows a microcrack being stabilized, and Figure 5b which shows a microcrack just becoming unstable and spread across the full cross-section of the specimen. The loading at this stage corresponds to the first crack strength (Figure 5b and Figure 6b). Between these two stages (A and B), distributed microcracking (not shown) can be expected. After stage B, multiple cracking (not shown) can occur, depending on the relative magnitude of the first crack strength σ_{fc} and on the post-crack strength σ_{pc} (Li and Leung, 1990). Subsequently, the load bearing capacity is fully due to the bridging fibers and the load-deformation behavior follows that of the softening σ_B-δ curve (Figure 6b). This discussion reveals that the pre-peak σ_B-δ relation plays a dominant role in determining the presence or absence of the first type of quasi-brittleness, i.e. the pseudo strain-hardening discussed earlier, whereas the post-peak σ_B-δ curve controls the second type of quasi-brittleness, i.e. the composite fracture energy. Theoretical and experimental studies of the σ_B-δ curve has traditionally focused on the post-peak part. Indeed, the pre-peak part is generally not accessible experimentally, although it can be studied theoretically.

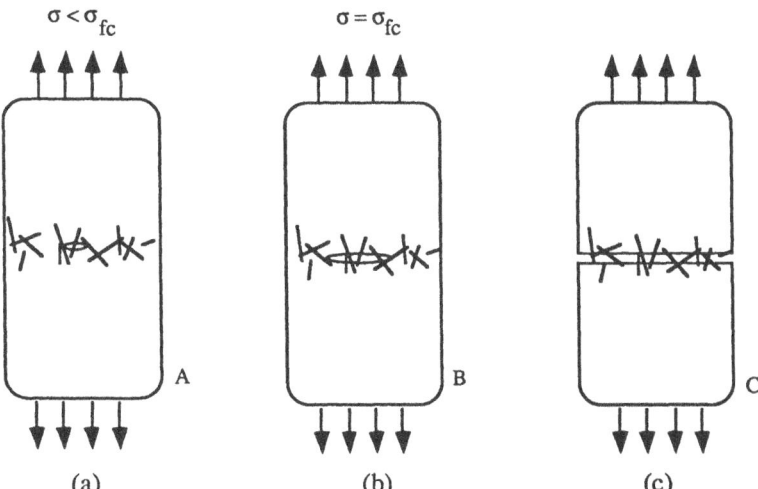

Fig. 5: Three Stages of Deformation Under Uniaxial Loading of a FRCC, (a) Prior to First Crack, (b) at First Crack Load, (c) Post-Peak

454

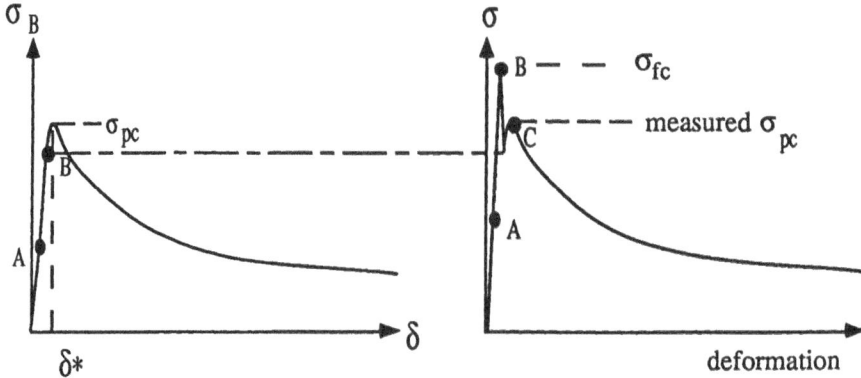

Fig. 6 (a): The Complete σ_B-δ curve and (b) the Corresponding Uniaxial Load-
Deformation Curve

5. A Micromechanical Model of σ_B-δ Curve

In the following, we review briefly a micromechanical model of the σ_B-δ curve proposed
by Li and co-workers. This model focuses on flexible fibers such as most steel and
polymeric fibers used in reinforcing cement matrixes. The material structure of FRCC
consists of fibers (volume fraction V_f, length L_f, diameter d_f, stiffness E_f and strength
σ_f), matrix and interface. For discontinuous fibers, they are typically randomly oriented,
although processing condition or structural boundaries may bias the distribution (Stroeven,
1988). The σ_B-δ relation may be derived from the single fiber bridging force P(δ)
through

$$\sigma_B(\delta) = \frac{V_f}{\pi d_f^2 / 4} \int_{\phi=0}^{\pi} \int_{z=0}^{(L_f/2)\cos\phi} P(\delta)p(\phi)p(z)dzd\phi \qquad (1)$$

where p(ϕ) and p(z) are probability density functions of the orientation angle (with respect
to the tensile loading direction) and centroidal distance of fibers from the crack plane. For
uniform random distributions, p(ϕ) = sin ϕ, and p(z) = 2/L_f (Li et al, (1990a)). The upper
integration limit for z in (1) is intended to discount those fibers not bridging across the
matrix crack in their initial position. That is, only fibers located or oriented in such a
direction as to have a positive initial embedded length would contribute to the composite
bridging stress. In particular, all fibers lying parallel to the matrix crack plane (at ϕ=90°)
are excluded in (1).

Eqn. (1) has been written in a modular form where the mechanisms of bridging (contained in the P(δ) term) and the distribution of fiber location and orientation are separated. This modularity enhances the flexibility of including a variety of mechanisms of bridging, such as fracture based fiber debonding (Gao et al, 1988) and/or fiber rupture (Li et al, 1990a).

To estimate the single fiber bridging force, mechanisms of fiber/matrix interaction has to be considered (Figure 7). The simplest model would be a frictional interface, with debonding and slippage resisted by a frictional bond strength of τ. Thus fiber bridging force would be a function of crack opening δ, embedded length *l* and the bond strength τ. Experimental test of Nylon and Polypropylene fibers, however, indicate that the pull-out load can be a strong function of the angle of inclination φ of the fiber to the loading axis (Figure 8). Morton and Grove (1976) and Li et al (1990) suggested that a flexible fiber being pulled from a matrix is similar to a rope pulled over a friction pulley. Thus the fiber bridging force may be modelled as

$$P(\delta) = \begin{cases} \pi\sqrt{E_f d_f^3 \tau \delta / 2}\, e^{f\phi} & for \quad \delta \le \delta_o \\ \pi\tau l d_f\left[1-(\delta-\delta_o)/l\right]e^{f\phi} & for \quad l \ge \delta > \delta_o \end{cases} \tag{2}$$

where $\delta_o \equiv \dfrac{2l^2\tau}{E_f d_f}$ corresponds to the crack width at which debonding of a fiber with

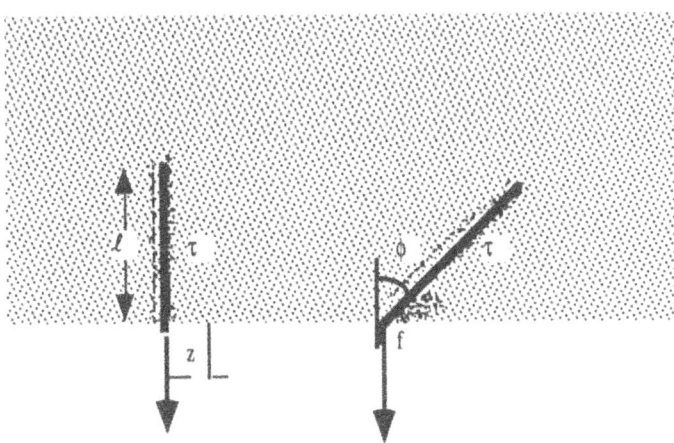

Fig. 7: Fiber Bridging Force and Mechanisms; **P** (δ; *l* (L$_f$, z, φ), d$_f$, E$_f$, τ, f, φ)

456

embedded length l is completed, and f is a snubbing coefficient. Apart from Nylon and Polypropylene fiber in normal strength cement data, steel wire pulled out at two different angles from a epoxy matrix was reported by Morton and Grove (1976), Figure 8. However, at the higher angle (40°), they reported that matrix yielding limited any increase in bridging force. In addition, Naaman and Shah (1976) reported inclined fiber pull out test in normal strength cement, but found that the peak load wavered about a medium value for ϕ between 0° and 75°. These data are perhaps indicative of matrix strength limiting the snubbing effect. The limit must be set by the fiber diameter and stiffness, as well as by the matrix spall resistant strength. Even for the polymer fibers which data follows the snubbing friction concept reasonably well, the data for high angle testing reveals significant amount of standard deviation associated with matrix surface spall. Matrix spalling can be observed under the SEM (Figure 9).

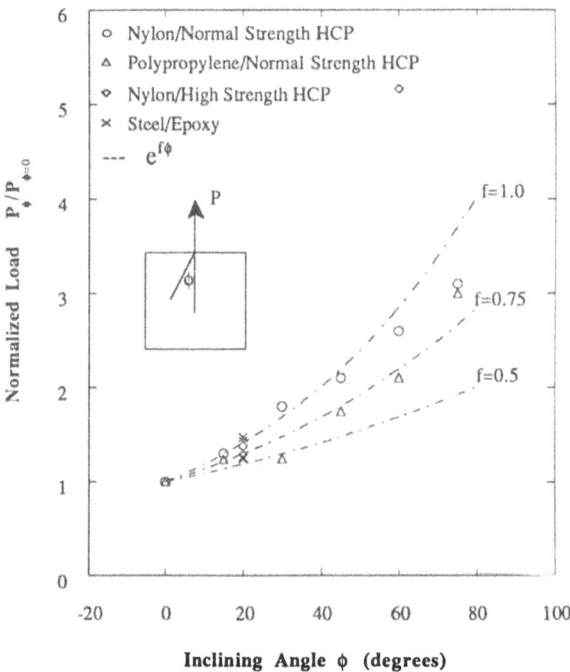

Fig. 8: Maximum Pull-out Load as a Function of Inclining Angle

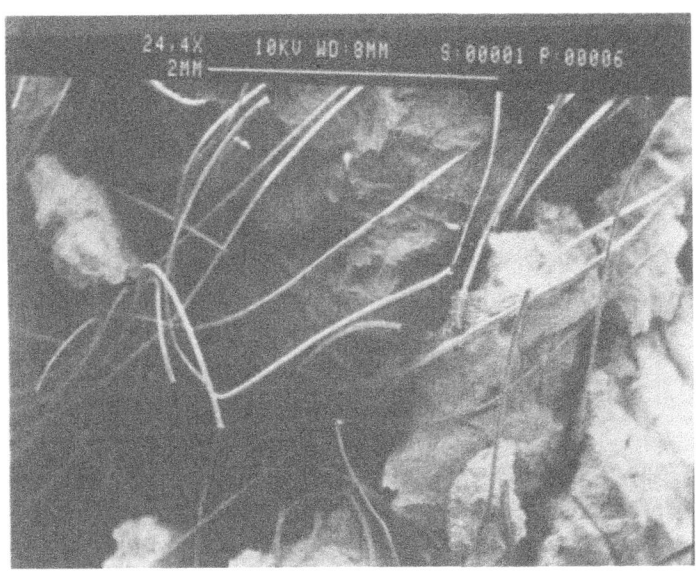

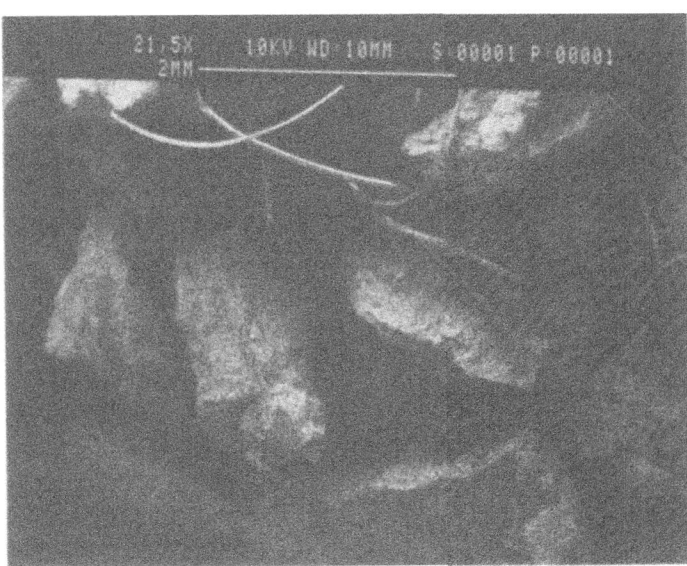

Fig 9: SEM of Surface Spalls from the Fracture Surface of Spectra FRCC

With (1) and (2), Li (1990) showed that the pre-peak and post-peak σ_B-δ can be expressed as:

$$\tilde{\sigma}_B(\tilde{\delta}) = \begin{cases} g\left[2(\tilde{\delta}/\tilde{\delta}^*)^{1/2} - \tilde{\delta}/\tilde{\delta}^*\right] & for\ \tilde{\delta} \le \tilde{\delta}^* \\ g(1-\tilde{\delta})^2 & for\ 1 > \tilde{\delta} > \tilde{\delta}^* \end{cases} \tag{3}$$

where $\tilde{\sigma}_B \equiv \sigma_B/\sigma_o$ and $\sigma_o \equiv V_f \tau (L_f/d_f)/2$, and $\tilde{\delta} \equiv \delta/(L_f/2)$. $\tilde{\delta}^* \equiv (\tau/E_f)(L_f/d_f)$ corresponds to the maximum attainable (normalized by $L_f/2$) value of δ_0 for the fiber with the longest embedment length of $L_f/2$. The snubbing factor is defined as

$$g \equiv \frac{2}{4+f^2}\left(1 + e^{\pi f/2}\right) \tag{4}$$

For the range of f between 0 and 1, g range from 1 to 2.32. Predicted pre-peak and post-peak σ_B-δ curves are shown in Figure 10 for three values of the snubbing coefficient f. A similar form of the post-peak $(1-\tilde{\delta})^2$ dependence was derived by Cotterell and Mai (1988) although their approach would not be able to account for the snubbing effect.

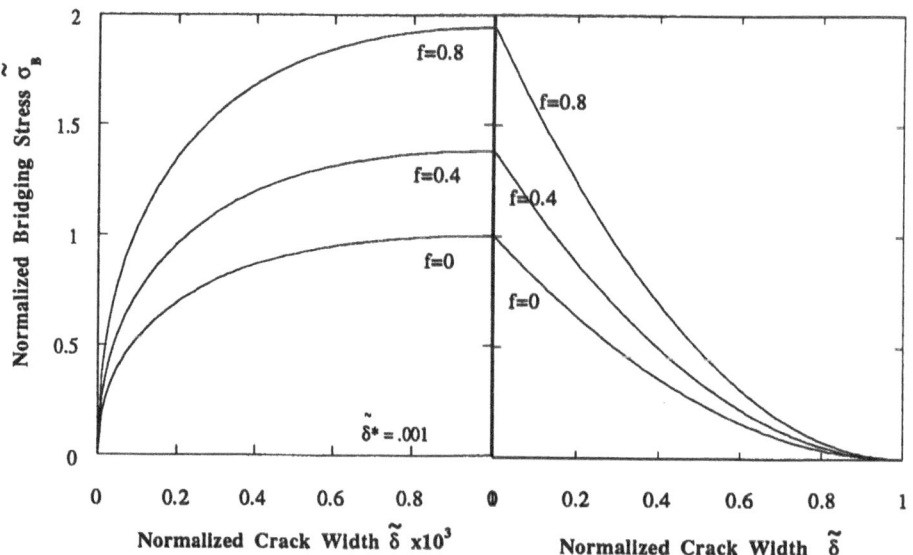

Fig. 10: Predicted Prepeak and Post-peak σ_B-δ curves

The pre-peak part of the σ_B-δ curve cannot be directly verified since there is no experimental data of this type. There is, however, plenty of data for the post-peak part of the σ_B-δ curve, for various FRCCs. Unfortunately, the values of g are typically not reported. Even so, it can be estimated from the peak value of the σ_B-δ curve and by noting that this peak value, often known as the post-crack strength σ_{pc}, is given (from (3)) by

$$\sigma_{pc} = \frac{1}{2} g \tau V_f \left(\frac{L_f}{d_f} \right) \tag{5}$$

By plotting σ_{pc}/τ versus the reinforcement index 1/2 V_f (L_f/d_f) (Li, 1990), data from Visalvanich and Naaman (1982) suggests a value of f ~1 for steel FRCC (a bond strength of 4 MPa is assumed) whereas data from Wang et al (1989) suggests a value of f~0.55 and f~0.77 for Spectra fibers (a high modulus polyethylene) in normal strength and high strength mortar, respectively (a measured bond strength of 1MPa is used, Li et al , 1990b). The well known effect of fiber plastic bending (Morton and Groves, 1974) is not accounted for in this calculation. For polymeric fibers, the bending stiffness is insignificant. Even in steel fibers, Morgan and Groves (1976) suggested that the bending effect is probably small compared to the snubbing effect based on limited pull-out test of steel fibers embedded in an epoxy matrix. However, contradictory data have been obtained by Naaman and Shah (1976).

Figure 11 shows the derived post peak σ_B-δ curve in normalized form (dashed line), and five sets of steel FRCC data (open symbols, Visalvanich and Naaman, 1982) made up of different fiber volume fractions, lengths and aspect ratios. Figure 12 shows a similar plot by Wecharantana and Shah (1983) of a different set of experimental data (except for the one labeled Visalvanich). They used a curve fitting procedure and obtained the same $(1-\tilde{\delta})^2$ dependence as theoretically derived (3). The post-peak σ_B-δ data of Spectra FRCC data (solid symbols, Wang et al, 1989) is also shown in Figure 11 . These plots suggest a definite universality of the shape of the post-peak σ_B-δ curve for FRCCs where fiber bridging composed of debonding and pull-out mechanisms controls the post-cracking behavior of the composite. When experimental data of concrete is examined, it is remarkable to find that a similar $(1-\tilde{\delta})^2$ dependence describes the data reasonably well. For example, Figure 13 shows a composite curve (Karihaloo and Nallathambi, 1990) and a $(1-\tilde{\delta})^2$ plot for the post-peak σ_B-δ of concrete. This consistency is indicative that in a material like concrete, the post-peak behavior may be governed by a pull-out mechanism of aggregates and /or unbroken ligaments bridging across the crack surface similar to the fiber bridging mechansim in FRCC.

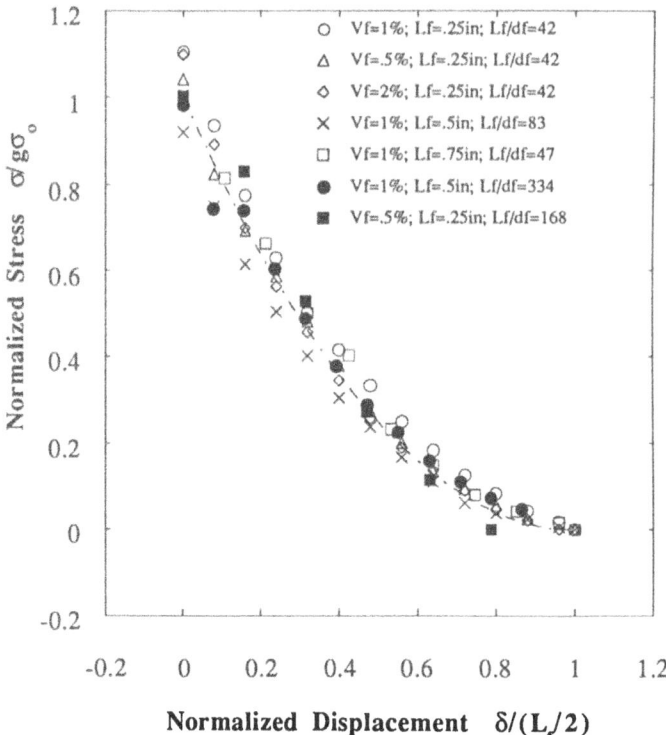

Fig. 11: Comparison Between Predicted (dashed line) and Experimentally Determined Post-peak σ_B-δ curve for Steel FRCC (open symbols) and for Spectra FRCC (solid symbols)

The fracture energy due to fiber pull-out can be derived from (3) and found to be (Li, 1990)

$$G_c = \frac{1}{12} g \tau V_f d_f \left(\frac{L_f}{d_f}\right)^2 \tag{6}$$

Figure 14 shows a log-log plot of (6) with both the steel and Spectra FRCC data included.

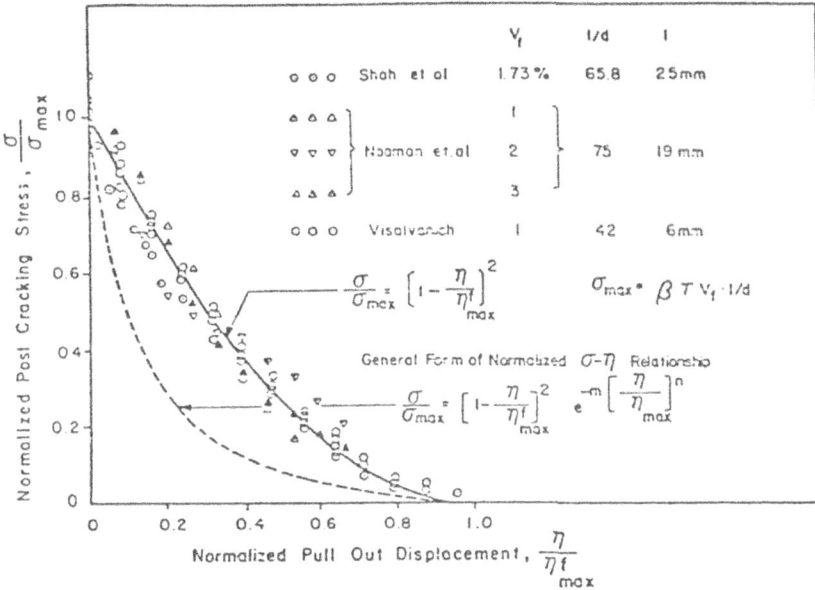

Fig. 12: Empirical Fit to Additional Steel FRCC Data by Wecharantana and Shah (1983)

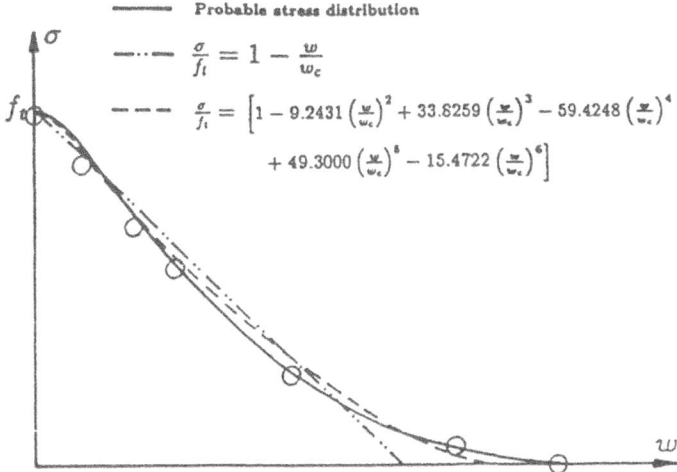

Fig. 13: Post-peak σ_B-δ curve for Concrete. Solid line represents "Probable Stress Distribution" (Karihaloo and Nallathambi, 1990). Circles denote that expressed by (3)

Eqn (5) and (6) suggest that the post-crack strength σ_{pc} and the fracture energy G_c scales with $g\tau$, V_f and $1/d_f$. To enhance σ_{pc} and G_c, therefore, it would be desirable to use a high loading of fibers with small diameters, good bonding and a high snubbing friction. This would be the case if the fibers do not rupture, in line with short fiber length for ease of processing. Otherwise the scaling laws of (5) and (6) no longer hold. Consideration of optimal fiber length for G_c with fiber rupture accounted for is given in Li et al (1990a).

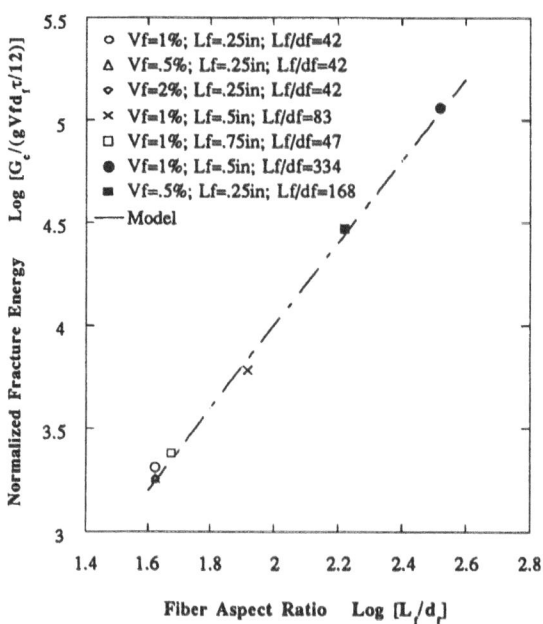

Fig. 14: Comparison of Predicted and Experimentally Determined Fracture Energy of Steel and Spectra FRCC

Because the snubbing friction term and the interface bonding appears as a product in (5) and (6), the snubbing friction has a multiplier effect rather than an additive effect. This implies that if the interface bond is doubled, say, due to surface finish treatment or due to mechanical deformation of the fiber, then σ_{pc} and G_c will quadruple if f~1. There is, however, significant differences in the physical origin of the snubbing friction and the interface bond strength. For example, Table 1 (Li et al, 1989) shows the effect of matrix strength on pull-out load of Nylon monofilaments. In one sample, the fiber was pulled out normal to the matrix crack plane, whereas the other sample involved the fiber pulled out at an angle of 60°. For the normal pull-out case, the average of 8 tests for the higher strength (achieved with use of superplasticizer and microsilica together with a low w/c ratio) matrix shows a 6.3% increase over that for the normal strength matrix, but the standard deviation was large enough not to attach much significance to this increase. For the 60° pull-out, the average of 8 tests for the higher strength matrix shows a 127% increase over that for the normal strength matrix. This increase was significant in comparison to the standard deviation recorded in the test data, and suggests that while the interface bond strength may not have increased much, if at all, the snubbing effect appears to have been enhanced for

Table 1: Effect of Matrix Strength on Pull-out of Nylon Monofilament

| | Pull-out Load (N) | |
	Normal strength HCP	High strength HCP
f = 0°	4.91	5.22
	(CV = 5.6%)	(CV = 22.8%)
f = 60°	11.85	26.94
	(CV = 27.1%)	(CV = 5.4%)
$P_{60°}/P_{0°}$ =	2.41	5.16

the fiber pulled out from the higher strength matrix, presumably due to the better spall resistance of the higher strength matrix. These observations are consistent with that of measured post-cracking strength for Spectra fiber reinforced normal and high strength concrete, discussed above. These preliminary results suggest plausibly different processing routes for the control of the snubbing friction and the interface bond strength.

Because the snubbing friction occurs only if the fiber is inclined to the loading direction, a high snubbing friction value would suggests a higher σ_{pc} and G_c than that of an aligned system. Again, this will be true only if no fiber rupture occurs. In an aligned fiber composite, more fibers tend to share the bridging load, whereas in a random fiber system, less number of fibers are sharing the bridging load, with each carrying a higher load. This implies a greater tendency for fiber rupture in the random case.

6. Conclusions

Although discussed only very briefly, and using extremely narrow and specific examples, the above presentation is meant to bring out the inter-dependencies between performance-properties-processing-structure for the development of cement based composites with pseudo-strain hardening and/or high fracture energy. It is shown that the fiber properties (distribution, length, aspect ratio, elastic stiffness and strength) and fiber/matrix interaction properties (interface bond strength, snubbing coefficient) directly govern the composite properties of post-crack strength and fracture energy which in turn controls the composite performance of pseudo strain-hardening and fracture energy. The need for novel processing routes of designing desirable bond strength, snubbing friction and fiber volume fraction is emphasized. Successful development of such an engineered material cannot be achieved without appreciating the inter-dependencies and constructing solutions which take advantage of the understanding of these inter-dependencies. In this process, it is recognized that micromechanics plays a significant role in quantifying the various links, and especially the link between material structure and properties.

The tetrahedron shown in Figure 1 also suggest an approach perhaps foreign to the development of construction materials -- the performance driven approach. If performance criteria (e.g. durability determined by the life expectancy of the structure; or energy absorption capacity determined by expected seismic load which may be imposed on the structure) can be specified, then the required material structure could be developed to achieve certain material properties using a certain processing route. In other words, a material could be engineered to satisfy the required performance of a given structure in a given environment. Certainly our current state of the art is far from this ideal, but that is precisely the challenge we must face in the research and development of high performance engineered FRCC.

7. Acknowledgement

The author would like to acknowledge the many stimulating discussions with S. Backer and Y. Wang. Part of this research has been supported with grants to the University of Michigan, Ann Arbor, from the Air Force Office of Sponsored Research (Program Manager: Dr. Spencer Wu), and from the National Science Foundation (Program Manager: Dr. Ken Chong).

8. Bibliography

Bache, H.H. (1987). "Compact Reinforced Composite Basic Principles", *CBL Report* No. 41, Aalborg Portland.

Cotterell, B. and Mai, Y-W. (1988). "Modelling Crack Growth in Fiber-Reinforced Cementitious Materials", *Materials Forum* 11, 341-351.

Gao, Y.C., Mai, Y.W. and Cotterell, B. (1988). "Fracture of Fiber-Reinforced Composites", *ZAMP* 39, 550-573.

Karihaloo, B. and Nallathambi, N. (1990). "Test Methods for Dterminating Mode I Fracture Toughness of Concrete", in *Proceedings of Toughening Mechanisms in Quasi-Brittle Materials*, NATO Advanced Research Workshop, Northwestern University, July 16-20, 1990.

Krenchel, H. and Stang, H. (1988). "Stable Microcracking in Cementitious Materials", paper presented at the *2nd International Symposium on Brittle Matrix Composites* -- BMC 2, Cedzyna, Poland.

Li, V.C. (1990). "Scaling Laws for the Post-Cracking Behavior of Fiber Reinforced Cementitious Composites", submitted for publication in the *J. of Materials in Civil Engineering*, 1990.

Li, V.C. and Leung, C. (1990). "First Crack Strength and Multiple Cracking of Flexible Fiber Reinforced Composites", in preparation.

Li, V.C., Wang, Y., and Backer S. (1990a). "A Micromechanical Model of Tension-Softening and Briding Toughening of Short Random Fiber Reinforced Brittle Matrix Composites", accepted for publication in *J. of Mechanics and Physics of Solids*.

Li, V.C., Wang, Y., and Backer S. (1990b). "Effect of Inclining Angle, Bundling, and Surface Treatment on Synthetic Fiber Pull-Out From a Cement Matrix", *Composites*, Vol. 21, 2, 132-140.

Mindess, S. and Shah, S.P., (1988). "Bonding in Cementitious Composites", *MRS Symposium Proceedings*, V. 114.

Morton, J. and Groves, G.W. (1974). "The Cracking of Composites Consisting of Discontinuous Ductile Fibers in a Brittle Matrix -- Effect of Fiber Orientation", *J. of Materials Science*, pp. 1436-1445.

Morgan, J. and Grove, G.W. (1976). "The Effect of Metal Wires on The Fracture of a Brittle Matrix Composite", *J. of Materials Science* 11, 617-622.

Naaman, A.E. and Shah, S.P., (1976). "Pull-out Mechanism in Steel Fiber Reinforced Concrete", *J of Structural Division*, ASCE, 1537-1548.

Naaman, A.E., Moavenzadeh, F., and McGarry, F.J. (1974). "Probabilistic Analysis of Fiber Reinforced Concrete", *J. of Engineering Mechanics*, ASCE, 100, 397-413.

National Research Council, (1990). *Materials Science and Engineering for the 1990s,*

Pub. National Research Council.

Stroeven, P. (1988). "Structural and Mechanical Characteristics of Steel Fiber Reinforced Concrete", *Report 25-88-70*, Faculty of Civil Engineering, Division of Mechanics and Structures, Section on Material Science, Delft University of Technology.

Visalvanich, K. and Naaman, A.E. (1982). "Fracture Modelling of Fiber Reinforced Cementitious Composites", *Program Report for NSF Grant ENG 77-23534*, Department of Materials Engineering, University of Illinois at Chicago Circle.

Wang, Y., V.C. Li, and S. Backer (1989) "Tensile Properties of Synthetic Fiber Reinforced Mortar", accepted for publication in *Int'l J. of Cement Composites and Lightweight Concrete*.

Wecharatana, M. and Shah, S.P. (1983). "A Model For Predicting Fracture Resistance of Fiber Reinforced Concrete", *Cement and Concrete Research*, Vol 13, 819-829.

FAILURE CHARACTERISATION OF FIBRE–REINFORCED CEMENT COMPOSITES WITH R–CURVE CHARACTERISTICS

YIU–WING MAI
Centre for Advanced Materials Technology
Department of Mechanical Engineering
University of Sydney
Sydney, NSW 2006
Australia

ABSTRACT. The failure of fibre–reinforced cement composites can be characterised by the crack–resistance (R) curve approach. In short fibre composites, where the fibres pullout rather than break, theoretical models are presented for the prediction of the R–curve based on the constitutive equation between the closure stress–crack face separation in the fibre–bridging zone at the wake of the crack tip. The influences of specimen size and geometry and the matrix fracture process zone on the R–curve are evaluated and compared to experimental results. Using this crack–face bridging concept the Weibull distribution of the tensile strength of short fibre composites is investigated both theoretically and by computer simulation experiments. It is shown that not only is the crack growth process stabilised but that the Weibull modulus is considerably increased due to the R–curve effect. By including the slow crack growth phenomenon in the theory the time–dependent strengths for short fibre composites with R–curve characteristics are also predicted.

1. Introduction

It is now generally recognised that the failure behaviour of fibre–reinforced cement composites cannot be adequately described by the conventional one–parameter fracture criterion such as the critical potential energy release rate G_c and the critical stress intensity factor K_c. When a crack develops in a short fibre–reinforced composite we have a small matrix fracture process zone (FPZ) at the crack tip region where microcracking activities take place and a fibre bridging zone (FBZ) at its wake where fibre–matrix debonding and fibre pullout occur, Figure 1. The size of these two zones depend largely on the fibre aspect ratio, fibre volume fraction and specimen configuration, Table 1. However, it is the development of the FBZ that gives rise to stable crack growth and the so–called crack–resistance (R) curve usually plotted in the form of stress intensity factor (K_R) versus crack growth (Δa). In these lecture notes only those short fibre cement composites whose failure mechanism is predominantly fibre pullout rather than fibre breakage are considered. Some of the experimental difficulties associated with R–curve and matrix fracture process zone size measurements are first highlighted (Section 2). Analytical modelling of R–curves consistent with the particular toughening mechanisms

S. P. Shah (ed.), Toughening Mechanisms in Quasi-Brittle Materials, 467–505.
© 1991 *Kluwer Academic Publishers.*

468

are then described in relation to effects of specimen size and geometry and influences of the matrix FPZ (Section 3). The application of R–curve in prediction of tensile strength of short fibre cement composites is finally studied both theoretically and by computer simulation (Section 4).

TABLE 1. Size of fibre bridging and matrix
fracture process zones for cementitious materials

Material	Fibre bridg— ing zone (mm)	Fracture process zone (mm)	Reference
Asbestos fibre — cement	~ 125	28	14.48
Cellulose fibre — cement	~ 80	28	4
Steel wire— concrete	> 610	—	17
Steel wire— cement	760	—	48
Glass fibre — cement	—	15	49

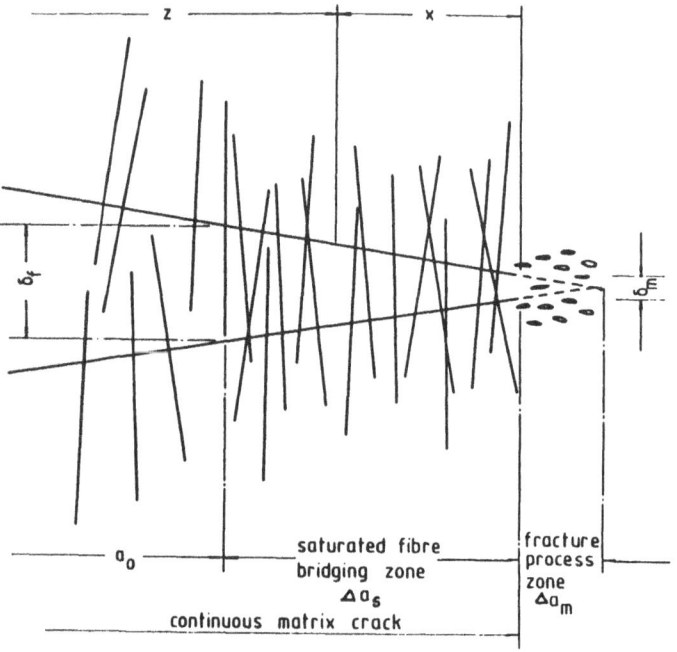

Figure 1. Fibre bridging (FBZ) and fracture process (FPZ) zones at crack tip of a fibre reinforced cement composite.

2. Measurements of FBZ and FPZ for Fibre Cement Composites

In the experimental evaluation of the crack–resistance (R) curve it is required to determine quite accurately the relative sizes of the fibre bridging zone (FBZ) and the matrix fracture process zone (FPZ). Depending on the relative size of the matrix FPZ it may or may not have to be included in the theoretical analysis. Distinction of the exact boundary between the FPZ and the FBZ is always difficult. Many direct and indirect methods have been used to measure either of these two zones. These include optical and scanning electron microscopy techniques, photography, staining and Moire' fringes, replicas, electrical potential difference methods, mercury porosimetry, acoustic emissions and compliance techniques. Foote [1] has given a review of these test methods and their relative merits and usefulness for measurements of the FPZ and FBZ. It is perhaps important to point out that the compliance method, so often employed by the majority of investigators to measure crack length, is not very accurate. The fibres bridging across the crack faces tend to reduce the compliance of a stress–free crack of the same length [2]. Evaluation of crack length from theoretical compliance calculations therefore underestimates its true length (see Figure 2). Recently, Hu and Wittmann [3] have developed a "multi–cutting" technique to measure the FPZ of cementitious matrices and the bridging stresses within this zone. Elegant and simple as this technique may be, it is difficult to extend to fibre cement composites for the separation and measurement of the FPZ and FBZ.

2.1 NEW METHODS FOR FRACTURE PROCESS ZONE SIZE MEASUREMENT AND CRACK GROWTH

The author and his co–workers have developed a new technique to detect the crack tip position and hence to distinguish the boundary of the FBZ and FPZ of fibre cements. An automated method of continuous crack length measurement using a screen–printed conductive grid and a micro–computer is also given. Details of these experimental methods have already been published elsewhere [4]. In the following only the essential features are described.

Figure 3 shows a schematic diagram of the computer aided crack growth monitor system and Figure 4 gives the conductive grid pattern that was used for a compact tension (CT) wood fibre cement composite specimen. There were 64 bars in eight blocks of eight allowing the crack to be measured over a distance of approximately 140 mm. The bars were nominally 1 mm in width and had a repeated distance of 2.14 mm. The conductive ink used consisted of finely divided graphite particles dispersed in a vinyl resin binder and a butyl cellosolve acetate solvent. In operation the computer scanned each bar on the grid serially and tested each for continuity. When a broken bar was detected, indicating the presence of a crack at that location, a voltage corresponding to its number was sent to a plotter and a graphic display of the grid on a monitor showed that bar as broken. With each scan the latest broken bar was detected, plotted and displayed. This permitted the crack growth process to be clearly seen. Both load–time and crack growth–time records could be simultaneously and continuously obtained with this technique until final failure. Figure 5 shows the results obtained for the wood fibre cement CT specimen. The complex nature of crack grwoth was revealed by the bars recording breaks and sometimes closures followed by a second break. The region of bars breaking, or the "activity zone", was approximately 20 mm over the period of crack growth.

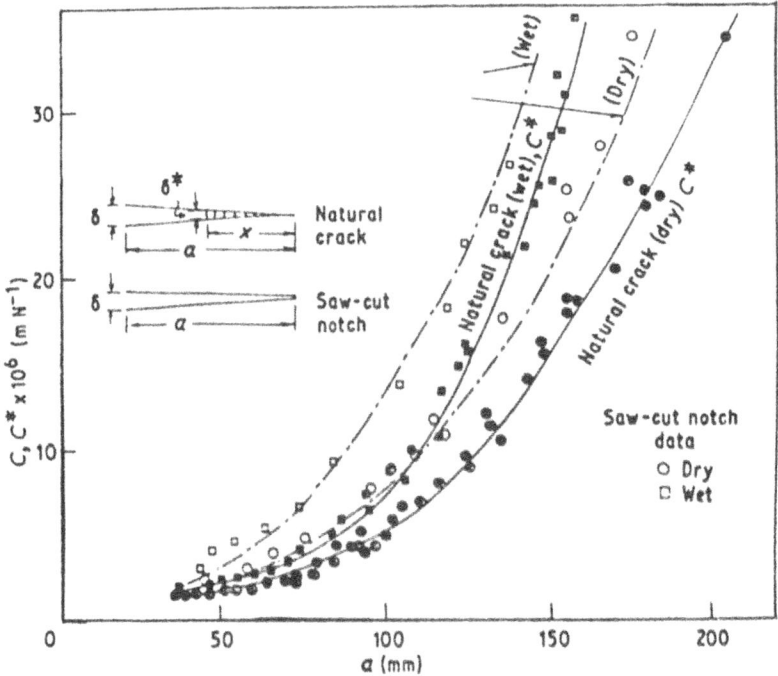

Figure 2. Fibre bridging effect on compliance measurements of a DCB cellulose fibre cement mortar. Compliances C^* and C refer to presence and absence of FBZ respectively.

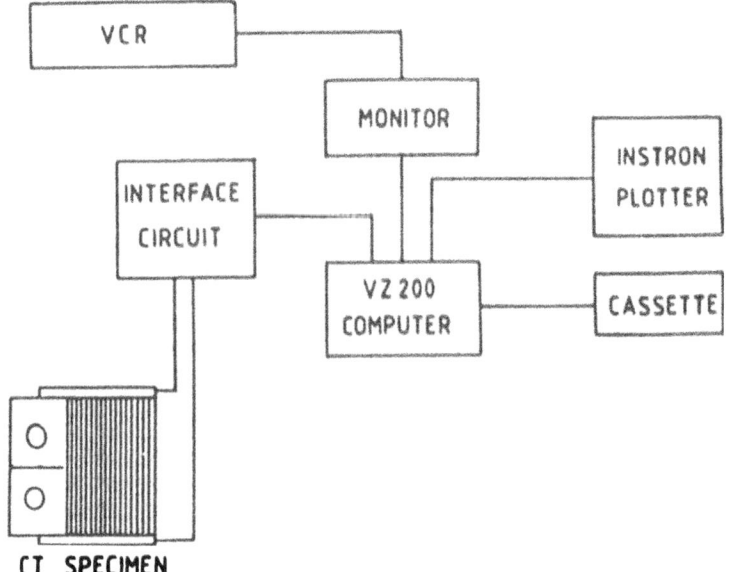

Figure 3. Computer aided crack growth monitoring system.

Table 2 compares the observed crack tip using optical microscopes and the location of the activity zone. Quite clearly, the leading edge of the activity zone gave a good measure of the real crack length and hence a plausible demarcation of the FBZ and FPZ.

In order to measure the size of the matrix fracture process zone narrow strips were cut from the CT specimen after crack growth had taken place. Strips away in the unstressed region were also cut to represent the undamaged material, Figure 6(a). The bending stiffness of crack strip was measured in pure bending with the compressive surface facing the crack growth direction, Figure 6(b), and the location of fracture noted. In strips cut from the cracked (or fibre bridging) region of the specimen, the failure sites would lie along the prolongation of the machined notch. However, in the matrix fracture process zone, the fracture sites were scattered. Figure 7 shows the variation of the normalised bending stiffness with distance from the crack origin and the locations of the failure sites. It is clear that these results support the general concept discussed above. The crack length indicated here is about 80 mm which agrees closely with the optical and computerised electrical methods shown in Table 2. Figure 8 plots the normalised bending stiffness results for four CT specimens as a distance from the crack tip. The matrix FPZ size deduced from this figure varies between 28 mm to 40 mm. Such a size is certainly not negligible and it would seem that it must be properly included in the R–curve modelling. However, as shown later in Section 3.5, its inclusion in the model is not really necessary.

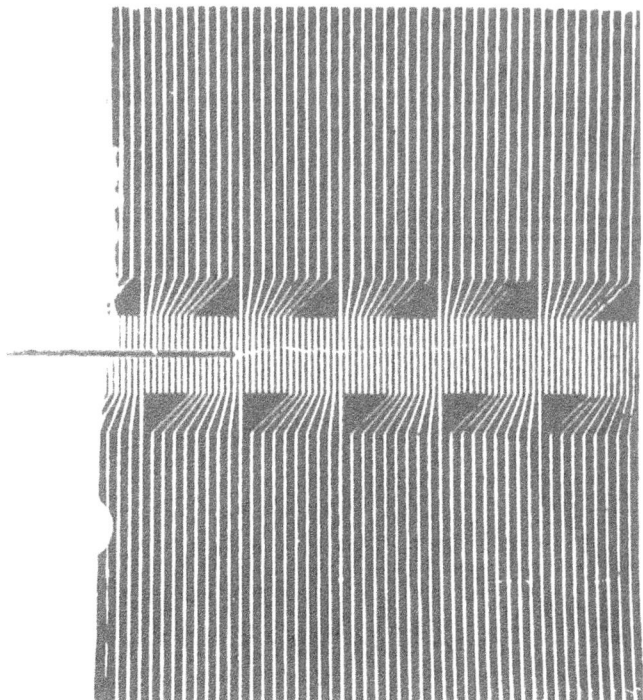

Figure 4.   Conductive grid pattern printed on CT specimen.

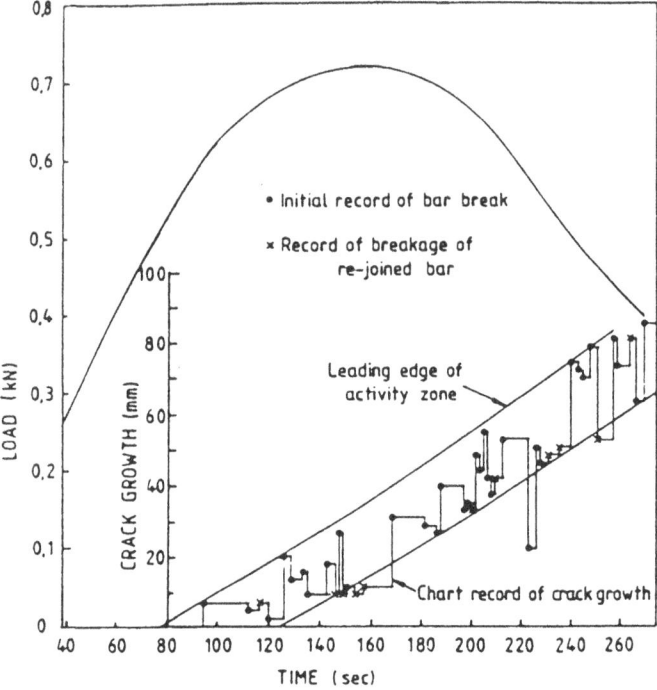

Figure 5.   Crack growth/load-time records using the computer-aided system.
Activity zone leading edge gives a measure of real crack length.

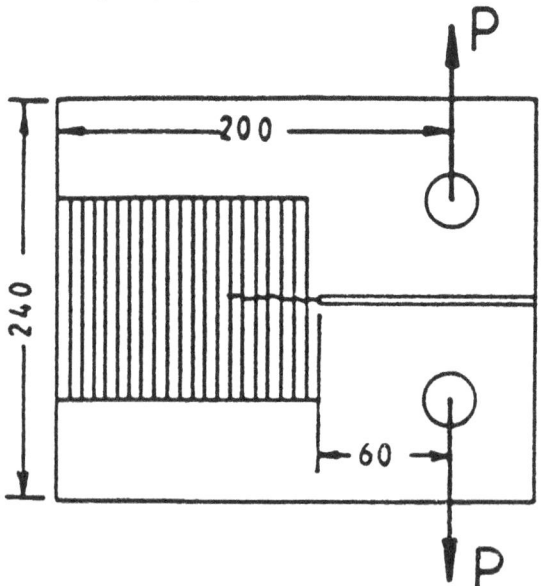

Figure 6(a).   Compact tension specimen dimensions in mm and
strips cut for bending stiffness tests.

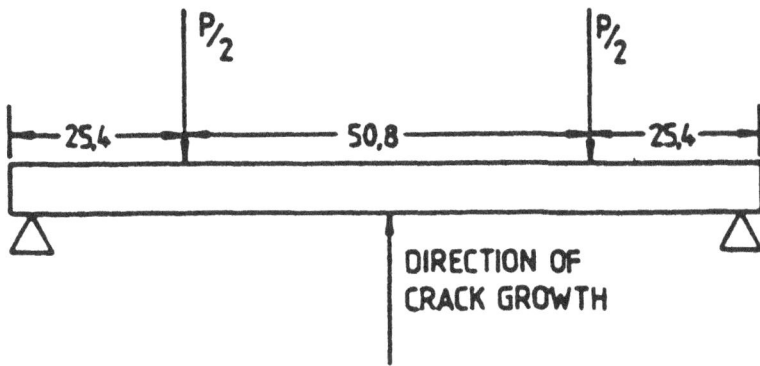

Figure 6(b). Four-point bend tests on damaged strips to evaluate size of the FPZ.

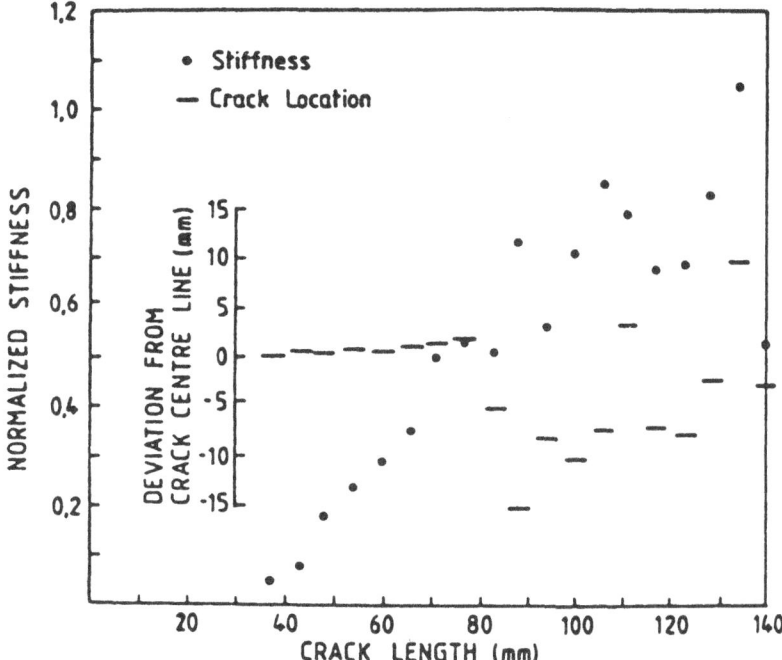

Figure 7. A plot of normalized stiffness versus crack length. Failure sites of sectioned strips are indicated by horizontal bars.

TABLE 2. Crack length measurement using the computerised electrical method and the section and bend stiffness test

Specimen No.	Crack length (mm)		
	Optical method	Electrical conductive bars/activity zone	Bend stiffness measurement
1	78	62–86	77
2	83	56–88	83
3	99	70–105	112

3. Crack–Resistance (R) Curve and Failure Characterisation

The crack–resistance curve is basically controlled by both "intrinsic" and "extrinsic" variables. Intrinsic variables are microstructural in nature and include the level of porosity, the fibre type and its surface treatment, the residual stresses and the matrix properties. By changing these variables it is possible to change the stress–displacement constitutive relations in the FBZ and FPZ. Extrinsic variables are those of specimen geometry, size and loading configuration which would affect the deformation of the FBZ and FPZ and hence the crack tip shielding mechanisms. In the following sub–sections the mechanics of crack–interface bridging and analytical modelling of specimen size, geometry and matrix FPZ effects on R–curves are presented.

3.1 MECHANICS OF CRACK–INTERFACE BRIDGING

In fibre cement composites the single most important variable on the crack–resistance curve is the closure stress (σ)–crack face separation (δ) relationship in the FBZ. Once this relation is known it is possible to predict R–curves for different effects of size and geometry and alternatively the load–deflection or moment–curvature diagrams for various types of structural components. For the very simple situation of fibre debond/pullout the σ–δ relationship can be worked out theoretically. If it can be assumed that the mechanical shear bond strength τ is a material constant then the force F to pullout a fibre is $F = \pi d \tau \ell$, where d and ℓ are the diameter and embedded length of the fibre. When the fracture surfaces have separated a distance δ the maximum possible pullout force is given by $F_m(1-2\delta/L)$ where $F_m = \pi d \tau L/2$ and L is the fibre finite length. The number of fibres per unit area of fracture surface (N) is $\eta V_f/(\pi d^2/4)$, where η is the orientation efficiency factor equal to unity for aligned fibres, $2/\pi$ for 2–D randomness and 1/2 for 3–D randomness; V_f is the fibre volume fraction. Thus, the average closure stress–displacement relationship is as expected linear, i.e.

$$\sigma = [\eta V_f \tau L/d][1 - 2\delta/L]. \tag{1}$$

However, if it is considered that when the fracture surfaces are separated by an amount δ, then a proportion $2\delta/L$ of the embedded fibres have already pulled out completely. Hence the average stress on the composite is

$$\sigma = [\eta V_f \tau L/d][1 - 2\delta/L]^2 \tag{2}$$

indicating that the σ–δ relation is parabolic rather than linear. Indeed Ballarine et al [5] have shown that equation (2) is valid for steel fibre reinforced concrete. It may be criticized that neither equations (1) and (2) considers the Poisson's contraction of the fibres and the non–constant shear stress at the fibre–matrix interface. Gao and co–workers [6] have developed σ–δ relationship that incorporates these additional factors from a simple shear lag model.

For many fibre cement composites the complex failure mechanisms in the FBZ make it difficult to derive a rigorous theoretical σ–δ relationship. In these cases experimental methods remain the only solution and both direct and indirect techniques have been developed to measure the σ–δ relationship. In the direct tensile test method a very stiff testing machine or special stiffening devices are required to produce a stable fracture and misalignment has to be eliminated in order to obtain an accurate σ–δ curve [7,8]. Figure 9 shows a schematic tensile stress–displacement curve for a fibre cement composite. In region I the specimen is essentially elastic. In region II the non–linear deformation is largely due to dispersed cracking within the gauge section.

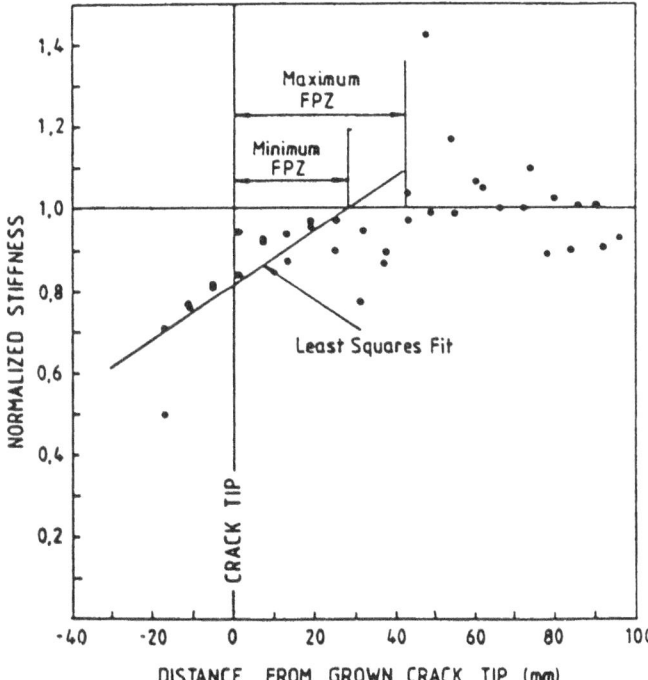

Figure 8. Evaluation of FPZ using normalised stiffness – crack tip distance plot.

476

When the cracking becomes localised on a future fracture plane the ultimate strength σ_m is achieved. Up till σ_m the deformation is uniform throughout the specimen; but in regions III and IV the deformation is localised in the fracture process and fibre bridging zones. During region III the matrix FPZ develops and when the displacement across it reaches δ_m a continuous matrix crack is established. (see Figure 1). Usually $\delta_m << \delta_f$ so that the stress σ_f is close to the maximum pullout stress. The difference between σ_f and σ_m depends upon V_f and F_m. In region IV the fibres gradually pullout and fracture. Complete fibre pull–out occurs when the crack face separation distance δ_f equals to half the fibre length. The σ–δ relation is therefore that given in regions III and IV as shown in Figure 9. The total specific work of fracture $J\infty$ in a tension test is given by the sum of the matrix fracture work J_{Ic} and the fibre pull–out work, i.e.

$$J_\infty = J_{Ic} + \eta V_f \tau L^2/6d \tag{3}$$

if $\delta_m << L/2$. In a sufficiently large notched specimen where the FBZ can be fully established and the crack profile remains unchanged during stable crack growth $J\infty$ can be obtained and may be identified with the plateau value $K\infty$ of the crack–resistance curve, i.e. $J_\infty E = K_\infty^2$. The crack resistance J_R prior to $J\infty$ is therefore

$$J_R(\delta) = \int_0^\delta \sigma(\delta)d\delta = J_{Ic} + \int_{\delta_m}^\delta \sigma(\delta)d\delta \tag{4}$$

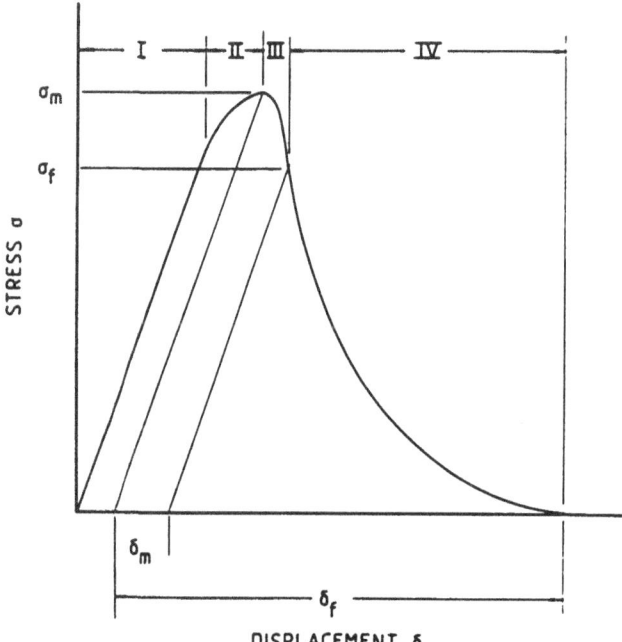

Figure 9. Schematic stress-displacement curve for a fibre cement composite.

Equation (4) has provided a theoretical basis for a simple indirect method to evaluate the $\sigma-\delta$ relationship if the crack-resistance curve $J_R(\delta)$ can be obtained. Using the energetic definition of J it is possible to determine J_R experimentally using specimens of different crack lengths. However, J_R is now evaluated at the load-point or crack mouth opening displacement Δ but not at the leading edge of the FBZ which is δ. For any given specimen geometry Δ and δ can be experimentally related to each other so that

$$\sigma(\delta) = \frac{\partial J_R(\Delta)}{\partial \Delta} \frac{d\Delta}{d\delta}. \tag{5}$$

Li and his co-workers [9,10] have successfully shown the use of equation (5) to determine the $\sigma(\delta)$ functional relationship for concrete and fibre cement composites. Figure 10 shows the $\sigma-\delta$ curves for a range of fibre cementitious materials obtained by Li and Ward [10]. It is interesting to observe that $\sigma-\delta$ is not linear but parabolic as predicted by equation (2). Since crack-resistance curves are usually represented in terms of crack growth Δa and not Δ or δ another version of equation (5) has been derived using the $K_R(\Delta a)$ curves. For the notched crack shown in Figure 11 where Δa is the fibre bridging zone and the initial crack length a_0 is much larger than Δa, then

$$K_R(\Delta a) = K_{Ic} + \left[\frac{2}{\pi}\right]^{\frac{1}{2}} \int_0^{\Delta a} \frac{\sigma(x)\,dx}{\sqrt{x}} \tag{6}$$

and it follows that

$$\sigma(\Delta a) = \left[\frac{\pi}{2}\right]^{\frac{1}{2}} (\Delta a)^{\frac{1}{2}} \frac{dK_R(\Delta a)}{d(\Delta a)} \tag{7}$$

in which $dK_R/d(\Delta a)$ is easily obtained from the $K_R(\Delta a)$ crack-resistance curve. To convert to $\sigma(\delta)$ it is necessary to determine the relationship between Δa and δ either theoretically or experimentally. If it is assumed that the crack profile is an approximate straight line [11] so that

$$\Delta a/\delta = 2\Delta a_s/L \tag{8}$$

where L/2 is the critical fibre pull-out length at the leading edge of the fully developed FBZ of size Δa_s, then

$$\sigma(\delta) = (\pi \Delta a_s \delta/L)^{\frac{1}{2}} dK_R(\Delta a)/d(\Delta a). \tag{9}$$

Since the direct tension test method is difficult to use, Chuang and Mai [12] have recently shown that for cementitious matrices the $\sigma-\delta$ relationship can be extracted from the load-displacement or moment-curvature curves of an

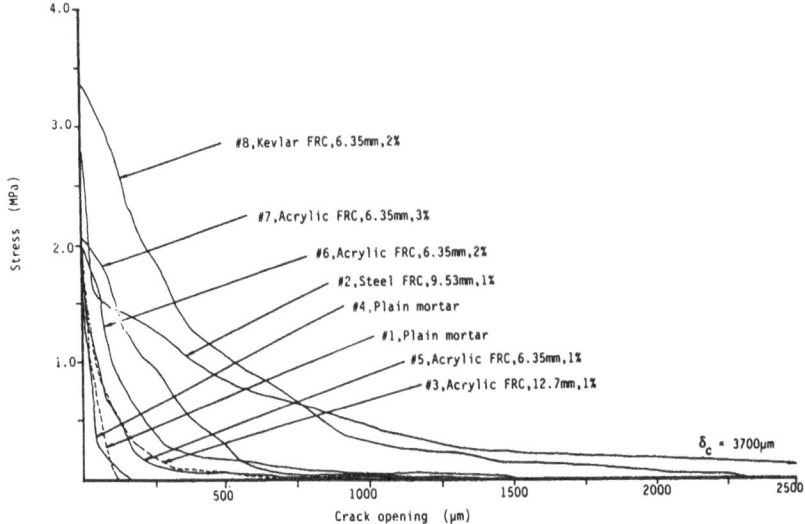

Figure 10.   Stress-displacement curves for various fibre cementitious composites.   (Taken from Li and Ward, ref. 10).

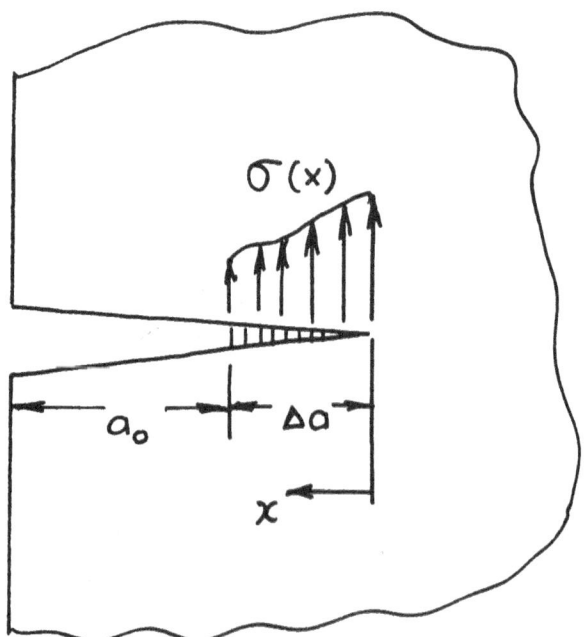

Figure 11.   FBZ and stress distribution at wake of crack tip.

unnotched beam in pure bending, provided it is possible to identify the points on these curves for the onset of the FPZ and the initiation of crack growth at the tensile surface. However, this method cannot be readily applied to fibre–reinforced cement composites.

3.2 UNIQUENESS OF CRACK–RESISTANCE (R) CURVE

It is of fundamental interest to ask if the crack resistance (R) curve is a unique material property for fibre cements and other cementitious matrices. If it were to be so then the characterisation of failure would be a simple matter of determining the tangency point on the R–curve due to the applied stress intensity factor K_a–curve (see Section 3.6). The evaluation of the instability point for non–unique R–curves however requires an iterative process. In terms of the crack face separation at the leading edge of the fibre bridging zone equation (4) describes the J_R–δ relationship as the crack resistance increases from a matrix toughness J_{Ic} to a plateau value J_∞ defined by equation (3). Thus, in between these two limits, i.e. for $\delta_m << L/2$ and $\delta > \delta_m$,

$$J_R(\delta) = J_{I_c} + \frac{\eta V_f \tau L^2}{6d} \left\{ 1 - \left[1 - \frac{2\delta}{L} \right]^3 \right\} \tag{10a}$$

or combining with equation (3) this becomes

$$J_R(\delta) = J_{I_c} + (J_\infty - J_{I_c}) \left\{ 1 - \left[1 - \frac{2\delta}{L} \right]^3 \right\}. \tag{10b}$$

It is quite obvious from equation (10) that when plotted in terms of δ the crack resistance curve J_R is a unique material property. However, since the relationship between δ and crack growth Δa (or extension of the FBZ) is both specimen geometry and size–dependent, J_R in terms of Δa is generally not invariant with these extrinsic variables. During the evolution of the crack resistance curve expressed in terms of Δa,

$$J_R(\Delta a) = J_{Ic} + \int_0^a \sigma(x) \frac{\partial \delta(x,a)}{\partial x} \, dx. \tag{11}$$

Except for a very large specimen the shape of the crack profile changes with crack growth so that $\partial \delta / \partial x \neq - \partial \delta / \partial a$ and $J_R(\Delta a)$ is not unique. A unique crack resistance curve can be obtained in theory if we consider a semi–infinite crack in an infinite sheet and this has indeed been studied by Foote et al [13] using Muskhelishville's method. In this way not only does the crack face profile remain unchanged but that the FBZ is very small compared to the crack length.

3.3 ANALYTICAL MODELLING OF CRACK—RESISTANCE CURVE

Many analytical models for crack growth in fibre cements follow the fictitious crack model of Hillerborg and co—workers [14,15] originally developed for cementitious matrices, e.g. Wecharatana and Shah [16], Ballarine et al [5], Visalvanich and Naaman [17] etc. Because the closure stresses in the FBZ and FPZ are dependent on the crack face separation the problem is a non—linear one and requires an iterative numerical solution [15,16]. To provide a simpler and more general model Jenq and Shah have extended their two—parameter fracture model for concrete [18] to fibre cements [19]. The two parameters are the critical value of δ_m already defined in Figure 1 and the critical stress intensity factor $K_{I_c}^s$ at the tip of the effective or fictitious crack. It is claimed that these parameters are independent of both specimen size and geometry and can be used to predict the crack resistance curve of fibre cementitious composites [20]. A review of these previous fracture models is given by Cotterell and Mai [21].

It appears that the simplest method of analysing crack growth in fibre cements is by superposition of the stress intensity factors due to the applied stresses K_a and the closure stresses in the FBZ, K_r, and the FPZ, K_m. Lenain and Bunsell [22] were the first to use the K—superposition method to model crack resistance in asbestos cements. They assumed that crack growth would commence when the effective K_e at the crack tip was equal to the matrix toughness K_{I_c}. Instead of calculating K_m they assumed K_r to prevail over an effective crack length which included a portion of the matrix fracture process zone. Thus, the equilibrium crack growth criterion becomes:

$$K_e = K_a + K_r = K_{I_c}. \tag{12}$$

The crack resistance is the term K_a which increases as the crack extends and it is usually described by K_R, i.e.

$$K_R(\Delta a) = K_{I_c} - K_r(\Delta a). \tag{13}$$

Lenain and Bunsell further assumed the bridging stresses to be constant and avoided solving the non—linear iterative problem. $K_r(\Delta a)$ can be obtained easily from the expression

$$K_r = - \int G(a,x)\sigma(x)dx \tag{14}$$

where the negative sign is to show that K_r is acting in an opposite sense to the applied K_a and the Green function $G(a,x)$ is dependent on specimen geometry and loading configuration. It is unfortunate that they have used the wrong Green function in their analysis making their results invalid.

The approximations used by Lenain and Bunsell are not realistic and unnecessary. Equations (13) and (14) can be used in conjunction with equation (2) for the $\sigma-\delta$ relationship in the FBZ to obtain an exact solution. Of course, this introduces a non—linear problem, since K_r depends on σ which in turn depends on δ that is the sum of the displacements due to the applied stresses and the fibre bridging forces. Iterations are required to obtain σ that is consistent with δ in the FBZ. For a bridged crack in a uniform stress field σ_∞, the solution of δ is relatively simple though iterations are still needed. In a non–uniform stress field, such as that in a notched bend (NB) or a double–cantilever–beam (DCB) geometry, the solution of δ is far more complicated. Foote et al [11] have derived a simple equation for δ from stress intensity factors by application of Castigliano's Theorem. Thus, the displacement $\delta(x)$ at a point x from the crack tip (Fig. 12) due to an applied load P acting at a point a away from the tip is given by [11, 23–25]

$$\delta(x) = \frac{2}{E} \int_{a-x}^{a} K_P(u) \left[\frac{\partial K_F}{\partial F} (u-a+x) \right]_{F=0} du \tag{15a}$$

where F is a fictitious force at x, K_P and K_F are stress intensity factors due to P and F. If the positions of P and F have been interchanged along the crack face $\delta(x)$ can be obtained from a similar expression

$$\delta(x) = \frac{2}{E} \int_{x-a}^{a} K_P(u-x+a) \left[\frac{\partial K_F(u)}{\partial F} \right]_{F=0} du. \tag{15b}$$

Note that the integration in equation (15) is carried out over the distance between the forces. If P is a unit force the integrals give the displacement coefficients along the crack face which can be used in the iterative scheme to evaluate the exact K_R–curves. It is important to point out here that equation (15) is a powerful tool for crack face separation calculations provided the K–solutions for the applied and fictitious loads P and F are known right to the crack tip for any given geometry.

Foote et al [11,21,23–26] have also developed an approximate method to calculate crack resistance curves for fibre cements. In common with Shah and his co–workers [5,16] the crack faces in the FBZ are assumed to remain straight so that using equations (2) and (8) it can be shown that:

$$\sigma(x)/\sigma_m = \left[1 - \frac{x}{\Delta a_s} \right]^n = \left[1 - \frac{2\delta}{L} \right]^n \tag{16}$$

where the maximum fibre pull out stress $\sigma_m = \eta V_f \tau L/d$ and n = 2 in this example but it may assume other values depending on the strain–softening characteristics of the material. An iterative solution is still required to determine Δa_s knowing K_{I_c}, E, σ_m and L while satisfying the equilibrium crack growth condition of equation (13). For crack growth less than Δa_s, $\sigma(x)$ is given by equation (16) and K_r calculated from equation (14). No further iterations are necessary. Justifications of the linear crack face profile and the approximate method are shown in Figures 13 and 14 respectively for a wood fibre cement composite and details of these calculations are given in [11,25].

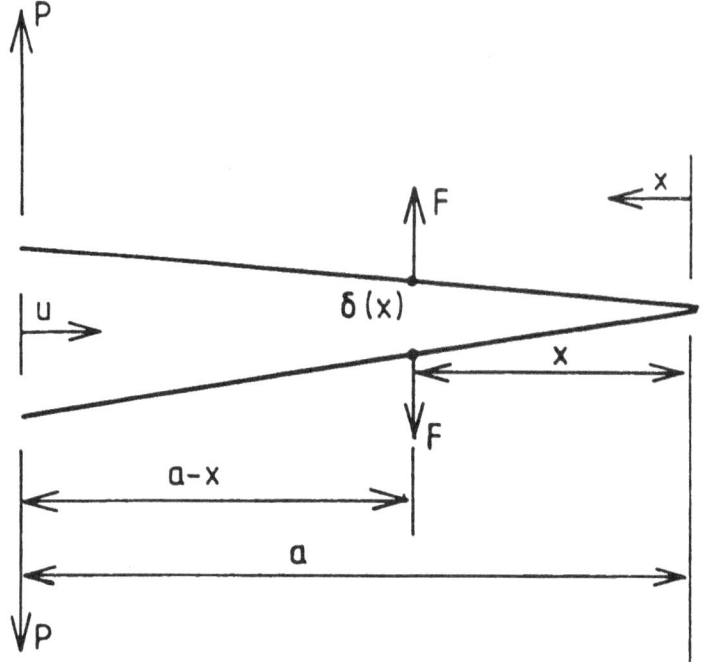

Figure 12. Evaluation of crack face separation (δ) by Castigliano's theorem at the location of the fictitious force (F).

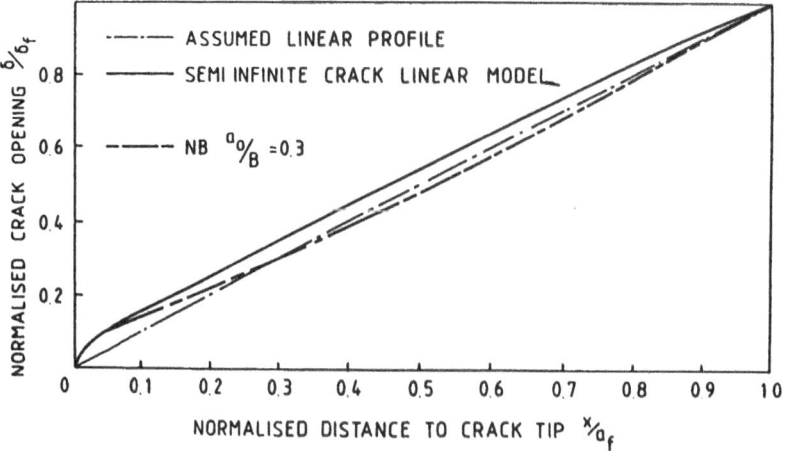

Figure 13. Shape of deformed crack faces for semi-infinite and NB specimens.

3.4 INFLUENCE OF SPECIMEN GEOMETRY AND SIZE ON CRACK RESISTANCE CURVE

It is discussed in Section 3.2 that the crack resistance curve is unlikely to be a unique material property because the FBZ is not small compared to the cracked or uncracked ligament lengths and the crack face profile is not invariant with crack growth. Consequently, specimen size and geometry must have some effects on the R–curve. Foote et al [11,25,27] have studied the size effect of the DCB and NB geometries for asbestos and wood fibre cement composites.

For ease of comparison non–dimensional crack resistance curves, where $\overline{K}_R = K_R/K_\infty$ and $\Delta\overline{a} = \Delta a/(K_\infty/\sigma_m)^2$, are plotted for the DCB geometry in Figure 15 and NB geometry in Figures 16 and 17. In these $\overline{K}_R$–curves a non–dimensional fracture toughness of the matrix $\overline{K}_{Ic} = 0.3$ is assumed which is typical of wood fibre reinforced mortars. K–solutions for the DCB and NB geometries are taken from Foote and Buchwald [28] and Tada et al [29] for use with equation (15) to calculate the crack face separation δ in the FBZ. In the DCB geometry, by keeping $a_o/H = 3$ constant, the crack resistance curves are dependent on the depth of the beam $\overline{H}(= H/(K_\infty/\sigma_m)^2)$ with the length of the fully developed FBZ $\Delta\overline{a}_s$ decreases as $\overline{H}$ decreases but the plateau values K_R are very close to the theoretical limit K_∞, Figure 15. The effect of the notch to depth ratio a_o/H on the crack resistance curve is practically zero, Figure 18. Some very interesting observations can be made for the NB geometry. Keeping the notch to beam depth a_o/B ratio of 0.30 constant it is shown that the size of the beam has a significant effect on the crack resistance curves. When the non–dimensional beam depth $\overline{B}(=B/(K_\infty/\sigma_m)^2)$ is large compared to the FBZ the $\overline{K}_R$ curve reaches a plateau value and indeed the whole crack resistance curve is identical to the DCB geometry when $\overline{H} \to \infty$. For smaller $\overline{B}$ the fully developed FBZ size decreases with specimen size and the crack resistance curve can be much larger than the plateau value K_∞, Figure 16. Also, unlike the DCB geometry, the notch to depth ratio a_o/B has a far greater effect on the $\overline{K}_R$ curves as shown in Figure 17.

It is worth noting that whenever the crack approaches the back–face the crack resistance curve rises rather sharply, such as in NB specimens in Figure 17. In this way small compact tension specimens give similar behaviour. There is no physical reason as to why the fracture resistance should suddenly increase near the back–face other than it is a consequence of the K–solution as the uncracked ligament approaches zero. In fact in cementitious matrices alone the crack resistance curve may decrease as the back–face is reached [30]. This result may be caused by the reduction in the width of the damage zone (a parameter which is not considered in a crack growth model presented in these lecture notes) and the prior damage of the material near the back–face due to precompression [31].

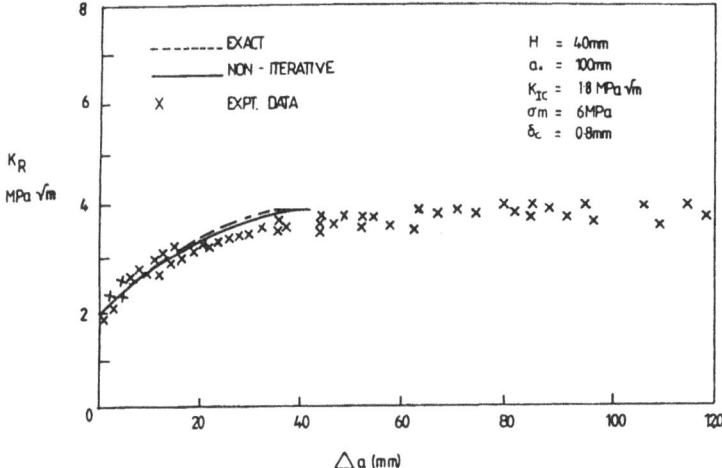

Figure 14. Comparison of 'exact' and approximate crack resistance curves for DCB geometry.

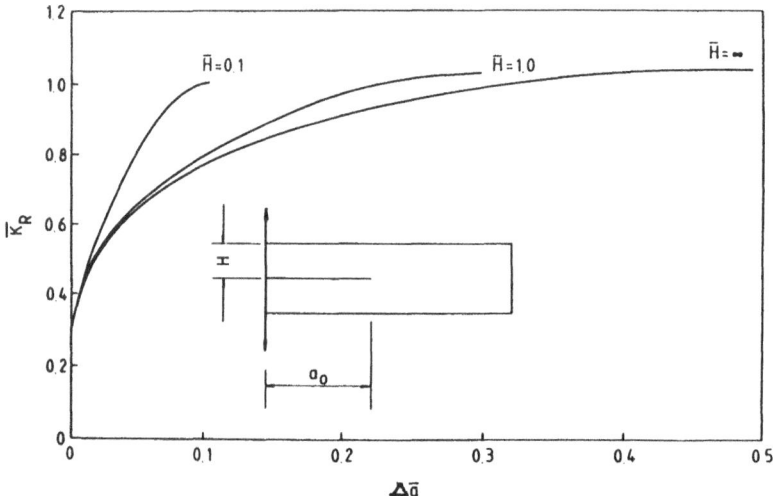

Figure 15. Non-dimensional crack resistance curves for DCB geometry for $a_o/H = 3.0$ and different $\bar{H}$.

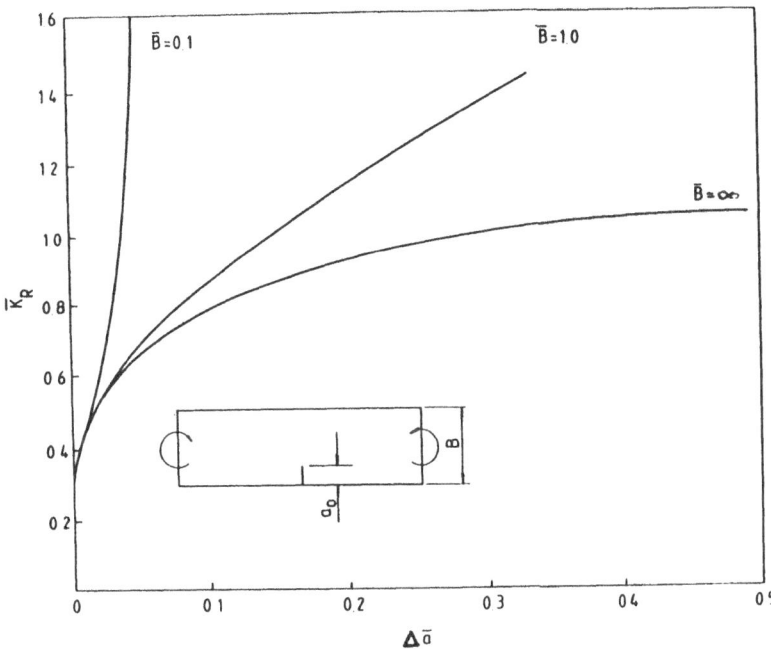

Figure 16. Non-dimensional crack resistance curves for NB geometry for $a_o/B = 0.3$ and different $\bar{B}$.

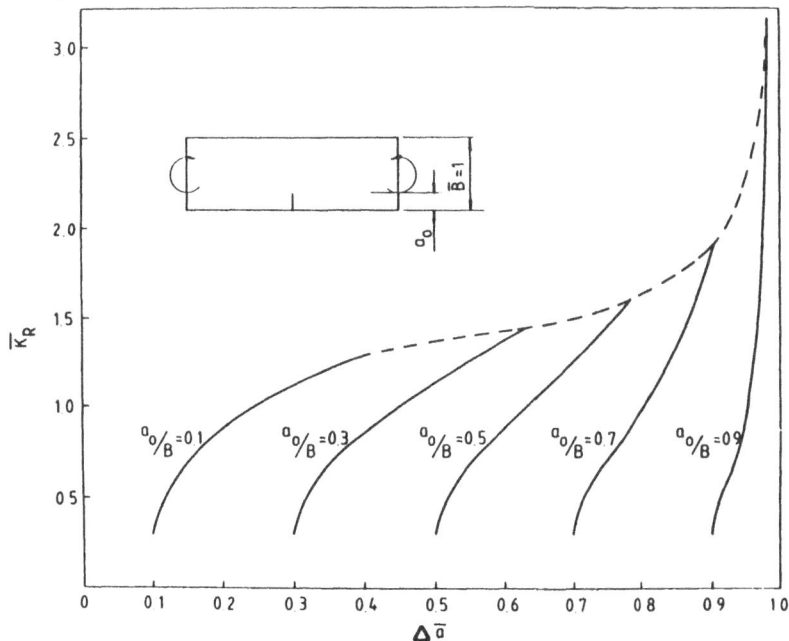

Figure 17. Non-dimensional crack resistance curves for NB geometry for $\bar{B} = 1.0$ and varying a_o/B.

Experimental crack resistance curves obtained for an asbestos/cellulose fibre cement mortar [27] for the NB geometry of varying depth and different initial notch length are shown in Figures 19 and 20. The crack growth was in the strong direction of the composite. Table 3 gives the composite and fibre properties. The bond strengths (τ) were not measured directly but selected to give good agreement with the experimental fracture strengths which were dependent on direction of the sheet. The crack resistance values $K_{Ic}(= 1.9$

MPa$\sqrt{m}$) and K_∞ (= 5 MPa$\sqrt{m}$) were chosen empirically to give the best fit to the K_R curve for the largest NB specimen with B = 200 mm. These parameters were then used to calculate the theoretical crack resistance curves for the smaller NB specimens in Figure 19 and varying initial notch length in Figure 20 using the approximate method outlined in Section 3.3. The agreement with experimental data is very good. An experimental K_R curve for a similar asbestos/cellulose cement mortar for the DCB geometry is given in Figure 14 in which theoretical K_R curves using both the exact iterative and approximate methods are superposed. Crack growth here was in the weak direction. The agreement with experimental data is again excellent.

In concluding this section it is important to point out that there are two levels of modelling. The fundamental level is to start from the closure stress (σ)– crack face separation (δ) relationship and use it to determine $K_r(\Delta a)$ as may be affected by size and geometry of specimen whilst still satisfying the crack growth criterion of equation (13). Since an accurate σ–δ relationship is difficult to obtain for a real composite material, it is appropriate to model crack resistance curves on a more practical level. This involves determining the most appropriate fracture parameters (K_{Ic}, K_∞, E, σ_m) or (K_{Ic}, σ_m, δ_f, E) that give the best fit to the K_R curve of a given geometry and size of specimen. It is also assumed, without any loss of accuracy, n = 1.0 in these calculations. With these parameters the behaviours of other geometries and even full–size structures can be theoretically predicted.

TABLE 3. Properties of asbestos/cellulose fibre cement mortars

(a) Composite properties		
Young's modulus (E)		6 GPa
Matrix toughness (K_{Ic})		1.9 MPa$\sqrt{m}$
Maximum toughness (K_∞)		5.0 MPa$\sqrt{m}$
Fracture strengths (σ_m)		
strong direction		10 MPa
weak direction		5 MPa
(b) Fibre properties		
	Cellulose	Asbestos
Aspect ratio (L/d)	135	80
Fibre length (L)	3.5 mm	2 mm
Volume fraction (V_f)	0.07	0.08
Bond strength (τ)	0.88 MPa	2.0 MPa

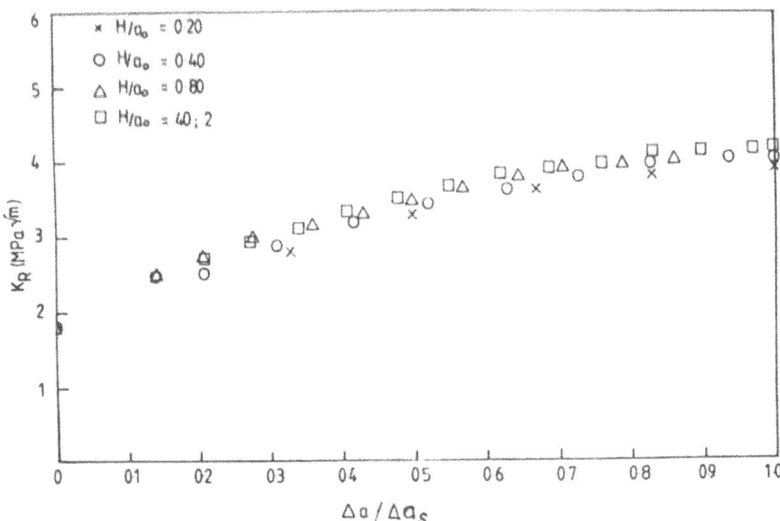

Figure 18. Normalised K_R - $\Delta a/\Delta a_s$ curves for DCB geometry for varying a_o/H ratios.

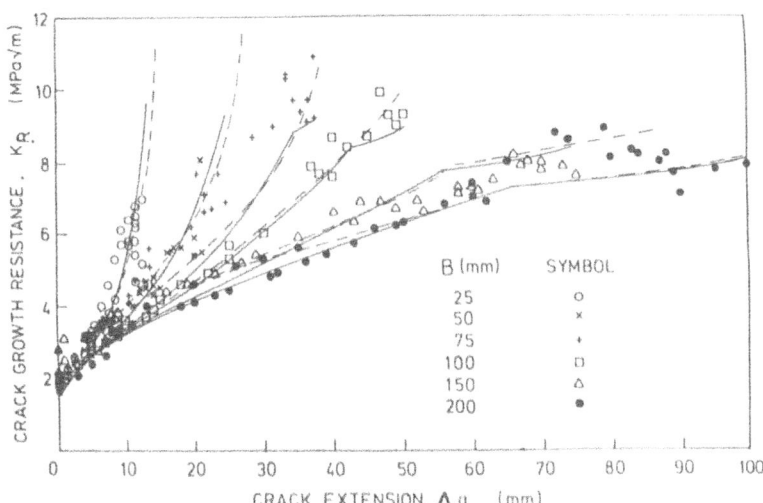

Figure 19. Experimental crack resistance curves for asbestos/ cellulose cement mortar NB specimens for $a_o/B = 0.3$ and different beam depth. Theoretical models based on $\sigma_m = 10$ MPa, $K_\infty = 5$ MPa$\sqrt{m}$, $K_{Ic} = 1.9$ MPa$\sqrt{m}$. ——— without FPZ; ----- with FPZ.

3.5 EFFECT OF MATRIX FRACTURE PROCESS ZONE ON CRACK RESISTANCE CURVE

In fibre cements the matrix fracture process zone is not small, (see Table 1 and Figure 8), it is hence necessary to consider what effect the FPZ will have on the crack resistance curve. To model this case equations (12) and (13) are not valid for crack growth. Instead the FPZ is considered as a fictitious extension of the continuous matrix crack and it carries a constant stress equal to σ_m. For equilibrium crack growth the crack opening displacement at the tip of the continuous matrix crack must be equal to the critical value δ_m. In addition since there is a finite stress at the fictitious extension of the continuous crack the sum of the stress intensity factors at the tip of the fictitious extension must be equal to zero, i.e.

$$K_e = K_a + K_r + K_m = 0 . \qquad (17)$$

Now K_a is not identical to the crack growth resistance K_R which is calculated at the continuous matrix crack tip but not the fictitious crack. A linearised $\sigma-\delta$ relationship and straight crack face profiles are assumed to hold in the FBZ in modelling crack growth. The method of solution is the same as that given in Section 3.3 except now both Δa_s and Δa_m (the matrix FPZ size) have to be determined by iteration. For Δa less than Δa_s it is still required to determine the current length of the FPZ Δa_m by iteration so that the crack face separation at the continuous matrix crack tip is δ_m. Crack resistances K_R evaluated at the continuous crack tip can now be obtained and plotted against crack extension Δa. These predicted K_R curves are also superposed in Figure 19 and, quite clearly, there is very little difference when they are compared to those K_R curves obtained from the earlier model without the inclusion of the matrix FPZ. In practice the scatter that would be obtained from experimental crack resistance curves is far greater than the difference between the two models. For simplicity therefore it is sufficient to model crack growth in fibre cement composites without considering the matrix FPZ.

3.6 FAILURE CHARACTERISATION OF FIBRE–CEMENTS WITH R–CURVE CHARACTERISTICS

In an earlier review [32] the author has discussed the usefulness of the crack resistance curves in the context of determining the maximum load that can be withstood by a structural component containing a well–defined crack. Certainly, the component may contain many micro–cracks then the application of the crack resistance curve instability analysis is somewhat complex and this is discussed in Section 4.1. The following discussion is limited to the single crack situation.

Mai and Cotterell [33] assumed that if the crack resistance curve were independent of specimen geometry and size it could be approximately represented by a power law function:

$$K_R = \beta(\Delta a)^\alpha = \beta(a - a_0)^\alpha \qquad (18)$$

in which α and β are constants and a_o is the initial crack length. The criteria for crack instability (at maximum load) under load–controlled conditions are:

$$K_a = K_R = K_c \qquad (19a)$$

$$\frac{\partial Ka}{\partial a} \geq \frac{dK_R}{d(\Delta a)} \qquad (19b)$$

and in general K_a can be expressed as:

$$K_a = PY(a) \qquad (20)$$

where P is the applied load and Y(a) is the geometry correction factor for a given specimen geometry and loading configuration [29]. Thus, combining equations (18) and (19) the maximum load K_c value can be obtained.

$$K_c = (\alpha\Phi)^{\alpha}\beta \qquad (21)$$

in which $\Phi = Y(a)/Y'(a)$ is a geometry parameter evaluated at the crack length a_c corresponding to the maximum load P_m. Experimentally, it has been established that there exists a linear relationship between a_o and a_c for a range of specimen geometries and sizes [33], i.e.

$$a_c = a_o + \Delta a_c = \gamma a_o \qquad (22)$$

so that $\Phi(a_c)$ can be determined easily knowing a_o. Using NB specimens of different sizes on an asbestos/cellulose fibre cement composite Mai and Cotterell [33] have confirmed the validity of equation (21). However, the compact tension geometry data do not fall within the scatter bend of the NB data, Figure 21. This means that α and β are geometry dependent and have to be evaluated separately for different geometries.

There are some fundamental problems with the above approach to obtain solutions for fracture loads. A major difficulty lies with the assumption of a geometry– and size–independent crack resistance curve as defined by equation (18). But as shown in Section 3.4 crack resistance curves generally vary with size and geometry. It appears that equation (18) may be approximately valid for a given geometry up to the maximum load and it is correct if small cracks in infinite plates are considered. The latter may be proven as follows. In general the geometry parameter $\Phi(a_c)$ can be written as the product of a_c and $\theta(a_c)$ which is a constant evaluated at a_c. Thus, using equations (18), (21) and (22), it can be shown that

$$\alpha = (\gamma-1)/\gamma\theta \qquad (23)$$

so that the crack resistance curve becomes

$$K_R = \beta(\Delta a)^{(\gamma-1)/\gamma\theta} = \beta(a-a_o)^{(\gamma-1)/\gamma\theta}. \qquad (24a)$$

490

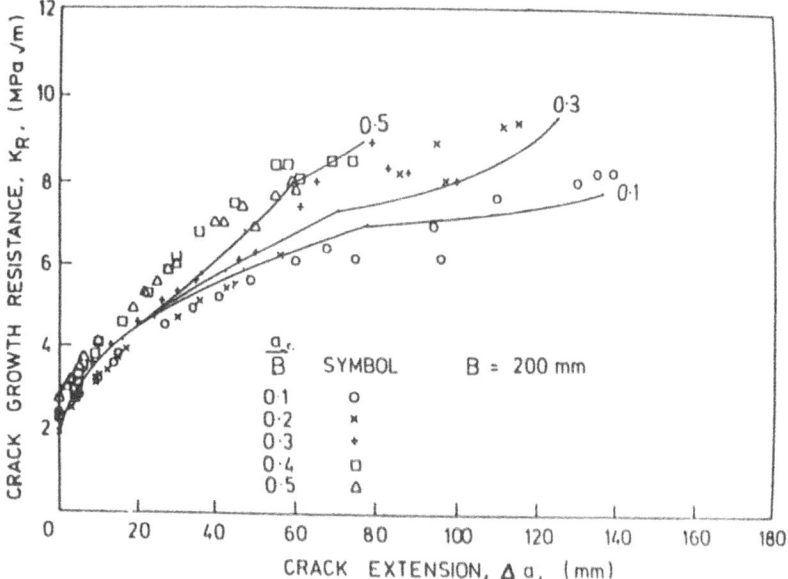

Figure 20. Experimental crack resistance curves for asbestos/ cellulose cement mortar specimens with B = 200 mm. Theoretical model without FPZ and same parameters as in Figure 19.

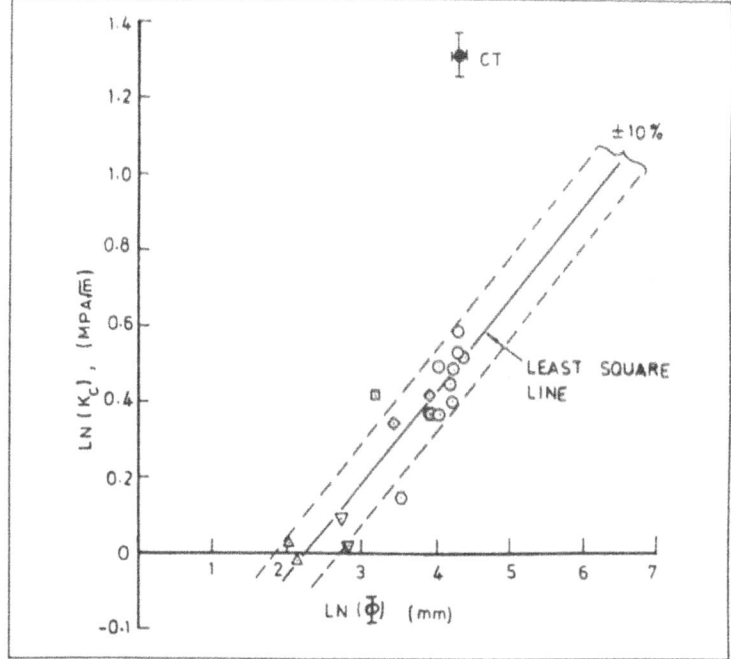

Figure 21. Variation of ℓn K_c versus ℓn Φ for NB and CT asbestos/ cellulose fibre cements according to equation (21).

$\theta(a_c)$ is not only geometry–dependent, but for a given geometry, it also varies with a_c. This suggests that the crack resistance curve is crack length–dependent. In this way the solution for K_c is not self–consistent. However, for small cracks in large sheets, θ is invariant with a_c and equals to 2, i.e.

$$K_R = \beta(\Delta a)^{(\gamma-1)/2\gamma} \tag{24b}$$

which is also obtained by Broek [34] and is now a unique material curve.

A similar approach has been used by Mobasher et al [20] for fracture load predictions in cement–based fibre composites. They also used an energy–based R–curve similar to the power–law equation (18) (in which K_R is replaced by R) and assumed equation (22) to apply. The constants β and γ (and hence α) are obtained by solving a set of non–linear integral equations satisfying the following conditions for fracture instability:

$$K_{Ic}^s = K_r(a_c) + K_a(a_c) \tag{25}$$

and

$$\delta_m = \delta_r + \delta_a. \tag{26}$$

The two fracture parameters K_{Ic}^s and δ_m are to be evaluated from the fracture properties of the unreinforced matrix. It seems that by using K_{Ic}^s instead of K_{Ic} in equation (25) the effect of the matrix FPZ is included. But as shown in Section 3.5 this does not really cause any difference in the predicted crack resistance curve. In addition the same comments made above with regard to the Mai and Cotterell approach [33] using a power law R–curve apply. It is also not obvious that K_{Ic}^s is geometry and size independent because the length of the matrix FPZ does depend on these variables [24].

In view of the absence of a unique material resistance curve a more rigorous and general approach is to start from the instability equation (19a) and compute the crack resistance curve as Δa is increased in accordance with the procedures outlined in Section 3.3 for any given specimen geometry and size and initial critical crack length. Fracture instability is reached if equation (19b) is satisfied. This solution gives the fracture load and the critical crack length. A numerical iterative scheme is required to solve this problem but it is not difficult. A demonstration of this technique has been given by Cotterell and Mai [35] for cement paste in NB specimens.

4.   Studies on Tensile Strength of Short Fibre Cement Composites

Practical cement–based short fibre composites contain many inherent defects which are bridged by fibres so that each individual crack essentially exhibits a R–curve behaviour. The size of these defects is not constant and in the absence of the fibres the Weibull weakest link theory applies, i.e. the tensile strength is predominantly controlled by the largest defect. When fibres are bridging these defects and because the fibres are not uniformly distributed some cracks will have more fibres and others fewer. The tensile strength cannot be determined by the simple weakest link theory and the crack–resistance curve characteristics for each crack has to be considered in any fracture model. Because the fibres exert closure stresses on the crack face a higher applied stress is needed to cause tensile failure of the matrix so that its strength can be increased. Hence, as shown by Andonian et al [36], low modulus fibres like polypropylene and cellulose can indeed reinforce cement mortar with a higher modulus (see Figure 22) because there are many crack–like defects from which fracture initiates. Both the first cracking and final tensile strengths are therefore increased due to the bridging effect of the fibres. In addition, if the matrix material exhibits a time–dependent strength degradation the same will apply to the strength of the fibre cements. These problems and others have been studied both theoretically and with computer simulations by Hu et al [37,38] and Hu [39]. In the following subsections only the essential features of the analytical and computer simulation studies are presented.

4.1   TENSILE STRENGTH OF SHORT FIBRE CEMENT COMPOSITES

Consider a rectangular plate specimen under uniform tension in which the fibre density ρ_f and the fibres of constant aspect ratio (L/d) are randomly distributed throughout the plate but are aligned in the direction of the applied stress. The size of the matrix cracks, which are considered as equivalent Griffith cracks lying normal to the applied stress, varies according to the Pareto distribution:

$$q(a) = (\rho_m \, m/2a_0)(a_0/a)^{(m+2)/2} \tag{27}$$

for $a > a_0$ and ρ_m is the density of matrix cracks, m is the Weibull modulus of matrix material and a_0 is the reference crack size. This choice of q(a) gives the usual Weibull strength distribution equation because

$$F(\sigma) \quad = 1 - \exp\left\{-V \int_{a(\sigma)}^{\infty} q(a)da\right\}$$

$$= 1 - \exp\left\{-V\rho_m \left[\frac{\sigma}{\sigma_0}\right]^m\right\} \tag{28}$$

where V is volume of material under tensile stress, $\sigma_0 = K_{Ic}\sqrt{2/\pi a_0}$ and K_{Ic} is the matrix toughness. Crack growth in each individual crack occurs if equation (12) is satisfied, i.e. $K_e = K_a + K_r = K_{Ic}$. The R–curve effect of the matrix

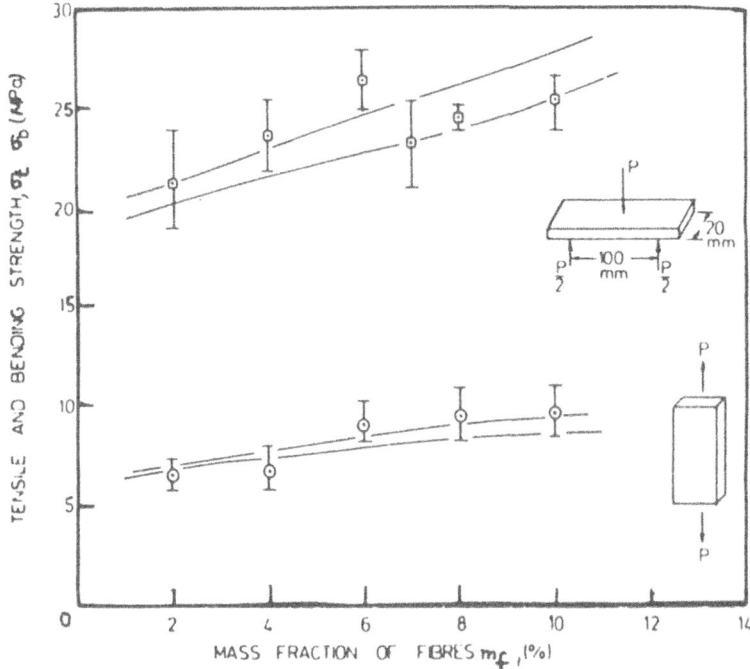

Figure 22. Tensile and bend strengths of cellulose cement mortar composites.

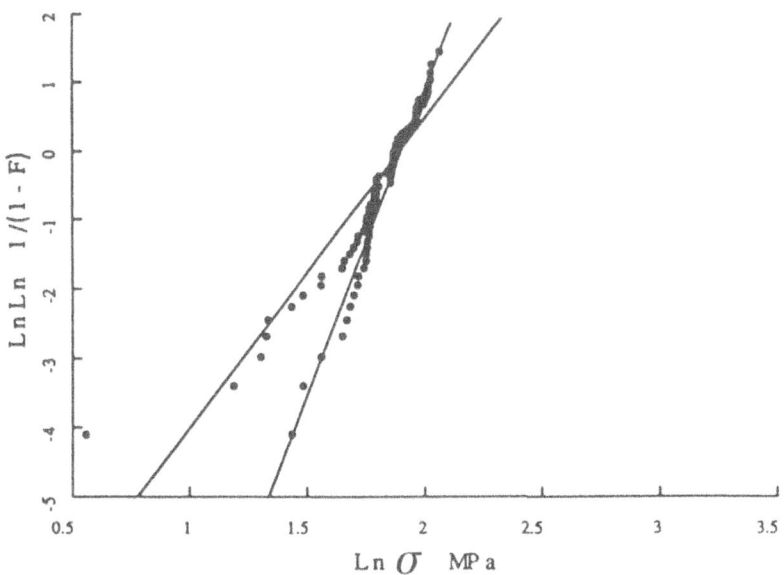

Figure 23. Weibull strength distribution of unreinforced cement matrix.

material is not considered in this crack growth criterion so that it is only necessary to evaluate K_r due to fibres bridging the individual cracks. Failure of the composite is complete when the cracks coalesce and spread across the plate width.

If the fibre density ρ_f is high, the number of fibres bridging a crack(s) is proportional to its length and the effective bridging stress (in this 2–D model)

$$\sigma_{fb} = \pi \tau d \rho_f L^2/4d \tag{29}$$

is the same in each crack being determined by the shortest of the two embedded lengths. Equation (28) then gives the strength distribution of these fibre cement composites if σ is replaced by the effective stress $\sigma_e(= \sigma - \sigma_{fb})$, i.e.

$$F(\sigma) = 1 - \exp\left\{-V\rho_m \left[\frac{\sigma - \sigma_{fb}}{\sigma_0}\right]^m\right\}. \tag{30}$$

In this way the Weibull weakest link theory applies because there is little variation of the bridging stress across the matrix cracks. However, when the fibre density is moderate the number of fibres in defects of a given length is not constant and the bridging stress changes. Unlike the high ρ_f case there will be considerable stable crack growth (Δa) prior to fracture instability due to the stabilization effect of the bridging fibres. Hu et al [38] derived the following failure probability equations:

$$F(\sigma, \Delta a_c) = 1 - \exp\left\{-V \int_{a(\sigma)}^{\infty} Q(a, \Delta a_c)da\right\} \tag{31}$$

for Δa equals Δa_c which is the statistically averaged maximum stable crack growth, and

$$F(\sigma) = 1 - \exp\left\{-V \int_{a(\sigma)}^{\infty} q(a) \sum_{s=0}^{\overline{s}} \frac{(\rho_f La)^s}{s!} \exp(-\rho_f La)da\right\} \tag{32}$$

for first cracking. Q is defined by

$$Q(a, \Delta a) = q(a) \sum_{s=0}^{\overline{n}} \frac{(\rho_f La)^s}{s!} \exp(-\rho La) \times$$

$$\sum_{\Delta s=0}^{\overline{s}-s} \frac{(\rho_f L\Delta a)^{\Delta s}}{\Delta s!} \exp(-\rho_f L\Delta a). \tag{33}$$

s̄ is the maximum number of fibres with which a crack of size a can propagate under a stress σ. S̄ is the equivalent of s̄ when a becomes $(a+\Delta a)$. It should be noted that when ρ_f becomes zero the Weibull equation (28) is recovered.

To illustrate the application of equations (31) and (32) the properties of a typical steel wire reinforced mortar as shown in Table 4 are used. Equation (32) shows that for $\rho_f = 0$ a constant strength of $\sigma(F=0.5)$ invariant with Δa is obtained. This is consistent with the weakest link theory. For $\rho_f = 0.1$ and 0.2 mm^{-2} equation (31) is used to calculate $\sigma(F=0.5)$ as Δa is increased to Δa_c. These strength results are tabulated in Table 5. For $\rho_f = 0.1$ mm^{-2} the tensile strength is about 9.6 MPa after $\Delta a_c \approx 10$ mm; and for $\rho_f = 0.2$ mm^{-2} this becomes about 13.22 MPa after $\Delta a_c \approx 40$ mm. These strength values should be compared with the matrix strength of 6.71 MPa. Even the first cracking strengths of the fibre–reinforced mortars are larger than the matrix strength. The amount of stable fracture prior to reaching Δa_c is enormous particularly for $\rho_f = 0.2$ mm^{-2}.

TABLE 4. Properties of a fibre reinforced cement mortar composite

(a) Matrix properties
Fracture toughness $K_{Ic} = 0.6$ MPa$\sqrt{m}$ Reference crack size $a_0 = 2$ mm Weibull modulus m = 8 Density of cracks $\rho_m = 0.003$ mm^{-2} Volume of specimen $V = L \times W = 100 \times 100$ mm^2
(b) Fibre properties
Fibre diameter d = 0.1 mm Fibre length L = 5 mm Bond strength $\tau = 4$ MPa Fibre density $\rho_f = 0.1, 0.22$ mm^{-2}

TABLE 5. Tensile strength of short fibre cement mortar composite

Fibre density ρ_f (mm^{-2})	Crack growth Δa (mm)	Tensile strength (F=0.5) σ(MPa)
0.00	—	6.71
0.10	0	8.51
	1	8.86
	2	9.14
	5	9.52
	10	9.60
	20	9.58
0.20	0	9.67
	1	10.42
	10	12.29
	20	12.74
	30	13.09
	40	13.22

In parallel with the theoretical calculations above computer simulation experiments have also been carried out for these fibre composites. Defects in accordance with the Pareto distribution, equation (27), and distribution of fibres are randomly generated. Incremental stresses are applied and the effective stress intensity factor K_e (due to the applied stress and the fibre closure forces) calculated for each crack. When $K_e = K_{Ic}$ of the mortar matrix crack growth is permitted from fibre to fibre until arrest. This procedure is repeated until complete failure of the specimen. Appropriate stress intensity factors for discrete forces acting on the crack face can be obtained from Tada et al [29]. To avoid stress singularity in K due to a point force acting at a crack tip the closure forces are assumed to act halfway along the shorter half of the bridging fibre and in a symmetrical position on the other side of the crack face. Figures 23 to 25 show the strength distributions for the matrix material and the two fibre cement mortars. Both the first cracking and final failure strengths are shown. Several comments may be made about these results. In Figure 23 the Weibull failure strength distribution obtained from computer simulations is given by:

$$\ell n \, \ell n \left[\frac{1}{1-F} \right] = 8.74 \, \ell n \sigma - 16.7 \tag{34}$$

which may be compared to that calculated from Weibull's equation (27) using the properties shown in Table 4, which is:

$$\ell n \, \ell n \left[\frac{1}{1-F} \right] = 8 \, \ell n \sigma - 15.56. \tag{35}$$

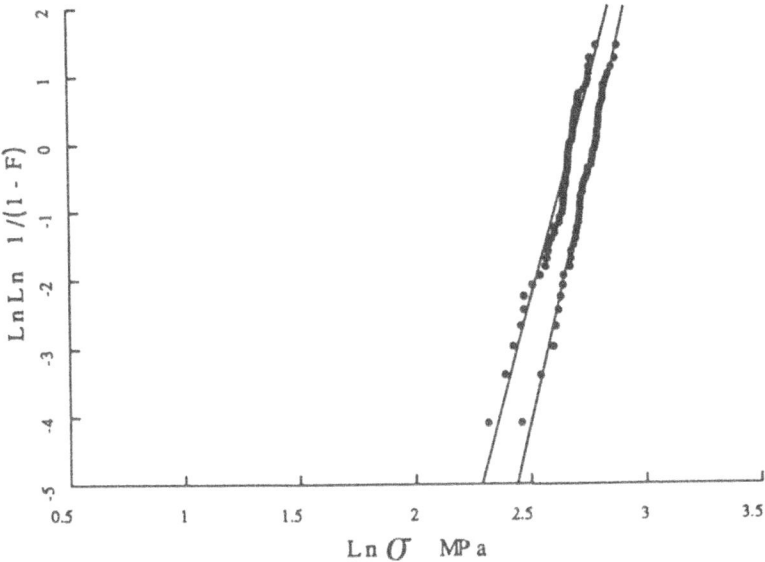

Figure 24. Weibull distribution of first-crack and final failure
strengths of fibre cement composites. $\rho_f = 0.1$ mm^{-2}

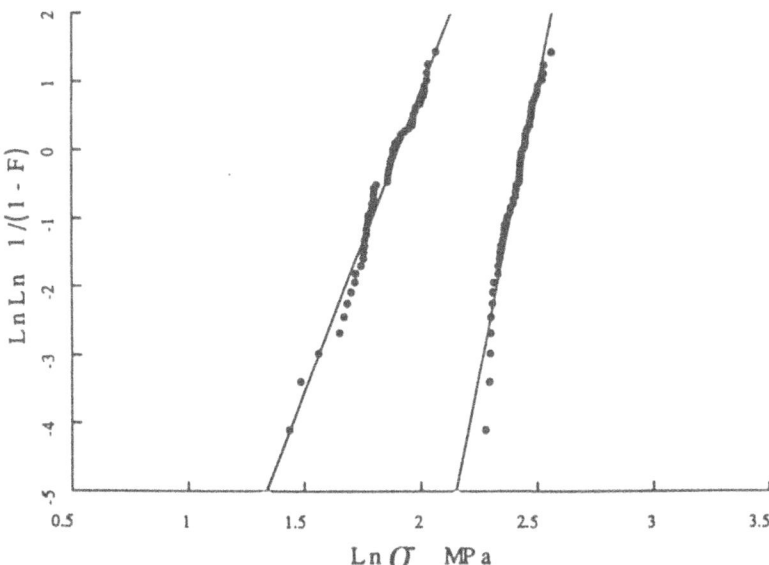

Figure 25. Weibull distribution of first crack and final failure
strengths of fibre cement composites. $\rho_f = 0.2$ mm^{-2}.

The agreement is hence excellent. In Figures 24 and 25 there are considerable increases in the Weibull moduli for the first cracking and final failure strengths, e.g. $m = 8.74$ and 16.7 for $\rho_f = 0.1$ mm^{-2}; and $m = 12.6$ and 14.8 for $\rho_f = 0.2$ mm^{-2}. Physically, this means that there is much less scatter in strength when the mortar matrix is reinforced with short fibres and this must be caused by the R–curve mechanism in the composite materials. Also, the tensile strengths as expected are increased several times that of the unreinforced mortar matrix.

4.2   TIME–DEPENDENT TENSILE STRENGTH OF SHORT FIBRE CEMENT COMPOSITES

Many cementitious matrices suffer slow crack growth due to the effects of a hostile environment such as water or moisture [40–42]. This phenomenon is usually manifested in terms of strength degradations in constant stressing rate and cyclic loading tests as well as the time under which a sustained stress is applied. The simplest equation to describe slow crack growth in brittle materials is [43]:

$$da/dt = AK_a^p \qquad (36)$$

where A and p are constants and K_a is the applied stress intensity factor at the crack tip. Hu et al [44,45] have studied the time–dependent strength behaviour of cementitious matrices by combining their statistical theory of fracture with equation (36) for constant sustained stresses, constant stressing rates and cyclic stresses. In short fibre composites the fibre bridging stresses σ_{fb} must be included in the analysis and are assumed to smear over the crack face. Hence in terms of the effective stress intensity factor K_e equation (36) is recast as:

$$da(t)/dt = AK_e^p = A\{(\sigma-\sigma_{fb})\psi\sqrt{a(t)}\}^p \qquad (37)$$

where a(t) is the crack size at time t and is used to distinguish from the initial value a at $t = 0$. The fibre closure stress σ_{fb} at time t is

$$\sigma_{fb}(t) = \{s(o) + \rho_f L[a(t)-a]\}(\pi L d\tau)/4a(t)d \qquad (38)$$

where s(o) is the initial number of bridging fibres. Thus, the time Δt required to extend a distance Δa may be obtained from equation (37), i.e.

$$\Delta t = \int_a^{a+\Delta a} \{A[(\sigma-\sigma_{fb}(t))\psi\sqrt{a(t)}]\}^{-p}da(t) . \qquad (39)$$

Hence, for any given initial values of s and a, Δt can be obtained for a given Δa or vice versa.

Computer simulation tensile experiments similar to those described in Section 4.1 for tensile strength can be conducted in accordance with equations (38) and (39). The properties of the fibre cement composite are the same as those given in Table 4 except the following changes: $\tau = 2$ MPa, $K_{Ic} = 0.34$ MPa$\sqrt{m}$, $p = 36$, $AK_{Ic}^p = 0.1359$ m/s. These variations are required because of a new matrix material is now selected [40]. Initially a time interval Δt is prescribed and the crack growth Δa is calculated according to equation (37) for all cracks. If no unstable crack propagation occurs and the specimen can still sustain the applied stress, another time interval is assigned and the crack growth calculated again. This procedure is repeated until a crack runs through the specimen. The lifetime to failure t_f at the given stress is the sum of the total time intervals.

Simulations are performed for the matrix material and the fibre composite with $\rho_f = 0.1$ and 0.2 mm^{-2}. These results are shown in Figure 26 for a 50% failure probability and for a given ρ_f the data can be described by the equation: $\sigma t_f^{p*} = $ constant [47]. The slope of the log σ–log t_f plot gives a measure of the effective slow crack growth exponent $p*$ of equation (36). Thus, when $\rho_f = 0$, $p* = 33$ which compares well with the input value of $p = 36$. When $\rho_f = 0.1$ and 0.2 mm^{-2}, $p*$ becomes 125 and 200 respectively. The observed changes in the effective slow crack growth exponent due to the R–curve effect are not uncommon and have also been reported for ceramics with residual stress effects [46,47].

Theoretical evaluation of the time–dependent strength may be obtained from an equation similar to (31), i.e.

$$F(\sigma, \Delta a_c, t_f) = 1 - \exp\left\{-V \int_{a(\sigma, t_f)}^{\infty} Q(a, \Delta a_c, t_f)\, da\right\}. \tag{40}$$

$Q(a, \Delta a_c, t_f)$ is more difficult to calculate and is given in [39]. Since no additional physical insight is to be gained from equation (40) the theoretical predictions of σ–t_f–F diagrams are not given here.

5.   Concluding Remarks

The fracture behaviour of fibre cementitious composites has been studied using the crack growth assistance curve concept. It is shown that the K_R–curve is not a material property but depends both on specimen geometry and size. In small NB beams the crack resistance does not tend to the limiting plateau value K_∞ as in large DCB specimens. A simple K–superposition method is proposed to model the crack resistance curve. This assumes that crack growth occurs when the effective stress intensity factors K_e at the tip of the continuous matrix crack is equal to the matrix toughness K_{Ic}. Since K_e is the sum of the applied and closure stress intensity factors, K_a and K_r, the modelling reduces to determining K_r due to the fibres bridging across the crack faces. With the

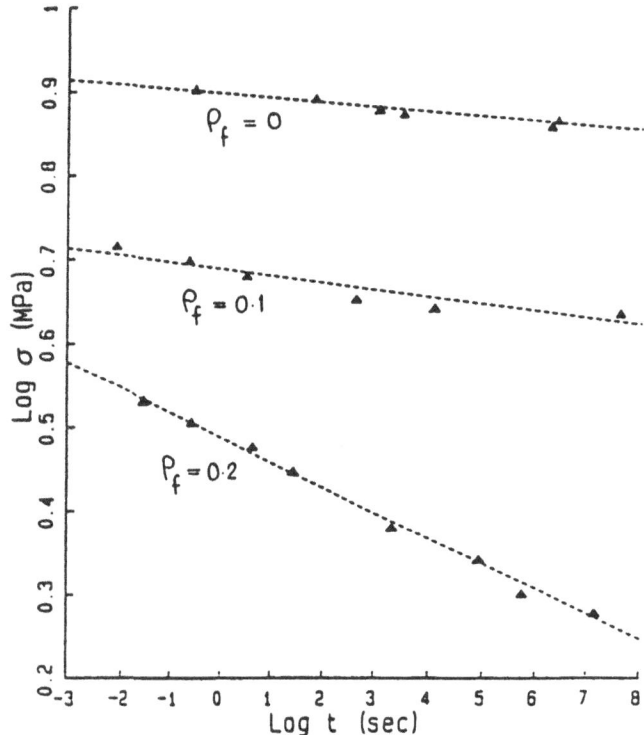

Figure 26. Failure lifetime versus applied stress for plain matrix and fibre cements with $\rho_f = 0.1$ and 0.2 mm^{-2}.

assumptions of a linearised closure stress (σ)–crack face separation (δ) relationship for the fibre bridges and straight crack flanks in the FBZ the crack resistance curve can be obtained by simple iteration routines. The effect of the exponent n in the stress–displacement equation (16) on the K_R curve is negligible [21]. What matters most are the fracture parameters K_{Ic}, σ_m and K_∞ (or J_∞). As shown in Section 3.4 these parameters can be determined by choosing such values that give the best fit to an experiment crack resistance curve for a given geometry and size. They can be used then to evaluate R–curves for other geometries and sizes.

Failure characterisation of fibre cement composites containing a single dominant crack with a R–curve characteristics may be effected by considering the interaction of the crack resistance curve and the applied stress intensity factor curve and determining the tangency point to evaluate the critical load and critical crack growth. Since the crack resistance curve is not unique iterations are required to solve this problem. Both theoretical and computer simulation studies on the tensile strength behaviour of short fibre cement composites containing many cracks show that as a consequence of the R–curve characteristics of each individual crack the tensile strength is increased even though low modulus fibres such as cellulose and polyethylene are used. Also, the Weibull modulus of the fibre composite is considerably increased compared to the unreinforced matrix. The time–dependent strength behaviours of short fibre cement composites when subjected to constant applied stresses are also investigated and they show that the effective stress corrosion exponent of the slow crack growth law is increased due to the R–curve effect of the bridging fibres.

A final remark is that whilst these lecture notes deal exclusively with the crack resistance curve – its theoretical modelling and its applications in failure characterisation of cement–based fibre composites – such a concept may not be readily comprehensible to the civil or structural enginner. For practical purposes, once the parameters K_{Ic}, K_∞ and σ_m are determined, it is desirable to calculate the load to extend a crack and hence to obtain a complete load–deflection curve or a moment–curvature diagram that are of more immediate use. Future work should be directed towards a sensitivity study of these fracture parameters on the shape of the load–deflection curves. Fracture mechanics models can also be extended to investigate the time–dependent strength behaviours of fibre cements particularly for cyclic loading situations.

Acknowledgements

The authors wishes to thank the Australian Research Council for financial support. The contributions in the form of original data and discussion from B. Cotterell, R.M.L. Foote and X–Z. Hu are much appreciated.

References

1. Foote, R.M.L. (1986) Crack Growth Resistance Curves in Fibre Cements, PhD Thesis, University of Sydney, Sydney.

2. Mai, Y–W. and Hakeem, M.I. (1984) 'Slow crack growth in cellulose fibre cements', J. Mater. Sci. 19, 501–508.

3. Hu, X–Z. and Wittmann, F.H. (1989) 'Fracture process zone and K_r–curve of hardened cement paste and mortar', in S.P. Shah, S.E. Swartz and B. Barr (eds.), Fracture of Concrete and Rock, Elsevier Applied Science, London, pp. 307–316.

4. Foote, R.M.L., Mai, Y–W. and Cotterell, B. (1987) 'Process zone size and slow crack growth measurements in fibre cements', in S.P. Shah and G.B. Barton (eds.), ACI Symposium on Fibre Reinforced Concrete, SP–105, Amer. Concrete Inst. Detroit, pp 55–70.

502

5. Ballarini, R., Shah, S.P. and Keer, L.M. (1984) 'Crack growth in cement–based composites', Engg. Fract. Mech. 20, 433–445.

6. Gao, Y–C., Mai, Y–W. and Cotterell, B. (1988) 'Fracture of fibre–reinforced composites', ZAMP 39, 550–573.

7. Petersson, P.E. (1981) 'Crack Growth Development of Fracture Zones in Plain Concrete and Similar Materials, Doctoral Dissertation, Lund Institute of Technology, published as Rept TVBM–1006 Division of Bldg. Mats., Sweden.

8. Reinhart, H. (1984) 'Fracture mechanics of an elastic softening material like concrete', Heron 29, 3–42.

9. Li, V.C. (1985) 'Fracture resistance parameters for cementitious materials and their experimental determinations', in S.P. Shah (ed.), Application of Fracture Mechanics to Cementitious Composites, Martinus Nijhoff Publishers, Dorsrecht, pp. 431–449.

10. Li, V.C. and Ward, R. (1988) 'A novel testing technique for post–peak tensile behaviour of cementitious materials', in H. Mihashi (ed.), Proc. Int. Workshop on Fracture Toughness and Fracture Energy: Test Methods for Concrete and Rock, pp. 139–156.

11. Foote, R.M.L., Mai, Y–W. and Cotterell, B. (1986) 'Crack growth resistance curves in strain–softening materials', J. Mech. Phys. Solids 34, 593–607.

12. Chuang, T–J. and Mai, Y–W. (1989) 'Flexural behaviour of strain–softening materials', Int. J. Solids & Structures 25, 1427–1443.

13. Foote, R.M.L., Cotterell, B. and Mai, Y–W. (1980) 'Crack growth resistance curves for cement composites', in D.M. Roy (ed.), Adv. Cement Matrix Coomposites, Materials Research Society, Penn., pp. 135–144.

14. Hillerborg, A., Modeer, M. and Petersson, P.E. (1976) 'Analysis of crack formation and crack growth in concrete by mean of fracture mechanics and finite elements', Cement & Conrete Research 6, 773–782.

15. Hillerborg, A. (1983) 'Analysis of one single crack', in F.H. Wittmann (ed.), Fracture Mechanics of Concrete, Elsevier, Amsterdam, pp. 223–249.

16. Wecharatana, M. and Shah, S.P. (1983) 'A model for predicting fracture resistance of fibre reinforced concrete', Cement & Concrete Research 13, 819–823.

17. Visalvanich, K. and Naaman, A.E. (1983) 'Fracture model for fibre reinforced concrete', J. Amer. Concrete Inst. 80, 128–138.

18. Jeng, Y. and Shah, S.P. (1985) 'Two parameter fracture model for concrete', J. Engg. Mech. Div. ASCE III, 1227–1241.

19. Jeng, Y. and Shah, S.P. (1986) 'Crack propagation in fibre–reinforced concrete', J. Struct. Engg. Div. ASCE 112, 19–24.

20. Mobasher, B., Ouyang C. and Shah, S.P. (1990) 'A R–curve approach to predict toughening of cement–based matrices due to fibre reinforcement', to be published.

21. Cotterell, B. and Mai, Y–W. (1988) 'Modelling crack growth in fibre–reinforced cementitious materials', Mater Forum 11, 341–351.

22. Lenain, J.C. and Bunsell, A.R. (1979) 'The resistance to crack growth of asbestos cement', J. Mater. Sci. 14, 321–332.

23. Foote, R.M.L., Cotterell, B. and Mai, Y–W. (1986) 'Analytical modelling of crack growth resistance curves in DCB fibre–reinforced cement specimens', in F.H. Wittmann (ed.), Fracture Toughness and Fracture Energy of Concrete, Elsevier, Amsterdam, 535–544.

24. Cotterell, B. and Mai, Y–W. (1988) 'The effect of a fracture process zone on a model for crack growth in fibre–reinforced cementitious composites', Adv. Cement Research 1, 75–83.

25. Cotterell, B., Mai, Y–W. and Foote, R.M.L. (1988) 'Bonding solutions for crack growth resistance curve in fibre reinforced cement composites", in S.a. Paipetis and G.C. Papanicolaou (eds.), Engineering Applications of New Composites, Omega Scientific, pp. 186–196.

26. Mai, Y–W. (1988) 'Fracture resistance and fracture mechanisms of engineering materials', Mater. Forum 11, 232–267.

27. Mai, Y–W., Foote, R.M.L. and Cotterell, B. (1980) 'Size effects and scaling laws of fracture in asbestos cements', Int. J. Cement Composites 2, 23–34.

28. Foote, R.M.L. and Buchwald, V.T. (1985) 'An exact solution for the stress intensity factor for a double cantilever beam', Int. J. Fract. 29, 125–134.

29. Tada, H., Paris, P.C. and Irwin, G.R. (1973) Stress Intensity Factor Handbook, Del Research Corporation, Hellertown, PA.

30. Hu, X–Z and Wittmann, F.H. (1990) 'Fracture energy and fracture process zone', Cement & Concrete Research, in press.

31. Hu, X–Z., Mai, Y–W. and Cotterell, B. (1990) 'Crack growth modelling in plain and fibre–reinforced cement composites', to be published in Proc 8th European Fracture Conference, Torino, Italy.

504

32. Mai, Y–W. 'Fracture measurements of cementitious composites', in S.P. Shah (ed.), Application of Fracture Mechanics to Cementitious Composites, Martinus Nijhoff Publishers, Dordrecht, pp. 399–429.

33. Mai, Y–W. and Cotterell, B. (1982) 'Slow crack growth and fracture instability in cement composites', Int. J. Cement Composites 4, 33–37.

34. Broek, D. (1986) Elementary Engineering Fracture Mechanics, Martinus Nijhoff Publishers, Dordrecht.

35. Cotterell, B. and Mai, Y–W. (1987) 'Crack growth resistance curve and size effect in fracture of cement paste', J. Mater. Sci. 22, 2734–2738.

36. Andonian, R., Mai, Y–W. and Cotterell, B. (1979) 'Strength and fracture properties of cellulose fibre reinforced cement composites', Int. J. Cement Composites 1, 151–158.

37. Hu, X–Z., Mai, Y–W. and Cotterell, B. (1988) 'Computer simulation of fracture behaviour of short–fibre reinforced cement', in R.N. Swamy et. al. (eds.), Proc. 3rd Int. Symp. on Developments in Fibre Reinforced Cement and Concrete, Sheffield, Vol. 2, Paper 6.4

38. Hu, X–Z., Mai, Y–W. and Cotterell, B. (1988) 'A statistical theory of time–dependent fracture for brittle materials', Phil. Mag. 58, 229–234.

39. Hu, X–Z. (1988) Statistical Fracture of Brittle Materials, PhD Thesis, University of Sydney, Sydney.

40. Mindess, S. and Nadeau, J.S. (1977) 'Effect of loading rate on the flexural strength of cement and mortar', Bull. Am. Ceram. Soc. 56, 429–430.

41. Nadeau, J.S., Mindess, S. and Hay, J.M. (1974) 'Slow crack growth in cement paste', J. Am. Ceram. Soc. 57, 51–54.

42. Mindess, S. (1985) 'Rate of loading effects on the fracture of cementitious materials', in S.P. Shah (ed.), Application of Fracture Mechanics to Cementitious Composites, Martinus Nijhoff Publishers, Dordrecht, pp. 617–636.

43. Atkins, A.G. and Mai, Y–W. (1985) Elastic and Plastic Fracture, Ellis Horwood/John Wiley, Chichester, U.K.

44. Hu, X–Z., Mai, Y–W. and Cotterell, B. (1989) 'A statistical theory of time–dependent fracture for cementitious materials', in S.P. Shah and S.E. Swartz (eds.), Fracture of Concrete and Rock, Springer–Verlag, Berlin, pp. 37–46.

45. Hu, X–Z., Mai, Y–W. and Cotterell, B. (1989) 'A statistical theory of time–dependent fracture for cementitious materials subjected to cyclic loading', J. Mater. Sci 24, 3118–3122.

46. Cook, R.F. and Lawn, B.R. (1984) 'Controlled indentation flaws for construction of toughness and fatigue master maps' in S.W. Friedman (ed.), Methods for Assessing the Structural Reliability of Brittle Materials, ASTM STP 844, pp. 22–42.

47. Fuller, E.R., Lawn, B.R. and Cook, R.F. (1983) 'Theory of fatigue for brittle flaws originating from residual stress concentrations', J. Am. Ceram. Soc. 66, 314–321.

48. Visalvanich, K. and Naaman, A.E. (1982) 'Fracture modelling of fibre reinforced cementitious composites', Program Rept for NSF Grant ENG 77–23534, Department of Materials Engineering, University of Illinois at Chicago Circle.

49. Patterson, W.A. and Chen, H.C. (1975) 'Fracture toughness of glass fibre reinforced cement', Composites 6, 102–104.

Characterization of Interfacial Bond in FRC Materials

Dr. H. Stang
Department of Structural Engineering,
Building 118,
Technical University of Denmark,
DK-2800 Lyngby, Denmark.

Professor S.P. Shah
Director of NSF Center for Science and Technology of
Advanced Cement-Based Materials
Northwestern University
Evanston, Ill. 60208-3109, USA.

ABSTRACT. A proper characterization of the fiber/matrix bond in fiber reinforced cementitious – FRC – composites is of great importance not only in the evaluation of the quality of a given fiber/matrix system. Different kinds of fiber/matrix debonding mechanisms are included in many models for the macroscopic mechanical behavior of FRC-materials and the applicability of such models naturally depends on the availability of experimentally determined fiber/matrix bond parameters. The present paper consist of a unified treatment of the fiber/matrix bond models proposed in the literature along with an evaluation of these models from a experimental and theoretical point of view. When dealing with perfectly bonded interfaces, basically two approaches are identified: the stress criterion and the fracture mechanical criterion approach. The results of these two approaches are examined and discussed. Finally some new research directions are proposed.

1   Introduction

Fiber Reinforced Cementitious – FRC – materials are a special type of the so-called *fiber reinforced brittle matrix composites* which again are a special type of *brittle matrix composites*. The latter group of materials consist of composites characterized by the fact that the matrix is very brittle compared to the other phases in the composite material.

Brittle matrix composite materials are designed in order to retain some of the matrix characteristics, typically the stiffness and the mechanical behavior in compression while other characteristics – typically the brittle behavior of the matrix material in tension – are modified.

Fiber reinforced brittle matrix composite materials are the most common of the brittle matrix composite materials. The fiber reinforcement modifies the brittle behavior of the matrix material by stabilizing cracks and microcracks. The crack stabilization takes place

S. P. Shah (ed.), Toughening Mechanisms in Quasi-Brittle Materials, 507–527.

both on the so-called *meso-level*, i.e. the level where the characteristic length is in the order of big pores, cement grains, preexisting cracks, and inclusions as well as on the *macro-level*, i.e. the level where the characteristic length is that of the structural element and where the cementitious material is treated as a homogeneous continuum.

Looking specifically at cementitious fiber reinforced materials a number of different mechanisms for crack stabilization on the meso- and the macro-level have been identified in the literature including:

- Crack blunting (meso-level), originating in the Cook-Gordon arrest mechanism, (Cook and Gordon, 1964)

- Crack path deviation (meso-level), presumably connected to the Cook-Gordon arrest mechanism, see e.g. Bentur *et al.* (1985a).

- Crack bridging (meso- and macro-level).Fibers crossing the crack introduce crack closing forces on the entire crack surface (meso-level) or part of the crack surface near the crack tip (macro-level). This effect is probably the most widely recognized crack stabilizing effect identified and modelled by numerous investigators, see e.g. Hillerborg (1980), Korczynskyj *et al.* (1981), Hannant *et al.* (1983), Selvadurai (1983), Mori and Mura (1984), Budiansky *et al.* (1986), Stang (1987), and Budiansky and Amazigo (1989).

- Crack shielding (macro-level), related to the increased formation of cracks on the meso-level. Crack shielding has been dealt with especially in ceramic systems, e.g. Evans (1984), and Hutchinson (1987) but is probably also a significant mechanism in cementitious systems.

In almost all fiber/cementitious matrix systems the bond between the fiber and the matrix is relatively week. It follows as a consequence that all the mechanisms mentioned above include fiber/matrix debonding to a certain extend. As a further consequence, the magnitude of the fiber/matrix bond – along with the mechanical characteristics of the fibers and the matrix material – has a significant influence on the macroscopical mechanical behavior of the composite material.

Numerus models have been introduced in the literatur dealing with the macroscopical behavior of cementitious fiber reinforced composite materials or brittle matrix composites in general as function of the interfacial fiber/matrix behavior, see e.g. Aveston *et al.* (1971), Hillerborg (1980), Korczynskyj *et al.* (1981), Hannant *et al.* (1983), Budiansky *et al.* (1986), Stang (1987), and Budiansky and Amazigo (1989), and Abudi (1989).

All these models include – implicitly or explicitly – a quantitative characterization of the fiber/matrix bond. The applicability of such models is of course limited if the fiber/matrix bond parameters cannot be determined independently by experiment.

Following this line of thought, the following requirements must be imposed on the parameters used to characterize the fiber/matrix bond:

- The bond parameters should be true "material" parameters in the sense that the magnitude of the parameters should be independent of the loading conditions and the geometrical configuration of the fiber/matrix system.

- The concept behind the parameters should be realistic enough to reflect the significant features of the bond mechanisms and simple enough to allow for application in simplified approaches to complicated composite material systems.

- It should be possible to evaluate the bond parameters directly from relatively simple (pullout) experiments or indirectly from the macroscopical behavior of simple composite material systems. In both cases the magnitude of the bond parameters should be determined by comparing the experimental results with a robust analytical model.

In the following models presented in the literatur will be examined and evaluated with respect to the requirements suggested above. First the different types of interface characterization will be identified from a general point of view. Secondly different types of debonding criteria will be examined and their application to the pullout problem will be investigated. Thirdly different types of experimental investigations will be mentioned and finally conclusions will be drawn on the basis of the findings in the previous sections. Finally some suggestions for future research and investigations will be given.

2 Interface characterization

The volume surrounding the surface separating the fiber and the matrix is usually denoted *the interfacial zone.*

This zone is often idealized as a surface i.e. as a zone with no extension perpendicular to the fiber. In this zone special requirements are then imposed on stresses and displacements – requirements which are different from those imposed in the matrix and fiber volume. These requirements reflect the nature of the interfacial zone as well as the fiber/matrix bond. (See fig.1.)

There is experimental evidence that this type of modelling *does* represent an idealization. Typically in steel fiber reinforced cementitious composite materials an interfacial *volume* or *transition zone* rather than a *surface* has been described. See Pinchin and Tabor (1978a), Barnes *et al.* (1978), Page (1982), Bentur *et al.* (1985b), and Wei *et al.* (1986). It has been shown that each fiber in the transition zone is surrounded by matrix material which is more porous and less stiff than the bulk matrix material.

Typically the interfacial zone is divided into two parts. One where the original bonding – which can be of a physical and/or chemical nature – between the fibers and the matrix is intact. This part is here denoted as the *perfectly bonded part*, and is usually described by a set of mathematical relations describing *continuity of stresses and displacements* across the interface.

The other part of the interfacial zone is denoted the *debonded interface*. This zone is usually thought of as a zone created by external thermal or mechanical loading. In the debonded part of the interface a completely new set of mathematical interfacial relations are set up which reflect *the ability of the fiber and the matrix volume to experience relative displacements.*

510

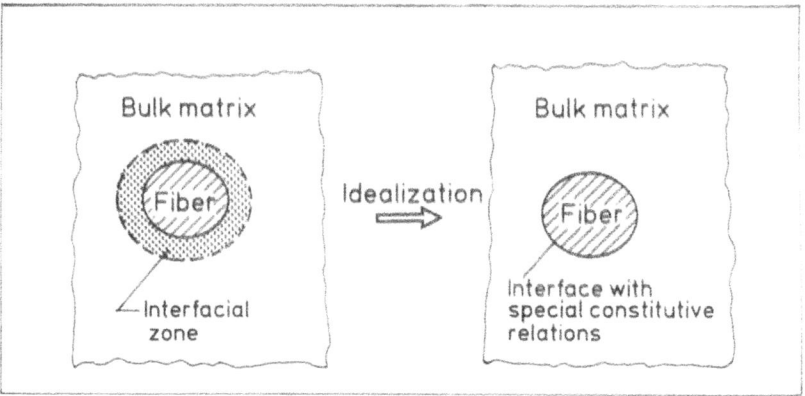

Figure 1: The physical reality: a fiber included in matrix material with a transition zone between the fiber and the bulk matrix material. The idealization: the transition zone is modelled by an interface with a special type of mechanical behavior.

2.1 The Perfectly bonded interface

From a continuum mechanics point of view – and accepting the surface idealization of the fiber/matrix interface – the perfectly bonded interface is characterized in the following way.

Using conventional index notation the stress field in the fiber is denoted σ_{ij}^f and the displacement field denoted u_i^f while the stress field and the displacement field in the matrix is denoted σ_{ij}^m and u_i^m. (See fig.2.)

The constitutive relation in the fiber and the matrix relates – along with the definition of strain – the stress and the displacement fields in each of the two volumes. However, on the perfectly bonded interface it is required that

$$\sigma_{ij}^f n_j = -\sigma_{ij}^m m_j \quad \text{on} \quad I^b \tag{1}$$

along with

$$u_i^f = u_i^m \quad \text{on} \quad I^b \tag{2}$$

where I^b represent all points on the perfectly bonded interfacial surface and n_i is the outward unit normal vector to the *fiber* surface while m_i is the outward unit normal vector to the *matrix* surface. (See also fig.2.)

Thus, the perfectly bonded interface is modelled by requiring continuity in the surface tractions and in the surface displacements on the interface.

Many analytical models are two dimensional and the analysis may not involve the complete displacement and stress tensor field, however, all models dealing with the perfectly

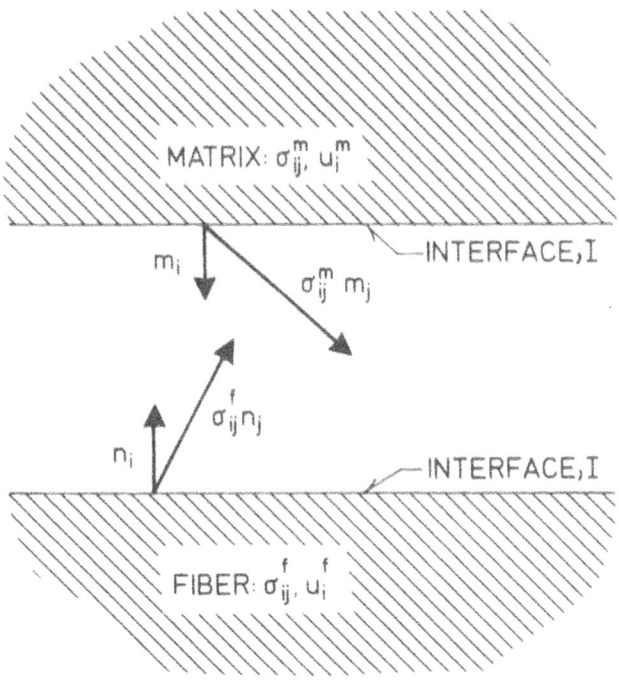

Figure 2: The notation used in order to describe conditions and constitutive relations on the interface, I. The fiber and the matrix part of the interface are moved apart for the sake of clarity.

bonded interface set up requirements which can be interpreted as simplifications or special cases of the equations (1) and (2).

2.2 The debonded interface

On the debonded interface the characterization is changed into a prescribed surface traction boundary condition which is applied to both the fiber surface and the matrix surface in the following way:

$$\sigma^f_{ij} n_j = f_i \quad \text{on} \quad I^d \tag{3}$$

and

$$\sigma^m_{ij} m_j = -f_i \quad \text{on} \quad I^d \tag{4}$$

512

where I^d represent all points on the debonded interfacial surface while f_i represent the prescribed surface traction.

This description obviously allows for displacement discontinuities along the debonded interface, however, it does not give any guarantee that surface overlapping will not occur. Thus the relations (3) and (4) are only sufficient if:

$$(u_i^m - u_i^f)n_i \geq 0. \tag{5}$$

The surface traction f_i represents frictional stresses. These frictional stresses can be assumed to be vanishing, Atkinson et al. (1982), Budiansky et al. (1986), Stang and Shah (1986), and Morrision et al. (1988). However, often the frictional surface tractions are assumed to be non-vanishing and constant, Lawrence (1972), Laws et al. (1973), Bartos (1981), Laws (1982), Gray (1984), Gopalaratman and Shah (1987), Palley and Stevans (1989), and Stang, Li, and Shah (1990).

If the surface traction is connected to surface roughness it is reasonable to assume some sort of relationship between the surface traction and the displacement discontinuity across the interface (see fig.3.):

$$f_i = f_i(u_j^m - u_j^f) \quad \text{on} \quad I^d \tag{6}$$

see e.g. Wang et al. (1988) who used this type of modelling in a fiber pullout investigation involving loading and unloading. In spite of the appealing concepts underlying equation (6), not much work has been done in this direction.

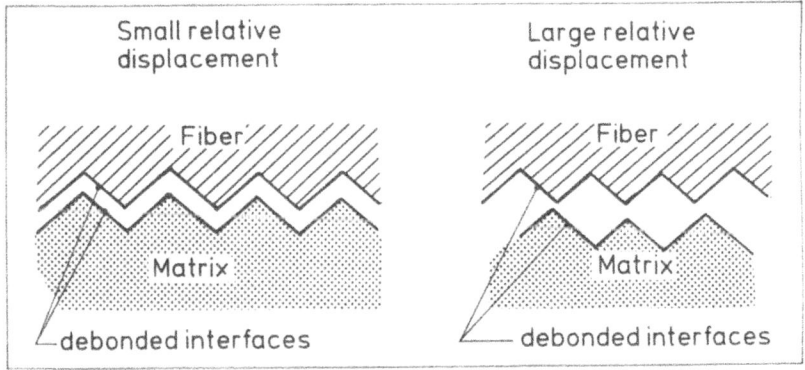

Figure 3: Schematic outline of different frictional conditions related to different magnitudes of relative interfacial displacements.

Some models, Pinchin and Tabor (1978b, 1978c), Beaumont and Aleszka (1978), Budiansky et al. (1986) and Gao (1987), Abudi (1989), and Sigl and Evans (1989) deal specifically with the case were the matrix exerts a *compressive* stress on the debonded interface due to thermal mismatch, mechanical loading, or matrix shrinkage. In that case the boundary condition on the debonded interface becomes a complicated mixed type of boundary condition requiring displacement continuity perpendicular to the interface:

$$n_i u_i^f = -m_i u_i^m \quad \text{on} \ I^d \tag{7}$$

(see fig.4.) along with stress continuity *perpendicular* to the interface:

$$n_p n_q \sigma_{pq}^f = m_p m_q \sigma_{pq}^m \quad \text{on} \ I^d \tag{8}$$

However, the surface traction *in* the interface plane (see fig.4.) is a prescribed frictional surface traction. This condition can be written as:

$$\sigma_{ip}^f n_p - n_i n_p n_q \sigma_{pq}^f = f_i \quad \text{on} \ I^d \tag{9}$$

and

$$\sigma_{ip}^m m_p - m_i m_p m_q \sigma_{pq}^m = -f_i \quad \text{on} \ I^d \tag{10}$$

assuming a symmetrical stress tensor.

The magnitude of the prescribed frictional surface traction can be assumed to depended on the normal stresses on the interface according to the Coulomb frictional law:

$$\|f\| = c - \mu n_p n_q \sigma_{pq}^f \tag{11}$$

where c is a measure for the cohesion, and μ is the frictional coefficient. The direction of the frictional surface traction can be determined from the direction of the displacement discontinuity rate, as suggested by e.g. Abudi (1989). Thus, a more general type of relationship for the surface traction can be written as:

$$f_i = f_i(u_j^d, \dot{u}_j^d, n_p n_q \sigma_{pq}^f) \quad \text{on} \ I^d \tag{12}$$

where

$$u_j^d = u_j^m - u_j^f \tag{13}$$

Again, many analytical models are two dimensional and the analysis may not involve the complete displacement and stress tensor field, however, all models dealing with the debonded interface set up equations which can be interpreted as parts of or special cases of the above equations.

514

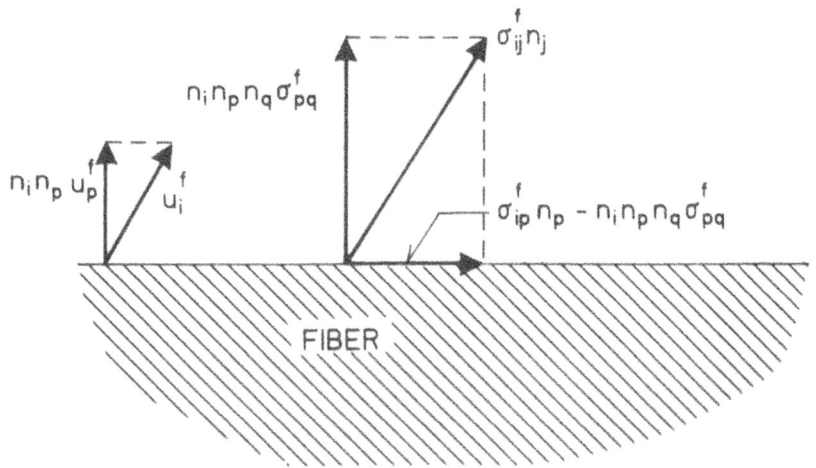

Figure 4: The normal component of interfacial displacement along with the normal and in-plane components of the surface traction on the fiber part of the interface.

2.3 Cohesive interface models

A very attractive interface model was suggested by Needleman (Needleman, 1987) originally for use in plastically deforming solids and composite materials. However, it seems obvious to use this kind of interface description in brittle matrix composites as well.

The basic idea of the Needleman interface model – which places it somewhere between the perfectly bonded interface and the debonded interface models – is that a separate constitutive relation is postulated relating interface surface traction with displacement discontinuity.

Thus, in this type of model there is no clear distinction between *bonded* and *debonded* interface since a relation of the kind (3), (4), and (6) is governing the interface at *all* times:

$$\sigma^f_{ij} n_j = f_i \quad \text{on} \quad I \tag{14}$$

and

$$\sigma^m_{ij} m_j = -f_i \quad \text{on} \quad I \tag{15}$$

and

$$f_i = f_i(u^m_j - u^f_j) \quad \text{on} \quad I \tag{16}$$

where I means all points on the total interface.

Using a potential formulation the response is specified in terms of three parameters: a maximum tensile surface traction σ_{max} corresponding to a positive interface separation, a maximum positive interface separation δ and the total work of separation, ϕ.

The potential is chosen so that the normal surface traction component reaches a maximum for increasing normal interface separation and then drops to zero when the normal interface separation exceeds δ. In case of negative interface separation (interface overlapping) compressive normal surface tractions build up. The in-plane surface tractions are assumed to depend linearly on the in-plane displacement discontinuety, however, the in-plane tractions also drop to zero when the normal separation exceeds δ.

This model has been used in a couple of papers for studying void formation at inclusion boundaries (Nutt and Needleman, 1987) and (Needleman and Nutt, 1989). Tvergaard (1989) modified the model in order to include Coulomb friction and used the modified model to study debonding in whisker-reinforced metals.

The model is somewhat similar to the cohesive crack models by Barenblatt (1962), and Hillerborg *et al.* (1976).

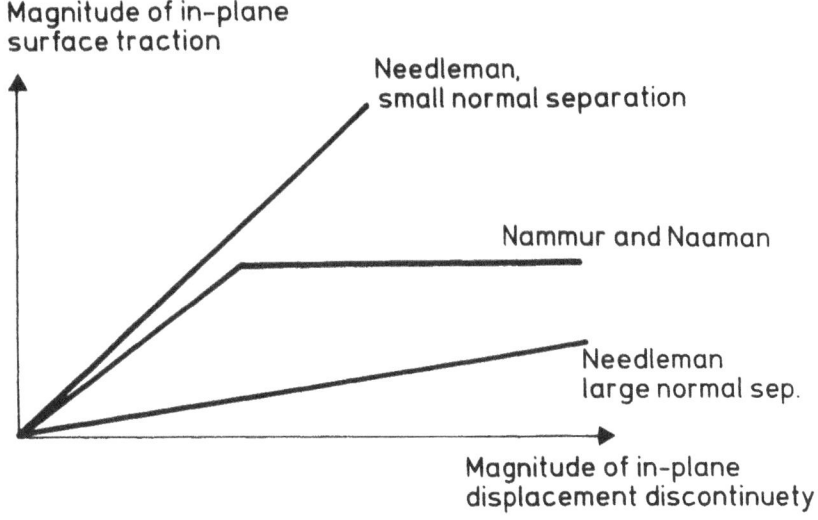

Figure 5: The normalized in-plane surface traction component as function of normalized in-plane displacement discontinuety as predicted by the Nammur and Naaman (1989) model and the Needleman (1987) model.

A Needleman type of model was applied to fiber reinforced cementitious matrix composites by Nammur and Naaman (1989) assuming a linear elastic perfectly plastic type of relationship between in-plane displacement discontinuety and in-plane surface traction at the interface. This relation was assumed to govern shear transfer at the total fiber/matrix interface at all times.

The Needleman and the Nammur and Naaman models are outlined in fig.5.

Finally, Leung and Li (1990) mentions the *possibility* of using cohesive crack models to describe the transition zone separating the debonded and perfectly bonded interface in a fiber pullout test.

3 Debonding criteria

Assume that a model for a fiber imbedded in a cementitious matrix has been established including an interface model derived from the general interface description above. If the interface is divided into a perfectly bonded and a debonded interface, a criterion is needed in order to determine weather a point on the perfectly bonded interface is about to change status and becoming a point on the debonded interface.

Basically debonding criteria can be divided into two classes: the stress based predicting onset of debonding when the interface surface traction reaches some critical value.

In general terms the stress based condition for debonding is

$$F(\sigma_{ij}^{f} n_j) = 1 \ \text{ on } \ I^b/I^d \tag{17}$$

where I^b/I^d means the transition point separating the perfectly bonded and the debonded interface.

The other approach is the fracture mechanical approach which was applied to the fiber pulllout problem by Gurney and Hunt (1967), Outwater and Murphy (1967), Bowling and Groves (1979), Wells and Beaumont (1982), Gray (1984), Piggott *et al.* (1985), Wells and Beaumont (1985), Stang and Shah (1986), Piggott (1987), Morrison *et al.* (1988), Gao *et al.* 1988.

Typically, the fracture mechanical postulate is related to the total energy release rate G in such a way that given a measure a for the debonded zone D, the debonded zone will increase in size if and only if

$$G = \Gamma \frac{\partial D}{\partial a} \tag{18}$$

where Γ is denoted *the work of fracture* and where

$$G = \frac{\partial W_{ex}}{\partial a} - \frac{\partial W_\epsilon}{\partial a} - \frac{\partial W_f}{\partial a} \tag{19}$$

Here W_{ex} is work done by prescribed external forces, W_ϵ is elastic strain energy, and W_f is dissipation in inelastic parts of the structure, e.g. work done by friction on the debonded interface.

Note that if the interface characterization is of the Needleman type no extra criterion is needed in order to determine weather debonding takes place or not.

4 The fiber pullout problem

The fiber pullout experiment where a single fiber or a number of fibers are loaded axially and pulled out of a piece of cementitious matrix material with a given geometry has been studied extensively in the literature.

Modelling of the fiber pullout problem involves modelling of the fiber/matrix interface along with a modelling of the matrix geometry and the fiber geometry.

The interface modelling identifies the bond parameters and the finalized model can be used for interpretation of fiber pullout experiments in order to determine the bond parameters experimentally.

A number of authors, Greszczuk (1969), Lawrence (1972), Laws et al. (1973), Bartos (1981), Laws (1982), Gray (1984), Gopalaratman and Shah (1987), Palley and Stevans (1989) Stang, Li, and Shah (1990) have used a simple shear lag type of analysis. The advantages of a simple shear lag approach is that simple closed form analytical expressions can be derived and at the same time the model is general enough to be able to identify some of the basic features of the fiber pullout problem.

A typical shear lag solution of the pullout problem based on the description (1), (2), (3), and (4) with a *constant* shear stress at the interface gives the following load/ displacement relation at the end of the fiber (see Stang Li and Shah, 1990):

$$U_f = \frac{P_f - q_f a}{E_f A \omega} \coth(\omega(L - a)) + \frac{P_f - \frac{1}{2}q_f a}{E_f A} a \tag{20}$$

Here, U_f is the displacement of the free fiber end while P_f is the pullout load at the free fiber end . The frictional force per length q_f is acting at the debonded interface which has a length of a. The stiffness and the cross sectional area of the fiber is denoted E_f and A respectively. The quantity ω is given by

$$\omega = \sqrt{\frac{k}{E_f A}} \tag{21}$$

where k is related to the shear stiffness and the geometry of the matrix material.

The total axial displacement field of the fiber $U(x)$ is given by:

$$U(x) = \frac{P_f - q_f a}{E_f A \omega} \frac{\cosh(\omega x)}{\sinh(\omega(L - a))} \qquad 0 \leq x \leq (L - a)$$

and

$$\tag{22}$$

$$U(x) = \frac{P_f - q_f a}{E_f A \omega} \coth(\omega(L - a)) - \frac{P_f - q_f L}{E_f A}(L - a)$$

$$- \frac{q_f}{2E_f A}(L - a)^2 + \frac{P_f - q_f L}{E_f A}x + \frac{q_f}{2E_f A}x^2 \qquad (L - a) < x \leq L$$

where L is the total fiber length, $x = L$ is the free fiber end, $x = 0$ is the embedded fiber end, while $x = (L - a)$ correspond to the perfectly bonded/ debonded interface transition. (See fig.6.)

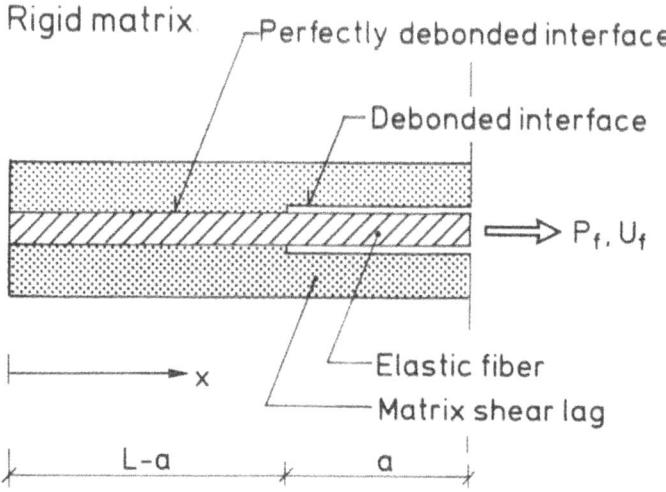

Figure 6: Schematic drawing showing the fiber pullout geometry modelled by Stang *et al.* (1990).

The fiber force as function of the position in the fiber is given by equations (22) in combined with the constitutive relation for the fiber:

$$P = E_f A U' \qquad (23)$$

Solutions of the same nature but introducing the shear stress as function of the normal stress (equation (11)) and furthermore relating the normal stress on the interface to a geometrical misfit due to e.g. matrix shrinkage have been presented by Pinchin and Tabor (1978b, 1978c) and by Beaumont and Aleszka (1978).

More detailed models than the shear lag type are models based on linear elastic analysis of a fiber embedded in an infinite half space, Muki and Sternberg (1970), Sternberg and Muki (1970), Luk and Keer (1979), Phan- Thien (1980), Phan-Thien and Goh (1981), Phan-Thien *et al.* (1982). These models take advantage of the axial symmetry of the pullout problem, thus reducing the 3D problem to 2D. To some extend analytical expressions for stresses and displacements are derived, but only in the case where the total interface is perfectly bonded is treated.

Applying finite element or boundary element solutions to the pullout problem allows for a more general type of analysis including modelling of the exact geometry and including more general types of interfacial models. Atkinson *et al.* (1982), Stang (1985), and Morrison *et al.* (1988) presented FEM and BEM solutions based on axial symmetric linear elastic models and used a perfectly bonded, stress free debonded interface model. Marmonier *et al.* (1988) presented an axial symmetrical FEM model but considered only a perfectly bonded interface. Steif and Hoysan (1986) considered a two dimensional FEM analysis of the pullout problem and introduced a interface relation of the kind (14), (15), and (16) governing the interface at *all* times.

4.1 Stress based criteria for debonding

As mentioned above the perfectly bonded, debonded interface type of models are not complete until a criterion for continued debonding is established. On the other hand models which do not include the perfectly bonded interface in the analysis, but assume some sort of slip/ stress transfer relation governing the interface at all times (e.g. Nammur and Naaman 1987) do not distinguish between *bonded* and *debonded* interface and all the bond parameters are already included in the model.

Consider a simple shear lag model for the pullout problem like the Stang, Li, and Shah (1990) model and apply a maximum shear stress criterion for continued debonding:

$$q = q_{max} \tag{24}$$

where q is the shear force per length at the interface. Then the pullout force as function of the debonded length a is given as

$$P_f = q_f a + \frac{q_{max}}{\omega} \tanh(\omega(L - a)) \tag{25}$$

The load that initiates debonding P_f^0 is given by eq. (25) for $a = 0$:

$$P_f^0 = q_{max} \frac{\tanh(\omega L)}{\omega} \tag{26}$$

Apart from a different interpretation of the factor ω similar expressions were obtained by Lawrence (1972), Laws (1982) and Bartos (1980).

The approach outlined above thus identified two parameters: the maximum shear stress at the perfectly bonded interface, q_{max} and the frictional shear stress at the debonded interface, q_f. However, it is important to note that if the shear lag approach was exchanged with a different type of stress analysis, e.g. a FEM analysis, then the magnitude of the maximum shear stresses at the perfectly bonded interface would change significantly.

This is pointed out by e.g. Atkinson et al. (1982) and Mamonier et al. (1988) reporting singular stresses at the interface in the complete linear elastic solution.

Thus by applying a maximum shear stress criterion to experimental results a bond parameter q_{max} is determined which depend on the type of analysis used in the analytical modelling.

4.2 Fracture mechanical criteria

The problem of dependency of the magnitude of the bond parameters on the analytical modelling is overcome if a fracture mechanical criterion for continued debonding is applied.

Applying e.g. the fracture mechanical criterion (18) and (19) to the shear lag analysis of Stang, Li and Shah (1990), and identifying the term W_f as the work done by the constant frictional stresses at the debonded interface then the solution for the pullout force as function of the debonded length a reads:

$$P_f = q_f a + \left(\frac{q_f}{2\omega} + \sqrt{\left(\frac{q_f}{2\omega} \right)^2 + 2 E_f A p \Gamma} \right) \tanh(\omega(L - a)) \tag{27}$$

or

$$P_f = q_f a + \left(\frac{q_f}{\omega} + \sqrt{2 E_f A p \Gamma} \right) \tanh(\omega(L - a)) \tag{28}$$

depending on the approximations done in order to estimate the work done by the frictional stresses. The quantity p designates the perimeter of the fiber.

The solution for the load that initiates the debonding process is readily obtained from the equations (27) and (28):

$$P_f^0 = \left(\frac{q_f}{2\omega} + \sqrt{\left(\frac{q_f}{2\omega} \right)^2 + 2 E_f A p \Gamma} \right) \tanh(\omega L) \tag{29}$$

or

$$P_f^0 = \left(\frac{q_f}{\omega} + \sqrt{2 E_f A p \Gamma} \right) \tanh(\omega L) \tag{30}$$

A similar solution was derived by Palley and Stevens (1989) for a different geometrical configuration, however, their solution was not presented on an explicit analytical form.

It is interesting to compare the solution obtained from the stress criterion with the two solutions obtained from the fracture mechanical criterion.

Identifying the latter approach as the "true" approach and the first approach as the "approximate" approach it is clear that the approximate parameter q_{max} contains contributions from two real parameters Γ, the work of fracture, and q_f, the interfacial friction. Since q_{max} includes a fracture mechanical parameter it is obvious that q_{max} will depend heavily on the approximations done in the modelling of the approximate stress distribution.

Furthermore, it is observed, that according to the fracture mechanical criterion approach:

$$q_f \leq q_{max} \tag{31}$$

a relation which is not included in the stress criterion approach.

5   Experimental investigations

The most commonly applied experimental setup used for the determination of bond parameters in FRC systems is the fiber pullout experiment where a single or number of fibers simultaneously are pulled out of a block of matrix material while the pullout load and displacement is measured.

Pullout experiments have been used by numerous investigators and a review paper summarizing some of the most commonly used experimental configurations has been prepared by Gray (1983).

In general it is not possible from a conventional pullout experiment to evaluate the validity of a given interfacial modelling scheme, i.e. it is not possible to deduct from a simple pullout test weather a perfect bonding/debonding approach or a more general type of modelling (14)-(16) should be used.

Bien and Stroven (Bień (1986), Bień and Stroeven (1988)) conducted some very detailed studies of a series of pullout tests using a holographic interferometry method combined with a special pullout setup. The experimental setup consisted of a 40x80x20 mmm block of concrete with an embedded steel strip modelling the pullout of a steel fiber from a cementitious matrix. This setup allowed the displacement fields in the fiber as well as in the matrix to be determined.

Bien and Stroeven clearly identified one interfacial zone with displacement *continuity* and another interfacial zone with displacement *discontinuity*, thus supporting a perfectly bonded/debonded approach to the interface modelling problem.

Applying a perfectly bonded/debonded approach to the interfacial modelling of the fiber pullout problem and furthermore applying either a stress criterion or a fracture mechanical criterion to predict continued debonding the following should be noted:

- According to the fracture mechanical approach the stress criterion parameters q_{max} and q_f are subjected to the requirement (31). Experimental observations seems to verify this, see Bentur and Mindess (1990), p.62. Bentur and Mindess applies a number of different stress criterion based models to different experimental pullout tests reported in the literature and find with one exception ($q_{max}/p = 9$ MPa and $q_f/p = 11.5$ MPa) that equation (31) holds.

- As pointed out by Leung and Li (1990) it is not possible from a single or a series of pullout tests with the same fiber radius to determine weather a stress criterion or a fracture mechanical criterion is the most appropriate one to use. To determine which of the two approaches to be used a series of tests with different fiber radii should be conducted to test weather the size effect predicted by the fracture mechanical approach can be observed.

- To the authors knowledge no experiments on cementitious matrix/fiber systems have been reported which directly indicate that the perfectly bonded/ debonded type of interface modelling is inadequate. However, only very few experiments allow for microscopic observation of the fiber/matrix interface during loading.

- Finally it should be noted that the displacement measurements done on conventional pullout tests include not only the displacements originating from the strains in the fiber/ matrix system but also the displacements originating from the strains in the free part of the fiber and sometimes part of the testing machine. Simple calculations show that the latter part often can one order of magnitude grater than the first part. This of cause makes a comparison with models which yield displacements in the fiber/matrix system difficult.

6 Conclusions

The literatur dealing with interface characterization in fiber reinforced brittle matrix composites has been investigated and the different types of interfacial modelling has been described.

Many different types of interfacial modelling is available at the present time. The models can roughly be divided into two groups: the perfectly bonded/ debonded type of modelling and the cohesive interface modelling. The first type of models need combination with a criterion for continued debonding. This criterion can be formulated either as a maximum stress criterion or as a critical energy release rate criterion. The cohesive interface models, on the other hand, are self contained models which do not need additional criteria.

At the present time no systematic work has been done to clarify the advantages and the disadvantages of the different types of interfacial modelling in relation to different FRC-systems.

The fiber pullout problem has been investigated with a perfectly bonded/ debonded interface characterization using a stress based as well as a fracture mechanical based criterion for continued debonding.

From a theoretical point of view the stress based criterion does not predict true material parameters since the magnitude of the parameters depend heavily on the type of modelling used.

From an experimental point of view a fracture mechanical criterion for continued debonding can only be verified using a test series designed especially to bring out the size effects involved.

7 Future research

On the basis of the investigations above the following suggestions for future research are made:

- A systematic investigation of the applicability of different interfacial models in different FRC-systems including further work on the implementation in pullout models of interface modelling of the type (14)-(16), either on the whole interface or on part of the interface serving as a transition zone between the debonded and the perfectly bonded interface.

- Further experimental investigations of the size effects involved in bond characterization (e.g. in connection with pullout tests) in order to verify of reject the size effect predicted by the fracture mechanical approach.

- Further development of advanced pullout test setups which are capable of determining loads and displacements near the matrix surface as well as debonded lengths at different load levels.

8 Acknowledgements

Dr. Henrik Stang acknowledge the support from the Research Program on Fiber Reinforced Cementitious Composites sponsored by The Danish Council for Scientific and Industrial Research and The Danish Ministry for Industry.

Professor Surendra P. Shah appreciates the support of the NSF Science and Technology Center for Advanced Cement-Based Materials.

9 References

Abudi, J. (1989) "Micromechanical Analysis of Fibrous Composites with Coulomb Frictional Slippage Between the Phases." *Mech. Mat.* 8(2 & 3), 103-115.

Aveston, J., Cooper, G. A., and Kelly, A. (1971) "Single and Multiple Fracture" in *The Properties of Fibre Composites, Conference Proceedings, National Physical Laboratory, 4 November 1971* IPC Science and Technology Press Ltd. 15-26.

Atkinson, C., Avila, J., Betz, E., and Smelser, R. E. (1982) "The Rod Pull Out Problem, Theory and Experiment." *J. Mech. Phys. Solids.* 30(3), 97-120.

Barenblatt, G. I. (1962) "The Mathematical Theory of Equilibrium Cracks in Brittle Fracture" *Adv. Appl. Mech.* 7, 56-129.

Barnes, B. D., Diamond, S., and Dolch, W. L. (1978) "The Contact Zone Between Portland Cement Paste and Glass 'Aggregate'Surfaces" *Cem. Concr. Res.* 8, 233-244.

Bartos, P. (1981) "Review Paper: Bond in Fibre Reinforced Cements and Concretes." *Int. J. Cem. Comp. Ltwt. Concr.* 3, 159-177.

Beaumont, P. W. R. and Aleszka (1978) "Polymer Concrete Dispersed with Short Steel Fibers." *J. Mater. Sci.* 13, 1749-1760.

Bentur, A., Diamond, S., and Mindess, S. (1985a) "Cracking Process in Steel Fiber Reinforced Cement Paste." *Cem. Concr. Res.* 15, 331-342.

Bentur, A., Diamond, S., and Mindess, S. (1985b) "The Microstructure of the Steel Fibre-Cement Interface" *J. Mater. Sci.* 20, 3610-3520.

Bentur, A. and Mindess, S. (1990) *"Fibre Reinforced Cementitious Composites"* Elsevier Applied Science, UK.

Bień, J. (1986) "Holographic Interferometry Study of the Steel Concrete Bond in Pull-Out Testing." *Report 1-86-9*, Delft University of Technology, Stevin Laboratory.

Bień, J. and Stroeven, P. (1988) "Holographic Interferometry Study of Debonding between Steel and Concrete." in *Engineering Applications of New Composites* (eds. S. A. Paipetis and G. C. Papanicolaou). Omega Scientific, 213-218.

Bowling, J. and Groves, G. W. (1979) "The Debonding and Pull-out of Ductile Wires from a Brittle Matrix." *J. Mater. Sci.* 14, 431-442.

Budiansky, B., Hutchinson, J. W., and Evans, A. G. (1986) "Matrix Fracture in Fiber-Reinforced Ceramics." *J. Mech. Phys. Solids* 34(2), 167-189.

Budiansky, B. and Amazigo, J. C. (1989) "Toughening by Aligned, Frictionally Constrained Fibers" *J. Mech. Phys. Solids* 37(1), 93-109.

Cook, J. and Gordon, J. E. (1964) "A Mechanism for the Control of Crack Propagation in all Brittle Systems." *Proc. Roy. Soc.* 282A, 508-520.

Evans, A. G. (1984) "Aspects of the Reliability of Ceramics" *Defect Properties and Processing of High-Technology Nonmetalic Materials* (eds. J. H. Crawford, Y. Chen, and W. A. Sibley) North-Holland, 63-80.

Gao, Y. C. (1987) "Debonding along the Interface of Composites." *Mech. Res. Com.* 14(2), 67-72.

Gao, Y., Mai, Y. W., and Cotterell, B. (1988) "Fracture of Fiber-Reinforced Materials" *J. Appl. Math. Phys. (ZAMP)* 39, 550.

Gopalaratnam, V. P. and Shah, S. P. (1987) "Tensile Failure of Steel Fiber Reinforced Mortar." *J. Eng. Mech. ASCE* 113, 635-652.

Gray, R. J. (1983) "Experimental Techniques for Measuring Fibre/ Matrix Interfacial Bond Strength." *Int. J. Adhesion and Adhesives* 3, 197-202.

Gray, R. J. (1984) "Analysis of the Effect of Embedded Fibre Length on Fibre Debonding and Pull-out from an Elastic Matrix. Part 1. Review of Theories." *J. Mater. Sci.* 19, 861-870.

Greszczuk, L. B. (1969) "Theoretical Studies of the Mechanics of the Fibre-Matrix Interface in Composites." In *Interfaces in Composites.* American Society of Testing and Materials, ASTM STP 452, Philadelphia, 42-58.

Gurney, C. and Hunt (1967) "Quasi-Static Crack Propagation" *Proc. Roy. Soc. Lond.* A299, 508.

Hannant, D. J., Highes, D. C., and Kelly, A. (1983) " Toughening of Cement and Other Brittle Solids with Fibres." *Phil. Trans. R. Soc. Lond.* A310, 175-190.

Hillerborg, A. Modéer, M., and Peterson, P.-E. (1976) "Analysis of Crack Formation and Crack Growth by Means of Fracture Mechanics and Finite Elements" *Cem. Concr. Res.* 6(6), 773-781.

Hillerborg, A. (1980) "Analysis of Fracture by Means of the Fictitious Crack Model, Particularly for Fibre Reinforced Concrete." *Int. J. Cem. Comp. Ltwt. Concr.* 2(4), 177-184.

Hutchinson, J. W. (1987) "Crack Tip Shielding by Micro-Cracking in Brittle Solids." *Acta Metall.* 35(7), 1605-1619.

Korczynskyj, Y., Harris, S. J., and Morley, J. G. (1981) "The Influence of Reinforcing Fibres on The Growth of Cracks in Brittle Matrix Composites." *J. Mater. Sci* 16, 1533.

Lawrence, P. (1972) "Some Theoretical Considerations of Fibre Pull-Out from an Elastic Matrix." *J. Mater. Sci.* 7, 1-6.

Laws, V., Lawrence, P., and Nurse, R. W. (1973) "Reinforcement of Brittle Matrices by Glass Fibers" *J. Phys. D.: Appl. Phys.* 6 523-537.

Laws, V. (1982) "Micromechanical Aspects of the Fibre-Cement Bond" *Composites* 13, 145-151.

Leung, K. Y. and Li, V. C. (1990) "Strength-based and Fracture-based Approaches in the Analysis of Fiber Debonding." To be published.

Luk, V.K. and Keer, L.M. (1979) "Stress Analysis for an Elastic Half Space Containing an Axially-Loaded Rigid Cylindrical Rod." *Int. J. Solids Struct.* 15(10), 805-827.

Marmonier, M.F., Desarmot, G., Barbier, B., and Letalenet, J.M. (1988) "A Study of the Pull-out Test by a Finite Element Method." (in French) *J. Theo. Appl. Mech.* 7, 741-765.

Mori, T. and Mura, T. (1984) "An Inclusion Model for Crack Arrest in Fiber Reinforced Materials" *Mech. Mat.* 3, 193-198.

Morrison J.K., Shah S.P., and Jenq Y.-S. (1988) "Analysis of Fiber Debonding and Pullout in Composites." *J. Eng. Mech.* 114(2), 277-294.

Muki, R. and Sternberg, E. (1970) "Elastostatic Load-Transfer to a Half-Space from a Partially Embedded Axially Loaded Rod." *Int. J. Solids and Struc.* 6(1), 69-90.

Nammur, G., and Naaman A.E. (1989) "Bond Stress Model for Fiber Reinforced Concrete Based on Bond Stress-Slip Relationship." *ACI Materials Journal* 86, 45-57.

Needleman, A. (1987) "A Continuum Model for Void Nucleation by Inclusion Debonding" *J. Appl. Mech.* 54, 525-531.

Needleman, A. and Nutt, S. R. (1989) "Void Formation in Short-fiber Composites" In *Advances in Fracture Research* (eds. K. Salama *et al.*) Pergamon Press, 2211-2220.

Nutt, S. R. and Needleman, A. (1987) "Void Nucleation at Fiber Ends in Al-SiC Composites." *Scripta Metall.* 21, 705-710.

Outwater J. D. and Murphy, M. C. (1967) "On the Fracture Energy of Uni-Directional Laminates." In *Proceedings of the 24th Annual Technical Conference of the Reinforced Plastics/Composites Division.* The Society of the Plastics Industry, Washington, D.C. 11-C.

Page, C. L. (1982) "Microstructural Features of Interfaces in Fibre Cement Composites." *Composites* 13, 140.

Palley, I. and Stevans, D. (1989) "A Fracture Mechanics Approach to the Single Fiber Pull-out Problem as Applied to the Evaluation of the Adhesion Strength Between the Fiber and the Matrix" *J. Adh. Sci. Techn.* 3(2), 141-153.

Phan-Thien, N. (1980) "A Contribution to the Rigid Fibre Pull-Out Problem." *Fibre Science and Technology* 13, 179-186.

Phan-Thien, N. and Goh, C. J. (1981) "On the Fibre Pull-out Problem." *J. Appl. Math. Mech. (ZAMM)* 61, 89-97.

Phan-Thien, N., Pantelis, G. and Bush, M. B. (1982) "On the Elastic Fibre Pull-Out Problem: Asymptotic and Numerical Results." *J. Appl. Math. Phys. (ZAMP)* 33, 251-265.

Piggott, M. R., Chua, P. S., and Andison (1985) *Polym. Composites* 6, 242-248.

Piggott, M. R. (1987) "Debonding and Friction at Fibre-Polymer Interfaces. I: Criteria for Failure and Sliding." *Comp. Sci. Techn.* 30, 295.

Pinchin, D. J. and Tabor, D. (1978a) "Interfacial Phenomena in Steel Fibre Reinforced Cement I. Structure and Strength of the Interfacial Region" *Cem. Concr. Res.* 8, 15-24.

Pinchin, D. J. and Tabor, D. (1978b) "Inelastic Behavior in Steel Wire Pull-out from Portland Cement Mortar" *J. Mater. Sci.* 13, 1261-1266.

Pinchin, D. J. and Tabor, D. (1978c) "Interfacial Contact Pressure and Frictional Stress Transfer in Steel Fibre Cement" In *Proc. RILEM Conference, Testing and Test Methods of Fibre Cement Composites.* (Ed. R. N. Swamy) The Construction Press, UK. 337-344.

Selvadurai, A. P. S. (1983) "Concentrated Body Force Loading of an Elastically Bridged Penny Shaped Flaw in a Unidirectional Fibre Reinforced Composite" *Int. J. Fracture* 21, 149-159.

Sigl, L. S. and Evans A. G. (1989) "Effects of Residual Stress and Frictional Sliding on Cracking and Pull-out in Brittle Matrix Composites" *Mech. Mat.* 8, 1-12.

Stang, H. (1985) "The Fibre Pull-Out Problem:; An Analytical Investigation." *Series R, No 204.* Department of Structural Engineering, Technical University of Denmark.

Stang, H. and Shah S. P. (1986) "Failure of Fibre-Reinforced Composites by Pull-Out Fracture." *J. Mater. Sci.* 21, 953-957.

Stang, H. (1987) "A Double Inclusion Model for Microcrack Arrest in Fibre Reinforced Brittle Materials" *J. Mech. Phys. Solids.* 35(3), 325-342.

Stang, H., Li, Z., and Shah, S. P. (1990) "The Pullout Problem. The Stress versus Fracture Mechanical Approach." To be published in *J. Eng. Mech. ASCE.*

Steif, P. S. and Hoysan, S. F. (1986) "On Load Transfer Between Imperfectly Bonded Constituents" *Mech. Mat.* 5, 375-382.

Sternberg, E. and Muki, R. (1970) "Load-Absorption by a Filament in a Fiber-Reinforced Material." *J. Appl. Math. Phys. (ZAMP)* 21, 553-569.

Tvergaard, V. (1989) "Effect of Fibre Debonding in a Whisker-reinforced Metal." *DCAMM* Tecnical University of Denmark, Report no. 400.

Wang, Y., Li, V. C., and Backer, S. (1988) "Modelling of Fibre Pull-out from a Cement Matrix." *Int. J. Cem. Comp. Ltwt. Concr.* 10(3), 143-149.

Wei, S., Mandel, J. A., and Said, S. (1986) "Study of the Interface Strength in Steel Fibre Reinforced Cement-based Composites." *J. Am. Concr. Inst.* 83, 597-605.

Wells J. K. and Beaumont P. W. R. (1982) "Fracture Energy Maps for Fibre Composites." *J. Mater. Sci.* 17, 397-405.

Wells J. K. and Beaumont P. W. R. (1985) "Debonding and Pull-Out Processes in Fibrous Composites." *J. Mater. Sci.* 20, 1275-1284.

SUMMARY OF SESSION 8: FRACTURE TOUGHNESS OF FIBER-REINFORCED CEMENT COMPOSITES

METHI WECHARATANA
Department of Civil and Environmental Engineering
New Jersey Institute of Technology
University Height
Newark, New Jersey 07102

There were three papers presented in this session. The summary and discussion during the workshop are summarized as followed:

1. "RESEARCH CHALLENGES IN FIBER REINFORCED CONCRETE"

Author: Victor C. Li
            University of Michigan

Summary and Discussion

The paper presented the concept of Performance-Property-Processing-Structure Inter-relationship. The author also called for more emphasis on the theoretical modelling of FRC since most of the existing models and formulae are largely obtained empirically. New type of high strength high toughness fiber reinforced cement composites with fiber volume fraction of up to 12-15 % can be obtained. There is a need to understand the physical and mechanical properties of these new products. With the increasing use of high strength concrete in the construction industry, the presence of fiber in the high strength concrete matrix can significantly improve the brittleness of these composites. One of the main objectives of this paper was the introduction of the snubbing mechanism of fiber in the pull-out process of fiber reinforced concrete composites. The snubbing effect which is basically derived from the pulley law varies with the type of matrix and fiber. It is not clear as how the snubbing mechanism affected by the type of fiber such as steel and synthetic. Although the proposed snubbing factor has accounted for the probability of fiber distribution, it is questionable as how the fiber distribution may affect the principal of the pulley law. It should be noted that there are two probability factors in equation [1], one in the term 1/2 due to the 3-D random fiber distribution and another in the

529

S. P. Shah (ed.), Toughening Mechanisms in Quasi-Brittle Materials, 529–537.
© 1991 Kluwer Academic Publishers.

snubbing factor g. Nonetheless, the snubbing effect is believed to improve the post-cracking strength of the fiber reinforced composites.

The introduction of the snubbing mechanism is certainly an interesting concept which might exist in certain type of fiber reinforced cement composites. However, many questions remain unclear as to the generalization of the snubbing mechanism. Some of these remarks are as follows.

1. Increasing maximum pull-out load and energy versus the inclined angle as shown in fig.3 may not always be true especially when the fiber inclined angle is 90 degree. Under this condition it seems that spalling of the matrix tends to dominate the failure.

2. Multiple fibers interaction may affect the snubbing law.

3. The stiffness ratio of the matrix and the fiber plays a significant role on the pull-out mechanism of fiber. This factor should certainly be incorporated into the snubbing law.

4. The statistical factor due to fiber distribution cannot be explicitly separated from the snubbing factor which by itself is also affected by the fiber distribution.

Discussion:

Mindess: The snubbing effect seems to apply only to the straight smooth fiber. For most of the new fibers which often have hooked end, it seems that the snubbing effect may be small as compared to the effect from the hooked end.

Li: This is a very valid point and for the hooked end or deformed fiber the snubbing effect may be small. However, this effect has not yet been measured for the hooked end fiber and it may never be measurable since this is a multiplicative effect.

Stroeven: The distribution of fiber should be easily simplified rather than using the integral concept as presented. Also, the snubbing effect should include fiber shielding and orientation.

Li: The distribution has been included in the integral by the a probability density function. Secondly, the shear and plastic bending effect were reported to be small as compared to the snubbing effect.

Shah: The effect of fiber inclination in the pull-out of fiber from the cement matrix is available and has been reported by Shah and Naaman.

2. "FAILURE CHARACTERIZATION OF FIBER-REINFORCED CEMENT COMPOSITES WITH R-CURVE CHARACTERISTICS"

Author: Yiu-Wing Mai
            University of Sydney

Summary and Discussion:

The paper provided a very detailed summary of the failure characteristics of fiber-reinforced cement composites with emphasis on the R-curve concept. Also discussed are different approaches of modelling the behavior of fiber reinforced composites. In general, the modelling process consists of fiber bridging zone (FBZ) and fracture process zone (FPZ) of the matrix. FBZ and FPZ are normally a function of the specimen size and geometry, the fiber aspect ratio, and the fiber volume fraction. There are many experimental techniques that have been used in detecting the FBZ and FPZ zones. One of these techniques that was developed by the author is the conductive grid method which measures the continuity of the printed circuit. In modelling the fracture characteristics of FRC, one of the most important relationships is the stress-displacement law in the FBZ. Many of the reported stress-displacement relationship in the literature are mostly for steel fiber. Very few are for other types of fiber. Typical stress-displacement law is frequently in the form of normalized stress-displacement relationship. The FPZ of the matrix or the aggregate debonding and interlocking is often neglected since its magnitude is believed to be small when compared to the contribution of the fiber bridging zone. Since the bridging force in the FBZ depends on the crack opening or the crack profile and in turns the crack profile is a function of the bridging force, the modelling process often involves tedious and lengthy iteration. As a result, certain simplifications need to be adopted such as the linear straight crack profile, etc. to simplify the computational process. It has been shown by many investigators that the assumed linear crack profile is more than adequate to achieve a reasonable prediction on the fracture behavior of FRC.

There are many approaches in accounting for the effect of FBZ and FPZ in FRC composites. One of the very promising approaches is the superposition of the stress intensity factor of the applied load and of the closing pressure which normally includes both the effect from the FBZ and FPZ. The resulted R-curve from this method is often recommended as the fracture criterion for FRC. However, the uniqueness of the obtained R-curve is frequently questionable as to its geometry and specimen size dependency. The authors and many other investigators have previously concluded that the K_R concept is definitely specimen size dependent. This dependency is very obvious for the case of notched beam specimen. In general, one can easily see that for the same crack length in a notched beam specimen and a large compact tension specimen, an obvious different in the shape of crack opening is certain in these two types of test specimen, leading to a different R-curve.

In conclusion, the author summarized that R-curve for FRC can be computed from the stress-displacement relationship if such a relationship is available. The author also expressed some doubts over the practical application of R-curve. Finally, the author concluded that there seems to have some similarity of R-curves in cement composites and in ceramics.

Discussion:

Li: Do you find the first crack strength dependent on the initial flaw sizes and with the high enough fiber volume fraction then it becomes independent of flaw size ?

Mai: It is possible. However, We have not done exhausted size effect study on the high fiber volume fraction. The study was only done on the general FRC.

Bazant: How good is the hypothesis of the linear crack profile over the fiber bridging zone and the fracture process zone ?

Mai: The assumption of linear crack profile has been used successfully by our study and many other investigators (Wecharatana and Shah). Some of the studies carried out on double cantilever specimen with the linear crack profile showed that the difference between the actual crack profile and the assumed linear profile is very small compared to the scattered of the experimental data. That is why I question why we have to go to such extent of measuring the fracture process zone.

Atkinson: With these arguments, why can't we just simply use the Dugdale zone with an average stress ?

Mai: Yes, certainly. But his method has no iteration. He just assumed a certain necessary boundary conditions.

3. "CHARACTERIZATION OF INTERFACIAL BOND IN FRC MATERIALS"

Author: H. Stang
Technical University of Denmark

S.P. Shah
Northwestern University

Summary and Discussion:

The paper provides a very thorough summary on different concepts of interfacial bond model. Two major assumptions discussed in the modelling of the interfacial zone are the concept of perfectly bonded interface and the debonded interface. Other models include the fracture mechanics based concept and the stress based debonding concept. Also discussed in the paper are different crack stabilization mechanisms which cover crack blunting, crack path deviation, crack bridging, and crack shielding. No matter what criteria to be used on the interfacial zone, the stress-pull-out displacement relationship of fiber from the cementitious matrix remains to be the most important factor. Some of these reported relationships are obtained from the pull-out of a single fiber from the matrix while others are from pull-out of randomly distributed fibers from cementitious matrix. The role of size effect on the bond characterization is not clear and remains to be investigated.

Among these reported models on the interfacial bond in FRC discussed in this paper, many questions remain unanswered. These problems include:

1.   Where do debonding first start in the fiber, either at the middle or at the end (tip) of the fiber ?   Either of these two places has been used in many reported models.

2.   Can the model developed from debonding of a single fiber be used for the practical randomly distributed fibers in FRC ? It is quite obvious that the statistical function due to fiber distribution definitely need to be included.   However, the existence of multiple fibers often changes the debonding characterization of each single fiber. Some may argue that these effects have been incorporated into the model by the statistical function.   Nonetheless, it is not clear, under that circumstances as how the pull-out process develops. The problem needs to be carefully addressed since the new type of high-performance fiber reinforced composite which consists of high fiber volume fraction of up to 15% is presently being investigated. For these new FRC products, the existing concepts on perfect bonded interface and the debonded interface may have to be revised. Each of the single fiber in the new FRC may not be completely bonded due to the high fiber volume content. Also questionable is the concerns over the commercially production feasibility of this new FRC product.

Discussion:

Mai: What is the effect of synthetic fiber reinforcing concrete ? Because synthetic fibers probably have much larger Poisson's effect than the matrix and you can imagine the situation when the fiber is drawn in such a way that it is completely detached from the surface and that will become like a frictionless debonding process.

Stang: For the case of synthetic fibers, the interfacial work of friction is probably the dominant parameter.

Stroeven: It seems that your approach in modelling the debonding is sophisticated. The more sophisticated the method, the more sensible the results. With a more sensitive method you used the earlier you see the debonding.

Stang: What you are saying is that may be there is no perfect debonding and it is just the sensitivity of the approach used.

SHORT PRESENTATION

Issac Daniel
Northwestern University

Most of the models discussed here assumed the existence of the macrocracks propagating through the fibers or being bridged by fibers. For ceramic composites, what we observed is multiple transverse macrocracks which increase in number. The first critical length is the failure of the matrix, subsequently the debonding of the matrix-fiber interface at the point of the matrix crack and then the fibers break. Experimental results show that it starts with matrix cracks. They do not start from the edge and propagate through the fibers. Actually, they start from the fiber-matrix interface and propagate outward. This is for the uniaxially loaded fiber composites. Soon after the matrix cracks are formed, we observed the debonding and subsequently we see the isolated fibers break. The average spacing can be measured as a function of loading and can be compared with the stress-strain curves. In most discussed behaviors of concrete and ceramic composites, it was assumed that there was a peak and the post-peak strain softening region. In our observation, we found that there is a continuously strain hardening. There is an initial linear portion, then a region which corresponds to matrix crack multiplication and a region of constant tangent modulus which corresponds to the saturation of the matrix cracks and also the debonding of fibers. We also used a shear lag model which we believed to be a more realistic assumption based on our observation. We obtained a closed form solution for the average stresses in the fiber and the matrix as a function only of material parameters and geometric parameters. Based on this model we can predict the applied stress and calculate crack spacing in the matrix and the minimum crack spacing. According to our study for this material the minimum crack spacing is about 35 micron. The observed one is 36 micron. Some other models reported a value of 5 micron and some as large as 400 micron. It is obvious that the interface plays a significant role. Also discussed are the prediction of the stress-strain curves, matrix crack density, debond length. It is not easy to measure debond length experimentally.

GENERAL DISCUSSION:

Rossi In most of the model of FRC discussed, the group effect is not included, is it practical and realistic ?

Mai: I was trying to ask the same kind of question. Since most of the proposed models are theorized and can not be used practically. For instance, the proposed shear lag model cannot be used for synthetic fibers because of the size effect. I wonder "Can we actually observe progressive debonding ?" Not many of these models when used in flexure can predict that the debond load increases as the debond fracture increases. But I do not know whether we can actually see this. In addition, how do we due with other factors such as the snubbing effect proposed by Victor Li.

Sidney: In real fiber reinforced concrete, you see a relatively few fibers sticking out of the fracture surface. It is hard to tell whether they are broken or pulled out. We have no idea what is happening behind most of the other fibers. We do not know whether there is debonding outside the fracture zone. It is enormously difficult for us, the experimentalists, to apply the single fiber models. It is rather difficult to verify these models experimentally for the real fiber reinforced concrete.

Shah: If you are trying to predict any macroscopic response, you will need some material parameters which characterize bond and one of them was the common test, whether it is reinforced concrete, fiber reinforced cement, or fiber reinforced ceramics, such as the pull-out test. Henry's point was that if you have material parameter such as a maximum bond stress. That is not a very valid parameter because it will depend on the geometry and size. In the simplest case, if you use the fracture criteria you need two. Henry also quoted Piet Stroeven's work that the progressive debonding was observed with the holography.

Mai: We can distinguish the two maximum energy based criteria. It does support the fracture mechanics approach.

Shah: For real materials, you need more variables and complexity.

Stroeven: If you look to fiber spacing, we will normally use volume fraction of steel fiber and if you follow the result of Rilem you see the spacing is too large on the average. The spacing in fiber reinforced concrete using 1 or 2% or even less is too large to be included in the model. The effect is small.

Mindess: How about polypropylene fibers ?

Stroeven: Single fiber pull-out test is not a representation of a more realistic random fiber contribution to tensile properties of the materials. There are pull-out characteristics and the shearing characteristics in the tensile test of FRC. Pull-out test with inclined fibers may be better for the tensile test because the single fiber pull-out test ignore the shearing effect.

Shah: Single fiber pull-out shows the snubbing effect.

Li: Concerning the fracture based debonding model or the strength based debonding model, I believe there are experimental data that supports both of these models. Different materials system will prefer different criteria.

Mindess: Every model has its support data.

Li: There are experimental data that do not fit some models.

Mai: Different criteria is good for different materials. For steel wire epoxy resin, the fracture based criteria is much better by using different steel wire diameters.

Stang:	The dependency on the fiber radius can be completely hidden even if the fracture mechanics approach applies if the contribution from the shear stress is fairly large from the friction. You can find both the size effect and no size effect.
Shah:	I think it depends on what you are comparing. For instance, the development of crack spacing, then you will see the influence of various models.
Pagano:	It seems that there is no perfect model.
Swartz:	For the use in engineering application, more experimental work needs to be done, followed by numerical modelling, parameter evaluation, and design recommendation.
Bennison:	We still need the basic science to better understand the problems.
Mai:	Is there a common approach among different materials to improve ductility, fracture toughness and strength of FRC composites?
Li:	Interface bond strength, interface fracture strength, and fiber diameter are generally factors that affect the pull-out force of fibers. If fibers are randomly oriented then there are bending action, fiber matrix interaction and those contained mechanical parameters that have to incorporated. If one can link these parameters to macroscopic composites properties, then we try to optimize these properties. Sometimes, they contradicted one another. For example, if we try to maximize fracture energy, we may not get very high first cracking strength. So one has to decide what one wants to maximize in terms of macroscopic properties of the composites.
Francois:	By looking at these natural mechanisms, we should try to find ways to improve the pull-out mechanism.
Becker:	In ceramic field, there are a number of examples that the model may not be completely correct, but in terms of the desired properties, for ceramic materials some of the microstructures has been changed. Although we do not know all the material parameters, but at least we have some guidance on how to process the materials. This is the step in the right direction.
Bennison:	I agree with Paul Becker's comment
Rick:	For the case of beam under bending, is it possible that there are other mechanisms such as shrinkage crack or something that can not be in the process zone and still shows up and that could be mistaken as the effect in the process zone?
Mai:	For the continuous cracks you referred to, all the break points are on one side of the crack. That was primarily because we have actually

overgrown the crack. So when the crack has opened up, the ligament behind the crack is actually precompressed. The precompression also causes damage. We do the four-point-bend test in uniform bending to locate where the path of the crack is. It can actually fall on either side of the crack. There are certainly a lot of crude assumptions. This is just to show that if the variation is so large, do we really need to do it since it can not be verified by the experimental data.

Bennison: Did you assume each crack (with different crack length) has different R-curve?

Mai: No, I assumed the same R-curve behavior. If you generate the cracks and put the fiber in, you just calculate the K field (the K of the crack tip) and then your condition is that when K is equal to K of the matrix and allow the crack to propagate to the next fiber. So you have the crack growth criteria and each will have the same resistance curve.

Bennison: Based on your experimental results cracks below the tendency point will grow to T_0, then should all crack grow to the same length?

Mai: This is exactly the reason why you get less scattered on the strength because of the high fiber modulus. The number of fibers bridging each crack are not the same and that is why you can not do a smearing model and you got to consider each of the individual fiber force. My idea is that when you have the R-curve effect, you have less scattered.

Shah: New type of fiber reinforced composites is presently being developed in Denmark with up to 6 or 7% of steel fiber volume fraction. The stress-strain behavior of this high volume fraction FRC is essentially linear and then yielding. This response is approaching structural steel. This new materials will allow the addition of a larger amount of conventional reinforcement. This is a completely new materials which might be the new direction of FRC.

Mindess: Would you like to state about to cost of this new materials compared to conventional reinforced concrete?

Shah: We should not compare the cost of the new materials with conventional reinforced concrete but rather with structural steel.

----: Do you have to use special extruding process to produce this new FRC materials?

Shah: This was made with conventional concrete-making process.

Li: This is a good example to understand the concept behind the high first crack strength properties in relation to the microstructure. They generally use very fine fibers. The first crack strength is related to the fiber diameter.

The session was adjourned.

Strain Rate, Thermal, Time and Fatigue Effects

GROWTH OF DISCRETE CRACKS IN CONCRETE UNDER FATIGUE LOADING

D. A. HORDIJK
TNO-Institute for Building
Materials and Structures
Lange Kleiweg 5
NL-2288 GH Rijswijk
The Netherlands

H. W. REINHARDT
Institute for Engineering
Materials
Stuttgart University
Pfaffenwaldring 4
D-7000 Stuttgart 80
West-Germany

ABSTRACT.   Based on the post-peak cyclic tensile behaviour of concrete,
a model for the fatigue behaviour of this material is proposed. In order
to verify the model, crack opening and closure of a discrete crack under
cyclic loading is studied by computer simulation. Input for the model is
an appropriate description for alternating stresses in the post-peak
tensile region. Results of deformation-controlled uniaxial tensile tests
have been used to develop a new model, which consists of continuous
functions. This constitutive model was implemented in the FE-code DIANA.
It is shown by numerical analysis of a beam under four-point bending,
how a crack or softening zone propagates under repeated loading. The
fatigue life of this beam for cyclic loading between load-levels 0 and
95 % of its maximum load bearing capacity, is predicted. The results of
this preliminary analysis are promising and show good agreement with
results that can usually be found in fatigue experiments.

1. INTRODUCTION

Fatigue of concrete has been studied for many years now. Especially,
after the oil crisis in the seventies, research activities in the
Netherlands concerning this topic increased considerably. All these
investigations yielded an enormous amount of data, like Wöhler curves or
S-N curves and Goodman diagrams. Nevertheless, despite all these
efforts, the cause and mechanism of the fatigue behaviour of concrete is
yet still not fully understood. Meanwhile, fracture mechanics had
entered the research field of concrete structures. Especially the
nonlinear fracture mechanics, in which a softening zone ahead of a
visible crack tip is assumed, has shed new light on the behaviour of
concrete structures (see for instance [1]). By using appropriate
material models, the behaviour of most structures can now very well be
predicted. It may be obvious that also new achievements in computational
techniques and the fact that much more powerful computers became
available, has contributed strongly to this progress.
    A fatigue model based on nonlinear fracture mechanics will be
presented. With the model it will be shown that a discrete crack in
concrete will grow under cyclic loading. For the sake of clarity, in

S. P. Shah (ed.), Toughening Mechanisms in Quasi-Brittle Materials, 541–554.

this paper, a discrete crack is defined to be a visible crack with a softening zone ahead of it, or just a softening zone alone. As soon as the concrete strain reaches the strain that belongs to the tensile strength, which amounts about 100 microstrain, a softening zone is created. It may be obvious that such discrete cracks exist in most concrete structures.

Input for the model is the post-peak cyclic behaviour of concrete under tensile loading. Therefore, deformation-controlled uniaxial tensile tests were performed in the Stevin Laboratory of the Delft University of Technology. Based on the results for post-peak cyclic loading, a new constitutive model was proposed [2]. This model was implemented in the DIANA finite element code and a first fatigue analysis was performed. In this paper, the material model will be presented and some model predictions will be compared with experimental results. Furthermore, the results of the preliminary numerical analysis will be presented and discussed.

2. APPROACH

From deformation-controlled uniaxial tensile tests on concrete, it is known that a loading cycle in the post-peak region of the σ-δ relation displays a behaviour as sketched in Fig. 1. It appears that after an unloading-reloading cycle, the curve will not return to the same point of the envelope curve where it started from, but to a point which belongs to a lower stress. This phenomenon is due to the damage which is caused in such an unloading-reloading loop. It may be clear that some mismatch of the crack surfaces will occur at unloading, resulting in a propagation of existing microcracks. From the experiments, it was furthermore concluded that the envelope curve is not significantly affected by the cyclic loading [2].

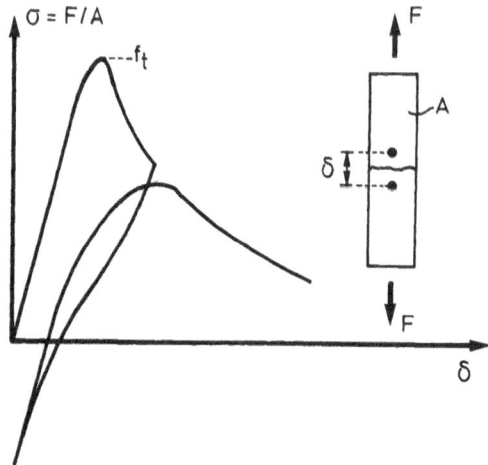

Figure 1. Post-peak cyclic tensile behaviour of concrete.

In Fig. 2 the stress-state in front of a visible crack after n loading cycles (n≥0) is plotted (solid line). Suppose this is part of a total structure which is loaded till a certain load F. If this structure is subjected to another unloading-reloading cycle, then we know by the post-peak tensile behaviour (Fig.1) that the different area in the softening zone cannot attain the same stress as they had at the beginning of this load cycle. This means that for the same maximum external load F, an internal stress-redistribution has to take place. By the dashed line in Fig. 2, a possible stress-distribution after the load cycle is plotted. The actual new stress-distribution depends on the total structure in combination with its load application. Nevertheless, the basic idea is that the softening zone propagates under cyclic or fatigue loading. Furthermore, it can be assumed that deformations increase with the number of load cycles and that this will continue till no longer equilibrium can be found. For the load-deformation relation of the structure, it means that in that case the descending branch is reached. To illustrate this, Fig. 3 shows an assumed load-deformation relation for a certain structure under a continuous increasing deformation. The maximum load bearing capacity is equal to F_{max}. If the same structure is loaded cyclically, probably loops as sketched in Fig. 3 will be found. The loops shift a little to the right, each time a loading cycle is performed. This will proceed till, in a certain cycle, the reloading curve meets the descending branch which is the boundary for combinations of load and deformation that fulfill the requirement of equilibrium. Then failure of the structure occurs.

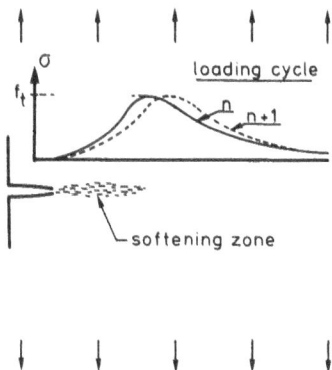

Figure 2. Assumed stress-distribution near a crack; before and after a loading cycle.

In Fig. 3, it can be seen that for a decreasing upper load level of the cycles, the maximum available increase of deformation, which is equal to the deformation at the descending branch minus the deformation at the ascending branch for that load level, increases. This points to an increasing fatigue life for a decreasing upper load level, which tendency corresponds to that one known from S-N curves (see Fig. 4a). Nevertheless it may be clear that it can only be a minor contribution to the increase of the fatigue life N for a decreasing upper load level.

This, namely, yields a S-N relation which is approximately linear, while in reality a logaritmic relation will be found (see Fig. 4a). Therefore, for a decreasing upper load-level of the cycles, the increase in deformation per cycle has to decrease considerably.

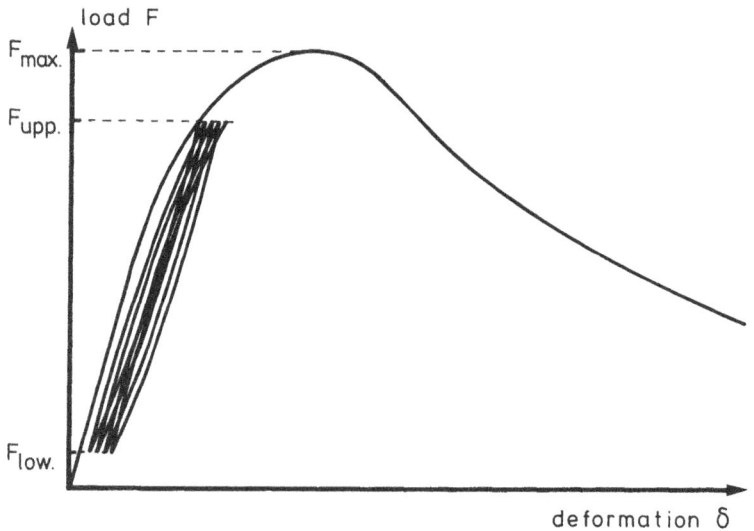

Figure 3. Schematic representation for the load-deformation relation of a structure loaded under continuous increasing deformation or cyclically loaded between $F_{upp.}$ and $F_{low.}$.

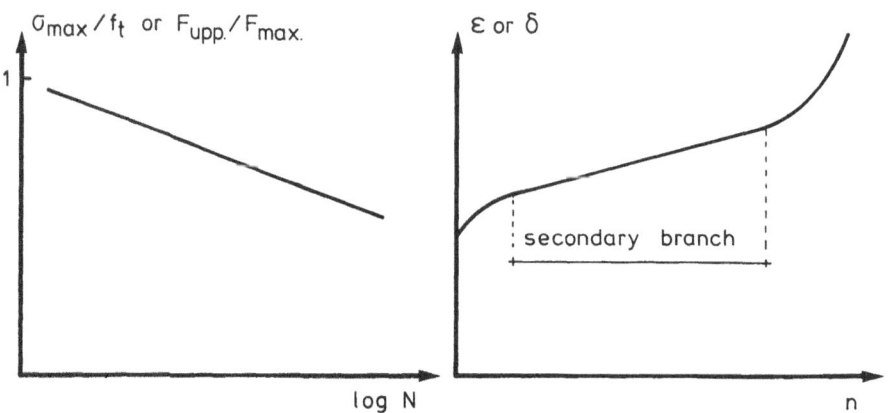

Figure 4. Typical results for fatigue tests on plain concrete; a) S-N curve or Wöhler curve, b) cyclic creep curve (see also [3]).

Experiments show for the increase of strain (or deformation) at the upper stress-level (or load-level) a relation as schematically plotted in Fig. 4b. It is now interesting to see if such a relation can also be found with a numerical analysis. Finally, the fact that experiments showed that there is a strong relation between the cyclic strain rate in the secondary branch of the ϵ-n relation and the number of cycles till failure [3], supports the existence of a failure criterion based on a maximum deformation as it is included in the proposed model.

The idea as presented above, is not completely new. A similar model was proposed in 1983 by Gylltoft [4]. The material model for the post-peak behaviour he used, however, deviated strongly from that one that can be found in experiments. In his material model, he did not assume a unique envelope curve. As a result, the deformation at the descending branch did not act as the failure criterion in his model. Nevertheless, the idea of tackling the fatigue behaviour of concrete, starting from a fracture mechanics point of view was the same. Then, at the Stevin Laboratory some preliminary investigations in this direction were performed (see [5]).

3. CONTINUOUS-FUNCTION-MODEL

For the approach of the fatigue behaviour, as presented above, it was necessary to have an appropriate model for the tensile post-peak cyclic behaviour of concrete. It can be mentioned, however, that such a model is not only required for the fatigue modelling, but for all analyses in which unloading (and reloading) occurs. In this respect it should be noted that a structure may partially unload due to stress-redistributions, while the overall loading increases continuously.

For the post-peak cyclic tensile behaviour of concrete, a limited number of models has been presented in recent years (for a review see [2]). Out of these models, that one by Yankelevsky and Reinhardt [6] simulates appropriately the real material behaviour. It exists of multilinear loops which are constructed by means of so-called "focal-points". A draw back of this model for the fatigue approach is that the increase in crack opening after a loading cycle is independent of the lower stress-level of this loop. As a result, the influence of the lower-stress level of loading cycles on the fatigue life cannot be studied with this material model. Furthermore, for convenience of implementation in numerical programmes, it was preferred to use continuous functions instead of multilinear descriptions. For these reasons, a new model, that can easily be implemented as a mathematical subroutine in numerical programmes, was proposed. The model gives a description for the stress-crack opening relation, while crack opening is defined according to the Fictitious-Crack-Model by Hillerborg et al. [7]. It consists basically of three continuous functions (see Fig. 5) and is therefore called "Continuous-Function-Model" (CFM). Here, the basic equations will be presented. For a complete description of the model, including descriptions for inner loops (reversals within a loop), the reader is referred to [2].

The three basic equations are, respectively, empirical expressions for the unloading curve (I), the gap in the envelope curve (II) and the reloading curve (III). It has been chosen to use only characteristic points in the σ-w relation $(f_t, w_c, w_{eu}, \sigma_{eu}, \sigma_L)$ as variables in the

546

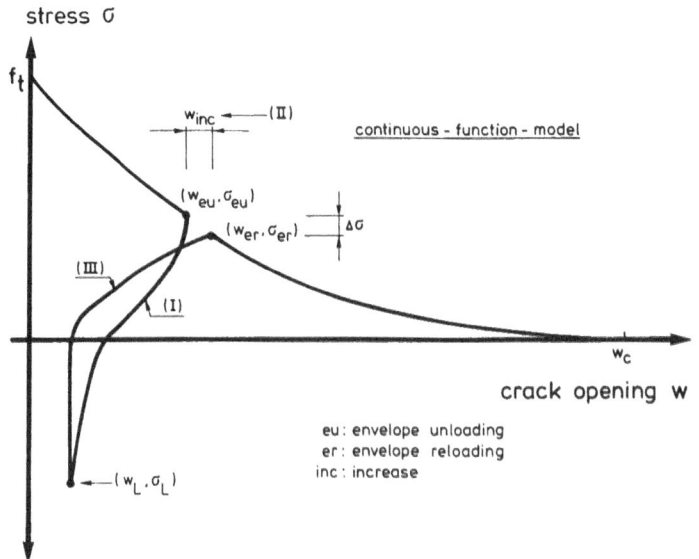

Figure 5. Set-up for the crack cyclic behaviour model [2].

expressions, while w_c is defined as $5.14G_F/f_t$. The expressions that were chosen are based on a close inspection of the experimental results. The softening relation is described by the following expression:

$$\frac{\sigma}{f_t} = \{1+(c_1\frac{w}{w_c})^3\}\exp(-c_2\frac{w}{w_c}) - \frac{w}{w_c}(1+c_1^3)\exp(-c_2) \qquad (1)$$

with $c_1=3$, $c_2=6.93$ and $w_c=5.14G_F/f_t$.

Starting from point (w_{eu},σ_{eu}) at the envelope curve, the unloading curve is determined by:

$$\frac{\sigma}{f_t} = \frac{\sigma_{eu}}{f_t} + \{\frac{1}{3(w_{eu}/w_c)+0.4}\}[0.014\{\ln(\frac{w}{w_{eu}})\}^5 - 0.57/(1-\frac{w}{w_{eu}})] \qquad (2)$$

When reloading starts from a lower stress level σ_L, the gap in the envelope curve is known by an expression for w_{inc}:

$$\frac{w_{inc}}{w_c} = 0.1\frac{w_{eu}}{w_c}\{\ln(1+3\frac{\sigma_{eu}-\sigma_L}{f_t})\} \qquad (3)$$

The coordinates of the returning point at the envelope curve (w_{er},σ_{er}) can now be found with

$$w_{er} = w_{eu} + w_{inc} \qquad (4)$$

and eq. 1. The reloading curve, starting from the point at the lower stress level (σ_L, w_L) up to point (w_{er}, σ_{er}) at the envelope curve, is determined by:

$$\frac{\sigma}{\sigma_L} = 1 + [\frac{1}{c_3}\{\frac{(w-w_L)}{(w_{er}-w_L)}\}^{0.2c_3} + \{1-(1-\frac{(w-w_L)}{(w_{er}-w_L)})^2\}^{c_4}](\frac{c_3}{c_3+1})(\frac{\sigma_{er}}{\sigma_L} - 1) \quad (5a)$$

with for the coefficients c_3 and c_4:

$$c_3 = 3(3\frac{f_t-\sigma_L}{f_t})^{(-1-0.5\frac{w_{eu}}{w_c})} \{1-(\frac{w_{eu}}{w_c})^{(\frac{0.71f_t}{f_t-\sigma_L})}\} \quad (5b)$$

$$c_4 = [2(3\frac{f_t-\sigma_L}{f_t})^{-3} + 0.5]^{-1} \quad (5c)$$

4. COMPARISONS BETWEEN THE MATERIAL MODEL AND EXPERIMENTAL RESULTS

Predictions by the Continuous-Function-Model have been compared with experimental results of deformation-controlled uniaxial tensile tests. Therefore a test series was performed in which post-peak loops between the envelope curve and five different lower stress-levels (σ_L is approximately 1, 0, -1, -3 and -15 MPa) were performed. For a complete description of these comparisons, the reader is referred to [2]. It could be concluded that the model represents the material behaviour very well. To illustrate this, Fig. 6 shows the results for one experiment. In this figure also the predictions by the Focal-Point-Model are plotted.

As stated above, the relation between the gap in the envelope curve and the lower stress-level should be modelled properly for studying the influence of the lower load-level of the loading cycles on the fatigue life of a structure. The gap in the envelope curve can be represented either by the increase in crack opening w_{inc} or the stress drop $\Delta\sigma$ (see Fig. 5). Fig. 7 shows the comparison between the model predictions and experimental results for the relative stress-drop as function of the relative crack opening. As can be seen, the experiments show the stress drop to increase for a decreasing lower stress level, which is also included in the Continuous-Function-Model. The fact that $\Delta\sigma$ is independent of σ_L in the Focal-Point-Model can also clearly be seen in this figure.

5. NUMERICAL ANALYSIS OF A BEAM UNDER FATIGUE LOADING

In order to verify the proposed material model, a number of four-point-bending tests on plain notched beams were performed and simulated with the finite element code DIANA. Results of these analyses are presented in [8]. One of these beams was used to perform a fatigue analysis. Specimen dimensions and loading arrangement are schematically plotted in Fig. 8a. The notch depth was 50 mm which is half of the beam height. Symmetry was used for the applied FE-idealization (see Fig. 8b). For the concrete, eight-noded and six-noded quadratic elements have been used, while six-noded interface-elements were applied to model the crack. For the interface-elements, the vertical displacements of each two nodes that were positioned besides each other were assigned to be equal. In horizontal direction, the Continuous-Function-Model becomes active as soon as the tensile strength is reached. For reasons of comparison with experiments, the vertical displacement of point A (see Fig. 8) was used for the deflection of the beam.

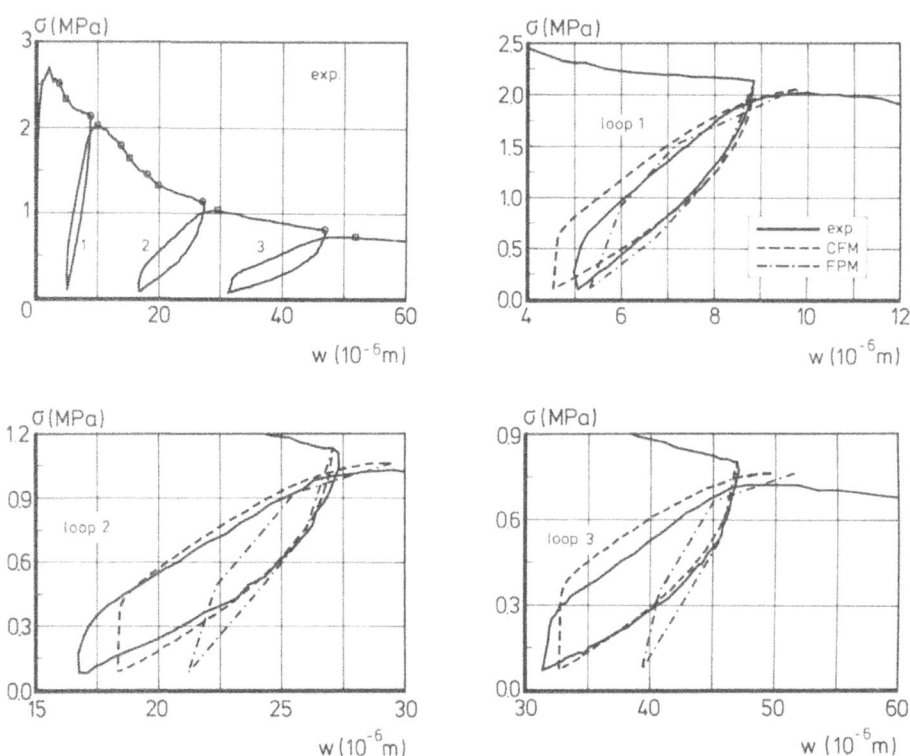

Figure 6. Experimental stress-crack opening relation and the predictions by the Continuous-Function-Model (CFM) and the Focal-Point-Model (FPM) [2].

The elastic concrete properties were taken as: Young's modulus E=38000 MPa and Poisson's ratio ν=0.2. The softening parameters, tensile strength and fracture energy were: f_t= 3.0 MPa and G_F=125 J/m^2.

5.1. Load-deflection relations

First an analysis for a continuous increasing deformation was performed. In this analysis the vertical deformation at the loading point was used as control-parameter. The load-deflection relation that was obtained is plotted in Fig. 9 by the dashed line. Here, only the result up to a deflection of 0.15 mm is plotted. For the force at peak load F_{max}, a value of 1292 N was found.

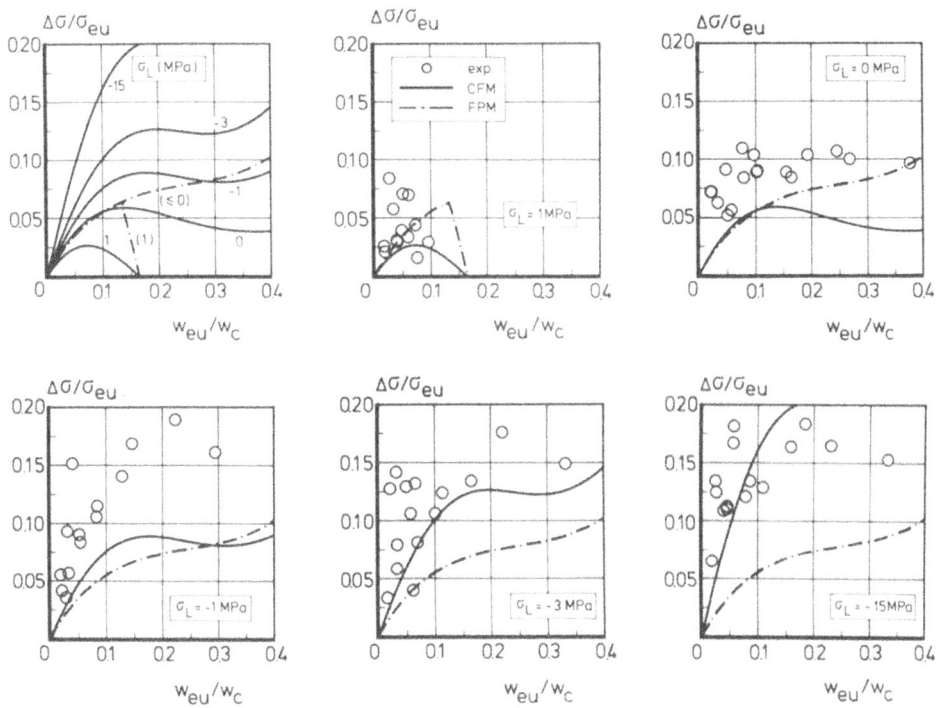

Figure 7. Experimental results and model predictions for the stress drop $\Delta\sigma$ at the envelope curve [2].

For the fatigue analysis, the same input was used as for the quasi-static analysis, while in this case the load was applied by load increments (load-controlled). First, load steps up to a maximum of 95% of the peak load were performed. Subsequently, the beam was unloaded till zero load again, followed by a reloading till the same upper value of 0.95 F_{max}. For the unloading as well as for the reloading path, nine load steps of equal size were performed. This procedure was repeated till it was no longer possible to find a new equilibrium. The result of this analysis as far as the load-deflection relation is concerned can be seen in Fig. 9. It appeared that 35 unloading-reloading loops could be performed, while in the subsequent reloading part it was not possible to find a solution at the upper load-level. It can be assumed that for smaller loading steps the descending branch as found in the quasi-static analysis would have been reached at a load level smaller than F_{upp}. Therefore, it can be concluded that the descending branch of a quasi-static analysis was an envelope curve and failure criterion for the fatigue loading.

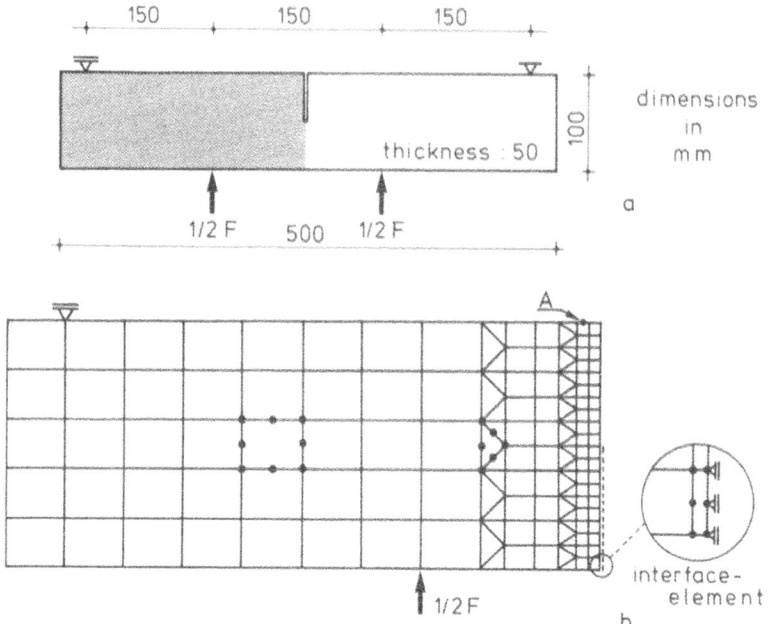

Figure 8. Specimen dimensions of the four-point bending specimen used for the analysis (a) and applied FE-idealization (b) [8].

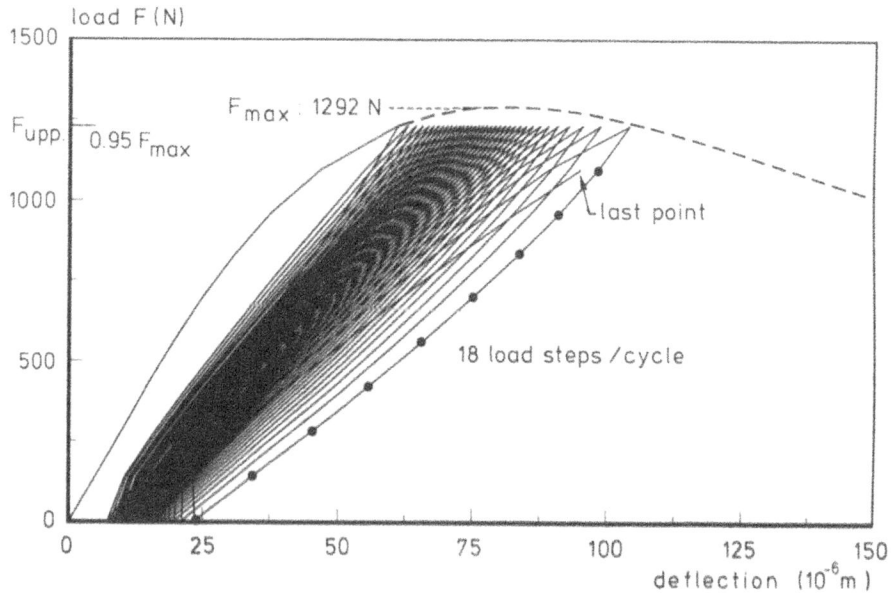

Figure 9. Load-deflection relations predicted by the FE analyses [8].

5.2. Stress-distributions

For a number of loops, the stress-distributions at the upper ($F=F_{upp}$) and lower ($F=0$) loading points are plotted in Fig. 10. First of all, it can be seen that the length of the softening zone increases with the number of load cycles. Furthermore, the stress-distributions at zero load show that important residual stresses are active after a preloading till 95% of the maximum load. It appears that the tensile residual stresses in the centre part of the cross-section increase with the number of cycles performed.

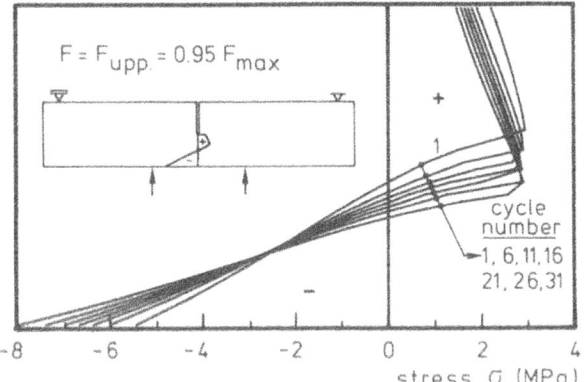 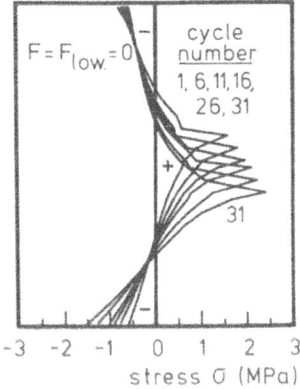

Figure 10. Stress-distributions at the upper and lower load level for a number of loading cycles [8]

5.3. Cyclic creep curve

As already mentioned before, fatigue tests usually show a particular shape for the increase of strain or deformation at the maximum load level as function of the number of cycles performed (Fig. 4b). Normally strain is plotted as function of number of cycles, but it may be obvious that the same holds true for deformation, or, as for this analysis, deflection. Characteristic for a cyclic creep curve is that first the increase of deformation per cycle decreases with increasing n. This first branch of the curve is followed by a secondary branch where this increase is constant. Just before failure occurs, the increase of deformation per cycle increases rapidly.

The deflection as function of number of loading cycles for the performed analysis is plotted in Fig. 11. As can be seen, the shape of the curve resembles very well the cyclic creep curve as found in experiments.

552

6. DISCUSSION

In the model for the fatigue behaviour of concrete as presented, starting point is the existence of a softening zone in the structure. This means that if the material of a structure behaves elastically till the upper load level and that no softening zone existed in advance, the fatigue model will not become active. First of all the authors believe that for most structures, softening zones, for instance, due to differential drying shrinkage, exist in concrete structures. Furthermore, the authors believe that a similar model can also be used to study the fatigue behaviour of a structure without a softening zone as defined above. For that purpose, however, the structure should be modelled at a lower level. In fact, the model was presented at a macro-level. Concrete is treated as a homogeneous isotropic continuum, in which a crack will arise after the strain belonging to the tensile strength is passed. At a lower level, the concrete can be modelled as a two-phase material, in which larger aggregates are embedded in a mortar matrix. In this matrix, softening zones will exist, for instance near aggregates, due to a difference in stiffness between the matrix material and the aggregate (see Fig. 12). It may be obvious that an analysis with a structure modelled in this way demands for extremely powerful computers. Here, however, it is only intended to show that the model can be used at different levels of modelling.

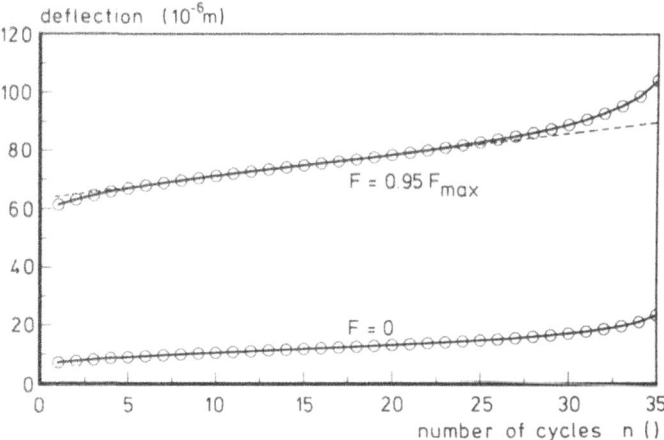

Figure 11. Deflection as function of number of cycles [8].

Now the model can further be used to study the relation between F_{max} and N (S-N curve) or the effect of the lower load level on N. The analysis which is presented, however, is very computer-time consuming. Therefore, for such a study a more simplified modelling of the concrete outside the crack area is required. Nevertheless, even then simulating all loadings cycles (thousands or millions) will not be possible. If, however, in one way or another, the secondary branch can be reached directly, then an estimation for the number of cycles till failure can be obtained by simulating only a limited number of cycles. Yet, it looks not to be possible to do this with the presented model. Further study is this direction is required.

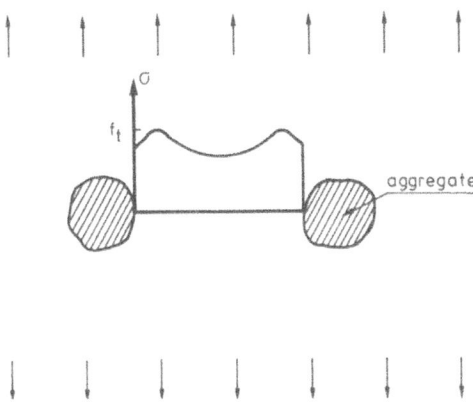

Figure 12. Assumed stress-distribution between the aggregates in a two-phase model of concrete.

7. CONCLUSIONS

1 The proposed approach for the fatigue behaviour of plain concrete looks promising.

2 The constitutive model for the stress-crack opening relation of concrete represents the actual material behaviour, including the relation between stress drop and the lower stress-level of the loading cycle, very well.

3 By numerical simulation, it has been demonstrated that the softening zone will propagate under repeated loading.

4 Failure criterion in the presented modelling is the deformation corresponding to the deformation at the descending branch obtained by a quasi-static analysis.

5 For the increase of deflection as function of number of cycles, a curve was obtained by the numerical analysis which resembles the cyclic creep curves, as usually found in fatigue experiments, very well.

554

ACKNOWLEDGEMENTS

The experiments were carried out at the Stevin Laboratory of the Delft University of Technology and the calculations have been carried out with the DIANA finite element code of the TNO Institute for Building Materials and Structures (TNO-IBBC). Regarding the numerical analyses, the authors are indebted to J.G. Janssen who carefully carried out the analyses and to dr. ir. J.G. Rots for his support. This investigation was partly supported by STW (Netherlands Technology Foundation) and by CUR (Netherlands Centre for Civil Engineering, Research, Codes and Specifications).

REFERENCES

1 Elfgren, L. (Ed.), Fracture mechanics of concrete structures; From theory to applications. Report of RILEM TC 90-FMA, Chapman and Hall, 1989, 407 pp.

2 Hordijk, D.A., Doctoral Thesis, Delft University of Technology, The Netherlands, in preparation.

3 Cornelissen, H.A.W., State-of-the-art report on fatigue of plain concrete. Stevin report 5-86-3, Delft University of Technology and Chapter 3, CEB Bulletin, 188, "Fatigue of concrete structures", pp. 85-142.

4 Gylltoft, K., Fracture mechanics models for fatigue in concrete structures. Doctoral thesis, Lulea University of Technology, 1983, 210 pp.

5 Reinhardt, H.W., Cornelissen, H.A.W. and Hordijk, D.A., Tensile tests and failure analysis of concrete. ASCE J. Structural Eng., 112(11), 1986 pp 2462-2477.

6 Yankelevsky, D.Z. and Reinhardt, H.W., Uniaxial behaviour of concrete in cyclic tension. ASCE J. of Structural Eng., 115(1), 1989, pp. 166-182.

7 Hillerborg, A., Modeer, M. and Petersson, P.E., Analysis of crack formation and crack growth in concrete by means of fracture mechanics and finite elements. Cement and Concrete Res., 6, 1976, pp. 773-782.

8 Janssen, J.G., Mode I fracture of plain concrete under monotonic and cyclic loading; Implementation and evaluation of a constitutive model in DIANA. Graduate Thesis, Delft University of Technology, The Netherlands, IBBC-TNO report nr. BI-90-110.

CREEP AND CREEP RUPTURE OF STRUCTURAL CERAMICS

S.M. WIEDERHORN, B.J. HOCKEY and T.-J. CHUANG
National Institute of Standards and Technology
Gaithersburg, MD 20899, USA

ABSTRACT. Structural ceramics are often two phase materials, in which rigid refractory grains, fibers, or whiskers are bonded by a less refractory matrix. At elevated temperatures, creep occurs by deformation of the matrix, resulting in the localization of stresses along grain boundaries, followed by cavitation and eventually by structural failure of the ceramic. The sequence of events prior to failure depends on the grain size and the amount of bonding matrix within the solid. Large amounts of matrix phase reduce constraints at grain boundaries, and thus the stresses that cause cavitation. Large grains, fibers or whiskers provide easy paths for crack growth, so that, once initiated, cracks readily propagate to failure. In structural ceramics, failure time can be expressed as a power function of the creep rate; the coefficient and exponent of the power function are determined by the failure mechanism. In this paper, models for creep rupture of ceramics are presented for cavity coalescence and for crack growth as primary failure processes. The model for cavity coalescence, in particular, is used to rationalize creep rupture data for a two phase ceramic that fails by cavity coalescence.

1. INTRODUCTION

Failure of structural ceramics at elevated temperatures can be classified into two general types depending on the time required for failure [1-5]. For relatively short failure times, failure is a consequence of crack growth from preexisting defects. For relatively long failure times, failure results from distributed cavities that are nucleated by the creep process. A threshold stress separates the two types of failure [2]. Most theoretical and experimental work on ceramics at elevated temperatures have dealt with crack growth as the principal mode of failure and a series of excellent reviews and research papers are available on this subject [6-8]. The subjects of cavity nucleation and cavity growth in ceramic materials have also been investigated and are similarly referenced in the literature [9-12]. With few exceptions [13], an area of the ceramics literature that has been neglected is cavity coalescence, which pertains to the sequence of events that transforms cavities into microcracks.

Cavity coalescence is the main subject of this paper. First, the steps leading to the formation of cavities and/or microcracks are discussed and are shown to be critically dependent on the microstructure of the ceramic. Whether microcracks, or isolated cavities are formed during cavitation is shown to depend on the ratio of the volume of bonding matrix to the volume of the refractory grains, and also on the size and shape of the refractory grains in the

555

S. P. Shah (ed.), Toughening Mechanisms in Quasi-Brittle Materials, 555–576.

structural ceramic. Next, it is shown that the creep life of two phase ceramics can be expressed as a power function of the strain rate, independent of temperature or applied stress. This relation between creep rate and failure time was first proposed by Monkman and Grant [14] for structural alloys. It also provides a convenient representation for analyzing creep rupture data in structural ceramics. Finally, two theories to rationalize the dependence of rupture life on creep rate are developed. Of these, a theory modelled on cavity coalescence is shown to provide a satisfactory description of the stress rupture behavior of a ceramic that forms isolated cavities during deformation.

2. CAVITY MORPHOLOGY

All materials discussed in this paper are two phase ceramics, in which refractory grains or whiskers are surrounded and bonded by a less refractory matrix. The grains are assumed to be completely rigid so that all deformation occurs within the matrix. This assumption is valid for a wide class of structural ceramics over a range of conditions[1]. Despite differences in the microstructures of these two phase materials, certain common features are apparent both in the mode of cavitation, and in the way cavities grow in these materials. First, cavities always seemed to form at interfaces between the rigid grains and the more ductile matrix. Most often, cavities are observed to form within the interfacial regions separating closely spaced grains, or at triple junctions. Interfacial cavitation is illustrated in figure 1 for two grades of alumina [16], and in figure 2 for a grade of siliconized silicon carbide [17-18]. Cavities form at such locations largely because of physical limitations on the rate of flow of the matrix between the grains as they move apart [19, 20]. If the matrix does not flow rapidly enough, high dilational stresses develop within the matrix, eventually resulting in the formation of cavities. Since the highest stresses develop at the midpoint between two planes that are being pulled apart, cavitation is expected to occur preferentially between grains that are closely spaced.

A quantitative estimate of the stresses that develop between two grain interfaces can be obtained by modeling the interfaces by two circular cylinders of radius G, separated by a distance h, figure 3. As the cylinders are forced to move apart, fluid is sucked into the space between the cylinders. Assuming no slip between the fluid and the disk, the pressure, p, that develops between the cylinders is given by the following relation [21]:

$$p = p_o - (3 \cdot \mu/h^3) \cdot (dh/dt) \cdot (G^2 - g^2) \tag{1}$$

---

[1] As the refractoriness of the matrix phase increases relative to that of the grains, higher temperatures have to be used to deform the solid. As a consequence, deformation and hence cavitation may also occur within the grains, at the grain boundaries. Such cavitation has been observed in one of the newer grades of silicon nitride, in which small lenticular shaped cavities decorate the grain boundaries of the solid after creep at elevated temperatures [15]. The morphologies of these cavities is essentially different from those discussed below.

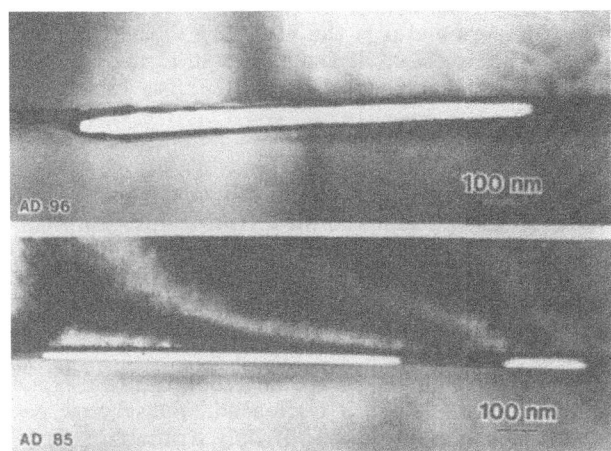

Figure 1. Transmission electron micrographs showing cavities formed within narrow grain boundaries in (a) AD-96 and (b) AD-85. In both grades of alumina, grain boundary cavitation represents the initial stage in creep crack formation. Microstructural analysis indicates grain sizes of ≈9μm and ≈3μm for AD-96 and AD-85, respectively. AD-96 contains 96 wght. % aluminum oxide, whereas AD-85 contains 85 wght. % aluminum oxide. The volume fraction of intergranular binding phase ranged from ≈0.2 to ≈0.25 for the AD-85 and from ≈0.06 to ≈0.08 for the AD-96. Data from reference 16.

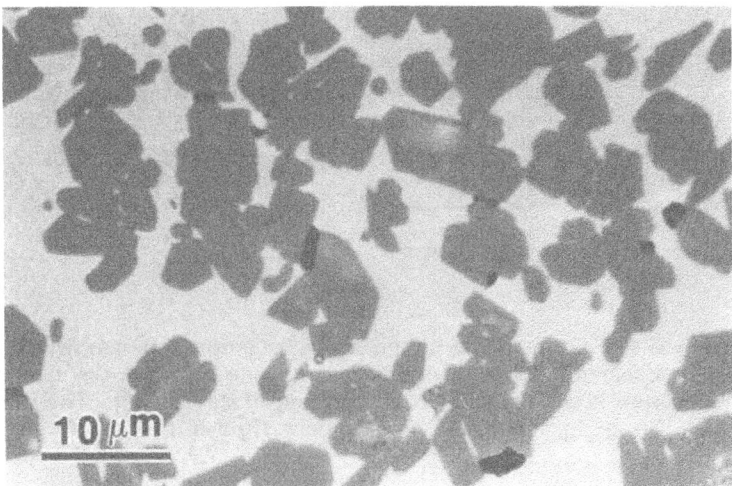

Figure 2. Cavity formation in Carborundum KX01. This material contains ≈33 vol. % silicon. Cavities are always located at Si/SiC interfaces, most often between two closely spaced SiC grains. As the cavities grow from the narrow space between the grains, they arrest on encountering a larger pool of silicon. The applied tensile stress was vertical (top to bottom) in the figure.

where g is the distance from the center of the cylinder, p_o is the pressure in the fluid surrounding the cylinders and μ is the viscosity of the fluid. The maximum absolute pressure occurs when g = 0. If the grains in a material are considered to be rigid cylinders of height 2G, then dh/dt $\approx$ 2G$\cdot\dot{\varepsilon}$, so that

$$p = p_o - 6\cdot\dot{\varepsilon}\cdot\mu\cdot(G/h)^3 \qquad (2)$$

For the grade of siliconized silicon carbide shown in figure 2, a viscosity of 10^{10} Pa·s is estimated at 1300°C and 100 MPa from the creep map for silicon[22]. For a creep rate of $\approx 10^{-8}$ s^{-1}, and a SiC grain size of 5µm, pressures >1 GPa are estimated for values of h <21nm. Grain boundary separations of this magnitude are not unusual for SiC bonded by Si, suggesting that cavitation should be relatively easy once the creep rate has exceeded a threshold value determined by the grain separation.

Once nucleated, a cavity tends to grow along the grain interface until it covers the facet of the grain on which it nucleated. Subsequent growth then depends on the microstructure of the solid. If only a small amount of bonding phase is at the grain interface, the cavity will continue to grow from one grain to the next along grain interfaces, thus forming a crack. Such crack formation

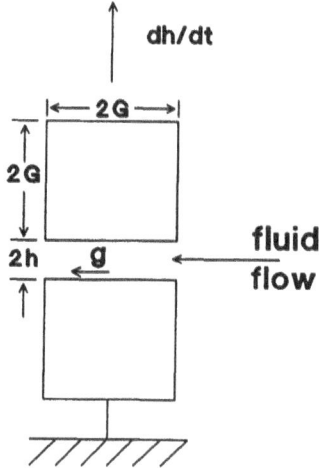

Figure 3. Model of two grains being pulled apart as a consequence of the creep process. Fluid is assumed to flow freely into the space between the two grains, establishing a pressure gradient, which is the cause of cavitation. The bonding matrix is assumed to be approximated by a viscous fluid in this model.

---

[2] In this estimate a grain size of 5 µm is assumed for the silicon carbide grains. The viscosity of silicon is calculated as the ratio of the applied stress to the creep rate at 1300°C. As silicon is a power-law material, it is recognized that equation 2 can only approximate the true situation.

via rapid interfacial cavity growth was typically observed in alumina containing ≈8 vol. % glass, figure 4. Similar creep crack formation due to small amounts of glass at grain boundaries has also been observed by Evans et al. [23] and by Wilkinson [24] in studies on nominally "pure" alumina containing isolated pockets of glass. In these materials, the glass wets the grain boundaries of the aluminum oxide grains only in the vicinity of the glass pockets. Under stress, cracks readily generated from the glass pockets, then grew into the bulk of the alumina where they arrested. Thus, in small quantities, glass apparently "embrittles" aluminum oxide at elevated temperatures, permitting the nucleation of sizable cracks within the material as a consequence of creep cavitation.

As the volume fraction of bonding phase within the solid is increased, cavities tend to stabilize once they have propagated across a facet. This is illustrated for a grade of siliconized silicon carbide containing 33 vol. % silicon (Carborundum KX01), in which isolated damage occurs as a consequence of cavitation, figure 2. After a nucleus is formed between two narrowly spaced grains, cavities generally grow until they intersect the larger pockets of silicon on each side of the boundary. The cavity then blunts and arrests. Cavities are formed at random throughout the gauge section and, with continued creep, the gauge section becomes filled with these cavities. For this type of cavitation, cavity coalescence is required to form a crack that then propagates through the material causing failure, figure 5.

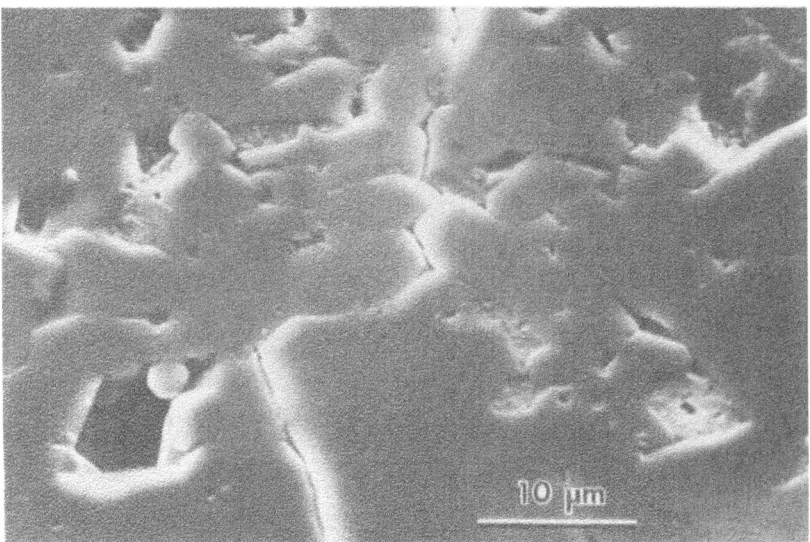

Figure 4. Scanning electron micrograph illustrating representative portion of a creep crack developed at the tensile surface of a flexure bar in AD-96 alumina. Cavities in this material are not arrested after nucleation, but grow as cracks along grain interfaces. Figure taken from the work presented in reference 16.

560

Large grains, fibers or whiskers in an otherwise fine grain material promote crack growth over cavity coalescence as a fracture mechanism. The process of crack growth is illustrated in a grade of siliconized silicon carbide (Coors SCRB 210) that contains a mixture of ≈ 50 µm and ≈ 3 µm grains of SiC bonded by ≈ 18 vol. % Si, figure 6 [25]. Generalized cavitation does not occur for this material. Cavities form along the larger grain boundaries and then link up to form a crack. Once the crack has been formed, cavitation occurs preferentially from the crack tip and the material fails by crack growth. Tensile specimens that fail by this process, show no evidence of generalized cavitation within the gauge section. Instead, one or two cracks such as those shown in figure 6 are observed; these usually form at the "higher" stress regions of the gauge section[3].

A similar crack nucleation process is observed in whisker reinforced silicon nitride [27]. In this material, cavities nucleate from the large whiskers and then propagate into the silicon nitride matrix forming a crack that eventually results in component failure, figure 7a. Below a stress threshold, micro-cracks, such as that show in figure 7a, do not grow in this material. Cavities are observed at the SiC-whisker interface, but these tend to have blunted morphologies, indicating limited growth. Above the crack growth threshold, high densities of

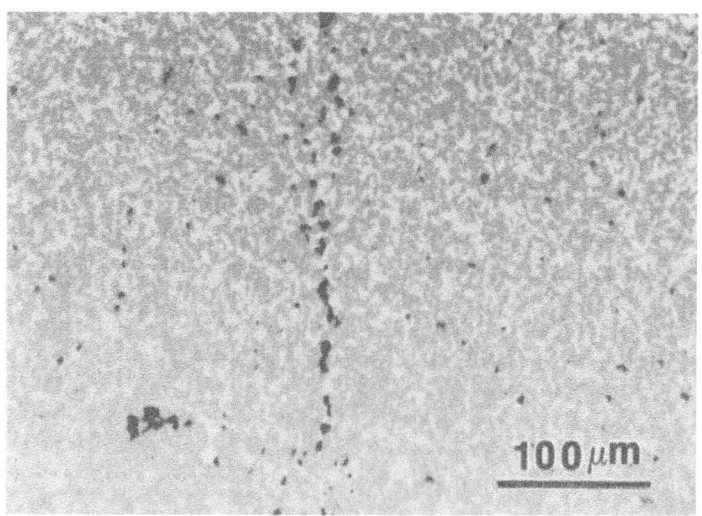

Figure 5. Cavity coalescence in Carborundum KX01. Very few of these coalesced features are observed in the gauge section after testing, suggesting a low probability event. The applied tensile stress is horizontal in this figure.

---

[3] To reduce stress concentrations in the gauge section, a gradual reduction of the section from the gripping area was used in the design. Never-the-less, an enhancement of stress of approximately 3% is expected [26]. Apparently, this enhancement is sufficient to nucleate cracks at the ends of the gauge section.

microcracks are observed, all of them slowly growing as a function of time, figure 7b. As the applied stress is increased, however, fewer microcracks are observed. Apparently, the crack growth sensitivity of the material to applied stress is greater than the cavity nucleation sensitivity of the material to applied stress, so that a crack, once nucleated, rapidly grows to failure, leaving very little damage behind in the rest of the gauge section.

3. CREEP RUPTURE LIFETIME

Although the lifetime of structural ceramics depends on the exact details of the mechanism responsible for failure, some characteristics of the creep rupture process seem to apply regardless of failure mechanism. Creep rupture studies on a number of structural ceramics indicate, for example, that failure strain is greatest for specimens tested at the slowest strain rates. This relation between failure strain and strain rate is illustrated in figure 8 for three, two phase ceramics. Strains to failure in tension for two phase ceramics tend to be <2%, which means that these materials do not exhibit extended tertiary creep behavior. In fact, primary creep in these materials can be exaggerated, lasting for thousands of hours, when processes such as devitrification occur at grain interfaces [15, 27].

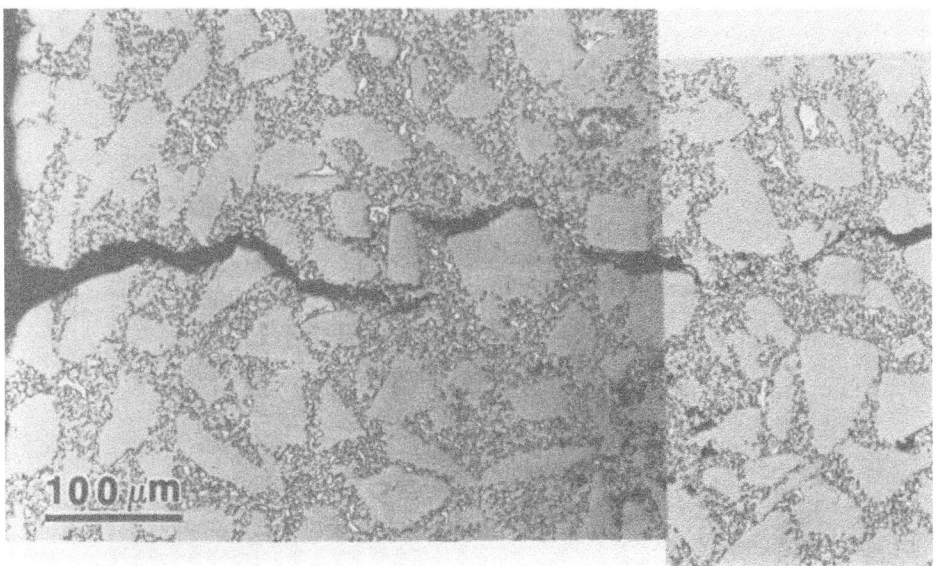

Figure 6. Microcrack formation in Coors SCRB 210. Cavitation appears to occur preferentially at the interfaces of large grains within the structure. This material contains ≈18 vol. % silicon in which are imbedded a mixture of ≈50μm and ≈3μm grains SiC.

562

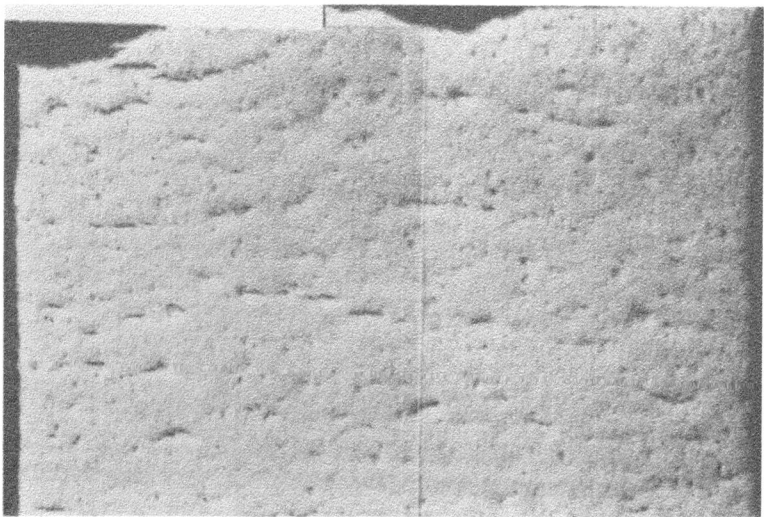

Figure 7. Cavity nucleation in the whisker reinforced materials occurs at the vitreous interfaces bordering the SiC whiskers (a). Cavities grow from the whisker interfaces into the bulk of the material, where they link-up with other cavities eventually forming life-limiting cracks. As can be seen in the polished gauge section of the whisker-reinforced Si_3N_4 specimen in (b), cavities form uniformly throughout the gauge section. The gauge section width is ≈2.5mm. Figure taken from reference 27.

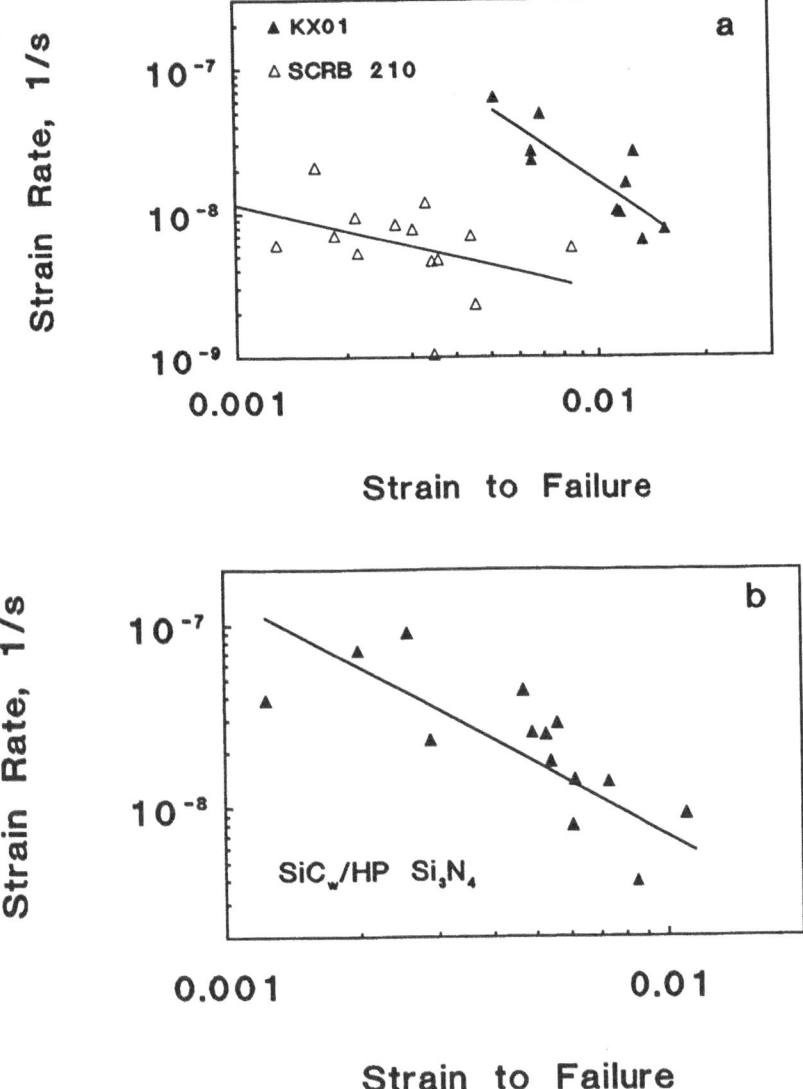

Figure 8. Relation between strain to failure, ε_f, and strain rate, $\dot{\varepsilon}$ (a) two commercial grades of siliconized silicon carbide; (b) SiC_w reinforced Si_3N_4. For all three materials, a greater strain is sustained at lower rates of strain.

With regard to the lifetime of two phase ceramics, a useful observation is that a logarithmic plot of the creep rate versus rupture time often yields a single curve for a material regardless of temperature or applied stress, figure 9. This type of plot was first suggested by Monkman and Grant [14] for the creep rupture of metals. Their data can be represented by a power law relation between rupture time, t_f, and minimum creep rate, $\dot{\varepsilon}$:

$$t_f = C \cdot \dot{\varepsilon}^{-m}, \tag{3}$$

where C and m are empirical constants. The strain rate exponent, m, had a value that ranged from 0.77 to 0.93 for metals studied by Monkman and Grant [14]. The curves for the data presented by Monkman and Grant were independent of stress and temperature, but exhibited a small dependence on the ductility of the metal at fracture. Monkman and Grant noted that most of their data could be represented adequately by setting m=1, which suggests a constant strain to failure, provided most of the creep strain can be accounted for by strain that develops at the minimum creep rate. A value of 1 for m was observed by Kossowsky et al. [26] in an early study of the tensile creep rupture behavior of hot-pressed silicon nitride. In studies conducted by the present authors, however, values of m were larger than 1. From the data for the three materials plotted in figure 9, values of m were 1.45, 2.39 and 1.69, respectively, for KX01, SCRB210 and SiC_w/Si_3N_4[4]. These values are consistent with the observation that the failure strain decreases with increasing strain rate. Other recent analyses of two phase ceramics show similar Monkman-Grant curves for silicon nitride and vitreous bonded aluminum oxide [29]. These results suggest that the Monkman-Grant curve is a common feature of creep rupture behavior in two phase ceramics.

The fact that data for ceramic materials fit an equation, such as the Monkman-Grant equation, simplifies the analysis of failure mechanisms and the prediction of lifetime. Two equations now describe creep-rupture behavior: the creep equation defines the creep process; the Monkman-Grant equation defines the rupture process. Therefore, the mode of deformation and the mode of failure can be analyzed separately. As the Monkman-Grant equation is independent of stress and temperature, the temperature and stress dependence of the creep rupture behavior will depend only on the creep behavior of the ceramic. Furthermore, if the Monkman-Grant curve is also a universal relation for a given type of material, then improvements in lifetime can be achieved solely by improving the creep resistance of the solid.

The Monkman-Grant equation and the creep equation can be combined to provide an estimate of the failure time as a function of applied stress and temperature. Thus, if $t_f = C \cdot \dot{\varepsilon}^{-m}$, and $\dot{\varepsilon} = \dot{\varepsilon}_o \cdot e^{-H/RT} \cdot (\sigma/\sigma_o)^n$, then $t_f = C \cdot \dot{\varepsilon}_o^{-m} \cdot e^{mH/RT} \cdot (\sigma/\sigma_o)^{-mn}$, so that both the apparent activation energy and the stress exponent for the failure time can be calculated from those for the creep rate. Obtaining the Monkman-Grant data and the creep data separately permit the failure time to be obtained as a function of stress more efficiently than if

---

[4] The identification of commercial grades of ceramics is not to be construed as endorsement by the National Institute of Standards and Technology.

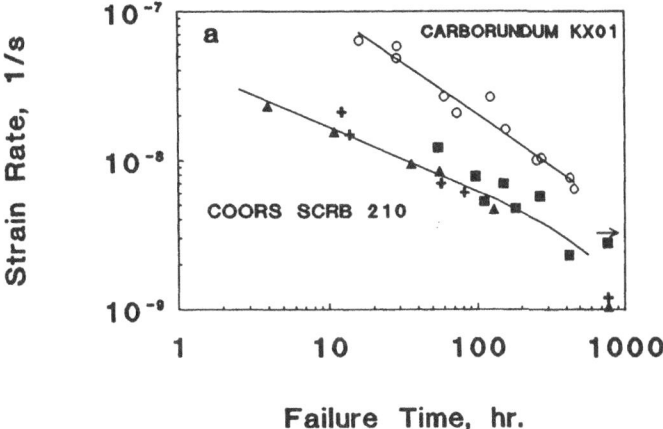

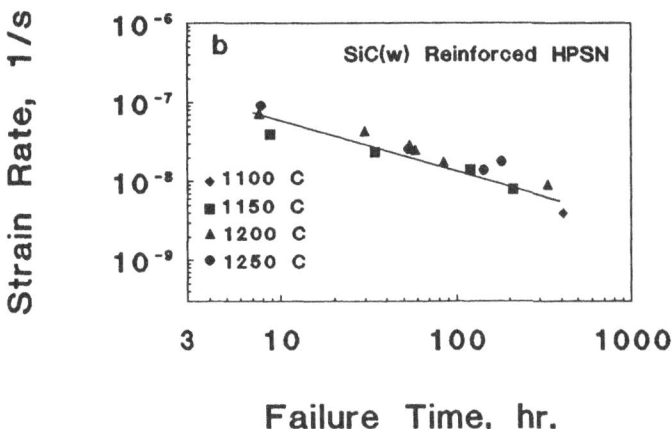

Figure 9. Creep rupture behavior of (a) siliconized silicon carbide and (b) SiC_w reinforced Si_3N_4. The three sets of data were taken over a range of stresses and temperatures. Regardless of the test conditions, data for each material cluster about a single curve. The strain exponents for the two grades of Si/SiC are 1.45 for the Carborundum KX01 and 2.39 for the Coors SCRB 210 (data from reference 25). The arrow over the data point in the lower right hand part of figure 9a indicates a specimen that was discontinued after ≈800 hr. The strain exponent for the SiC_w/Si_3N_4 is 1.69 (data from reference 27).

they were collected directly. This conclusion follows from the fact that only a small number (e.g. 6) of creep-rupture tests are required to define the Monkman-Grant curve. If multiple creep experiments can be used on individual specimens, an even smaller set of specimens (e.g. 2) can be used to define the stress and temperature dependence of the creep rate. Thus, a relatively rapid initial assessment of the creep behavior of a solid can be obtained from a limited number of samples.

4. MECHANISMS OF FAILURE IN TWO PHASE CERAMICS

Data presented in this paper suggest that the failure of two phase ceramics can be characterized by two extremes in behavior. In materials that contain a large volume fraction of bonding phase, cavities tend to nucleate at grain interfaces and then stabilize after growing into full facet cavities. Failure in this mode is by cavity coalescence to form a critical size crack. Once a critical crack is formed, crack growth is relatively rapid, so that failure is determined by the time required for the cavities to coalesce into cracks.

At the other extreme, failure is caused by creep crack nucleation and propagation. In these microstructures, interfacial cavitation followed by rapid interfacial cavity growth to form crack nuclei is favored. This occurs for materials that have relatively small amounts of second phase, or in materials that contain whiskers or large grains. The rate limiting step for component lifetime can be crack nucleation, or crack growth, or a combination of the two. As an example, in vitreous bonded alumina containing 8 vol. % vitreous phase (Coors AD96), $\approx 33\%$ of the failure time is required for cavity nucleation, indicating that both processes determine the lifetime [30]. Similar studies have yet to be conducted on other materials.

4.1. Failure by crack Growth

A relatively simple derivation of a Monkman-Grant equation can be obtained if the time for nucleating cracks is assumed to be short relative to the time for crack growth[5]. As an extreme of this condition, we consider creep failure from pre-existing flaws. If the crack growth rate is a power function of the applied stress intensity factor, i.e. $v = v_0 \cdot \exp(-Q/RT) \cdot (K_I/K_o)^p$, then the time to failure, t_f, is given by $t_f = B \cdot \exp(Q/RT) \cdot S_i^{n-2} \cdot \sigma^{-p}$, where B is a materials/environmental constant and S_i is the strength determined from the initial flaw size [4]. Combining this equation with the Norton equation for creep, $\dot{\varepsilon} = \dot{\varepsilon}_0 \cdot e^{-H/RT} \cdot (\sigma/\sigma_o)^n$, the following equation is obtained for the time to failure[6]:

$$t_f = \dot{\varepsilon}^{-p/n} \cdot B \cdot e^{[-(p/n)H+Q]/RT} \cdot S_i^{p-2} \cdot [\sigma_o^n/\dot{\varepsilon}_o]^{-p/n} \qquad (4)$$

---

[5] For crack growth to be an important mechanism of failure, either crack growth must initiate from pre-existing defects, in which case the nucleation time is zero, or a fixed but small number of nucleation sites must be present. Once these sites are activated, the remaining lifetime is determined by crack growth from these sites.

[6] This approach to component lifetime was first suggested by Lange [32].

From this equation, the strain rate exponent for the Monkman-Grant equation is m = p/n. Furthermore, the coefficient C in the Monkman-Grant equation (equation 3) is temperature independent only if H/n = Q/p. Otherwise, a series of parallel lines will be obtained on a Monkman-Grant type of plot.

Equation 4 provides a means of estimating the Monkman-Grant exponent directly from creep and crack growth measurements. The exponent for the creep rate, n, can be obtained from tensile creep measurements. The crack growth exponent can be obtained from indentation cracks in tensile bars, or from fracture mechanics specimens. However, as indentation cracks may not propagate in the same way as those that nucleate naturally during creep, it is probably better to obtain the crack growth exponent from naturally nucleated cracks. The fact that these two types of cracks may behave differently during creep was demonstrated on flexural bars of vitreous bonded alumina (Coors AD96) by Jakus et al. [30]. As shown in figure 10, the stress-intensity-factor exponent of the crack propagation rate was 4 for artificially introduced indentation cracks, but only 1 for natural cracks generated by the creep process[7]. Recently, Jakus and Nair [31] demonstrated the same type of behavior on SiC whisker-reinforced Al_2O_3. Results shown in figure 10 should be confirmed on tensile specimens because of the greater simplicity of the stress distribution in this type of loading. Despite this precaution, equation 4 provides a useful representation for creep rupture under conditions when lifetime is determined primarily by crack propagation parameters.

4.2. Failure by Cavity Coalescence

The following sequence of events is consistent with observations of cavitation damage in Carborundum KX01 [17, 18]. Prior to coalescence, cavities form at random throughout the gauge section. At small strains, cavities are isolated, but stable. With increasing strain, the cavity density increases and some cavities are nucleated adjacent to others, thus, forming small cracks. When a sufficient number of cavities are linked together, a self sustaining crack is formed which then propagates to failure. At any time, the size of the crack that forms depends on the density of the cavities and therefore on the creep strain. If a relationship between creep strain and critical crack size can be developed, then, as shown below, the time to failure can be determined solely from the creep behavior of the material.

An estimate of the crack size can be obtained from total creep strain by a two step process. First, the cavity density is determined experimentally as a

---

[7] This kind of behavior cannot be rationalized on the basis of stress relaxation in the surface of the flexure beam. Stresses in the beams that contain the indentation cracks are expected to be higher during crack growth, since the indentation cracks are present for the entire period of testing, whereas the naturally grown cracks form after the stress field in the flexure bar has relaxed. With the higher stress, a higher crack velocity would be expected for a given applied stress intensity factor, in contrast to what is observed in figure 10.

568

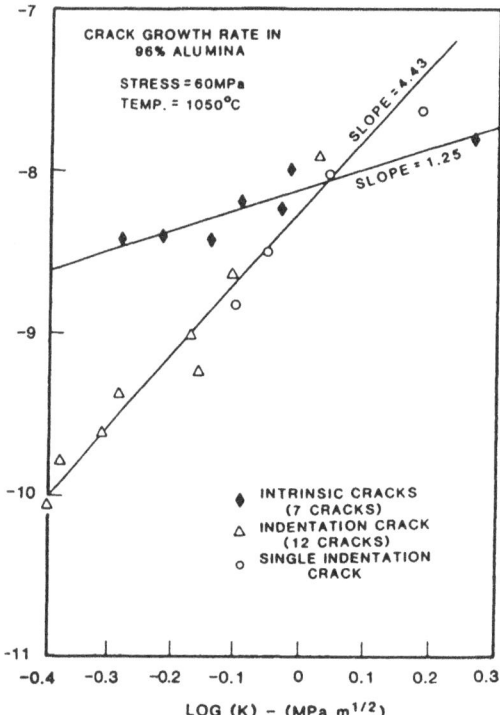

Figure 10. Crack propagation curves for both creep-nucleated and indentation-nucleated cracks in vitreous-bonded aluminum oxide (AD-96). The slope of the curve for the indentation-nucleated cracks is about 4 times that of the curve for creep-nucleated cracks. Data taken from reference 30.

function of total strain, either by density measurements or by ceramographic techniques. Then, the crack size is estimated from the cavity density by using statistical theory. Recently, Carroll and Tressler [18] determined the cavity density in siliconized silicon carbide specimens (Carborundum KX01) that were strained in tensile experiments conducted at 1100°C (versus 1225°C to 1400°C for the creep rupture data shown in figure 9). The cavity density was found to be linearly related to the total strain in the gauge section of the specimen, figure 11. A positive intercept on the strain axis of the data suggests a finite strain for cavitation. A linear relation between cavity density and creep strain has also been reported in other studies on ceramic materials [33 ,34]. Using the cavity density data in figure 11 and the strain to failure date in figure 8, it can be concluded that the density of cavities at failure ranged from $\approx680/mm^2$ (ε=0.004) to $\approx3000/mm^2$ (ε=0.014).

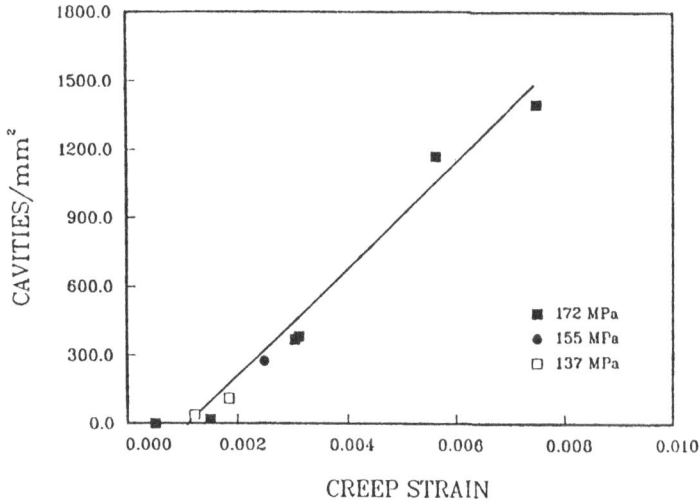

Figure 11. Area density of cavities as a function of creep strain at 1100°C. The fraction of cavitated grains, p, was estimated as the density of cavities divided by density of grains per unit area of surface.

The relation between cavity density and crack size can be determined from the statistical analysis of Lindborg[8] [35]:

$$f = 0.2(2 \cdot q/N)^{1/q}, \tag{5}$$

where f is the fraction of grains cavitated in the solid and N is the total number of grain boundaries normal to the applied stress. q is the cluster size, i.e. the number of cavities aggregated together to form a crack. If the gain size, G, is assumed to be uniform throughout the solid, and if the grains are arranged in a cubic lattice, then $N = (W/G)^2 \cdot (L/G)$, where W is the width of the gauge section, which is assumed to be square, and L is the length of the gauge section . The area occupied by a cluster is just $G^2 \cdot q$, and the crack size, 2a, is assumed proportional the square root of the cluster area, $G \cdot \sqrt{q}$.

---

[8] See the paper by Hunt [36] for a critical discussion of Lindborg's work. In deriving equation 5 the implicit assumption is made that failure occurs once a critical failure probably is reached, ≈60-70%. This same assumption was used in a paper by Rana and Evans [13] in their discussion of failure by cavity coalescence.

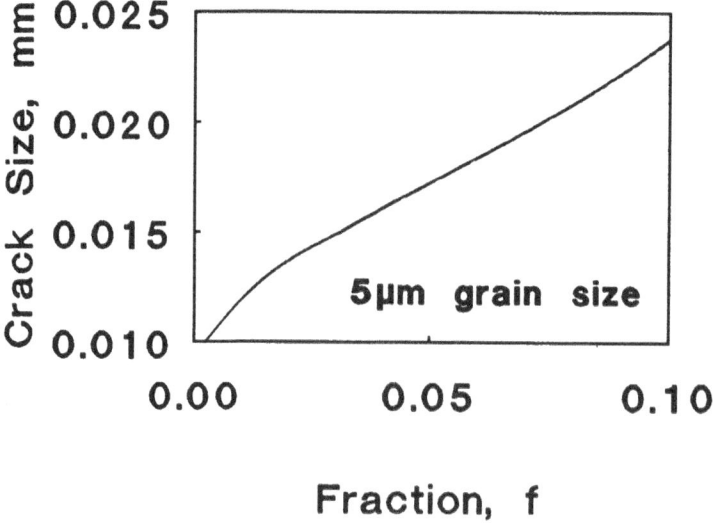

Figure 12. Calculated crack size as a function of the fraction of cavitated grains. This curve is a plot of equation 5, for which the crack size, 2a, is equal to $G \cdot \sqrt{q}$, the grain size, G, is equal to 5μm and the total number of boundaries, N, normal to the applied stress is equal to 1×10^9 (assuming W=3mm and L=15mm).

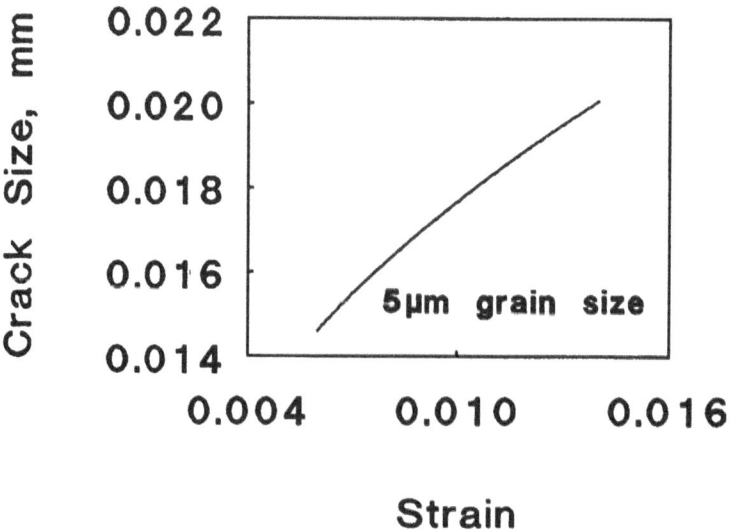

Figure 13. Calculated crack size as a function of specimen strain. This figure is obtained by using the curves in figures 11 and 12 to eliminate the fraction of cavitated grains as a variable.

Using the above definitions, equation 5 is plotted in figure 12, where the crack size, 2a, is the ordinate and the fraction of cavitated boundaries, f, is the abscissa. This equation is nonlinear; the crack size, 2a, approaches infinity as f approaches 0.2. Therefore, f=0.2 is an upper bound for the fraction of ruptured boundaries in a solid. In the set of experiments on Carborundum KX01, the cluster size was much smaller than this value. The range of values estimated above for the cavity density at failure is $\approx 680/mm^2$ to $\approx 3000/mm^2$. Assuming the grains may be represented by stacked cubes 5μm on a side, the grain boundary density is $40x10^3/mm^2$. Combining these two estimates, the fraction of cavitated boundaries at fracture ranged from ≈ 0.017 to ≈ 0.076. For this range of f, the critical crack size ranges from ≈ 13 to ≈ 20μm, suggesting a cluster size of from ≈ 7 to ≈ 16 for a self sustaining crack.

Combining the data in figure 11 with the theoretical curve in figure 12, an estimate of the dependence of crack size on strain is obtained, figure 13. A power law fit to the points in figure 13 yields the following equation for the estimated crack size as a function of strain:

$$2a = b \cdot \varepsilon^{0.38} \tag{6}$$

where b has the value $1x10^{-4}$μm.

Equation 6 may now be combined with empirical representations of the creep data to obtain an equation for the creep rupture behavior. The creep data shown in figure 14 for Carborundum KX01 is linearized, $\varepsilon_{cr} = \varepsilon \cdot t_f$, since the linear equation represents the creep data over approximately 80 percent of the

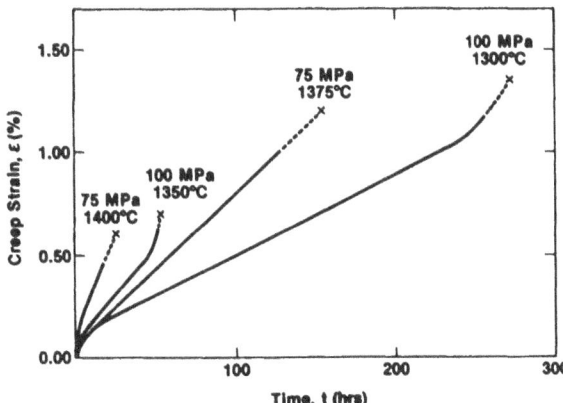

Figure 14. Uniaxial creep curves for KX01 at various thermal and mechanical loadings. Three important observations can be made from this data: (1) over most of the data range, creep strain can be represented as a linear function of time; (2) rupture strain decreases with increasing creep rate; (3) the maximum rupture strain is $\approx 1.5\%$, which indicates brittle behavior even at low rates of creep. Figure from reference 38.

lifetime of each specimen. The stress dependence of the creep behavior is represented by the Norton equation: $\dot{\varepsilon} = \dot{\varepsilon}_o \cdot (\sigma/\sigma_o)^n$, where $\dot{\varepsilon}_o$ now contains the temperature dependence of the creep rate. For experiments on KX01, n = 11. If σ_o is selected as 100 MPa, then $\dot{\varepsilon}_o = 1 \times 10^{-8} s^{-1}$.

Finally, at some threshold stress intensity factor, K_{th}, a crack is assumed to grow from a coalesced nucleus. The crack nucleus is defined by: $K_{th} = (2/\sqrt{\pi}) \cdot \sigma \cdot \sqrt{a}$, where a is the critical crack radius. Combining these equations, the relationship between creep rate and rupture time is given by the following equation:

$$\dot{\varepsilon}^{1+5.3/n} \cdot t_f = [(\pi/2b)^{1/2} \cdot (K_{th}/\sigma_o) \cdot \dot{\varepsilon}_o^{1/n})]^{5.3} \tag{7}$$

The form of this equation is identical to that of a Monkman-Grant equation, with m = 1+5.3/n. Since n ≈ 11 in the present experiment [18], m ≈ 1.48, which compares favorably with the value of 1.45 obtained from the creep rupture data shown in figure 9.

Once the value of m is determined, the right hand side of equation 7 can be evaluated and K_{th} can be determined. In the current experiment, K_{th} ≈ 0.4 MPa-m$^{1/2}$, which is consistent with a value of 2a ≈ 22 μm for the critical crack size. This value of K_{th} is much lower than the value of K_{Ic} measured on chevron notched specimens of this material tested at elevated temperatures. At 1300°C, for example, K_{Ic} for KX01 is ≈15 MPa-m$^{1/2}$ [37]. This value of K_{Ic} is a consequence of a substantial cavitation zone that develops around the crack tip in these materials at elevated temperatures. These zones are readily observed in siliconized silicon carbide at the tips of cracks at freshly introduced indentations, figure 15. For these cracks, propagation is preceded by the

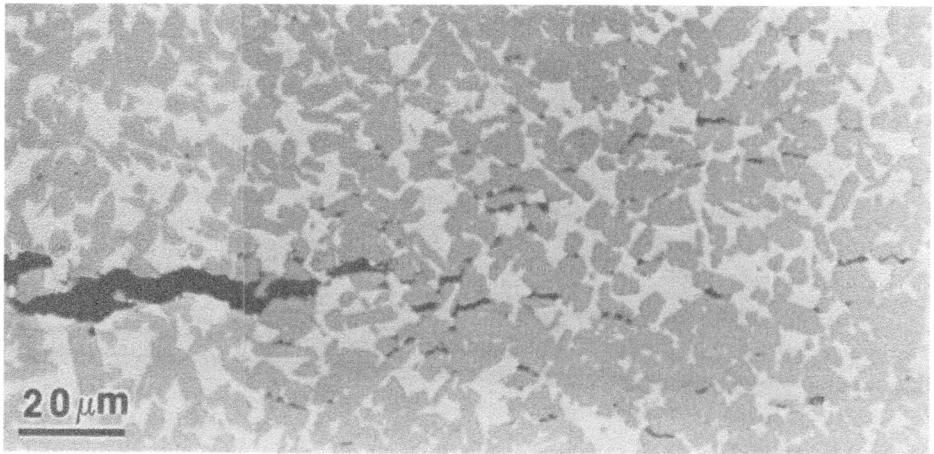

Figure 15. Cavity distribution around the tip of an indentation crack in KX01. Cavities form a cloud around the crack tip, which later link up as the crack propagates (Vickers indentation, 10 kg; temperature 1300°C; stress 120 MPa; time under load 19 hr).

linkage of cavities with the crack tip. Thus, plastic work at the crack tip, and volume expansion accompanying cavitation probably account for the high toughness of this material at elevated temperature. By contrast, cracks that nucleate within the creeping material as a consequence of creep , do not develop large cavitated zones as they propagate, figure 5, which again suggests a basic difference in behavior between cracks freshly introduced into an unstrained material, and cracks that develop as a consequence of the creep behavior of the materials. Creep apparently "work hardens" ceramic materials, embrittling them and reducing resistance to crack growth. As in the case of the alumina tested in flexure [30], self nucleated cracks appear to propagate differently from ones that are introduced by indentation.

5. SUMMARY

Two types of creep-rupture behavior are observed in two phase structural ceramics. Ceramics that contain large amounts of bonding phase form distributed isolated cavities during creep. With continuing creep, the density of cavities increases. Cavities link to form critical size cracks that eventually cause failure. As the amount of bonding phase is decreased, constraints are such that cavities, once nucleated, propagate as cracks. Small amounts of bonding phase in the material effectively embrittle ceramics at elevated temperatures reducing strain to failure. The strain to failure is also reduced if large grains or fibers are included within the solid, as these are preferred sites of cavity nucleation and rapid growth into cracks. Regardless of fracture mechanism, however, lifetime can be expressed as a power function of the creep rate, independent of temperature or applied stress. This separation of creep rupture behavior into creep behavior and rupture behavior simplifies the analysis of creep rupture data and permits a more rapid evaluation of material performance for high temperature application.

Acknowledgement: Support of the Ceramic Technology for Advanced Heat Engines Program, U.S. Department of Energy, and Test Methodology for Tubular Components Program of the Gas Research Institute is gratefully acknowledged.

REFERENCES

[1] A.G. Evans (1985) "Engineering Property Requirements for High Performance Ceramics," Mat. Sci. and Eng. 71 3-21.
[2] A.G. Evans and B.J. Dagleish, "Some Aspects of the High Temperature Performance of Ceramics and Ceramic Composites," pp. 929-955 in Creep and Fracture of Engineering Materials and Structures, B. Wilshire and R.W. Evans Eds. The Institute of Metals, London (1987).
[3] G. Grathwohl, "Creep and Fracture of Hot-Pressed Silicon Nitride with Natural and Artificial Flaws," pp. 565-577 in Creep and Fracture of Engineering Materials and Structures, Edited by B. Wilshire and R. Owen, Pineridge, Swansea, U.K. (1984).
[4] S.M. Wiederhorn and E.R. Fuller, "Structural Reliability of Ceramic Materials," Mater. Sci. and Eng. 71 169-186 (1985).
[5] G. Quinn (1984) "Static Fatigue in High-Performance Ceramics," pp.177-193 in Methods for Assessing the Structural Reliability of Brittle Materials, ASTM STP 844, S.W. Freiman and C.M. Hudson, Eds.

574

[6] M.D Thouless and A.G. Evans (1986) "On Creep Rupture in Materials Containing an Amorphous Phase," Acta Metall. **34** 23-31.

[7] M.D. Thouless (1987) "A Review of Creep Rupture in Materials Containing an Amorphous Phase," Res Mechanica **22** 213-242.

[8] T.L. Tsai and R. Raj (1982) "Creep Fracture in Ceramics Containing Small Amounts of a Liquid Phase," Acta Metall. **30** 1043-1058.

[9] R. Raj and M.F. Ashby (1975) "Intergranular Fracture at Elevated Temperature," Acta Metall. **23** 653-666.

[10] J.R. Porter, W. Blumenthal and A.G. Evans (1981) "Creep Fracture in Ceramic Polycrystals - I. Creep Cavitation Effects in Polycrystalline Alumina," Acta Metall. **29** 1899-1906.

[11] C.H. Hsueh and A.G. Evans (1981) "Creep Fracture in Ceramic Polycrystals - II. Effects of Inhomogeneity on Creep Rupture," Acta Metall. **29** 1907-1917.

[12] M. Thouless and A.G. Evans (1984) "Nucleation of Cavities during Creep of Liquid-Phase-Sintered Materials," J. Am. Ceram. Soc. **67** 721-727.

[13] A.G. Evans and A. Rana (1980) "High Temperature Failure Mechanisms in Ceramics," Acta Metall. **28** 129-141.

[14] F.C. Monkman and N.J. Grant (1956) "An Empirical Relationship between Rupture Life and Minimum Creep Rate in Creep-Rupture Tests," Proc. ASTM **56** 593-620.

[15] D.S. Cramner, B.J. Hockey, and S.M. Wiederhorn, "Creep and Creep-Rupture of Hot Isostatically Pressed Silicon Nitride," to be published.

[16] S.M. Wiederhorn, B.J. Hockey and R.F. Krause, Jr., (1987) "Influence of Microstructure on Creep Rupture"; pp. 795-806 in **Ceramic Microstructures '86: Role of Interfaces**, edited by J.A. Pask and A.G. Evans, Plenum Press, New York.

[17] S.M. Wiederhorn, D.E. Roberts, T.-J. Chuang, and L. Chuck (1988) "Damage Enhanced Creep in Siliconized Silicon Carbide: Phenomenology," J. Am. Ceram. Soc., **71** 602-608.

[18] D.F. Carroll, and R.E. Tressler (1988) "Accumulation of Creep Damage in a Siliconized Silicon Carbide," J. Am. Ceram. Soc. **71** 472-477.

[19] D.C. Drucker (1965) "Engineering and Continuum Aspects of High-Strength Materials"; pp. 795-833 in **High Strength Materials**, edited by V.F. Zackay, Wiley, New York.

[20] F.F. Lange (1972) "Non-Elastic Deformation of Polycrystals with a Liquid Boundary Phase"; pp. 361-81 in **Deformation of Ceramic Materials**, edited by R.C. Bradt and R.E. Tressler, Plenum Press, New York.

[21] A.H. Cottrell (1964) **The Mechanical Properties of Matter**, John Wiley and Sons, New York.

[22] H.J. Frost and M.F. Ashby (1982) **Deformation-Mechanism Maps: The Plasticity and Creep of Metals and Ceramics**, Pergamon Press, Oxford.

[23] B.J. Dalgleish, S.M. Johnson and A.G. Evans (1984) "High-Temperature Failure of Polycrystalline Alumina: I, Crack Nucleation," J. Am. Ceram. Soc. **67** 741-750.

[24] D.S. Wilkinson, A.G. Robertson and C.H. Cáceres, "Damage Mechanisms during High Temperature Creep in Hot-Pressed Alumina," to be published.

[25] S.M. Wiederhorn, W. Liu, D.F. Carroll, D.F., and T.-J. Chuang (1989) "Creep Rupture of Two Phase Ceramics," Presented at the 91st Annual Meeting and Exposition of the American Ceramic Society, April 23- 27, Paper 7-JIII-89.

[26] R.E. Peterson (1974) **Stress Concentration Factors,** Wiley-Interscience, New York.

[27] B.J. Hockey S.M. Wiederhorn, W. Liu, J.G. Baldoni and S.-T. Buljan, "Tensile Creep of Whisker-Reinforced Silicon Nitride," J. Mat. Sci., in press.

[28] R. Kossowsky, D.G. Miller, and E.S. Diaz (1975) "Tensile and Creep Strengths of Hot-Pressed Si_3N_4," J. Mat. Sci. **10** 983-997.

[29] M.K. Ferber, M.G. Jenkins and V.J. Tennery (1990) "Comparison of Tension, Compression, and Flexure Creep for Alumina and Silicon Nitride Ceramics," Ceram. Eng. Sci. Proc. **11** 1028-45.

[30] K. Jakus, S.M. Wiederhorn and B.J. Hockey (1986) "Nucleation and Growth of Cracks in Vitreous Bonded Aluminum Oxide at Elevated Temperatures," J. Am. Ceram. Soc. **69** 725-31.

[31] K. Jakus and S.V. Nair (1990) "Nucleation and Growth of Cracks in SiC/Al_2O_3 Composites," Comp. Sci. and Technol. **37** 279-97

[32] F.F. Lange (1976) "Interrelations between Creep and Slow Crack Growth for Tensile Loading Conditions," Int. J. Fracture **12** 739-744.

[33] E. Messner (1990) "Kriechverhalten und Kriechporenbildung von heissgepresstem Siliciumnitrid," Ph.D. Thesis, Technischen Universität Hamburg-Harburg.

[34] J. Stark (1988) "Verfomungsverhalten und Kriechporenbildung von reinem und glasphasehaltigem Aluminiumoxid," Ph.D. Thesis, Technischen Universität Hamburg-Harburg.

[35] U. Lindborg (1969) "A Statistical Model for the Linking of Microcracks," Acta Metall. **17** 521-526.

[36] R.A. Hunt (1978) "A Theory of the Statistical Linking of Microcracks Consistent with Classical Reliability Theory," Acta Metall. **26** 1443-1452.

[37] K. Kromp, T. Haug, R.F. Pabst and V. Gerold (1987) "C^* for Ceramic Materials? Creep Crack Growth at Extremely Low Loading Rates at high Temperatures using Two-Phase Ceramic Materials," pp. 1021-1033 in **Creep and Fracture of Engineering Materials and Structures,** B. Wilshire and R.W. Evans, Eds., The Institute of Metals, London.

[38] T.-J. Chuang, D.F. Carroll and S.M. Wiederhorn (1989) "Creep Rupture of a Metal-Ceramic Particulate Composite," pp. 2965-2976 in **Advances in Fracture Research,** Proceedings of the 7th International Conference on Fracture (ICF7), K. Salama, K. Ravi-Chandar, D.M.T. Taplin, P. Rama Rao, Eds., Houston, Texas, 20-24 March 1989, Pergamon Press, New York.

FRACTURE OF CONCRETE AT HIGH STRAIN-RATE

C. Allen Ross

Air Force Engineering and Services Center

Tyndall AFB, Florida 32403-6001

ABSTRACT

This paper is a review of high strain-rate effects on concrete strength, fracture and fracture mechanics parameters. Strain-rate effects on unsaturated concrete, where considerable strength and crack density increases are evident, begins to occur at approximately 5/sec in tension and approximately 100/sec in compression. Modulus of rupture data falls between that of tension and compression. High strain-rate shear strength data is not available but is expected to occur at strains above that of compression.

INTRODUCTION

Fracture of concrete at high strain rates, as well as other nonmetallic materials, show a rather similar dependence of tensile fracture strength σ_f upon the strain-rate $\dot{\varepsilon}$ to the one-third power. Here the high strain term is used to mean a strain-rate above 1.0/sec. This rather interesting similarity amongst several brittle materials is shown in Figure 1. These several sets of data were experimentally determined by different researchers and collected for comparison by Grady and Lipkin [1]. An analytical approach to determining fracture strength of brittle materials either from a inherent flaw size [1] or an energy method [1-4] leads to the same result. i.e. fracture strength of brittle materials at high strain rates is a function of strain-rate to the one-third power. The relationship, based on an inherent flaw size of a penny shaped crack, presented by Grady [1-3] and discussed in detail by Reinhardt [5] is given as

$$\sigma_f = 0.77\pi^{1/3}\left[\frac{EK_{IC}^2}{C_s}\right]^{1/3} \dot{\varepsilon}^{1/3} \qquad (1)$$

where E is Young's Modulus, K_{IC} is the fracture toughness, C_s is the elastic shear wave velocity, σ_f is the fracture stress and $\dot{\varepsilon}$ is the strain-rate. It is worth noting

S. P. Shah (ed.), Toughening Mechanisms in Quasi-Brittle Materials, 577–596.

here that Equation (1) is limited to dynamic loadings or high strain-rates. Apparently the relationship given by Equation (1) holds, regardless of the quasistatic cylinder compressive strength f'_{cs}. However, at strain-rates below approximately 1.0/sec and at quasistatic strain-rates, fracture strength is also dependent upon f'_{cs}. Semiempirical curves of various concrete mixes and strengths are presented by Reinhardt [6]. However, the slope of the fracture strength versus strain-rate for the dynamic and static regimes are not sufficient information to determine where strain-rate effects become very large.

Using data presented by Reinhardt [6] a simplified fracture strength versus strain-rate curve may be drawn as shown in Figure 2. For this curve a critical strain-rate $\dot{\varepsilon}_{cr}$ may be defined, above which strain rate effects are very important because of the rather large increases in fracture strength with relatively small changes in strain-rate. The critical strain-rate is different for different types of loadings. The normalized strength comparision of tensile and compressive loading of concrete by Ross [7] shows this rather distinct difference in Figure 3,. The critical strain rate for tensile fracture of a concrete (f'_{cs} = 48.3 MPa, 7000 psi for 6-inch diameter cylinder) was found to be approximately two orders of magnitude less than that for compressive fracture of the same material. Differences will also exist between flexure and shear, however very little data are available for these comparison but John and Shah [23] show some estimates for various loadings.

Following the same assumptions and development for Equation (1) an expression for concrete fragment size α_f as a function of strain-rate is given by Reinhardt [6] as

$$\alpha_f = 0.147\pi^{1/3}[\frac{C_s K_{IC}}{E}]^{2/3} \dot{\varepsilon}^{-2/3} \tag{2}$$

Again, the expression of Equation (2) is for high strain-rates. For compressive fracture of a split-Hopkinson pressure bar (SHPB) concrete specimens, the fragment distribution of a sieve analysis by Ross and Kuennen [8] showed considerable difference between a low strain-rate test of 25/sec and high strain rate tests of 100/sec and 200/sec. The size distribution for the SHPB tests is shown in Figure 4. Equation (2), based on an inherent flaw size distribution gives a good estimate of the average experimentally determined high-strain-rate fragment size of Figure 4, but underpredicts the fragment size of the 25/sec test. Fragment size based on an energy criterion equation, similar to Equation 1 predicted much higher post-test average fragments for the SHPB tests of Reference [8].

EXPERIMENTS

Initial studies and experiments of dynamic concrete fracture and increases in compression strength were conducted using impact hammers or dropped weights against short concrete specimens and long bars [9-15]. These early studies (1950-1980) may be classified as dynamic tests as opposed to quasistatic tests, but were mainly conducted in the low to intermediate strain-rates 10^{-3}/sec up to 10^{-1}/sec. The disadvantage of dropped weights is they require large drop heights to achieve impact velocities necessary for high strain-rates. Probably the first use of the higher strain-rates generated by a SHPB to test concrete is reported by Sierakowski at al [16], as early as 1977. Gas gun driven or torsional spring loaded SHPB test devices offer an advantage of high velocity impact and high strain rates for compression strength test and fracture of concrete. Several large diameter SHPB devices have been built for compression testing of concrete. A 76-mm (3.0 inch) diameter SHPB is located at the University of Florida and is described in detail by Malvern and Ross [17]. A 64-mm (2.5 inch) diameter SHPB, described by Felice [18], is located at Los Alamos National Laboratory. A dual mode compression-tension 51-mm (2.0 inch) diameter SHPB is located at the Air Force Engineering and Services Center, Tyndall AFB, FL. and details of this device are given by Ross [7].

Experiments on compression of cement paste at strain-rates up to 0.3/sec were conducted by Harsh et al [45]. The effect of strain-rate on cement paste and mortar show a nonlinear increase up to the maximum strain-rate of 0.3.sec. Harsh also reports increases in Poisson's ratio with strain-rate and decreases in strain capacity with increases in water-cement ratio. The presence of pore fluid and it's movement in saturated cement is shown by Harsh to contribute to strain-rate sensitivity. The higher the porosity the higher the strain-rate sensitivity.

Fracture of concrete in tension at strain-rates above 1.0/sec, by gas gun impact of steel projectiles on concrete rods was reported by Mellinger and Birkimer [19] in 1966. These data showed very large increases in concrete tensile strength; much larger than the 100% and 150% increases in compressive strength that had been reported earlier. Similar experiments were conducted by Griner [20] but the large increases in concrete tensile strength were not evident in the Griner data. A falling weight driven tensile SHPB was developed at Delft University, Netherlands and is described in some detail by Komeling et al [21]. Tensile fracture of uniform cross section concrete bars was accomplished in References [19-20] whereas short notched specimens were fractured in References [7,21] by cementing the specimens to the SHPB. The tensile pulse in the concrete bar specimens [19-20] was generated by reflection of a compressive pulse from the concrete bar free end, as opposed to the direct tension SHPB loading devices of [7 and 21]. High strain-rate tensile fracture of splitting-tensile specimens is also reported by Ross [7] and this data along with other dynamic tensile strength data are shown in Figure 3. A rather novel technique developed by Gran [22] uses axial release waves from an initial compression of a cylindrical concrete rod. The superposition of the two relief waves at the rod midlength place the rod in tension and the rod fails in tension. Tensile strengths found by Gran are not nearly as high as those reported by Millinger and Birkimir [19] and Ross [7].

Experiments specifically designed to study effect of high strain-rate on fracture mechanics parameters are reported by John and Shah [23]. For their experiments crack mouth opening displacement (CMOD) and modulus of rupture (MOR) were measured in three point bending test specimen mounted in a modified instrumented Charpy impact test device. The highest strain-rate obtained in these tests was 0.4/sec. Using linear elastic fracture mechanics (LEFM) the simplified fracture toughness $K_{IC}{}^s$ and the crack tip opening displacement CTOD can be determined from the experimentally determined CMOD and MOR of Reference [23]. In similar experiments John and Shah [24] measured crack velocity of concrete at high strain-rates using foil Krak gages, developed by the TTI Division, Hartrun Corp., St. Augustine, FL. These same gages were used by Ross et al [25] to measure crack velocity of splitting tensile concrete specimens in a SHPB.

Bentur et al [26] examined the effect of concrete strength on fracture energy and energy absorption capacity of impact tested flexural beams made of reinforced concrete. Discussion on the operation of this device and analysis of experiment al data is given by Banthia et al [27]. Other dynamic modulus of rupture and flexure tests are reported by Gopalaratnam et al [28-29], as well as the modified Charpy impact data of Sauris and Shah [30], and additional flexural impact experiments by Mindess and Nadeau [31].

It appears that there are no laboratory experiments designed specifically for gathering dynamic shear strength. This was pointed out by Sierakowski [32] in a paper written for this NATO Advanced Research Workshop of 1984 and there still appears to be no specific research in this area. Field tests of scaled reinforced concrete slabs failing by direct shear at the edges are reported by Slawson and Kiger [33]. For these field tests direct shear occurs for very short rise time blast loads and occur at approximate strain-rates of 10/sec. For rise times which are longer, the response and subsequent failure of the slabs is that of flexure.

Laboratory SHPB compression tests of concrete with confining pressure is reported by Malvern and Jenkins [34]. These tests, were performed on concrete cylinders of unconfined static $f'_{cs} = 48.3$ MPa using confining pressures of approximately 3.3 to 10.3 MPa (485-15000 psi). Dynamic tensile tests on concrete with confining pressures are reported by Gran et al [22,35].

DISCUSSION

Fracture of concrete at high strain-rates is significantly different than that found at the lower rates. Figure 5 shows a schematic of the initial compressive fracture of concrete taken from a high speed (10,000 f/sec) film for research reported in Reference [7]. For Figure 5a the strain-rate is approximately 25/sec and the strain-rate of Figure 5b is approximately 130/sec. For both strain-rates the fracture appears as wavy cracks running almost parallel to the longitudinal axis of the specimen. For the lower rate of Figure 5a the number of cracks are less than that of the higher rate of Figure 5b. In addition there appears to be more transverse cracks in Figure 5b which gives smaller fragments which is in agreement with the fragment distribution of similar tests shown in Figure 4. These rather longitudinal surface cracks appear to be similar to the internal crack patterns, Figure 6, found by Malvern and Jenkins [34] when testing concrete in a SHPB with a limiting strain collar which is slightly shorter than the specimen. The crack patterns of Reference [34] were obtained by a special crack enhancement process after sectioning of the cylindrical specimen. These internal cracks of Reference [34] and the surface cracks of Figure 5 are probably the result of tensile straining caused by Poisson's effect and tensile stress cracking due to boundary conditions at the interface of the specimen and SHPB. Experimental evidence by Malvern and Ross [17], using circumferential strain gages on a dynamically compressed concrete cylinder in a SHPB, shows that a compressive stress greater than the static compressive strength is evident prior to failure in the circumferential direction. Also, for Reference [17] the ratio of the circumferential strain to the longitudinal strain compared to Poisson's ratio for the concrete material. It is interesting to note here that the onset of high strain-rate effect $\dot{\varepsilon}_{cr}$ for tensile fracture is much less than that of compressive fracture, see Figure 3.

The effect of a static confining pressure on the compressive fracture strength is to increase the compressive strength for a given strain-rate. The combined effect of strain-rate and static confining pressure of 19 confined and 8 unconfined SHPB tests is given in Reference [34] as

$$\sigma_f = A + B \ \dot{\varepsilon}_f + Cp_f$$
$$A = 64.4 \ \text{MPa}$$
$$B = 0.388 \ \text{MPa's}$$
$$C = 7.59 \tag{3}$$

where σ_f is the compressive failure stress, $\dot{\varepsilon}_f$ is the strain-rate at failure and p_f is the confining pressure at failure. The expression, Equation 3, was obtained for an unconfined concrete strength of 48 MPa (7,000 psi) at strain-rates of approximately 50/sec to 150/sec and confining pressures of 3.3 MPa (485 psi) to 10.3 MPa (1500 psi). Extrapolation outside these ranges is not recommended. For unconfined tests of References (7,8,17,34) the stress-strain curve rises to a peak and decays very quickly to zero with no elastic recovery and the specimen is reduced to rubble. For the confined tests, the stress-strain curve rises to a plateau and stays relatively constant (resembling a ductile curve) throughout the loading pulse then falls to zero with an apparent elastic unloading. The effect of confinement in the SHPB tests on specimen fracture is to produce an almost intact specimen with one major crack running diagonally across the specimen from one end to the other as opposed to the highly fracture rubble of the unconfined SHPB tests.

Tensile fracture of direct tension SHPB specimen by Ross [7] showed one or more rather clean fractures across the test specimen. Multiple fractures occur as strain-rate is increased. The experimental data indicates that fracture occurs during the first passage of the tensile stress wave across the specimen. The tensile fracture surface of the 48 MPa compressive strength concrete showed both aggregate pull out and fracture in an approximate one to one proportion. Splitting-tensile SHPB tests [7], of the 48 MPa concrete gave slightly higher tensile strength than that of the direct tension tests. However both the splitting-tensile tests and direct tensions tests show similar strain-rate effects. Large increases in tensile strength show up at lower strain-rates than the compressive data (see Figure 3).

Both photographic and Krak gage data were obtained for crack propagation in the splitting-tensile tests of Reference [7]. Figure 7 shows the crack pattern and the stress associated with these patterns for a SHPB splitting-tensile test. The Figure 7 crack pattern was copied from a Polaroid film taken using an image converter camera running at an equivalent rate of 200,000 f/sec. The initial crack of Figure 7 appears to start just to the left of the center of the specimen on the side of the incident pulse at a time equivalent to the twice the transit time of the specimen. The pulse then travels in both directions and the specimen fails into two major pieces. Crack velocity measurements for these splitting tension specimens at strain-rates of 1/sec to 10/sec were found to be of 10 to 50% of the 3620 m/sec acoustic velocity of the specimen. Crack velocities obtained for the center notch concrete splitting tensile specimens agree with crack velocities obtained using concrete notched beams and are plotted as a function of strain-rate in Figure 8.

Dynamic analysis, using an ADINA finite element code, of the splitting-tensile cylinder by Tedesco et al [36] shows a build-up of stress distribution very similar to the static stress distribution. A computer generated crack pattern using the specimen loadings of Reference [7] is shown in Figure 9. The bifurcation of the crack shown in Figure 9 was also observed in the experiment. Bifurcation of the crack occurs at the higher strain-rates and appears to be a result of the biaxial compression-tension that occurs off the centerline of the specimen. The failure surface model used in Reference [36] degrades the tensile strength linearly with the presence a compressive principal stress. It is interesting how well the fracture pattern generated using a structural analysis code resembles the fracture pattern observed in the experiments. The differences in times of events of Figures 7 and 9 is due to the difference in crack velocities in the analysis and experiment. In the analysis of Reference 36 the acoustic velocity of a linear elastic material is used in the calculations and in the experiment the crack velocity may be only 10% of the acoustic velocity.

Linear elastic fracture mechanics (LEFM) parameters have been used to predict the effects of strain-rate on the tensile strength and modulus of rupture. John and Shah [23] measured crack mouth opening displacements (CMOD) and used LEFM relationships to predict crack tip opening displacement, (CTOD) and fracture toughness for Mode I fracture, K_{IC}. These values and Young's modulus E may then be used to predict uniaxial tensile strength and splitting-tensile strength. For the higher strain-rate regimes the CTOD is assumed to be the only strain-rate dependent parameter in their calculation. The ratio of dynamic CTOD and quasistatic CTOD of [23] is based on an exponential function using the ratio of dynamic to quasistatic strain-rates. The elastic modulus E and the fracture toughness K_{IC} are assumed to be strain-rate independent but vary with concrete compressive strength f'_{cs}. Strain-rate independence of compressive Young's modulus has been observed in Reference [28] and in experiments of Reference [7]. The compressive modulus obtained in SHPB work such as [7] is not usually reported as the initial linear portion of the stress-strain occurs during the rise time of the loading pulse, before uniform stress along the specimen length is obtained. Use of the quasistatic fracture toughness in [23] was justified on the basis of the very low crack velocity observed at the higher strain-rates.

The two parameter fracture model of [28] appears to predict the low strain-rate tensile strength data of Cowell [37], Takeda et al [38], Kormeling et al [21] and data presented by Oh [39]. However, it underpredicts experimental data at strain-rates above 1/sec [4,7]. This underprediction of the tensile fracture strength of concrete at high strain-rates, for a predictor based on slightly lower strain-rate data, may be attributed to the steep rise in strain-rate sensitivity on tensile strength shown by Reinhardt [6].

The two parameter fracture model of John and Shah [23] was also used to predict modulus of rupture (MOR) by using CMOD data taken from fracture of three point bend specimens in a modified Charpy device. Again the model agrees well with MOR data taken at strain-rates below 1/sec, but would not be expected to predict MOR for the higher strain-rates.

Probably the highest dynamic tensile strength data reported is by McVay [40]. Dynamic tensile strength increases of over 700% at strain-rates of 100/sec were calculated from spall of a concrete slab subjected to blast from a conventional weapon. These stresses were obtained by back calculating the failure stress from the measured spall plane and the stress wave shape measured at the back of the slab. These data are shown in Figure 3 and fall in the high strain regime predicted by Equation (1).

The observations and data for both low and high strain-rates, point out the importance of knowing the fracture strength and fracture processes for the entire strain-rate region of the loading environment. This is especially true for the tensile dominated fractures of uniaxial tension, transverse tensile fracture of direct compression, splitting-tensile fracture and the tensile fracture of flexure. Models predicting tensile fracture must include the nonlinear portion of the low strain-rate regime and the rather linear and steep slope of the higher strain-rates.

A model presented by Weerheijm and Reinhardt [41], based on an array of penny shaped cracks agrees very well with the concrete tensile date shown in Figure 3. This model, shown as solid lines of Figure 3 predict both the lower strain-rate concrete strength as well as changing slope to predict the concrete strength at the higher strain-rates. This model takes into account the strain-rate dependence of crack speed and fracture energy as well as the effect of compressive concrete strength on tensile strength.

The micromechanical modelling of concrete is based mainly on some distribution of microcracks, microdefects or weak planes. Chen [42] uses a Weibull statistical crack per unit volume distribution which is activated by bulk or volumetric strain based on the mean stress and bulk modulus. The crack distribution is defined in terms of constants taken from experimentally determined fracture stress versus strain-rate data or estimated from Equation (1). In addition the average crack dimension used by Chen [42] is based on a fragment dimension, similar to Equation 2. The interaction of the cracks in Chen's model is treated as an internal state variable which represents the accumulation of damage in the material. This damage is assumed to degrade the material stiffness based on equations by Budinasky and O'Connell [43], for an array of penny shape cracks in an isotropic medium. The point here is that modelling of fracture of concrete at strain-rate, whether it be a complicated code such as [42], a single formula for dynamic tensile strength as given by Oh [39] or empirical equations such as Soroushian et al [48], are all based on some form of experimental data.

SUMMARY

Based on experimental observations, fracture strengths of concrete for most all types of loading are dominated by tensile fractures. At the quasistatic and low strain-rates the fracture strength versus strain-rates is nonlinear up to some critical strain-rate. Beyond this critical strain-rate the strength increases rather drastically as a function of the strain-rate to the one-third power. It appears that the critical strain-rate occurs at different strain-rates for the different loadings. It is not that any type of loading is more strain-rate sensitive than another, but that the high strain-rate sensitivity occurs at different strain-rates for different types of loading.

All the modelling of concrete fracture at high strain-rates at some point in the analysis requires time dependent fracture data. Many times this data is not available and care must be exercised in extrapolating beyond the strain-rate range in which the experimental data was taken.

For the tensile dominated concrete fractures, Mode I fracture appears to occur in the majority of failures. In the low strain-rate compression loading the inherent flaws at the higher stress concentrations grow and relieve the stresses at other sites. Under the low loading rates these activated cracks have time to grow into the weakest areas of the concrete matrix and aggregate. The result is a post-test specimen with very few cracks and large broken pieces. As the compression load rate increases many more of the inherent flaws are activated and forced to grow into the high strength areas of the mortar and aggregates. This results in increased strength, a larger number of much shorter cracks and smaller fragments. At much higher rates the fragments are much smaller and the concrete is reduced to rubble. In the case of tension/compression biaxial or triaxial stress states the tensile failure stress is reduced by the presence of the compressive stress. Hydrostatic confining pressure on concrete inhibit crack growth and fracture at high strain-rates. Continued research into high strain-rate compression of concrete is needed to characterize the crack area and crack pattern for both unconfined and confined loadings.

High strain-rate uniaxial tension specimens usually fail by a single crack plane across the specimen. At high strain-rates multiple cracks form. Experimental direct tension fracture strength at strain-rates higher than 1.0/sec are needed to completely define this area of strain-rate. Data on the effects of confining pressure on tensile strength at higher strain-rates are needed to better understand fracture processes of the compression loading.

Strain-rate effects data on fracture mechanics parameters are scarce and further experimentation is needed. Fracture toughness is a vary fundamental part of the prediction of fracture strength of concrete. Some fundamental high strain-rate experiments in Mode I and Mixed Mode fracture, with and without confining pressures, are needed to determine the extent of strain-rate effects on fracture toughness.

REFERENCES

1. Grady, E. E. and Lipkin, J., "Criteria for Impulsive Rock Fracture", Geophysical Research Letters, Vol. 7, No. 4, April 1980, pp 255-58.

2. Grady, D. E., "The Mechanics of Fracture Under High-Rate Stress Loading", Sandia National Laboratories Report SAND-82-1148C, 1983.

3. Grady, D. E. and Lipkin J., "Mechanisms of Dynamic Fragmentation: Factors Governing Fragment Size", Sandia National Laboratories Report SAND 84-2304C, 1984.

4. Birkimer, D. L., "A Possible Fracture Criterion for the Dynamic Tensile Strength of Rock", Procedures 12th Symposoum on Rock Mechanics, G. B. Clark, ed., 573, 1971.

5. Reinhardt, H. W., "Tensile Fracture of Concrete at High Rates of Loading", Application of Fracture Mechanics to Cementation Composites (NATO-ARW), S. P. Shah, ed., 1984 pp 559-90.

6. Reinhardt, H. W., "Dynamic Loading", Fracture Mechanics of Concrete Structures, L. Elfgren, ed. Champman and Hall, Ltd. London, 1989 pp 188-90.

7. Ross, C. A., "Split-Hopkinson Pressure Bar Tests", ESL-TR-88-82, Engineering and Services Laboratory, Air Force Engineering and Services Center Tyndall AFB, FL, 1989.

8. Ross, C. A. and Kuennen, S. T., "Fracture of Concrete at High Strain Rates", Fracture of Concrete and Rock: Recent Developments, S. P. Shah et al, eds. Elsevier Applied Science, London, 1989 pp 152-61

9. Watstein, D. "Effect of Straining Rate on the Compressive Strength and Elastic Properties of Concrete", ACI Journal, Vol 24, 1953, pp 729-44.

10. Goldsmith, W., Kenner, V. H. and Ricketts, T. E., "Dynamic Behavior of Concrete", Experimental Mechanics, Vol. 66, 1966, pp 65-69.

11. Green, H., "Impact Strength of Concrete" Institution of Civil Engineers, Proc., Vol. 28, 1968, pp 731-40.

12. Hughes, B. P., and Gregory, R., "The Impact Strength of Concrete Using Green's Ballistic Pendulum", Institution of Civil Engineers, Proc., Vol 81, 1968, pp 731-40.

13. Atchley, B. L. and Furr, H. L., "Strength and Energy Absorption Capabilities of Plain Concrete Under Dynamic and Static Load", ACI Journal, Nov. 1967, pp 745-56.

14. Hughes, B. P. and Gregory, R., "Concrete Subjected to High Rates of Loading in Compression", Mag. of Concrete Res., Vol 24, 1972, pp 25-36.

15. Hughes, B. P. and Watson, A. J., "Compressive Strength and Ultimate Strain of Concrete Under Impact Loading", Mag. of Concrete Res., Vol. 30, 1978, pp 189-99.

16. Sierakowski, R. L., Malvern, L. E., Collins, J. A., Milton, J. E. and Ross, C. A., "Penetration Impact Studies of Soil/Concrete", Final Report, AFOSR Grant No.

77-3029 and AFATL Tech. Report TR-78-9, University of Florida Gainesville, FL, Nov. 1977.

17. Malvern, L. E. and Ross, C. A., "Dynamic Response of Concrete and Concrete Structures", Two annual reports and a final report for AFOSR Contract F49620-83-K007. University of Florida, Gainesville, FL, Feb. 1984, Feb. 1985, May 1986.

18. Felice, C. W., The Response of Soil to Impulse Loads Using the Split-Hopkinson's Pressure Bar Technique, Ph.D. Dissertation, The University of Utah, 1985.

19. Mellinger, F. M. and Birkimer, D. L., "Measurements of Stress and Strain on Cylindrical Test Specimens of Rock and Concrete Under Impact Loading", Tech. Rep. 4-46 Dept. of Army Ohio River Div. Lab., Apr. 1966.

20. Griner, R. C., Dynamic Properties of Concrete, Thesis for Master of Science Degree in Engineering, University of Florida, Gainesville, FL 1974.

21. Kormeling, H. A., Zielinski, A. J., and Reinhardt, H. W.,"Experiments on Concrete Under Single and Repeated Uniaxial Impact Tensile Loading", Stevin Lab. Rept. 5-80-3, Delft University of Technology, 2nd Print, 1981.

22. Gran, J. K., and Seaman, L., "Observations and Analysis of Microcracks Produced in Dynamic Tension Tests of Concrete", Final Rept. AFOSR Contract No. F49620-82-15-0021. SRI International, Merlo Park, CA, 1986.

23. John, R. and Shah, S. P., "Effect of High Strain-Rate of Loading on Fracture Parameters of Concrete", Proc. of Fracture of Concrete and Rock, SEM-RILEM Int. Conf., S. P. Shah and S. E. Swartz, eds Published by Soc. of Ecp. Mech. (SEM), Bethel, CT., 1987, pp 35-52

24. John, R. and Shah, S. P., "Fracture of Concrete Subjected to Impact Loading", Cement, Concrete and Aggregates, CCAGDP, Vol. 8, No. 1, 1986, pp 24-32.

25. Ross, C. A., Kuennen, S. T., and Tedesco, J. W., "Experimental and Numerical Analysis of High Strain-Rate Concrete Tensile Tests", Micromechanics of Failure of Quasi-Brittle Materials, S. P. Shah et al, eds, Elsevier Applied Science, London, 1990, pp 353-364.

26. Benture, A., Mindess, S. and Banthia, N., "The Behavior of Reinforced Concrete Beams Under Impact: The Effect of Concrete Strength", Proc. of Fracture of Concrete and Rock, SEM-RILEM Int. Conf., S. P. Shah and S. E. Swartz, eds, Published by Soc. of Exp. Mech. (SEM), Bethel, CT, 1987, pp 449-58.

27. Banthia, N. P., Mindess, S., and Benturc, A., "Energy Balance in Instrumented Impact Tests on Plain Concrete Beams", ibid pp 21-34.

28. Gopalaratnam, V. S. and Shah, S. P., "Properties of Steel Fiber Reinforced Concrete Subjected to Impact Loading", ACI J., Proceedings, V. 83, No. 1,

1985, pp 117-2

29. Gopalaratnam, V. S., Shah, S. P., and John, R., "A Modified Instrumented Charpy Test for Cement-Based Composites", Exp. Mech., SEM, V. 24, No. 2, 1984, pp 102-11.

30. Sauris, W. and Shah, S. P., "Properties of Concrete Subjected to Impact Loading", Jour. of Struct. Eng., ASCE, Vol. 109, No. 7, 1983, pp 1727-41.

31. Mindess, S. and Nadeau, J. S., "Effect of Loading Rate on the Flexural Strength of Cement and Mortar", Bull. of Amec. Ceramic Soc., Vol. 56, No. 44, 1977, pp 429-30.

32. Sierakowski, R. L., "Dynamic Effects in Concrete Materials", Application of Fracture Mechanics to Cementitious Composites, S. P. Shah, ed., Martinus Nijhoff Publ, Dordrecht, The Netherlands, 1985.

33. Slawson, T. R. and Kiger, S. A., "Dynamic Shear Failure of Shallow-Buried Flat-Roofed Reinforced Concrete Structures Subjected to Blast Loading", Waterways Experiment Station, Vicksburg, MS, Tech. Dept., SL-84-7, 1984.

34. Malvern, L. E. and Jenkins, D. A., "Dynamic Testing of Laterally Confined Concrete", ESL-TR-89-47, Engineering and Services Laboratory, Air Force Engineering and Services Laboratory, Tyndall AFB, FL, 1990.

35. Gran, J. K., Florence, A. L. and Colton, J. D., "Dynamic: Traxial Tests of High-Strength Concrete", Jour. of Eng. Mech., Vol. 115, 1989, pp 891-904.

36. Tedesco, J. W., Ross, C. A. and Brunair, "Numerical Analysis of Dynamic Split Cylinder Tests", Computers and Structures, Vol. 32, No. 3/4, 1989, pp 609-24.

37. Cowell, W., "Dynamic Properties of Plain Portland Cement Concrete", Tech. Dept. No. R447, U. S. Naval Engineering Laboratory Post Hueneme, CA, 1966.

38. Takada, J., Tachikawa, H. and Fujimoto, K., "Mechanical Behavior of Concrete Under Higher Rate Loading than in Static Test", in Mechanical Behavior of Materials, Proc. of the Soc. of Material Sciences, Vol. II, 1974.

39. Oh, B. H., "Behavior cf Concrete Under Dynamic Tensile Loads", ACI Materials Journal, Jan-Feb 1987, pp 8-13.

40. McVay, M. K., "Spall Damage of Concrete Structures", Tech. Rept. SL-88-22, Waterways Experiment, Corps of Eng., Vicksburg, MS, 1988.

41. Weerheijm, J. and Reinhardt, H., "Modelling of Concrete Fracture Under Dynamic Tensile Loading", Fracture of Concrete and Rock: Recent Developments, S. P. Shah et al, eds. Elsevier Applied Science, London, 1989, pp 721-28.

42. Chen, E. P., "A Strain Dependent Fracture Model Based on Continuum Damage Mechanics", in Dynamic Constitute/Failure Models, A. M. Rojendran and T. Nicholas eds. AFWAL-TR-88-4229, Univ. of Dayton Res. Inst., Dayton, OH,

590

1988, pp 212-238.

43. Budiansky, B. and O,Connell, R. J., "Elastic Modulus of a Cracked Solid", International Journal of Solids and Structures, 12, 81, 1976.

44. Soroushian, P., Choi, K., and Alhamad, A., "Dynamic Constitutive Behavior of Concrete", ACI Journal, Mar-Apr 1986, pp 251-59.

45. Harsh, S., Shen, A. and Darwin, D., "Strain-Rate Sensitive Deformation of Cement Paste and Mortar in Compression", Final Report AFOSR Research Grant, AFOSR-85-0194, University of Kansas, Lawrence, K. S., 1989.

46. Malvern, L. E. and Jenkins, D. A., "Strength and Deformanton of Confined and Unconfined Concrete Under Axial Dynamic Loading", Final Report AFOSR Contract, No. 87-0201, University of Florida, Gainesville, FL, 1988.

ACKNOWLEDGEMENT

The author would like to acknowledge the financial support of the Air Force Office of Scientific Research and the Air Force Engineering and Services Center. Thanks to Mrs. Mary Young for typing of the manuscript.

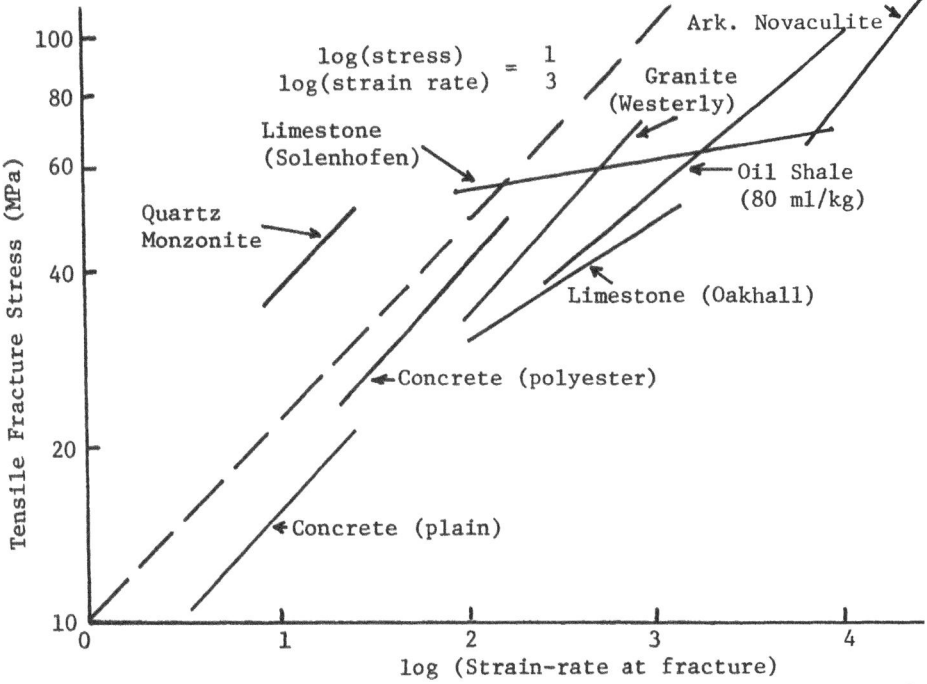

Figure 1. Strain-rate dependence of fracture stress. Fig. 1 of Ref. [1]
Dotted line added by the author.

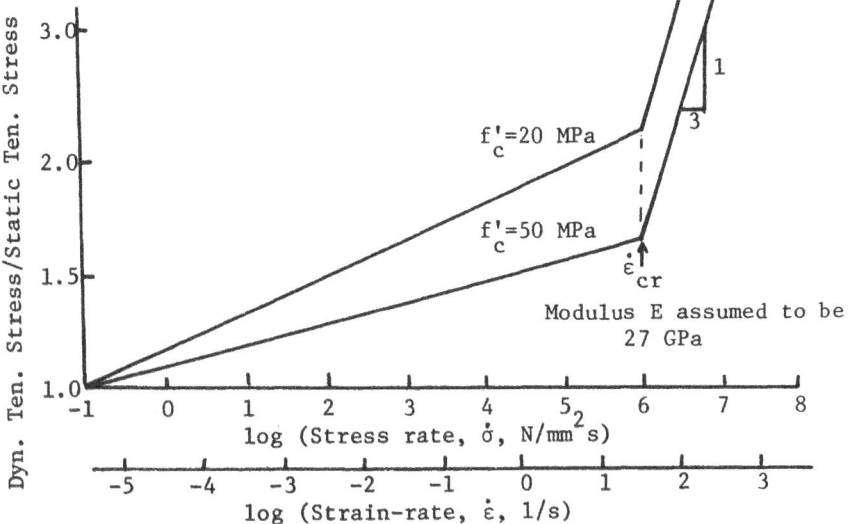

Figure 2. Stress-rate dependence of fracture stress. Fig. 2 of Ref. [6]
Critical strain-rate $\dot{\varepsilon}_{cr}$ and modulus E added by the author.

592

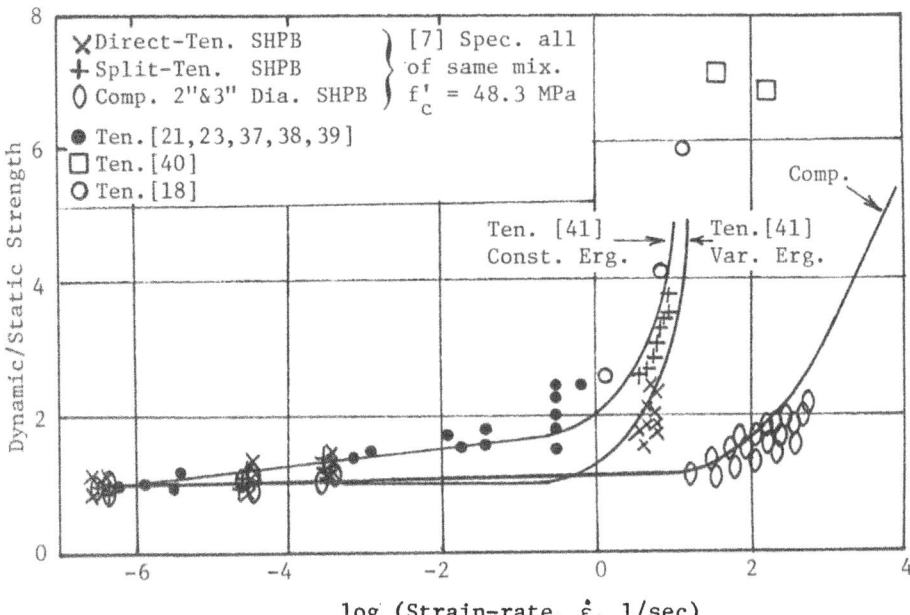

Figure 3. Strain-rate dependence of fracture stress. [7]

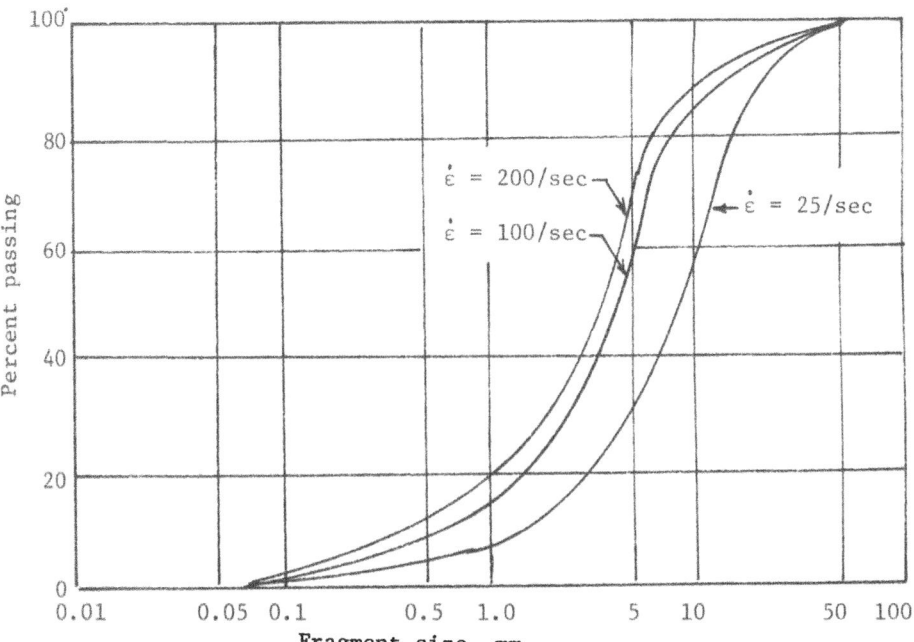

Figure 4. Fragment size distribution from sieve analysis of a post-test
concrete specimen tested in the compressive SHPB. [8]

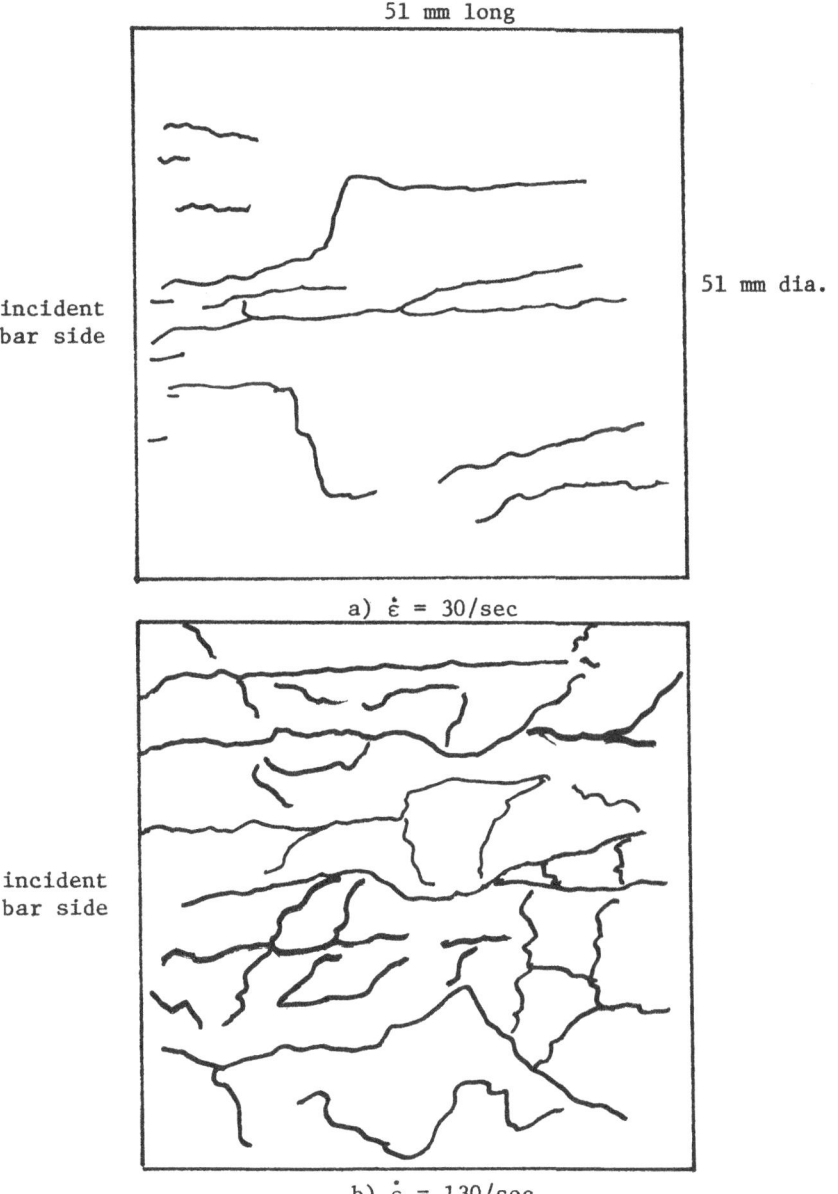

51 mm long

incident
bar side

51 mm dia.

a) $\dot{\epsilon}$ = 30/sec

incident
bar side

b) $\dot{\epsilon}$ = 130/sec

Figure 5. Tracing of surface cracks of SHPB concrete compressive
specimen taken from high speed film. Time is approx-
200 microsec after arrival of the incident pulse.

594

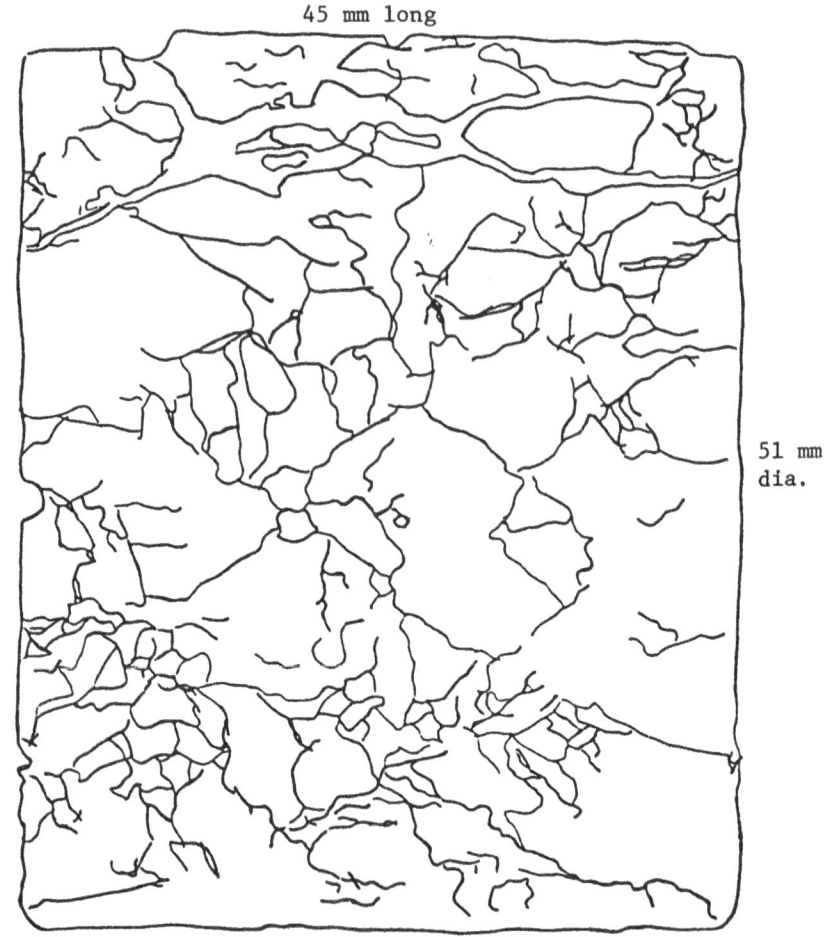

45 mm long

incident
bar side

51 mm
dia.

Figure 6. Internal crack pattern of SHPB concrete compressive test with
strain limiting collar. Strain-rate 150/sec. Maximum strain of
0.0128 at approximately 150 microsec after arrival of the
incident pulse. (Fig. 24 Ref. [46])

595

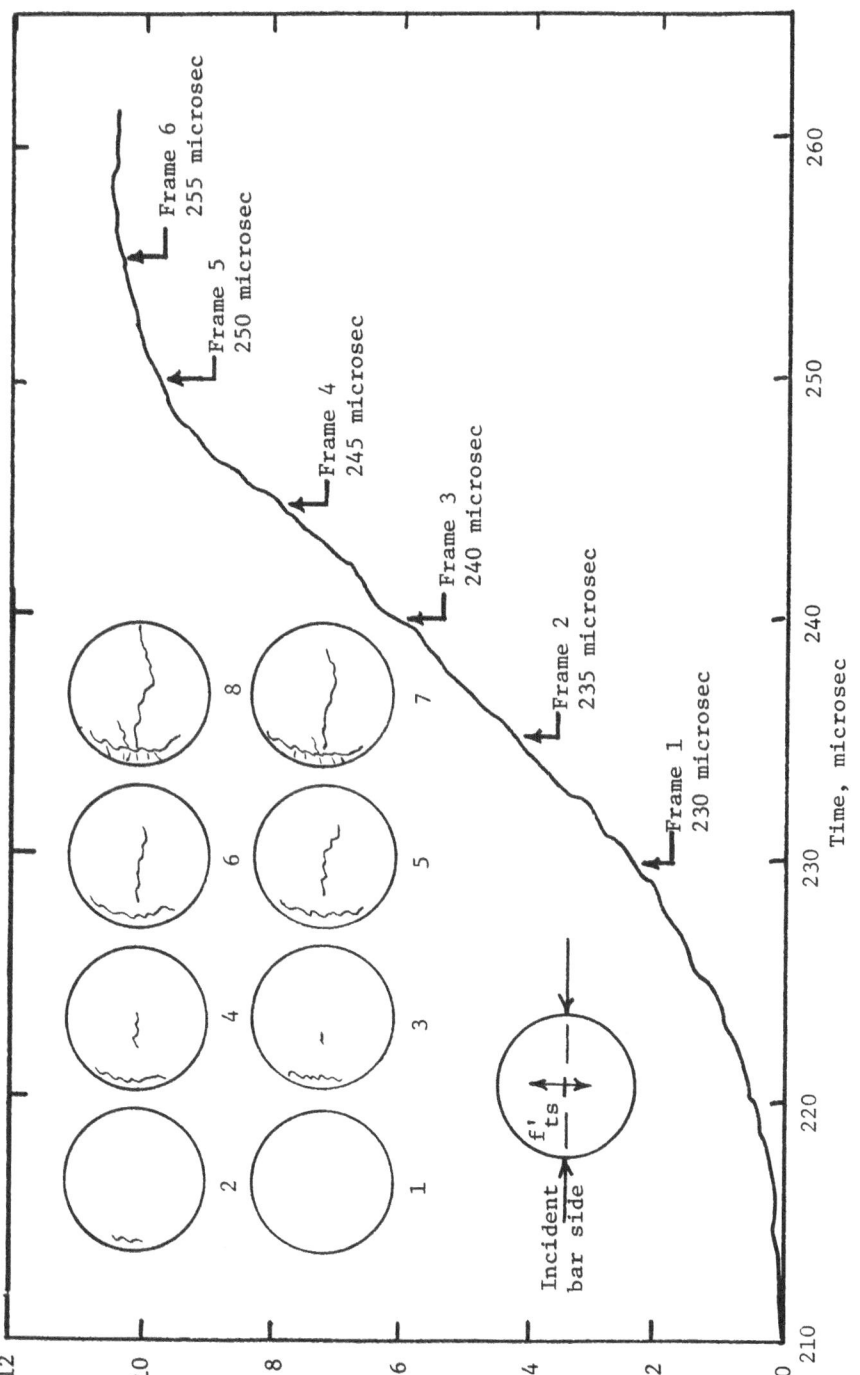

Figure 7. Crack pattern and associated tensile stress for a splitting-tensile SHPB test. Strain-rate approximately 5/sec. 51 mm diameter x 51 mm long specimen. f'_c = 48 MPa. [8]

596

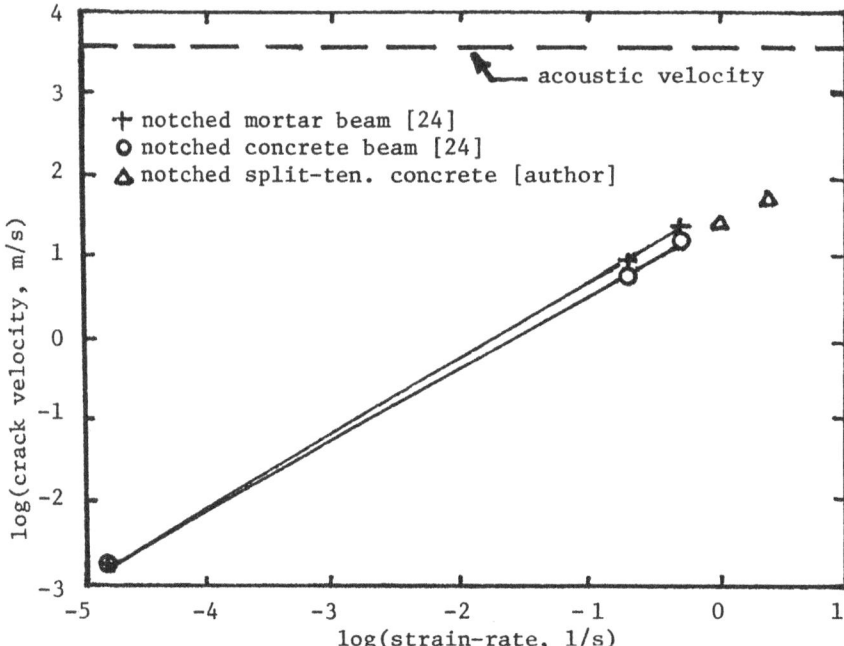

Figure 8. Crack velocity shown as a function of strain-rate in concrete and mortar.

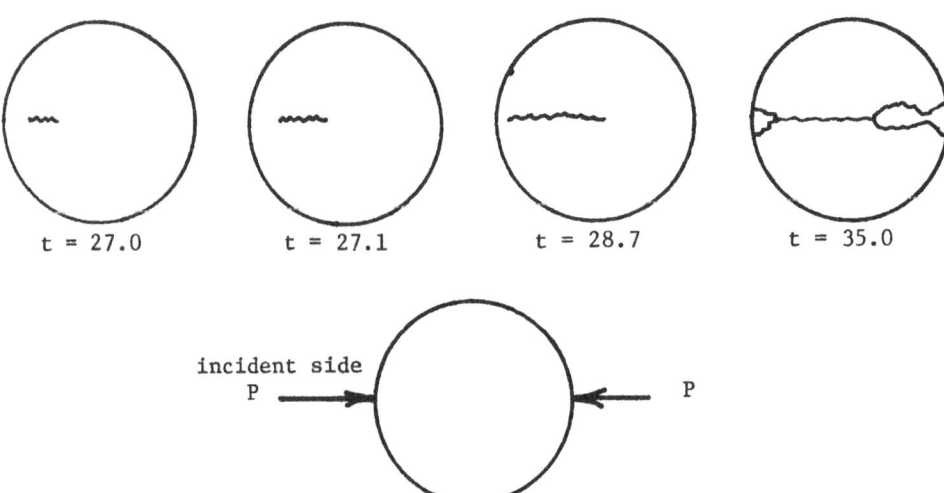

t = 27.0   t = 27.1   t = 28.7   t = 35.0

incident side
P →   ← P

Figure 9. Computer generated crack patterns for splitting-tensile concrete SHPB specimen [36]. Time given in microseconds.

SUMMARY OF SESSION 9: STRAIN RATE, THERMAL, TIME AND FATIGUE EFFECTS

S. E. SWARTZ
Department of Civil Engineering
Kansas State University
Manhattan, KS, USA

1. Papers Presented

"Growth of Discrete Cracks in Concrete Under Fatigue Loading" by Dirk A. Hordijk, TNO-Institute of Building Materials and Structures, The Netherlands, and Hans W. Reinhardt, Stuttgart University, Federal Republic of Germany.

"Crack Growth, Creep, and Creep Rupture at High Temperatures" by Sheldon Wiederhorn, National Institute of Standards and Technology, Gaithersburg, Maryland, USA.

"Fracture of Concrete at High Strain Rate" by C. Allen Ross, Air Force Engineering and Services Center, Tyndall AFB, Florida, USA.

2. Summary of Paper by Hordijk and Reinhardt

The paper, presented by Reinhardt, exploited the concept of using the static, tensile stress-softening curve as a failure envelope denoting the limit of damage effects induced by fatigue growth of a single crack in concrete. If one examines the problem of fatigue this can be divided into: material behavior – nucleation of existing flaws into cracks, crack extension; structural behavior – high strain concentrations ("hot spots"), crack initiation, stable and unstable crack propagation; and type of fatigue – low cycle, high amplitude or high cycle, low amplitude. Issues, or questions, which can be raised which are particularly pertinent to concrete include: presence of existing flaws – curing procedures, finishing procedures, moisture content, mix variables; continued hydration processes – influence of crack size, service/test environment, testing time or service life including time-rate effects (creep); growth of a visible (macro) crack – is the pro-

597

S. P. Shah (ed.), Toughening Mechanisms in Quasi-Brittle Materials, 597–602.
© 1991 *Kluwer Academic Publishers.*

598

cess the same for static growth versus fatigue growth? (This question has been addressed earlier by the Reporter on the basis of calculated K_{IC} using LEFM[1.]).

The strategy presented to model fatigue crack growth in concrete is;

1. Obtain stress-displacement data in static, uniaxial tension with a number of unload-reload cycles;

2. Develop a constitutive model of the softening curve (envelope), unload curve, reload curve and "gap" which develops in the envelope curve upon subsequent re- loading;

3. Use this model to develop an interface element which is then implemented in a finite element program (DIANA);

4. Verify the model on the basis of the uniaxial data;

5. Then implement the model for other structures, eg., beam in bending.

The resulting model developed in step 2 is called the Continuous-Func- tion-Model (FCM) and uses parameters of stress and displacement at the unload point, the minimum load point (which may or may not be at re- versed stress) and the point where the reload curve rejoins the enve- lope curve. The "gap" is modeled in terms of stress drop.

An example of applying this method of analysis to a beam in four point bending was presented which demonstrated the change in stress distribution with load cycle and showed the development of eigen- stresses due to crack closure associated with the minimum load. Deflections versus load cycles were presented which resembled well experimental cyclic-creep data for short-time, high-displacement fatigue (maximum of 35 cycles, the maximum load was 95% of the peak, static load).

2.1 DISCUSSION

R. Steinbrech (Research Center Jülich, Inst. for Reactor Materials, Jülich, FRG): Fatigue crack growth in ceramics is related to the deformation control and especially time aspects.

---

1. "Crack Growth and Fatigue in Plain Concrete - Static Versus Fatigue Loading" by Stuart E. Swartz, James Huang and K. K. Hu, Fatigue of Concrete Structures, ed. S. P. Shah, ACI SP-75, American Concrete Inst., Detroit, MI, 1982, pp. 47-69.

P. Rossi (Laboratoire Central des Ponts et Chaussees, Paris, France):
Can the softening branch at the micro level be described? Also, what
was the criterion to stop the numerical analysis (beam problem)?
Answer: The softening curve is based on the observed macro response
and the numerical analysis is stopped when the reload cycle intersects
the softening/ failure envelope.

S. P. Shah (NSF Center for Science and Technology of Advanced Cement-
Based Materials, Northwestern University, Evanston, USA): How applicable is the envelope curve for low-cycle, high-amplitude fatigue? –
Are there creep effects? Answer: Creep could affect results in this
region.

Z. P. Bazant (Northwestern University, Evanston, USA): Were rate and
irreversibility effects on the gap considered? Answer: Creep effects
were not investigated but testing was done using different rates.

3. Summary of Paper by Wiederhorn

The paper focused on mechanisms leading to creep rupture in
ceramic materials tested in tension at high temperatures. At high
temperatures the material is no longer brittle, there is some ductility, and the failure mechanism is different from that obtained at
low temperatures. In particular, the matrix plays a major role in the
failure. The discussion centered on two-phase materials:

1. vitreous – bonded aluminum oxide; 1a. 20% glass with large
   grains of aluminum oxide embedded in the vitreous matrix,
   1b. 8% glass with narrow grain boundaries and large facets;

2. silicon-disilicon carbide, 33% silicon with a grain size of
   about five microns.

In these materials, the second phase (matrix) softens at high
temperature and flows when subjected to stress. If flow is prevented
by the first phase material then cavitation occurs. How (or if) the
cavities coalesce depends on the first phase inclusions. For material
1a. (20% glass) the cavities grow into large holes that are stabilized
without coalescing. When a crack finally starts to propagate it does
so with a process zone and multiple cracking. For material 1b. (8%)
the cavities nucleate and propagate without arrest in a sharp crack.
For material 2, flow is difficult because of the closeness of the
grains and cavities tend to arrest and not form cracks. Basically,
the time to failure is correlated with the time necessary to coalesce
cavities into the formation of a crack. Thus, there are two major
mechanisms: cavity formation (flow) - coalescence - crack; and crack
propagation through the material.

Based on these ideas and physical observations a time rate model was presented which correlates strain level with cavity density and crack size. Knowing this leads to relations predicting the time it takes to initiate crack growth and creep rupture.

An interesting - and important - point was made that the effect of indentation on the specimen surface was to cause the subsequent crack to grow more rapidly than if the crack had formed without an indentation. Since it is useful to indent the specimen so that the crack location will be known (from a testing standpoint) this is an important effect to consider.

3.1 DISCUSSION

A. Kobayashi (Mechanical Engineering Department, University of Washington, Seattle, USA): What happens in two-or three-dimensional testing situations? Answer: Testing was done only under uniaxial conditions. However, on the micro level the cracks formed surfaces not necessarily perpendicular to the main crack line and so local triaxiality was present.

D. Krajcinovic (Mechanical and Aerospace Engineering Department, Arizona State University, Tempe, USA): Was the tertiary stage of creep not considered in the model? Answer: Most of the creep observed occurred in stage one or the plateau (stage two).

Krajcinovic: How were the indentations made? Answer: At room temperature - then the test was conducted at high temperature.

4. Summary of Paper by Ross

It was noted that at high strain rates (> 1.0/sec.) the tensile fracture stress for brittle materials - including concrete - varies approximately as the strain rate to the 1/3 power (and independently of the strength of the material). At lower rates, work presented by Reinhardt indicates this to be dependent on the strength of the material. The critical strain rate at which this transition in behavior occurs apparently depends on the type of load - tension giving this at about two orders of magnitude lower than compression.

The presentation was centered around impact tests using the Split-Hopkinson Pressure Bar (SHPB) on concrete specimens. These included direct compression impact, direct tension and split-cylinder indirect tension at different strain rates. The presentation (and paper) showed the development of the crack patterns for the various loading cases. These generally appeared to be dominated by tensile, mode I, failure on the macro scale. It was also verified that the transitional - or critical - strain rate varied for the different types of loading.

4.1 DISCUSSION

A. Kobayashi: How was the strain rate computed? Answer: It was obtained from the slope of the output curve (not at the point of fracture).

A. Kobayashi: How was K_I obtained? Answer: It was obtained from the stress and crack length for the split cylinder test.

S. Wu (Air Force Office of Scientific Research, Washington, DC, USA): Were the different failure mechanisms associated with low and high strain rates examined microscopically? Answer: An associate (Dr. Malvern) is doing this.

S. Wu: Was pore collapse examined? Answer: No.

S. Wu: Was a damping mechanism included in the finite element analysis of the test? Answer: Yes.

H. Reinhardt: Was there a pad between the specimen and loading platens in the splitting test. Answer: No, but now a thin pad is being tried. However, this causes a serious delay in the response signal.

P. Rossi: The influence of moisture content an the strain rate effect is important. Dry specimens do not have the rate effect as presented here but wet specimens do. Answer: The concrete tests presented did not have humidity control.

S. Shah: One could apply a static fatigue or stress corrosion model to predict strain rate effects.

R. Miller (Dept. of Civil and Environmental Engineering, University of Cincinnati, USA): If the tensile specimen is glued, what effect does this interface have? Answer: In the finite element (ADINA) results, at high load rates failure is next to the interface; for low load rates failure is at the center but cracks form at both places.

5. General Discussion

The discussion revolved around the failure mechanisms, primarily in concrete, associated with creep, low cycle fatigue, high cycle fatigue and the interaction of these mechanisms. With respect to damage in concrete a five-minute presentation was given by H. Horii (University of Tokyo, Japan) of results of a recent study on fatigue tests using the compact tension specimen. The crack profile and damage zone (at the specimen surface) were measured using the laser speckle method. These were measured for both loading (to a maximum

load of 80% of the static failure load) and unloading (to zero load).
For low cycle fatigue he used a model of the compression upon unloading giving rise to eigenstresses which induce tension at the
crack tip. For this process, a degradation parameter was defined.
For high cycle fatigue a different mechanism must be considered.
B. Karihaloo (University of Sydney, Australia) commented that they had
investigated fracture processes for high cycle fatigue in ceramics and
that these were due to crack bridging and attraction of cracks into
pores.

Further discussion of fatigue elicited the key points that creep
is also present; the moisture conditions are important; the build up
of residual stresses – or eigenstresses – are a major factor in,
especially, long cycle fatigue; and the stress state is important with
regard to distributed cracking.

Discussion on creep mechanisms again mentioned these points and
indeed it seems that there is a variety of effects present at all
times, some of which will dominate for a certain strain rate and
others for a different strain rate. Further, these effects seem to be
quite similar between cementitious materials and ceramics. The work
in the field of ceramics in identifying these damage mechanisms and
ways to toughen the material appears to be ahead of related work in
the field of cement-based materials.

Author Index

The numbers following each name indicate where an author is referenced in this volume. An italicized number indicates a contribution to this volume.

606

608

Subject Index

The page numbers following each key word indicate where the subject is covered by a paper in this volume.

611

612